Ecohydrology

Ecohydrology is an interdisciplinary field studying the interactions between water and ecosystems and utilizing that knowledge for enhancing environmental sustainability. With the growing concerns about nature conservation with increasing human activity/urban growth, there is a dramatic increase in research activities studying the changes in our ecosystems and related impacts. Apart from the geosciences point of view, these studies also come from fundamental scientific fields like physics, biology, and chemistry, as well as from engineering and social perspectives. The planned handbook consists of five individual volumes aiming to provide a comprehensive overview of the state of the art of current ecohydrological studies: fundamental concepts, practical monitoring methods, modeling studies, and special topics in water-limited environments. Each volume is edited by well-known expert(s) in the field, bringing in international authors for each chapter. The book is structured in a way that is appropriate for advanced graduate students and professionals in diverse scientific and engineering communities devoted to relevant fields, including geoscience, chemistry biochemistry, and biology, as well as various engineering disciplines. Although the handbook is planned to be published in series, each volume provides a self-contained description of its topic, with a standard and consistent format across each volume: Vol. 1, Water and Ecosystems; Vol. 2, Observation and Measurement of Ecohydrological Processes; Vol. 3, Water-Limited Environments; Vol. 4, Integrated Ecohydrological Modeling; and Vol. 5, River Basin Management.

More information about this series at http://www.springer.com/series/13407

Xin Li • Harry Vereecken
Editors

Observation and Measurement of Ecohydrological Processes

With 173 Figures and 32 Tables

Springer

Editors
Xin Li
Institute of Tibetan Plateau Research
Chinese Academy of Sciences
Beijing, China

CAS Center for Excellence
in Tibetan Plateau Earth Sciences Chinese
Academy of Sciences
Beijing, China

Harry Vereecken
Agrosphere Institute (IBG-3)
Forschungszentrum Jülich GmbH
Jülich, Germany

ISSN 2522-8242 ISSN 2522-8250 (electronic)
ISBN 978-3-662-48296-4 ISBN 978-3-662-48297-1 (eBook)
ISBN 978-3-662-48298-8 (print and electronic bundle)
https://doi.org/10.1007/978-3-662-48297-1

Library of Congress Control Number: 2019930844

This Springer imprint is published by the registered company Springer-Verlag GmbH, DE part of Springer Nature.
The registered company address is: Heidelberger Platz 3, 14197 Berlin, Germany

Series Preface

In the broadest sense, ecohydrology is an interdisciplinary field studying the interactions between water and life. With the growing concerns about nature conservation and environmental sustainability under intensifying global change and human activities, there has been a dramatic increase in research activities studying the ecohydrological processes and how they shape and impact the ecosystems we live in. Apart from the geosciences point of view, these studies also involve fundamental scientific fields like physics, biology, and chemistry, as well as engineering and social perspectives.

The handbook has been planned to consist of five individual volumes aiming to provide a comprehensive overview about the state of the art of contemporary ecohydrological studies: fundamental concepts and theories, field observation and monitoring methods, integrated ecohydrological modeling, river basin management, and special topics in water-limited environments. Each volume is edited by well-known experts in the field, bringing in international authors for each chapter.

Individual volumes are intended to include specific topics, such as interactions and feedbacks between ecological and hydrologic processes under different vegetation types and landscape patterns; soil seasonal freezing/thawing processes and their ecohydrological effects; groundwater-surface water interactions and their controlling factors; mechanisms of water utilization and transpiration by vegetation canopies across multiple scales; system complexity and scaling of coupled ecological-hydrologic processes; integrating ecohydrological models with economic models for decision support; field observatories to support integrated ecohydrological studies; assessment of ecosystem services and development of decision-making frameworks; and basin-scale biogeochemical processes and water quality issues.

The book is structured in a way that is appropriate for advanced graduate students and professionals in diverse scientific and engineering communities devoted to relevant fields, including geoscience, hydrology and water resources management, biogeochemistry, biology, and various engineering disciplines. Although the handbook is planned to be published in series, each volume provides a self-contained description of its topic, with standard and consistent format across each volume.

It is noteworthy that the motivation for this handbook stems from a large ecohydrological research program sponsored by the National Natural Science

Foundation of China (NSFC) from 2010 to 2018. The research program, titled "An Integrative Study of Ecological and Hydrologic Processes in the Heihe River Basin," utilizes a well-instrumented inland river basin in northwest China to explore fundamental ecohydrological processes across multiple spatiotemporal scales and to develop sustainable water management and ecosystem preservation strategies. Many of the chapters in the handbook draw from the data, findings, and perspectives of the Heihe Program. We gratefully acknowledge the financial support by the NSFC.

Chunmiao Zheng
Guodong Cheng
Bojie Fu
Editors

Volume Preface

Water is among the most essential resources that sustain Earth's ecosystems, but these systems are currently under threat due to land use and climate change. Assessing the impact of this global change on water security and ecosystem's functions is not only the key concern of the United Nation's Sustainable Development Goals, playing a key role in sustainable development of society, but also the most important research focus in "Future Earth" and Earth critical zone research. In view of global change, assessing the role of water in the functioning of ecosystems has become of utmost importance, and it is a key challenge in the field of "ecohydrology". With the continuous support of various international organizations, ecohydrology has gradually become a topic of considerable research and a frontier discipline in Earth and environmental sciences.

This book, Observation and Measurement of Ecohydrological Processes, systematically presents and summarizes the novel ideas, methodologies, and techniques for observations of ecohydrological processes at multiple scales, e.g., leaf, field, and watershed, regional and global scales that contribute to better understanding of the functioning of ecosystems. Many cutting-edge observing techniques, e.g., stable isotope, wireless sensor network, cosmic ray probe, and multisource remote sensing, are extensively introduced and their applications are discussed. Several key but very challenging issues, including scaling effect, heterogeneity, and uncertainty, are deeply explored and discussed.

Observation and Measurement of Ecohydrological Processes condenses the collective knowledge of scientific teams in various disciplines associated with ecohydrological processes' observations, including remote sensing, hydrology, ecology, soil, and agricultural sciences. Representative experimental studies and data are collected from the newest achievements of internationally large-scale ecohydrological experiments and observing networks, e.g., Heihe Watershed Allied Telemetry Experimental Research in China and Terrestrial Environmental Observatories in Europe. These studies and data are valuable in helping to understand and quantify various states and fluxes of ecohydrological processes from multiple scopes. Thus, this book is intended as a reference book for scientists, engineers, and graduate students of related disciplines.

Great progress has been made in the observation of ecohydrological processes, and new observing possibilities and platforms are constantly emerging and present new opportunities to better understand the hydrological processes, ecosystem functions, and their interactions. As the observational database is progressively increasing, Big Data analysis methods will gain importance and offer unprecedented opportunities to extract new knowledge and to develop and validate ecohydrological models. We hope that this book proves to be informative and instructive for all science disciplines involved in ecohydrological research.

Institute of Tibetan Plateau Research
Chinese Academy of Sciences
Beijing, China

CAS Center for Excellence in Tibetan
Plateau Earth Sciences
Chinese Academy of Sciences
Beijing, China

Xin Li

Agrosphere Institute (IBG-3)
Forschungszentrum Jülich GmbH
Jülich, Germany

Harry Vereecken

February 2019

Contents

About the Series Editors

Prof. Chunmiao Zheng
School of Environmental Science and Engineering
South University of Science and Technology of China
Shenzhen, China

Dr. Chunmiao Zheng currently holds the position of chair professor and dean of the School of Environmental Science and Engineering at the Southern University of Science and Technology in Shenzhen, China. He has also been chair professor and director of the Institute of Water Sciences at Peking University, and the George Lindahl III Endowed Professor of Hydrogeology at the University of Alabama. The primary areas of his academic research are contaminant transport in the subsurface, groundwater management, and ecohydrological processes in large watersheds. He is developer of the widely used MT3D and MT3DMS series of contaminant transport models and author of the textbook *Applied Contaminant Transport Modeling* (Wiley). He has served as associate editor for five leading hydrology journals and as president of the International Commission on Groundwater of the International Association of Hydrological Sciences. Among the numerous honors and awards he has received, Zheng is a recipient of the O.E. Meinzer Award from the Geological Society of America and the M. King Hubbert Award from the National Groundwater Association (USA). Zheng received a Ph.D. in hydrogeology with a minor in environmental engineering from the University of Wisconsin-Madison.

Prof. Guodong Cheng
Cold and Arid Regions Environmental and Engineering Research Institute
Chinese Academy of Sciences
Lanzhou, China

Institute of Urban Studies
Shanghai Normal University
Shanghai, China

Prof. Guodong Cheng is an internationally renowned scientist at the Cold and Arid Regions Environmental and Engineering Research Institute of Chinese Academy of Sciences in Lanzhou, Gansu Province, China. He is a member of the Chinese Academy of Sciences and a foreign member of the Russian Academy of Engineering. Because of his research accomplishments in geocryology, he has received numerous prestigious awards, including the Outstanding Achievement Award of Chinese Academy of Sciences in 2005, the Top Class National Science and Technology Progress Award in 2008, the Special Award for Outstanding Academic Papers from the Chinese Association for Science and Technology in 2007, and the Lifetime Achievement Award from the International Permafrost Association. He has served as vice-president and president of the International Permafrost Association from 1988 to 1993 and from 1993 to 1998, respectively. Professor Cheng has published more than 200 papers in peer-reviewed professional journals and several books.

Prof. Bojie Fu
Research Center for Eco-Environmental Sciences
Chinese Academy of Sciences
Beijing, China

Dr. Bojie Fu is a distinguished professor of geography and landscape ecology at the State key Laboratory of Urban and Regional Ecology, Research Centre for Eco-Environmental Sciences, Chinese Academy of Sciences (CAS). He is dean of the Faculty of Geographical Science, Beijing Normal University. He is member of the Chinese Academy of Sciences, fellow of the Academy of Sciences for Developing World (TWAS), and corresponding fellow of the Royal Society Edinburgh, UK. He is president of the Geographical Society of China, director-general of the Department of Earth Sciences, National Natural Science Foundation of China, and director of the Division of Geosciences of Chinese Academy of Sciences. His research areas are land use and land cover change, landscape pattern and ecological processes, and ecosystem services and management. He has published more than 400 scientific papers and 10 books, including *Science*, *Nature Geoscience*, and *Nature Climate Change*. His awards and honors include China National Natural Science Prize, Outstanding Science and Technology Achievement Prize of CAS, The Ho Leung Ho Lee Science and Technology Prize-Geosciences, and Award of Distinguished Service of the International Association of Landscape Ecology.

About the Editors

Xin Li
Institute of Tibetan Plateau Research
Chinese Academy of Sciences
Beijing, China

CAS Center for Excellence in Tibetan
Plateau Earth Sciences
Chinese Academy of Sciences
Beijing, China

Xin Li received his Ph.D. in remote sensing and a GIS from the Chinese Academy of Science (CAS) in 1998. He has been a full professor at Cold and Arid Regions of Environmental and Engineering Research Institute, CAS, since 2000. He is presently a professor at the Institute of Tibetan Plateau Research, CAS. He is a member of the WCRP GEWEX scientific steering committee, a member of the International Science Advisory Panel of Global Water Futures program, and the secretary general of the UNESCO IHP G-WADI Asia network. He is on the editorial boards of *Journal of Hydrology*, *Science Bulletin*, *Vadose Zone Journal*, and *Remote Sensing*. He is the lead scientist of WATER (Watershed Allied Telemetry Experimental Research, 2007–2010) and HiWATER (2012–2016). His primary research interests include land data assimilation, application of remote sensing and GIS in hydrological and cryospheric sciences, and integrated watershed study.

Harry Vereecken
Agrosphere Institute (IBG-3)
Forschungszentrum Jülich GmbH
Jülich, Germany

Harry Vereecken received his Ph.D. in agricultural sciences from the Katholieke Universiteit Leuven, Belgium (1988). He obtained his habilitation in hydrogeology at Bonn University, Germany (1997). Since 2000 he is full professor at the Faculty of Agriculture, Bonn University, and director of the Agrosphere Institute (IBG-3), Forschungszentrum Jülich GmbH. He is editor of *Vadose Zone Journal* and member of the Nordrhein-Westfälische Akademie der Wissenschaften und der Künste (North Rhine-Westphalia Academy for Sciences and Arts, Germany). He is coordinator of TERENO, which is a network of environmental observatories in Germany, director of the Geo Alliance ABC/J (Germany), and chair of the International Soil Modeling Consortium (ISMC). His research interests include hydrology, hydrogeophysics, and soil science.

Contributors

Alberto Alonso-González Microwave and Radar Institute, German Aerospace Center (DLR), Wessling, Germany

Heye Bogena Agrosphere Institute (IBG-3), Forschungszentrum Jülich GmbH, Jülich, Germany

David Caterina Institute of Bio- and Geosciences, Forschungszentrum Jülich GmbH, Jülich, Germany

Guodong Cheng Cold and Arid Regions Environmental and Engineering Research Institute, Chinese Academy of Sciences, Lanzhou, China

Institute of Urban Studies, Shanghai Normal University, Shanghai, China

Qian Cui State Key Laboratory of Remote Sensing Science, Institute of Remote Sensing and Digital Earth, Chinese Academy of Science, Beijing, China

Jie Ding Key Laboratory of Tibetan Environment Changes and Land Surface Processes, Institute of Tibetan Plateau Research, Chinese Academy of Sciences, Beijing, China

Haifeng Gao Key Laboratory of Tibetan Environment Changes and Land Surface Processes, Institute of Tibetan Plateau Research, Chinese Academy of Sciences, Beijing, China

Sarah Garré Gembloux Agro-Bio Tech, UR TERRA, Université de Liège, Gembloux, Belgium

Yingchun Ge Cold and Arid Regions Environmental and Engineering Research Institute, Chinese Academy of Sciences, Lanzhou, China

Xiaolin Guo State Key Laboratory of Hydroscience and Engineering, Department of Hydraulic Engineering, Tsinghua University, Beijing, China

Yanhong Guo Key Laboratory of Tibetan Environment Changes and Land Surface Processes, Institute of Tibetan Plateau Research, Chinese Academy of Sciences, Beijing, China

Irena Hajnsek Microwave and Radar Institute, German Aerospace Center (DLR), Wessling, Germany

Institute of Environmental Engineering, Swiss Federal Institute of Technology (ETH), Zurich, Switzerland

Zhongying Han State Key Laboratory of Hydroscience and Engineering, Department of Hydraulic Engineering, Tsinghua University, Beijing, China

Xiaobo He State Key Laboratory of Cryosphere Sciences, Cold and Arid Regions Environment and Engineering Research Institute, Chinese Academy of Sciences, Lanzhou, Gansu, China

Yang Hong State Key Laboratory of Hydroscience and Engineering, Department of Hydraulic Engineering, Tsinghua University, Beijing, China

School of Civil Engineering and Environmental Science, University of Oklahoma, Norman, OK, USA

Guangrong Hu State Key Laboratory of Earth Surface Processes and Resource Ecology, Faculty of Geographical Science, Beijing Normal University, Beijing, China

School of Natural Resources, Faculty of Geographical Science, Beijing Normal University, Beijing, China

Qi Huang State Key Laboratory of Hydroscience and Engineering, Department of Hydraulic Engineering, Tsinghua University, Beijing, China

Mathieu Javaux Earth and Life Institute/Environmental Sciences, Université catholique de Louvain, Louvain-la-Neuve, Belgium

Agrosphere Institute (IBG-3), Forschungszentrum Jülich GmbH, Juelich, Germany

François Jonard Institute of Bio- and Geosciences, Forschungszentrum Jülich GmbH, Jülich, Germany

Earth and Life Institute, Université catholique de Louvain, Louvain-la-Neuve, Belgium

Anja Klotzsche Institute of Bio- and Geosciences, Forschungszentrum Jülich GmbH, Jülich, Germany

Axelle Koch Earth and Life Institute/Environmental Sciences, Université catholique de Louvain, Louvain-la-Neuve, Belgium

Xiaoyan Li State Key Laboratory of Earth Surface Processes and Resource Ecology, Faculty of Geographical Science, Beijing Normal University, Beijing, China

School of Natural Resources, Faculty of Geographical Science, Beijing Normal University, Beijing, China

Xin Li Institute of Tibetan Plateau Research, Chinese Academy of Sciences, Beijing, China

CAS Center for Excellence in Tibetan Plateau Earth Sciences, Chinese Academy of Sciences, Beijing, China

Shaomin Liu State Key Laboratory of Earth Surface Processes and Resource Ecology, School of Natural Resources, Faculty of Geographical Science, Beijing Normal University, Beijing, China

Chunfeng Ma Key Laboratory of Remote Sensing of Gansu Province, Cold and Arid Regions Environmental and Engineering Research Institute, Chinese Academy of Sciences, Lanzhou, China

Yingzhao Ma State Key Laboratory of Hydroscience and Engineering, Department of Hydraulic Engineering, Tsinghua University, Beijing, China

Félicien Meunier Earth and Life Institute/Environmental Sciences, Université catholique de Louvain, Louvain-la-Neuve, Belgium

Alessandra Monerris Department of Infrastructure Engineering, University of Melbourne, Melbourne, Australia

Carsten Montzka Agrosphere Institute (IBG-3), Forschungszentrum Jülich GmbH, Jülich, Germany

Yakov Pachepsky Environmental Microbial and Food Safety Laboratory, USDA ARS Beltsville Agricultural Research Center, Beltsville, MD, USA

Yonghua Qu State Key Laboratory of Remote Sensing Science, Beijing Normal University and Institute of Remote Sensing and Digital Earth of Chinese Academy of Sciences, Beijing, China

Beijing Engineering Research Center for Global Land Remote Sensing Products, Institute of Remote Sensing Science and Engineering, Faculty of Geographical Science, Beijing Normal University, Beijing, China

Youhua Ran Heihe Remote Sensing Experimental Research Station, Northwest Institute of Eco-Environment and Resources, Chinese Academy of Sciences, Lanzhou, China

Mike Schwank Mountain Hydrology and Torrents, Swiss Federal Institute for Forest, Snow and Landscape Research WS, Birmensdorf, Switzerland

GAMMA Remote Sensing AG, Gümligen, Switzerland

Jiancheng Shi State Key Laboratory of Remote Sensing Science, Institute of Remote Sensing and Digital Earth, Chinese Academy of Science, Beijing, China

Antonina Simonenko State Hydrologic Institute, St. Petersburg, Russia

Guoqiang Tang State Key Laboratory of Hydroscience and Engineering, Department of Hydraulic Engineering, Tsinghua University, Beijing, China

Harry Vereecken Agrosphere Institute (IBG-3), Forschungszentrum Jülich GmbH, Jülich, Germany

Christian von Hebel Institute of Bio- and Geosciences, Forschungszentrum Jülich GmbH, Jülich, Germany

Cunguang Wang State Key Laboratory of Hydroscience and Engineering, Department of Hydraulic Engineering, Tsinghua University, Beijing, China

Haibo Wang Heihe Remote Sensing Experimental Research Station, Northwest Institute of Eco-Environment and Resources, Chinese Academy of Sciences, Lanzhou, China

Kunxin Wang Key Laboratory of Tibetan Environment Changes and Land Surface Processes, Institute of Tibetan Plateau Research, Chinese Academy of Sciences, Beijing, China

Xufeng Wang Heihe Remote Sensing Experimental Research Station, Northwest Institute of Eco-Environment and Resources, Chinese Academy of Sciences, Lanzhou, China

Ziwei Xu State Key Laboratory of Earth Surface Processes and Resource Ecology, School of Natural Resources, Faculty of Geographical Science, Beijing Normal University, Beijing, China

Daqing Yang Watershed Hydrology and Ecology Research Division, Water Science and Technology, Environment and Climate Change Canada, Victoria, Canada

Yuan Yang State Key Laboratory of Hydroscience and Engineering, Department of Hydraulic Engineering, Tsinghua University, Beijing, China

Yuting Yang State Key Laboratory of Hydroscience and Engineering, Department of Hydraulic Engineering, Tsinghua University, Beijing, China

Panpan Yao State Key Laboratory of Remote Sensing Science, Institute of Remote Sensing and Digital Earth, Chinese Academy of Science, Beijing, China

Ziyue Zeng State Key Laboratory of Hydroscience and Engineering, Department of Hydraulic Engineering, Tsinghua University, Beijing, China

Teng Zhang Key Laboratory of Tibetan Environment Changes and Land Surface Processes, Institute of Tibetan Plateau Research, Chinese Academy of Sciences, Beijing, China

Yinsheng Zhang Key Laboratory of Tibetan Environment Changes and Land Surface Processes, Institute of Tibetan Plateau Research, Chinese Academy of Sciences, Beijing, China

CAS Center for Excellence in Tibetan Plateau Earth Sciences, Beijing, China

Zhonghao Zhang Institute of Urban Studies, Shanghai Normal University, Shanghai, China

Tianjie Zhao State Key Laboratory of Remote Sensing Science, Institute of Remote Sensing and Digital Earth, Chinese Academy of Science, Beijing, China

Observing Ecohydrological Processes: Challenges and Perspectives

Xin Li, Harry Vereecken, and Chunfeng Ma

Contents

X. Li (✉)
Institute of Tibetan Plateau Research, Chinese Academy of Sciences, Beijing, China

CAS Center for Excellence in Tibetan Plateau Earth Sciences, Chinese Academy of Sciences, Beijing, China
e-mail: xinli@itpcas.ac.cn

H. Vereecken
Agrosphere Institute (IBG-3), Forschungszentrum Jülich GmbH, Jülich, Germany
e-mail: h.vereecken@fz-juelich.de

C. Ma
Key Laboratory of Remote Sensing of Gansu Province, Cold and Arid Regions Environmental and Engineering Research Institute, Chinese Academy of Sciences, Lanzhou, China

X. Li, H. Vereecken (eds.), *Observation and Measurement of Ecohydrological Processes*, Ecohydrology, https://doi.org/10.1007/978-3-662-48297-1_1

Abstract

The observation and measurement of ecohydrological processes have been witnessed a huge progress in terms of novel ideas, methodologies, and techniques. Many cutting-edge observing techniques, e.g., stable isotope, wireless sensor network, cosmic ray probe, multi-source remote sensing, are continuously introduced and widely applied. As the first chapter of this book, this chapter introduces the progresses, challenges, and perspectives of observing ecohydrological processes. We first introduced the key states and fluxes that control the ecohydrological processes and novel techniques that allow those controlling factors to be quantified. However, we found that knowledge gap remains, including: (1) improving the observation ability to understand and quantify the ecohydrological processes, (2) integrating multisource observations into a dynamics model to accurately estimate the state and flux variables of ecohydrological processes, (3) developing upscaling approaches through system observations to understand the scaling issue, and (4) estimating representativeness error to quantify the uncertainties. To this end, we pointed out the potential directions for filling these gaps, including: (1) to better translate remotely sensed data into information that helps us better understand ecohydrological processes and better inform land-surface models, (2) to better quantify the roles of subsurface processes in ecohydrological processes, (3) to develop observational systems that allow ecohydrological processes to be captured across different scales and across compartments, (4) to use well-instrumented watersheds as test beds of new concept for ecohydrological observations, (5) to combine monitoring and controllable and synthetic observation experiments, (6) to utilize technical advancements in new models, and (7) to integrate observation systems with integrated models, data services, and decision making. Overall, this chapter provides an insight into the-state-of-art of observing ecohydrological processes.

Keywords

Ecohydrological processes · Remote sensing · Uncertainty · Heterogeneity · Scaling

Introduction

Water is among the most essential resources that sustain Earth's ecosystems, and it supports essential ecosystem services, such as food, feed, fiber, and energy. Thus, water can be considered a strategic resource for mankind (Liu et al. 2010). As such, a better understanding of the hydrological processes that control ecosystem services is of great significance. Moreover, our ecosystems are under threat due to a lack of either sufficient water or water of good quality. The need for a better understanding of the roles of water in ecosystem functioning requires an integrated approach between hydrologists and ecologists (Bonacci et al. 2009). A new discipline, ecohydrology, has been proposed to demonstrate the importance of this integrated approach. Ecohydrology was proposed as an independent discipline at the International Conference on Water and the Environment in Dublin in 1992. The results of ecohydrological research have contributed to the better management and preservation of fragile ecosystems (Bonacci et al. 2009; Abbott et al. 2016). Under the continuous support of the International Geosphere-Biosphere Program (IGBP), United Nations Educational, Scientific and Cultural Organization-International Hydrologic Programme (UNESCO-IHP), and other related international organizations, ecohydrology gradually has become a topic of considerable research and a frontier discipline in earth and environmental sciences. Ecohydrology is a key discipline in "Future Earth" (Liverman et al. 2013) and Earth critical zone research (Anderson et al. 2008; Richter and Mobley 2009; Lin et al. 2010; Guo and Lin 2016).

Modeling and observation are the fundamental research methodologies of ecohydrological processes, and both have developed rapidly over the past two decades. This book focuses on the observation of ecohydrological processes.

Progress in the Observation of Ecohydrological Processes

Overview

Ecohydrological processes are controlled by, for example, soil moisture (SM), precipitation, evapotranspiration (ET), runoff, and vegetation through photosynthesis, and the statuses of these processes are characterized by, for example, the leaf area index (LAI). In this chapter, we present different observation technologies that allow these controlling factors to be quantified, including ground-based sensing, airborne remote sensing, and spaceborne remote sensing. Table 1 lists the key variables of ecohydrological processes and observational techniques that will be addressed in this chapter.

Key Factors Controlling Ecohydrological Processes

Precipitation

Precipitation is defined as condensed atmospheric water vapor that falls to Earth's surface in liquid, solid, or combined forms, i.e., rain, snow, drizzle, sleet, and hail. Precipitation is the driving flux for many processes that take place on land surface and is considered the most sensitive but also uncertain input in many ecohydrological models (Chen and Frauenfeld 2014; Liu et al. 2011). Ground-based measurements of precipitation can be obtained using various types of rain gauges and weather radars, and precipitation can be sensed remotely using satellite sensors operating in the near-infrared, thermal infrared, passive microwave, and active microwave bands (Prigent 2010). The most successful satellite missions include the Tropical Rainfall Measuring Mission (TRMM) and Global Precipitation Measurement (GPM) mission (Hou et al. 2014; Skofronick-Jackson et al. 2017).

Table 1 Key ecohydrological variables and observation techniques and platforms

Variables	Ground measurement	Airborne	Satellite mission
Precipitation	Doppler radar, rain gauge, precipitation particle drop size analyzer	N/A	FY-3, TRMM, GPM
ET	EC, LAS, Lysimeter	Imaging spectrometer, LiDAR, airborne EC	ASTER, MODIS, AATSR, FY-3
Runoff	ADCP, hydrological gauging station, runoff plot	InSAR, Imaging spectrometer	OSTM/Jason-2 Radar Altimeter, ASAR, Altimeter-2, TSX, SWOT
Soil moisture	Site-scale (e.g., TDR, FDR) Footprint scale (e.g., COSMOS, GPR)	Active (e.g., PLIS, AirSAR), passive (e.g., PLMR, PSR)	Active (e.g., SMAP, ENVISAT, ALOS), passive (e.g., SMAP, SMOS, FY-3 MWRI)
Vegetation type	Sampling strip, plot survey	Imaging spectrometer	PROBA CHRIS, HYSPIRI, HJ-1, CBERS
LUCC	Land use survey	Multispectral sensor, imaging spectrometer	Landsat, SPOT, MODIS, HJ-1, CBERS and other multispectral sensors
Photosynthesis and respiration	LI-6400/XT, EC	Airborne-EC	FLEX
LAI	LAI-2000, TRAC, LI-3100	VNIR, LiDAR, low frequency SAR	Landsat, SPOT, MODIS, HJ-1, CBERS and other multispectral sensors

There are several challenges in precipitation measurement. First, in situ observations of precipitation should be bias corrected; in particular, gauge measurements of snowfall are significantly underestimated and require the development of standard correction methods. Therefore, WMO initialized the Solid Precipitation Intercomparison Experiment (SPICE) to undertake a systematic assessment of the reliability of accurately measuring solid precipitation. Second, obtaining high-resolution and reliable precipitation forcings in complex terrains, particularly mountainous regions, remains challenging (Alemohammad et al. 2015; Gottardi et al. 2012; Ward et al. 2011). One promising approach is to combine regional climate modeling with various observations, including those from rain gauges, Doppler radars, and satellite remote sensing, via multiscale data assimilation (Pan et al. 2012, 2017).

Evapotranspiration

Terrestrial ET is composed of the evaporation of water from soils, canopy interception, and waterbodies, whereas transpiration represents the loss of water through stomata. ET can be expressed as either part of the energy balance (latent heat) or a flux in the soil water balance. ET is a key process linking hydrology with ecosystem dynamics from stoma to landscape scales (Wang and Dickinson 2012; Zhu et al.

2016). Ground-based ET measurement techniques include weighing lysimeters, eddy covariance techniques, and scintillometers (Liu et al. 2011). For satellite remote sensing, direct measurements of ET as a water flux remain impossible. ET can be estimated indirectly using the energy balance and Penman-Monteith-based methods (P-M) (for more details, refer to chapters ▶ "Remotely Sensed Evapotranspiration" and ▶ "Micrometeorological Methods to Determine Evapotranspiration"). Various algorithms, such as energy balance based algorithms, including SEBAL (Bastiaanssen et al. 1998), SEBS (Su 2002), and their variants, and algorithms based on plant physiology and ecology, such as P-M, have been developed.

Accurate quantification of ET is challenging due to its strong spatial heterogeneity and scale dependence, which encompasses leaf, canopy, field, and regional scales. Multiscale observational experiments on ET that provide insight into the spatial heterogeneities of ET, explore the energy balance closure problem, and identify scaling effects are needed to address those challenges. These experiments require the provision of ground truth data that correspond to the development of remote sensing models and scale transformation approaches. For example, HiWATER (Heihe Watershed Allied Telemetry Experiment Research) is a multiscale observation experiment designed to capture the heterogeneity of land-surface water and energy fluxes (Li et al. 2013; Xu et al. 2013; Liu et al. 2016).

Streamflow

Streamflow, another significant component of the water cycle and water balance, can be described as the overland flow in streams, rivers, and channels generated by rainfall, snowfall, meltwater, or irrigation water due to gravity and groundwater supplements (Beven et al. 2004). Streamflow feeds rivers and lakes with water that supports many ecosystem services. Streamflow is measured at gauging stations by measuring depth, width, and velocity using traditional methods or Acoustic Doppler Current Profiles (ADCPs). Satellite remote sensing that aims to quantify streamflows shows great potential but is still not operational. The challenge with this technique is determining how to measure streamflow velocity using remote sensing. This challenge will be overcome by the new Surface Water Ocean Topography (SWOT) satellite mission, which uses a Ka-band radar interferometer as the core technology for measuring velocity. The water level will be measured simultaneously by radar altimetry. The spatial resolution (tens of meters) and vertical precision (a few centimeters) of SWOT are also unprecedented (Durand et al. 2010). Therefore, streamflow measurements will be revolutionized through the implementation of the SWOT mission (chapter ▶ "Surface Runoff").

Soil Moisture

Soil moisture is a physical soil state variable that is defined as the water contained in the unsaturated (vadose) soil zone. It represents a key variable in many hydrological, climatological, environmental, and ecohydrological processes (Vereecken et al. 2008). The important role of SM in multidiscipline applications is well recognized (NRC 2007). The approaches for measuring soil moisture include point-measurements, field-scale measurements, ground-based networks, and remotely sensed methods (including satellite-based and airborne-based remote sensing) (Zacharias et al. 2008; Bogena et al. 2012). Soil moisture sensing techniques have progressed considerably over the last 50 years. In particular, the European Spatial Agency (ESA) Soil Moisture and Ocean Salinity (SMOS) (Kerr et al. 2001) and National Aeronautics and Space (NASA) Soil Moisture Active and Passive (SMAP) (Entekhabi et al. 2010) missions are designed to map global soil moisture under the background of climate change.

Soil moisture measurements are challenged by the issues of quantifying the spatial representativeness of in situ measurements and scaling micro-observations to macroscales, such as the pixel resolutions of satellite remote sensing products. The new technologies, which include footprint-scale in situ measurements, such as cosmic-ray probes (Zreda et al. 2012; Andreasen et al. 2017) and wireless sensor networks (Bogena et al. 2010; Jin et al. 2014) provide new opportunities because they can capture spatial variability in soil moisture at multiple scales. By using proper scaling methods, in situ measurements can be scaled to pixel-scale truth data and used to validate remote sensing products (Kang et al. 2015; Jin et al. 2017).

Photosynthesis and Respiration

Photosynthesis is the process by which green plants and other organisms transform sunlight energy into chemical energy. It is a primary life process for green plants and contributes significantly to making terrestrial ecosystems large carbon sinks (Canadell et al. 2000; Eagleson 2002). Through respiration, plants break down carbohydrates to provide energy for metabolism and release CO_2 into the atmosphere (Eagleson 2002). Photosynthesis and respiration are strongly controlled by hydrological states and fluxes. Understanding those relationships can aid in predictions of how hydrological changes affect vegetation and thus both photosynthesis and plant respiration (Running and Gower 1991).

The photosynthesis components of terrestrial plants can be measured by destructive dry matter, gas exchange, isotope, and fluorescence methods (Hunt 2003). Additional details are provided in chapter entitled ▶ "Photosynthesis (NPP, NEP, Respiration)". Photosynthesis components are estimated using light use efficiency or process-based biogeochemical models from satellite remote sensing. The challenge is determining how to measure light use efficiency directly. Solar-induced chlorophyll fluorescences retrieved by high-spectral-resolution (<0.5 nm) remote sensing could shed new light on this challenge (Meroni et al. 2009). In particular, the planned ESA FLuorescence EXplorer (FLEX) mission will usher in a new era of the photosynthesis remote sensing (Drusch et al. 2017).

Leaf Area Index

The leaf area index (LAI) is defined as single-sided green leaf area per unit ground area (Chen and Black 1991) and is a critical parameter in many climate, hydrology, and ecology models. The LAI has been widely used to investigate crop growth, hydrology processes, energy balances, and land-atmosphere exchanges and in other climate change studies (Sellers 1997). The LAI can be measured directly by collecting samples from a plant canopy or indirectly using optical methods. Typical indirect methods include canopy analyzers and the hemispherical photographic method (Jonckheere et al. 2004). Remote sensing is an important and feasible way to acquire the LAI at regional and global scales. However, LAIs estimated from remote sensing should be calibrated and validated with ground observations to ensure reliability (Qu et al. 2012). Additional details are provided in chapter ▶ "Leaf Area Index: Advances in Ground-Based Measurement".

Main Observation Platforms for Ecohydrological Processes

Ecohydrological processes can be studied using satellite, airborne, and ground-based platforms, as shown in Fig. 1, which represents the HiWATER observational system.

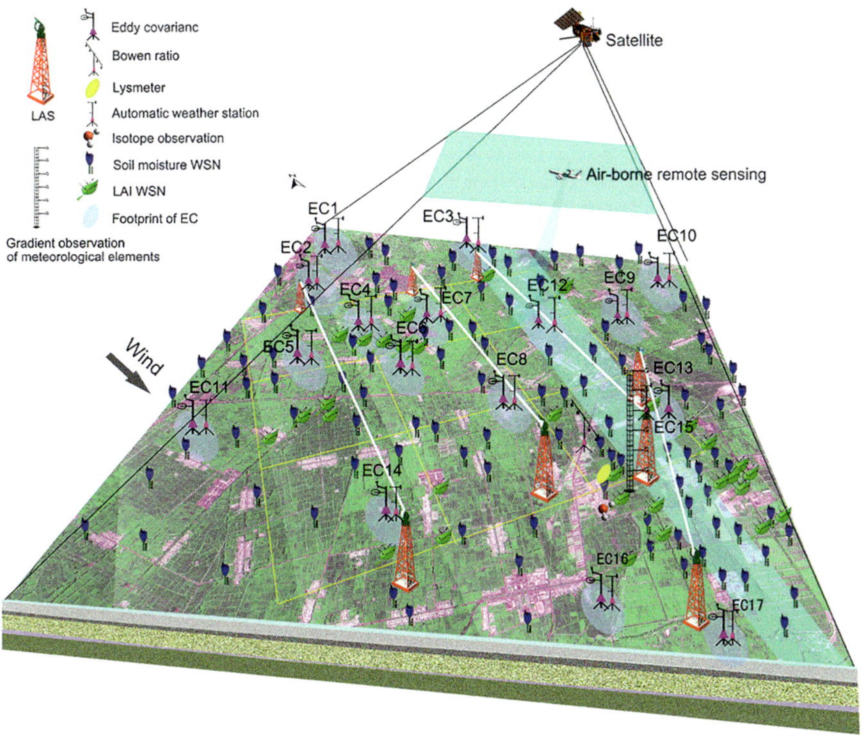

Fig. 1 Multi-platform observing system for ecohydrological processes deployed in the HiWATER observational system, taken as an example. The HiWATER observing system is composed of a flux matrix, a wireless sensor network for measuring soil moisture and the LAI, isotope observations, and airborne remote sensing. Details can be found in Li et al. (2013, 2017); Liu et al. (2016, 2018)

Satellite Observation Platforms

At present, ecohydrological variables largely can be observed or monitored by satellite remote sensing. The main optical-thermal, active and passive microwave satellite/sensors or satellite platforms that can provide observations on ecohydrological variables are listed in Tables 2, 3, and 4, respectively.

Since 2000, several new satellite observation missions have been implemented and can be used to monitor ecohydrological variables globally.

- GPM (Global Precipitation Measurement). The core satellite of the GPM, which was designed and developed jointly by NASA and JAXA, was equipped with both a dual-frequency Ku/Ka-band precipitation radar and a high-resolution, multichannel passive microwave rain radiometer. In addition, through international collaboration, several non-sun-synchronous and sun-synchronous orbit satellites were launched with passive microwave radiometers onboard, making up a satellite constellation. The temporal resolutions of global precipitation observations can reach 3 h (Hou et al. 2014).
- SMOS (Soil Moisture and Ocean Salinity). The SMOS satellite was successfully launched on November 2, 2009, with an L-band passive microwave radiometer. One of the main scientific objectives of SMOS is to monitor surface soil moisture. The spatial resolution is 40 km (Kerr et al. 2010).

Table 2 Main optical-thermal satellite platforms for ecohydrological processes observation

Satellite (Sensors)	Agency (First launch year)	Resolution (m)	Variables that have been observed
TIROs-N/NOAA (AVHRR)	TIROS-n/NOAA, USA (1978)	1100	ET, SM, Chlorophyll, NPP, etc.
Terra/Aqua (MODIS)	NASA, USA (1999)	250–1000	ET, SM, Chlorophyll, biomass, VWC, snow
Landsat (MSS/TM/ETM+/ OLI -TIRS)	USGS, USA (1972)	10–30	SM, biomass, NPP, VWC, snow.
SPOT (VEGETATION)	ESA, Europe (1986)	1.5–20	SM, VWC, biomass, NPP, LUCC, snow.
CBERS	China-Brazil (1999)	2–120	SM, VWC, biomass, NPP.
ALOS/AVNIR	JAXA, Japan (2006)	10–100	SM, VWC, snow.
PROBA-V	ESA, Europe (2013)	3–100	SM, VWC biomass, NPP, LUCC.
FY	China (1988)	250–1250	SM, VWC, biomass, NPP, LUCC, snow.
HJ	China (2008)	1–100	SM, VWC, biomass, NPP.
GF	China (2013)	0.8–16	SM, VWC, biomass, NPP, snow.
Sentine2-3	ESA, Europe (2014/2015)	5–40	ET, SM, VWC, biomass.

- SMAP (Soil Moisture Active Passive). The SMAP satellite, which was developed by NASA, was launched on January 31, 2015 with an integrated active and passive L-band radar and radiometer. As with SMOS, the scientific objectives of SMAP are to map soil moisture and freeze/thaw states. However, the spatial and temporal resolutions of SMAP are 3/9/36 km and 2–3 days, respectively (Entekhabi et al. 2010).

Table 3 Spaceborne SAR for ecohydrological process observations

Satellite (Band)	Agency (Launch year)	Frequency (GHz)	Resolution (m)	Variables that have been observed
ERS-1/2 (C)	ESA, Europe (1991/1995)	5.25	30	ET, SM, Chlorophyll, NPP, etc.
JERS-1 (L)	JAXA, Japan (1992)	1.275	20	Precipitation, ET, SM, Chlorophyll, biomass, VWC
SIR-C/X-SAR (L, C, X)	NASA/JPL/DLR (1994)	1.25, 5.3, 9.6	30	Precipitation, SM, biomass, NPP, VWC.
Radarsat-1 (C)	Canadian Space Agency (1995)	5.3	10–100	Precipitation, SM, VWC, biomass, NPP, LUCC.
SRTM (C, X)	NASA/JPL/DLR (2000)	5.25, 9.6	30	Snow, SM, biomass
ENVISAT/ ASAR (C)	ESA (2002)	5.25	20–100	Precipitation, SM, VWC, biomass, NPP.
ALOS/ PALSAR (L)	JAXA, Japan (2006)	1.27	10–100	Precipitation, SM, VWC, snow.
Radarsat-2 (C)	Canadian Space Agency (2007)	5.4	3–100	Precipitation, SM, VWC biomass, NPP, LUCC.
TerraSAR-X (X)	DLR/Astrium, Germany (2007)	9.65	0.24–260	SM, VWC, biomass, NPP, LUCC, snow.
COSMO-SkyMed (X)	ASI, Italy (2007/2010)	9.65	1–100	SM, VWC, biomass, NPP.
TanDEM-X (X)	DLR/Astrium, Germany (2010)	9.65	1–16	SM, VWC, biomass, NPP, snow.
RISAT-1 (C)	Indian Space Agency (2012)	5.35	3/6/9/25/50	SM, VWC, snow, biomass, NPP.
HJ-1C-SAR (S)	CRESDA/CAST/NRSCC, China (2012)	3.2	5/20	SM, VWC, biomass.
ALOS-2/ PALSAR-2 (L)	JAXA, Japan (2014)	1.20	1–100	Precipitation, SM, VWC, glacier, snow.
SMAP (L)	NASA/JPL (2015)	1.26	1000–3000	SM, biomass, NPP, freeze and thaw.
Sentinel-1a/b (C)	ESA, Europe (2014/2015)	5.4	5–40	Precipitation, ET, SM, VWC, biomass.
GF-3/SAR (C)	China (2016)	5.4	1–700	SM, VWC, LUCC, biomass.

- Biomass. The Biomass mission was selected as ESA's seventh Earth Explorer in May 2013 (Scipal et al. 2010; Toan et al. 2011). Biomass provides crucial information about the state of forests and how they are changing. The mission also explores the unique sensitivity of P-band SAR together with advanced retrieval methods and generates maps of forest biomass and forest height at a resolution of 200 m. The main goal of the Biomass mission is to understand the carbon cycle and related changes taking place in Earth's ecosystems, e.g., absorption, storage, and release of carbon in forests.

Table 4 Spaceborne radiometers for ecohydrological process observation

Satellite platform (Sensors)	Start-end (Year)	Frequency (GHz)	Spatial resolution (degree)	Temporal resolution (day)	Variables that have been observed
NIMBUS-7 (SMMR)	1978–1987	6.63, 10.69, 18.0, 21.0, 37.0	0.25	2	Precipitation, SM, snow
DMSP-F08 (F08 SSM/I)	1987–2009	19.35, 22.235, 37.0, 85.5	0.25	1	Precipitation, SM, snow.
DMSP-F19 (F19 SSMIS)	2014–	19.35, 22.235, 37.0, 85.5	0.25	1	Precipitation, SM, snow.
TRMM (TMI)	1997–2015	10.7, 19.4, 21.3, 37, 85.5	0.25	1	Precipitation, SM, snow.
WindSat	2003–	6.8, 10.7, 18.7, 23.8, 37.0	0.25	1	Rain rate, SM, snow, cloud liquid water.
EOS-Aqua (AMSR-E)	2002–2011	6.93, 10.65, 18.7, 23.8, 36.5, 89.0	0.25	1	Precipitation, SM, VWC, snow, biomass.
FY-3 (MWRI)	2010–				Precipitation, cloud liquid water, vapor, sea surface temperature.
GCOM (AMSR2)	2012–	6.93, 7.3, 10.65, 18.7, 23.8, 36.5, 89.0	0.25	1	Precipitation, SM, VWC, snow, biomass.
SMOS	2009–	1.4	0.36	1	SM, ocean salinity.
Aquarius	2014–	1.4	0.36	1	SM, SST, SSW.
SMAP	2015–	1.4	0.36	1	ET, SM, VWC, biomass
WCOM	2020	FPIR: 1.2, 2.4, 6.7 PMI: 6.8, 10.65, 18.7, 23.8, 37, 89, 50	FPIR: 50, 30, 15 (km) PMI: 4–50 (km)	1	Precipitation & vapor, ET, SM, sea surface salinity, froze/thaw, SWE.

- FLEX (FLuorescence EXplorer). FLEX is a mission to map vegetation fluorescence to quantify photosynthetic activity initiated by ESA. FLEX will be the first mission designed to observe fluorescence using a novel technique measuring the main part of the chlorophyll fluorescence spectrum that originates from the core of the photosynthetic machinery. The main scientific objectives of FLEX are to assess the quality of the fluorescence-derived photosynthesis data against a classical optical-based method, address temporal and spatial scaling issues in more detail, and indicate potential applications of the novel fluorescence observations.
- GCOM (Global Change Observation Mission). GCOM is a project for the global and long-term observation of Earth initiated by JAXA. GCOM is expected to play an important role in monitoring global water circulation and climate change (Imaoka et al. 2010). The mission consists of two satellite series, GCOM-W and GCOM-C. GCOM-W carries Advanced Microwave Scanning Radiometer 2 (AMSR2), an instrument for observing water-related targets, such as precipitation, water vapor, sea surface wind speed, sea surface temperature, soil moisture, and snow depth. GCOM-C carries a Second Generation Global Imager (SGGI), an instrument for gathering surface and atmospheric measurements of phenomena involved in the carbon cycle and radiation budget, such as clouds, aerosols, ocean color, vegetation, and snow and ice. Global and long-term (10- to 15-year) observations by GCOM will help scientists understand the mechanisms of water circulation and climate change (Imaoka et al. 2010).
- WCOM: The Water Cycle Observation Mission has been proposed by the Chinese Academy of Sciences to improve the capability of synergetic observations of key water cycle variables (Dong et al. 2014; Shi et al. 2014), including soil moisture, ocean salinity, freeze-thaw, and snow water equivalent. The payload configuration of the WCOM satellite is a combination of active and passive wide-frequency-coverage microwave remote sensors with innovative designs, including the L-S-C tri-frequency, Full-Polarized Interferometric synthetic aperture microwave Radiometer (FPIR), the Polarized Microwave radiometric Imager (PMI), which covers a band from 6.6 to 150 GHz, and the X-Ku Dual-Frequency Polarized SCATterometer (DFPSCAT), which has a high spatial resolution. The integration of those payloads can provide global mapping of soil moisture, snow water equivalent, freeze-thaw state, and ET.

Airborne-Based Observation Platforms

Compared to satellite-based remote sensing, airborne-based remote sensing enables a higher spatial resolution and more sensitive observation ability; for example, a super-high spectral resolution can be obtained using airborne sensors. Airborne-based remote sensing is an important complement of satellite-based remote sensing, and the rapid development of UAV remote sensing has enabled lower-cost, more flexible, and more efficient remote sensing experiments. Airborne-based remote sensing plays an important role in ecohydrological process research at the regional

Table 5 Main airborne system used for ecohydrological processes observation

Type	Sensor	Agency/Counry	Band	SR[a] (m)
SAR	C/X-SAR	CCRS/Canada	X/C	0.9,6
	AIRSAR	NASA/USA	C/X/L	0.6,3
	GeoSAR	NASA/JPL	P/X	1.25–3
	PLIS	ARRC/Australia	L	0.2–29
	UAVSAR	NASA/USA	L	1.0,1.8
Radiometer	PSR	NOAA/USA	X/K/Ka	–
	ESTAR	NASA/USA	L	–
	PLMR	NASA/USA	L	–
Spectrometer	AVIRIS	NASA, JPL/USA	360–2500 nm	1-4
	PRISM	NASA/USA	349.9–1053.5 nm	–
	GIFS	NASA/USA	–(multiple sensors)	–

[a]SR = Spatial resolution

and watershed scales. Here we list several representative airborne-based platforms/sensors to illustrate the characteristics of them (Table 5), almost all of the sensors could provide measurements for key ecohydrological variables.

The main characteristics of airborne-based remote sensing are summarized below.

Controllability

More controllable simultaneous airborne-ground or satellite-airborne-ground observing experiments, in which various observations can be precisely matched in space and time and various parameters can be obtained simultaneously, can be conducted based on airborne remote sensing. Furthermore, airborne-based remote sensing can be used to precisely calibrate satellite-based sensors. These special characteristics of airborne-based remote sensing can provide irreplaceable data for research on ecohydrological processes, model validation, and scale transformations. Therefore, airborne-based remote sensing has played key roles in important observational experiments of the past several decades, including FIFE (Sellers et al. 1987), SMEX (Jackson 2002), CLPX (Cline et al. 2009), WATER (Li et al. 2009), HiWATER (Li et al. 2013), SMAPEx (Panciera et al. 2014), and SMAPVEX (McNairn et al. 2015), and will continue to play more important roles in future observational experiments.

Enabling Scale Transformations

Compared to airborne-based remote sensing, the resolution of satellite-based remote sensing is relatively coarse; therefore, the within-pixel heterogeneity is stronger, and it is relatively difficult to develop accurate and detailed validations for satellite remote sensing. Airborne-based remote sensing allows more flexibility to configure spatial, spectral, and temporal resolutions according to researchers' requirements, and finer observations, with resolutions from centimeters to tens of kilometers, can be obtained. Therefore, airborne-based remote sensing data can be used as homogeneous pixel data for remote sensing validation and can thus play an important role in bridging scaling transformations.

Testing the Observing Ability of Satellite-Based Remote Sensing

Nearly all operational satellite-based sensors and the majority of forward model and inversion approaches are first calibrated and verified using airborne-based remote sensing data. In hydrological applications, comparisons between airborne-based remote sensing data and ground observations are preferred due to the coarse spatial resolutions of satellite-based remote sensing. Thus, airborne-based remote sensing is particularly important for testing the performances of satellite-based sensors. For example, airborne-based sensors have been widely applied in several large-scale ground-based remote sensing experiments.

Important Component of Watershed Observation Systems

Although the resolution of satellite-based remote sensing has increased considerably, particularly for satellite-based multispectral and SAR remote sensing techniques, the same resolutions cannot yet be achieved by hyperspectral/hyperspectral and laser radar satellites. The latter play a critical role in research in watershed hydrological and ecological sciences, such as in measurements of forest and crop structure parameters. Therefore, airborne-based remote sensing is not only a scientific experimental approach but also an important component of watershed observing systems.

Ground-Based Observation Platforms

Traditionally, the key ecohydrological elements have mainly been observed using point-scale equipment that does not adequately capture scale-related heterogeneities in the variables. The upscaling of ecohydrological observations is described and analyzed in detail in chapter entitled ▶ "Upscaling Issues in Ecohydrological Observations." The next generation of ground-based observation networks is presented below.

- WSNs (Wireless sensor networks): WSNs integrate sensor, automatic control, wireless transmission, data storage, pre-processing, and visualizing technologies. Each node in a network consists of one or more sensors, data loggers, transceivers, and energy supply systems (solar, wind or battery) that in turn constitute an intelligent and automate observation network (Gong 2010; Delin et al. 2005). In hydrological applications, various types of sensors, such as soil moisture and temperature sensors, have become increasingly more sophisticated, lower cost, and more reliable and have lower power consumptions, making deployments of large numbers of these sensors possible and applicable. Unprecedented increases in sensor networks have enabled us to capture spatial variations in observational variables at the river basin scale. However, observations from WSNs are more objective than manual observations. Thus, WSNs have become a bridge to fill the gaps between the traditional single-point observations and remote sensing pixel observations, thus playing an important role in remote sensing validation and scaling-related research (Kang et al. 2014; Jin et al. 2014; Hart and Martinez 2006; van Zyl et al. 2009; Qu et al. 2014).

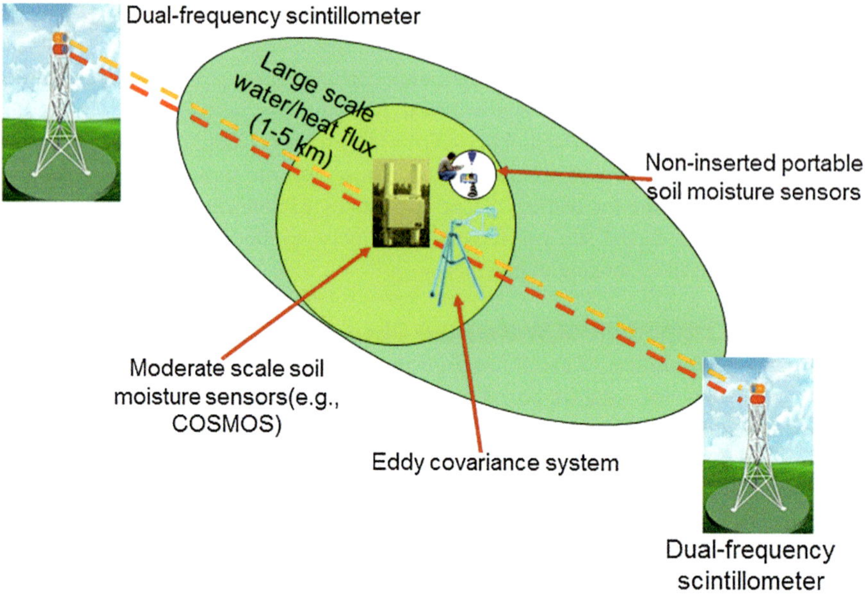

Fig. 2 Typical configuration of a multiscale ground-based observation system for energy balance and soil moisture (Liu et al. 2016; Li et al. 2017). Dual-frequency scintillometers (an infrared scintillometer for measuring sensible heat flux and a microwave scintillometer for measuring latent heat flux) are installed with a separation of approximately 1–5 km. An eddy covariance system is mounted between them, and a wireless sensor network is installed around the EC system, with a cosmic-ray probe in the center

- Large-scale flux and footprint-scale observations (Fig. 2): An eddy covariance (EC) system is a main approach for directly measuring sensible heat and latent heat flux at measurement scales ranging from tens of meters to hundreds of meters. However, EC energy disclosures have been found in nearly all field experiments, possibly because EC systems could not capture large-scale eddies using a single EC set (Foken 2008; Liu et al. 2013). Since the 1990s, optical large aperture scintillometers (LASs) have attracted considerable attention and have been widely used in field experiments (Kleissl et al. 2008; DeBruin 2009; Liu et al. 2011, 2013). LASs can measure sensible heat fluxes on scales of a hundred meters to several kilometers. In recent years, combinations of optical and microwave scintillometers that can measure area sensible heat and latent heat flux directly have been commercialized (Ludi et al. 2005). Their use also makes flux observations easier than medium-resolution satellite remote sensing, such as MODIS and MERIS. In addition, various footprint-scale ground-based observation techniques are widely used in observations of ecohydrological processes. Cosmic-ray probes are the most typical type of equipment; they utilize fast neutrons to detect variations in hydrogen atoms in the soil and atmosphere, thus enabling calculations of soil moisture and snow water equivalent at large scales (Zreda et al. 2012; Han et al. 2014, 2016).

- Many of the newer sensors have developed rapidly and have been widely used for ecohydrological observations. These new sensors include distributed water temperature observation techniques, GPS/GNSS hydrological observations, cosmic rays, isotope methods, new-generation weather radars, ground-based lidar technology, Acoustic Doppler current profiles (ADCPs), the time domain transient electromagnetic method (TDEM), and many other new technologies.

These new observation methods greatly enhance the observability of large-scale ecohydrological processes and enable distributed ground observations.

Challenges in Observing Ecohydrological Processes

Improving the Observation Ability for Ecohydrological Processes

It is critical to continue to develop remotely sensed observation systems that allow for the observation and quantification of ecohydrological processes. There are several challenges to improving observational capability. First, high-resolution remote sensing in space and time is needed to better quantify land-surface heterogeneities and processes occurring at short time scales. Second, better ground-based observational platform designs are needed in terms of measuring ecohydrological processes. In many cases, only variables in specific compartments are measured without considering the complex interactions between hydrology and ecology across all compartments of the land surface. Third, continuous improvements in our ground-based measuring capabilities are needed to allow ecohydrological processes to be quantified.

Integrating Multisource Observations into a Dynamic Model

Integrating multisource observations, e.g., ground- and remote sensing-based observations, into dynamic models to accurately estimate the state and flux variables of ecohydrological processes remains a challenge. Data assimilation is the most promising approach to combine the strengths of multisource observations and dynamic simulations to improve estimations of ecohydrological processes (Han et al. 2013; Huang et al. 2016). However, new assimilation approaches and strategies that are more effective and efficient for improving the estimation of ecohydrological processes and that account for the inherent heterogeneities of land surfaces and upscaling must be developed.

Deriving Upscaling Approaches Through System Observations

Observations of ecohydrological processes must address the scaling issue, a complex problem that is intertwined with the nonlinearities and heterogeneities of ecohydrological processes (Li 2014). Scaling is challenging within all branches of land-surface sciences, including hydrology (Vereecken et al. 2007; Blöschl and Sivapalan 1995; Blöschl 2001), ecology (Wiens 1989), soil science (Lin et al. 2005), and boundary layer meteorology (Raupach and Finnigan 1995), and becomes increasingly prominent as increasing amounts of observational data become available. Multiscale observations must be obtained to further improve our understanding of the scaling issue and validate upscaling methods.

Over the last few years, data availability has increased considerably due to state-of-the-art in situ and remote sensing observations and data acquisition techniques. Moreover, multiscale land-surface observation experiments have been implemented globally within the last decade (Vereecken et al. 2008; Jensen and Illangasekare 2011; Debeer et al. 2015; Li et al. 2013; Yang et al. 2013; Bogena et al. 2012; Zacharias et al. 2008). Those experiments have provided a promising method for bridging knowledge gaps among microscopic-, mesoscopic-, and macroscopic-scale understandings.

Ecohydrological processes are highly controlled by heterogeneities in states and parameters, and assessing the impacts of small-scale heterogeneity on larger-scale behaviors remains a challenge. Thus, how to capture and quantify spatiotemporal heterogeneities of states and parameters across scales is an important issue to resolve.

Overall, upscaling and heterogeneity issues are related to the quantification of uncertainty. Such quantification is extremely complex when observing ecohydrological processes at multiple scales. Thus, the scaling and heterogeneity issues are among the most challenging problems in ecohydrological observation.

Estimating Representativeness Error

Measurements are subject to instrument and representativeness errors, which are independent of each other. However, representativeness error, which is scale dependent, is considerably more difficult to quantify (Li 2014). From a mathematical perspective, the spatial representativeness errors of observations should be made statistically unbiased, and their variances should be as small as possible.

New-generation observing techniques, e.g., those introduced in the section above, are reliable for direct measurements. However, in terms of spatial representativeness, which requires appropriate scaling, both remote sensing and in situ observations exhibit greater uncertainties. In terms of the reliabilities of retrieval and estimation models, which transform direct measurements into the necessary hydrological and ecological variables or parameters, those observations also exhibit considerable uncertainties. Thus, although the original measurements capture the heterogeneities, translating those measurements into useful information for understanding ecohydrological processes requires appropriate scaling schemes and carefully developed models that can provide reliable retrievals. In turn, optimal observation network designs and sampling strategies as well as suitable approaches to quantify the uncertainties associated with scaling and estimation/retrieval models are needed.

Outlook and Perspectives

Advances in ecohydrological observation, both ground-based and remotely sensed, have contributed significantly to understandings of ecohydrological processes. In the future, these new observational technologies and methods should be fully employed to further benefit ecohydrological science.

- To better translate remotely sensed data into information that helps us better understand ecohydrological processes and better inform land-surface models. We should move toward mechanistic models that allow remote sensing measurements to be predicted, e.g., to mechanistically predict vegetation canopy fluorescence, and use those models to estimate parameters that are critical in ecohydrological processes through inverse estimation.
- To better quantify the roles of subsurface processes in ecohydrological processes. Data assimilation can play an important role by integrating different types of data with varying spatial and temporal resolutions. Remotely sensed data are critical because they provide global spatial coverage at different spatial resolutions. Remote sensing techniques that allow ecohydrological processes to be better captured are urgently needed. Recent steps have been undertaken, but considerable work remains to be done.
- To develop observational systems that allow ecohydrological processes to be captured across different scales and across compartments. This goal will require the integration of different sensing systems at selected locations to capture non-linearities and feedback processes among the subsurface, vegetation, and atmosphere.
- To use well-instrumented watersheds as test beds of new concept for ecohydrological observations. In designing this type of a watershed observation system, multiscale heterogeneity must be considered such that the knowledge obtained can eventually be scaled up to the basin, regional, and global scales.
- To combine monitoring and controllable and synthetic observation experiments. Earth science begins with observations rather than controlled experiments. Nevertheless, as earth science became an experimental science, controllable experiments, the goals of which are to verify scientific hypotheses, were advocated. Accordingly, synthetic observation experiments consisting of monitoring and controlled experiments have become more important.
- To utilize technical advancements in new models. Modeling is lagging behind new observational techniques. Modeling strategies must shift from being spatially explicit to being scale explicit. Data assimilation methods will play a key role in this shift. However, determining how to effectively combine modeling and observation using data assimilation methods remains a considerable challenge that must be addressed.
- To integrate observation systems with integrated models, data services, and decision-making. Observation systems should be fully coupled with data systems and fully used in ecohydrological modeling to gain understandings of ecohydrological processes and to support decision-making by water resources

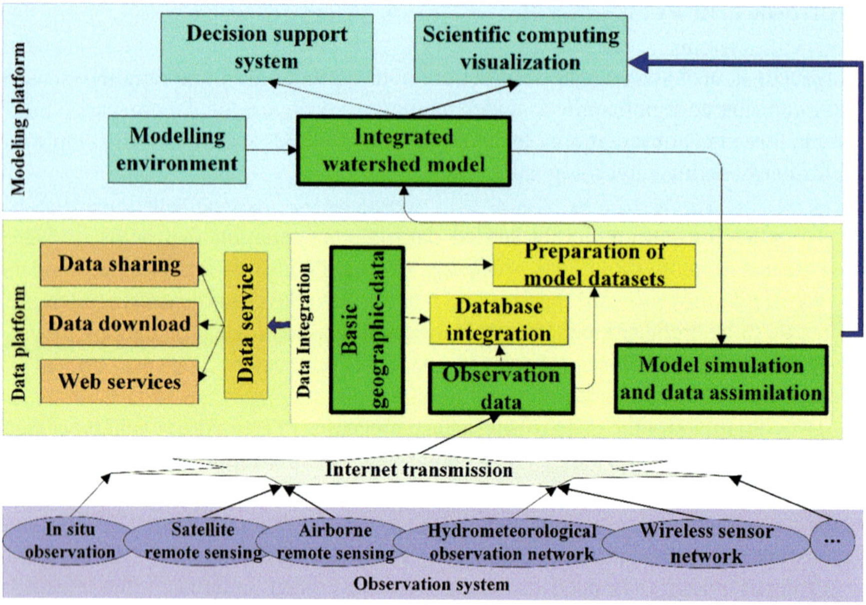

Fig. 3 Integration of multisource observations, integrated models, and information technologies to understand ecohydrological processes

and ecosystem management (Fig. 3). Automatic data collection, data transfer, and the remote control of observational nodes can be realized using the Internet of Things technologies. In addition, various observations at different scales and from different sources should be assimilated into integrated ecohydrological models to provide better estimates of ecohydrological state variables.

Summary of the Contents of this Book

This book is organized with the aim of introducing the state of the art of technologies and measurement methods of ecohydrological processes.

The first chapter provides an overview of the progress, challenges, and perspectives of observing ecohydrological processes. It demonstrates that observations of ecohydrological processes have developed considerably in terms of observation techniques, sensor design and application, and observing platforms (ground-, airborne-, and satellite-based). In particular, applications of next-generation information, communications, and sensor techniques have improved our ability to observe ecohydrological processes considerably. For example, the development and application of wireless sensor networks make it easier for scientists and engineers to collect, control, monitor, and visualize observational data. Nevertheless, problems and challenges remain and must be resolved. The mechanism of the scaling issue remains unclear, and methodologies for scaling transformations must be investigated. Thus, observational experiments must be conducted, and world-class observing systems that can capture heterogeneity, quantify uncertainty, and combine observations and modeling to develop advanced ecohydrological data assimilation systems must be developed.

Chapter ▶ "Ground-Based Soil Moisture Determination" introduces the estimation of soil moisture by ground-based approaches and provides an analysis of the advantages and limitations of each technique. The state of the art in ground-based observation sensors and estimation approaches for soil moisture are highlighted. Spatial scale remains the main issue throughout the chapter. Taking different spatial scales as the breakthrough point, the ground-based techniques of soil moisture are discussed from the point scale to the field scale. Point-scale measurements provide observations with high temporal resolutions, but they have limited spatial representativeness. The application of wireless sensor networks provides multitudes of soil moisture measurements that can capture spatial heterogeneity and allow estimates of soil moisture uncertainty to some extent. Several field-scale observational techniques, such as ground-penetrating radar, ground-based L-band radiometry, cosmic-ray probes, and global navigation satellite system reflectometry, are introduced in detail. The authors also expect to develop new instruments with higher accuracies and spatiotemporal resolutions for soil moisture measurements.

Chapter ▶ "Airborne and Spaceborne Passive Microwave Measurements of Soil Moisture" introduces the basic theories and methodologies of airborne and spaceborne microwave remote sensing for soil moisture, surface roughness, and vegetation parameters. Numerous international missions designed for soil moisture monitoring and mapping continue to be carried out, including the ESA/SMOS, NASA/SMAP, JAXA/GCOM-W, and China/WCOM missions. Surface roughness and vegetation effects are the main uncertainty sources for soil moisture estimation; thus, determining how to estimate those effects remains a challenge. The chapter describes related models and algorithms that attempt to solve the problem by proposing parameterized models and microwave vegetation indices (MVIs). The newly launched SMOS and SMAP satellites provide global soil moisture products for hydrological, meteorological, and climatological applications. However, current soil moisture products continue to have large uncertainties, and their spatial resolutions are coarse. The retrieval accuracy can be increased by combining the L-and S-bands; this finding will be utilized in the upcoming WCOM mission.

Chapter ▶ "Remote Sensing Precipitation: Sensors, Retrievals, Validations, and Applications" systematically introduces the ground-based sensing, retrieval, validation, and remote sensing of precipitation. The authors review the sensors and precipitation retrievals from two aspects of ground- and satellite-based platforms. Full-band remote sensing from visible/infrared to microwave satellite sensors and ground-based weather radars are widely used in precipitation sensing and retrieval. The former provide rainfall estimates best from space, whereas the latter accurately provide estimates of precipitation on the ground; thus, high-quality precipitation estimates can be generated by combining satellite-based sensors with ground-based radars and rain gauges. Global precipitation products, validation, and application are introduced in detail. Single sensors or individual satellites cannot provide adequate precipitation estimates; thus, numerous multi-satellite precipitation products have been developed and applied. The products can typically provide both near-real-time and post-real-time datasets for climatology, meteorology, and hydrology applications. Validations of remotely sensed precipitation products suggest that large uncertainties remain and that the products must be further improved.

Chapter ▶ "Inhomogeneity in Winter Precipitation Measurements" introduces precipitation measurement-related topics, mainly focusing on ground-based measuring methods. The precipitation gauge is regarded as one of the most important and direct measuring tools; this tool is not only the standard of ground-based precipitation measurements but also provides ground truth for validating remotely sensed precipitation estimates. Various types of precipitation gauges as well as their advantages and disadvantages are comprehensively introduced. With the advances of sensor techniques, various new instruments have been developed and applied in precipitation measurements, such as snow pillow, gamma radiation system, and disdrometer methods. The authors demonstrate that identifying the precipitation phase is very important for the functioning of models in many discipline applications, and thus, they provide several methods to identify the precipitation type. Finally, the authors point out the biases that exist in the precipitation observations and propose several methods to correct these biases.

Chapter ▶ "Remotely Sensed Evapotranspiration" introduces the quantification of land surface evapotranspiration based on remote sensing approaches. Three typical methods for evapotranspiration estimation, including surface energy balance models, vegetation index-land surface temperature space approaches, and the Penman-Monteith equation, are presented. Based on an actual experimental example, the authors comprehensively compare the advantages and drawbacks of the three methods. The spatio-temporal scaling and global remote sensing products of evapotranspiration are introduced. The authors believe that great achievements have been made in evapotranspiration estimation, but a tremendous challenge remains. These researchers propose one potential solution to incorporate the evapotranspiration equation into hydrological models to close the water balance within a given domain.

Chapter ▶ "Micrometeorological Methods to Determine Evapotranspiration" introduces the observation of evapotranspiration by micrometeorological methods. The Bowen ratio, eddy covariance, and scintillometer methods are described in detail from the aspects of theory, installation, maintenance, data processing, and footprint source analysis. The Bowen ratio is simple and can integrate latent heat fluxes over large areas and observe fluxes on fine time scales; however, it is overly sensitive to the instrument bias and requires an adequate upwind source area to establish an equilibrium. The eddy covariance method, which is based on the premise of Taylor's frozen turbulence hypothesis, is the most accurate method for measuring water, heat, and carbon dioxide fluxes at fine spatial and temporal resolutions. However, the method requires careful footprint source analysis and data processing. Scintillometers measure areal average sensible and latent heat fluxes based on the Monin-Obukhov similarity theory. Scintillometers have the potential to bridge the gap between point-scale observations and satellite pixel or model grid scales because they can provide average fluxes measurements over hundreds of meters to several kilometers. Furthermore, typical evapotranspiration observation systems and experiments, such as FLUXNET, NEON, CZO, TERENO HOBE, FIFE, WATER, and HiWATER, are reviewed. The scaling mismatch between ground- and remotely sensed-based measurements is the main problem and challenge to be overcome. Scientists and engineers should focus on developing large-scale ET observational methods, such as airborne EC and multiband LAS.

Chapter ► "Surface Runoff" introduces four kinds of measurement techniques for surface runoff, namely, the runoff plot method, curve number method, isotopic tracer method, and salt solution method. First, the definition and formation mechanism of surface runoff are introduced. The authors focus on the analysis of four observation methods for surface runoff and demonstrate the following: (1) the runoff plot method is a prerequisite of watershed scale investigations and is widely used to evaluate the rainfall–runoff processes with better controllability and is also a prerequisite for developing the regional hydrological model; (2) the curve number method depends on using the measured watershed runoff and rainfall data and has been widely applied by the engineers and hydrologists as a simple watershed model; (3) the isotopic tracer method is based on the mass balance of stable isotopes, which could provide comprehensive insights into runoff processes; and (4) the principle of the salt solution method is to measure the shallow water flow by detecting the movement of salt. The modeling methods and vegetation effects of surface runoff are also briefly introduced. Finally, challenges regarding the measurement and simulation of surface runoff are presented.

Chapter ► "Subsurface Flow" provides an overview of current research on subsurface flow, including the concept, theoretical development, classification, controlling factors, and main measurement methods of subsurface flow. The connection between subsurface flow and ecohydrological processes is highlighted, and future research directions are proposed. Physical-based direct observations, isotope tracers, and model simulations are the main methods for subsurface runoff estimation. Direct observation is highly effective but based on collecting water. Isotope tracers and other novel techniques are extensively used but remain underdeveloped in terms of detailed processes, and the precise transport method remains poorly understood. Thus, future directions in subsurface flow research must focus on the understanding of subsurface flow processes, obtaining high-precision and systematic field datasets, extracting commonalities from multiple different well-instrumented sites, testing models, and assessing uncertainties.

Chapter ► "Photosynthesis (NPP, NEP, Respiration)" systematically introduces the measurement and modeling methods for carbon fluxes from field to regional or global scales. The state of the art in carbon flux estimation is reviewed, and challenges are analyzed. Remote sensing-based light use efficient models, process-based ecosystem models, and upscaling approaches are the main methods for estimating terrestrial carbon fluxes. The authors demonstrate that assimilating multisource observations, including remote sensing and field measurements, into dynamic ecosystem models could allow for accurate carbon flux estimation at regional or global scales.

Chapter ► "Leaf Area Index: Advances in Ground-Based Measurement" systematically reviews the main methods for LAI measurements and emphasizes newly advanced methods, namely, LAINet and LAISmart, which are based either on WSNs or mobile computing platform techniques for measuring LAI at high efficiencies and low costs. The former can reduce field data collection costs, and the latter can provide automatic measurements using mobile applications deployed on smartphones. Integrating passive and active optical signals for LAI measurement may be a future trend in this field.

Chapter ▶ "Radar Remote Sensing of Land Surface Parameters" reviews and analyzes the advantages and limitations of radar remote sensing in land cover and agriculture applications. In particular, the use of PolSAR and InSAR techniques is useful in land cover monitoring, classification, parameter estimation, and crop and forest height retrieval. SAR polarimetry and interferometry should also be employed to separate different elemental scattering processes, which will improve classification and allow for the development of temporal series analysis to obtain multisource information for boosting classification.

Chapter ▶ "Root Processes Affecting Soil Moisture Patterns in Ecohydrology" introduces the effects of vegetation (especially the plant root processes) on the soil moisture distribution pattern. The root growth, root water uptake and transpiration, plant competitions and rhizosphere properties are identified as the potential drivers of the soil moisture content variability. The authors find that high transpiration, root growth, and root water uptake generally tend to increase the vertical variability for drying conditions in wet soils. These authors also indicate that mechanistic models might help investigate how such interactions control the shape of soil moisture relations in a quantitative manner.

Chapter ▶ "Upscaling Issues in Ecohydrological Observations" describes the upscaling of soil water processes and hydraulic properties in the vadose zone as well as the upscaling of soil water-plant processes. Applications of data assimilation to estimate the ecohydrologically relevant parameters and novel observational platforms and techniques are also introduced in detail. The integration of novel upscaling approaches and sensing techniques will provide insights into understandings of ecohydrological processes.

Chapter ▶ "Field Experimental Design for a Watershed Observing System" introduces a field experimental design and watershed observation system for ecohydrological processes. The watershed observation system should be capable of observing water and energy cycles, ecological processes, and socioeconomic activities with state of science. The observation system should be able to capture spatial and temporal heterogeneities and variabilities in ecohydrological variables as well as quantify uncertainties. Taking the WATER and HiWATER experiments as case studies, the design and conduct of a field experiment and establishment of a watershed observation system are described in detail from the perspectives of thematic experiments, observation system establishment, scaling transformation, remote sensing products, data information systems, and data curation and sharing systems.

Acknowledgments This work was supported by the National Natural Science Foundation of China (Grant No. 91425303), and the Strategic Priority Research Program of the Chinese Academy of Sciences, Grant No. XDA20100104.

References

B.W. Abbott, V. Baranov, C. Mendoza-Lera, M. Nikolakopoulou, A. Harjung, T. Kolbe, M.N. Balasubramanian, T.N. Vaessen, F. Ciocca, A. Campeau, M.B. Wallin, P. Romeijn, M. Antonelli, J. Goncalves, T. Datry, A.M. Laverman, J.R. de Dreuzy, D.M. Hannah, S. Krause, C. Oldham, G. Pinay, Using multi-tracer inference to move beyond single-catchment ecohydrology. Earth Sci. Rev. **160**, 19–42 (2016)

S.H. Alemohammad, D.B. McLaughlin, D. Entekhabi, Quantifying precipitation uncertainty for land data assimilation applications. Mon. Weather Rev. **143**, 3276–3299 (2015)

S.P. Anderson, R.C. Bales, C.J. Duffy, Critical zone observatories: building a network to advance interdisciplinary study of Earth surface processes. Mineral. Mag. **72**, 7–10 (2008)

M. Andreasen, K.H. Jensen, D. Desilets, T.E. Franz, M. Zreda, H.R. Bogena, M.C. Looms, Status and perspectives on the cosmic-ray neutron method for soil moisture estimation and other environmental science applications. Vadose Zone J. **16** (2017). https://doi.org/10.2136/vzj2017.04.0086

W.G.M. Bastiaanssen, M. Menenti, R.A. Feddes, A.A.M. Holtslag, A remote sensing surface energy balance algorithm for land (SEBAL) - 1. Formulation. J. Hydrol. **212**, 198–212 (1998)

K. Beven, Robert E. Horton's perceptual model of infiltration processes. Hydrol. Process. **18**, 3447–3460 (2004)

G. Blöschl, Scaling in hydrology. Hydrol. Process. **15**, 709–711 (2001)

G. Blöschl, M. Sivapalan, Scale issues in hydrological modelling: A review. Hydrol. Process. **9**, 251–290 (1995)

H.R. Bogena, M. Herbst, J.A. Huisman, U. Rosenbaum, A. Weuthen, H. Vereecken, Potential of wireless sensor networks for measuring soil water content variability. Vadose Zone J. **9**, 1002–1013 (2010)

H. Bogena, R. Kunkel, T. Puetz, H. Vereecken, E. Kruger, S. Zacharias, P. Dietrich, U. Wollschlager, H. Kunstmann, H. Papen, H.P. Schmid, J.C. Munch, E. Priesack, M. Schwank, O. Bens, A. Brauer, E. Borg, I. Hajnsek, TERENO – long-term monitoring network for terrestrial environmental research. Hydrol. Wasserbewirtsch. **56**, 138–143 (2012)

O. Bonacci, T. Pipan, D.C. Culver, A framework for karst ecohydrology. Environ. Geol. **56**, 891–900 (2009)

J.G. Canadell, H.A. Mooney, D.D. Baldocchi, J.A. Berry, J.R. Ehleringer, C.B. Field, S.T. Gower, D.Y. Hollinger, J.E. Hunt, R.B. Jackson, S.W. Running, G.R. Shaver, W. Steffen, S.E. Trumbore, R. Valentini, B.Y. Bond, Carbon metabolism of the terrestrial biosphere: A multitechnique approach for improved understanding. Ecosystems **3**, 115–130 (2000)

J.M. Chen, T.A. Black, Measuring leaf area index of plant canopies with branch architecture. Agric. For. Meteorol. **57**, 1–12 (1991)

L. Chen, O.W. Frauenfeld, A comprehensive evaluation of precipitation simulations over China based on CMIP5 multimodel ensemble projections. J. Geophys. Res.-Atmos. **119**, 5767–5786 (2014)

D. Cline, S. Yueh, B. Chapman, B. Stankov, A. Gasiewski, D. Masters, K. Elder, R. Kelly, T.H. Painter, S. Miller, NASA cold land processes experiment (CLPX 2002/03): airborne remote sensing. J. Hydrometeorol. **9**, 1427–1433 (2009)

C.M. Debeer, H.S. Wheater, W.L. Quinton, S.K. Carey, R.E. Stewart, M.D. Mackay, P. Marsh, The changing cold regions network: observation, diagnosis and prediction of environmental change in the Saskatchewan and Mackenzie River Basins, Canada. Sci. China Earth Sci. **58**, 46–60 (2015)

H.A.R. DeBruin, Time to think: reflections of a pre-pensioned scintillometer researcher. Bull. Am. Meteorol. Soc. **90**, Es17–Es26 (2009)

K.A. Delin, S.P. Jackson, D.W. Johnson, S.C. Burleigh, R.R. Woodrow, J. Michael Mcauley, J.M. Dohm, F. Ip, T.P.A. Ferré, D.F. Rucker, Environmental studies with the sensor web: principles and practice. Sensors **5**, 103–117 (2005)

Dong, X., H. Liu, Z. Wang, J. Shi, and T. Zhao, COM: the mission concept and payloads of a global water cycle observation mission, in *2014 IEEE Geoscience and Remote Sensing Symposium* (2014), pp. 3338–3341

M. Drusch, J. Moreno, U. Del Bello, R. Franco, Y. Goulas, A. Huth, S. Kraft, E.M. Middleton, F. Miglietta, G. Mohammed, L. Nedbal, U. Rascher, D. Schttemeyer, W. Verhoef, The FLuorescence EXplorer mission concept-ESA's Earth Explorer 8. IEEE Trans. Geosci. Remote Sens. **55**, 1273–1284 (2017)

M. Durand, L.L. Fu, D.P. Lettenmaier, D.E. Alsdorf, E. Rodriguez, D. Esteban-Fernandez, The surface water and ocean topography mission: observing terrestrial surface water and oceanic submesoscale eddies. Proc. IEEE **98**, 766–779 (2010)

P.S. Eagleson, *Ecohydrology: Darwinian expression of vegetation form and function* (Cambridge University Press, Cambridge, 2002)

D. Entekhabi, E.G. Njoku, P.E. O'Neill, K.H. Kellogg, W.T. Crow, W.N. Edelstein, J.K. Entin, S.D. Goodman, T.J. Jackson, J. Johnson, J. Kimball, J.R. Piepmeier, R.D. Koster, N. Martin, K.C. McDonald, M. Moghaddam, S. Moran, R. Reichle, J.C. Shi, M.W. Spencer, S.W. Thurman, L. Tsang, J. Van Zyl, The soil moisture active passive (SMAP) mission. Proc. IEEE **98**, 704–716 (2010)

T. Foken, The energy balance closure problem: an overview. Ecol. Appl. **18**, 1351–1367 (2008)

P. Gong, Progress in recent environmental applications of wireless sensor networks. J. Remote Sens. **14**, 153–157 (2010)

F. Gottardi, C. Obled, J. Gailhard, E. Paquet, Statistical reanalysis of precipitation fields based on ground network data and weather patterns: application over French mountains. J. Hydrol. **432**, 154–167 (2012)

L. Guo, H. Lin, Critical zone research and observatories: current status and future perspectives. Vadose Zone J. **15**(9) (2016). https://doi.org/10.2136/vzj2016.06.0050

X.J. Han, H.J.H. Franssen, X. Li, Y.L. Zhang, C. Montzka, H. Vereecken, Joint assimilation of surface temperature and L-band microwave brightness temperature in land data assimilation. Vadose Zone J. **12**(3), 155–175 (2013)

X.J. Han, R. Jin, X. Li, S.G. Wang, Soil moisture estimation using cosmic-ray soil moisture sensing at heterogeneous farmland. IEEE Geosci. Remote Sens. Lett. **11**, 1659–1663 (2014)

X.J. Han, H.J.H. Franssen, M.Á.J. Bello, R. Rosolem, H. Bogena, F.M. Alzamora, Simultaneous soil moisture and properties estimation for a drip irrigated field by assimilating cosmic-ray neutron intensity. J. Hydrol. **539**, 611–624 (2016)

J.K. Hart, K. Martinez, Environmental sensor networks: a revolution in the earth system science? Earth Sci. Rev. **78**, 177–191 (2006)

A.Y. Hou, R.K. Kakar, S. Neeck, A.A. Azarbarzin, C.D. Kummerow, M. Kojima, R. Oki, K. Nakamura, T. Iguchi, The global precipitation measurement mission. Bull. Am. Meteorol. Soc. **95**, 701–722 (2014)

C.L. Huang, W.J. Chen, Y. Li, H.F. Shen, X. Li, Assimilating multi-source data into land surface model to simultaneously improve estimations of soil moisture, soil temperature, and surface turbulent fluxes in irrigated fields. Agric. For. Meteorol. **230–231**, 142–156 (2016)

S. Hunt, Measurements of photosynthesis and respiration in plants. Physiol. Plant. **117**, 314–325 (2003)

K. Imaoka, M. Kachi, H. Fujii, H. Murakami, M. Hori, A. Ono, T. Igarashi, K. Nakagawa, T. Oki, Y. Honda, H. Shimoda, Global change observation mission (GCOM) for monitoring carbon, water cycles, and climate change. Proc. IEEE **98**, 717–734 (2010)

T. Jackson, Soil Moisture Experiments in 2002 and 2003. American Geophysical Union Meeting. May 28–31, 2002. Washington, DC, EOS Trans. of AGU,83:S191

K.H. Jensen, T.H. Illangasekare, HOBE: a hydrological observatory. Vadose Zone J. **10**, 1–7 (2011)

R. Jin, X. Li, B.P. Yan, X.H. Li, W.M. Luo, M.G. Ma, J.W. Guo, J. Kang, Z.L. Zhu, S.J. Zhao, A nested ecohydrological wireless sensor network for capturing the surface heterogeneity in the midstream areas of the Heihe River Basin, China. IEEE Geosci. Remote Sens. Lett. **11**, 2015–2019 (2014)

R. Jin, X. Li, S.M. Liu, Understanding the heterogeneity of soil moisture and evapotranspiration using multiscale observations from satellites, airborne sensors, and a ground-based observation matrix. IEEE Geosci. Remote Sens. Lett. **14**, 2132–2135 (2017)

I. Jonckheere, S. Fleck, K. Nackaerts, B. Muys, P. Coppin, M. Weiss, F. Baret, Review of methods for in situ leaf area index determination – part I. theories, sensors and hemispherical photography. Agric. For. Meteorol. **121**, 19–35 (2004)

J. Kang, X. Li, R. Jin, Y. Ge, J.F. Wang, J.H. Wang, Hybrid optimal design of the eco-hydrological wireless sensor metwork in the middle reach of the Heihe River Basin, China. Sensors **14**, 19095–19114 (2014)

J. Kang, R. Jin, X. Li, Regression Kriging-based upscaling of soil moisture measurements from a wireless sensor network and multiresource remote sensing information over heterogeneous cropland. IEEE Geosci. Remote Sens. Lett. **12**, 92–96 (2015)

Y.H. Kerr, P. Waldteufel, J.P. Wigneron, J. Martinuzzi, J. Font, M. Berger, Soil moisture retrieval from space: the soil moisture and ocean salinity (SMOS) mission. IEEE Trans. Geosci. Remote Sens. **39**, 1729–1735 (2001)

Y.H. Kerr, P. Waldteufel, J.P. Wigneron, S. Delwart, F. Cabot, J. Boutin, M.J. Escorihuela, J. Font, N. Reul, C. Gruhier, S.E. Juglea, M.R. Drinkwater, A. Hahne, M. Martin-Neira, S. Mecklenburg, M. Martı, The SMOS mission: new tool for monitoring key elements of the global water cycle. Proc. IEEE **98**, 666–687 (2010)

J. Kleissl, J. Gomez, S.H. Hong, J.M.H. Hendrickx, T. Rahn, W.L. Defoor, Large aperture scintillometer intercomparison study. Bound.-Layer Meteorol. **128**, 133–150 (2008)

X. Li, Characterization, controlling, and reduction of uncertainties in the modeling and observation of land-surface systems. Sci. China Earth Sci. **57**, 80–87 (2014)

X. Li, X.W. Li, Z.Y. Li, M.G. Ma, J. Wang, Q. Xiao, Q. Liu, T. Che, E.X. Chen, G.J. Yan, Z.Y. Hu, L.X. Zhang, R.Z. Chu, P.X. Su, Q.H. Liu, S.M. Liu, J.D. Wang, Z. Niu, Y. Chen, R. Jin, W.Z. Wang, Y.H. Ran, X.Z. Xin, H.Z. Ren, Watershed allied telemetry experimental research. J. Geophys. Res.-Atmos. **114** (2009). https://doi.org/10.1029/2008JD011590

X. Li, G.D. Cheng, S.M. Liu, Q. Xiao, M.G. Ma, R. Jin, T. Che, Q.H. Liu, W.Z. Wang, Y. Qi, J.G. Wen, H.Y. Li, G.F. Zhu, J.W. Guo, Y.H. Ran, S.G. Wang, Z.L. Zhu, J. Zhou, X.L. Hu, Z.W. Xu, Heihe Watershed Allied Telemetry Experimental Research (HiWATER): scientific objectives and experimental design. Bull. Am. Meteorol. Soc. **94**, 1145–1160 (2013)

X. Li, S.M. Liu, Q. Xiao, M.G. Ma, R. Jin, T. Che, W.Z. Wang, X.L. Hu, Z.W. Xu, J.G. Wen, L.X. Wang, A multiscale dataset for understanding complex eco-hydrological processes in a heterogeneous oasis system. Sci Data **4**, 170083 (2017). https://doi.org/10.1038/sdata.2017.83

H.S. Lin, D. Wheeler, J. Bell, L. Wilding, Assessment of soil spatial variability at multiple scales. Ecol. Model. **182**, 271–290 (2005)

H. Lin, H.J. Vogel, J. Seibert, Towards holistic studies of the Earth's critical zone: hydropedology perspectives' preface. Hydrol. Earth Syst. Sci. **14**, 479–480 (2010)

C.M. Liu, Y.N. Chen, Z.X. Xu, Eco-hydrology and sustainable development in the arid regions of China preface. Hydrol. Process. **24**, 127–128 (2010)

S.M. Liu, Z.W. Xu, W.Z. Wang, Z.Z. Jia, M.J. Zhu, J. Bai, J.M. Wang, A comparison of eddy-covariance and large aperture scintillometer measurements with respect to the energy balance closure problem. Hydrol. Earth Syst. Sci. **15**, 1291–1306 (2011)

S.M. Liu, Z.W. Xu, Z.L. Zhu, Z.Z. Jia, M.J. Zhu, Measurements of evapotranspiration from eddy-covariance systems and large aperture scintillometers in the Hai River Basin, China. J. Hydrol. **487**, 24–38 (2013)

S.M. Liu, Z.W. Xu, Q.Y. Zhao, T.R. Xu, Y. Ge, Y.F. Ma, L.S. Song, Z.L. Zhu, Z.Z. Jia, F. Zhang, Upscaling evapotranspiration measurements from multi-site to the satellite pixel scale over heterogeneous land surfaces. Agric. For. Meteorol. **230**, 97–113 (2016)

S.M. Li, X. Li, Z.W. Xu, The Heihe integrated observatory network: a basin-scale land surface processes observatory in China. Vadose Zone J., under review (2018)

D. Liverman, J. Rockström, M. Visbek, R. Leemans, T. Abrahamse, B. Becker, R. D'Sousa, K. Jones, H. Mooney, I. Niang, *Future Earth initial design* (International Council for Science, Future Earth, Interim Secretariat, Paris, 2013)

A. Ludi, F. Beyrich, C. Matzler, Determination of the turbulent temperature-humidity correlation from scintillometric measurements. Bound.-Layer Meteorol. **117**, 525–550 (2005)

H. McNairn, T.J. Jackson, G. Wiseman, S. Belair, A. Berg, P. Bullock, A. Colliander, M.H. Cosh, S.B. Kim, R. Magagi, M. Moghaddam, E.G. Njoku, J.R. Adams, S. Homayouni, E.R. Ojo, T.L. Rowlandson, J.L. Shang, K. Goita, M. Hosseini, The soil moisture active passive validation experiment 2012 (SMAPVEX12): prelaunch calibration and validation of the SMAP soil moisture algorithms. IEEE Trans. Geosci. Remote Sens. **53**, 2784–2801 (2015)

M. Meroni, M. Rossini, L. Guanter, L. Alonso, U. Rascher, R. Colombo, J. Moreno, Remote sensing of solar-induced chlorophyll fluorescence: review of methods and applications. Remote Sens. Environ. **113**, 2037–2051 (2009)

NRC, *Earth science and applications from space: national imperatives for the next decade and beyond* (National Academy Press, Washington, DC, 2007)

X.D. Pan, X.J. Tian, X. Li, Z.H. Xie, A.M. Shao, C.Y. Lu, Assimilating Doppler radar radial velocity and reflectivity observations in the weather research and forecasting model by a proper orthogonal-decomposition-based ensemble, three-dimensional variational assimilation method. J. Geophys. Res.-Atmos. **117** (2012). https://doi.org/10.1029/2012jd017684

X.D. Pan, X. Li, G.D. Cheng, Y. Hong, Effects of 4D-Var data assimilation using remote sensing precipitation products in a WRF model over the complex terrain of an arid region river basin. Remote Sens. **9**, 963 (2017). https://doi.org/10.3390/rs9090963

R. Panciera, J.P. Walker, T.J. Jackson, D.A. Gray, M.A. Tanase, D. Ryu, A. Monerris, H. Yardley, C. Rudiger, X.L. Wu, Y. Gao, J.M. Hacker, The soil moisture active passive experiments (SMAPEx): toward soil moisture retrieval from the SMAP mission. IEEE Trans. Geosci. Remote Sens. **52**, 490–507 (2014)

C. Prigent, Precipitation retrieval from space: an overview. Compt. Rendus Geosci. **342**, 380–389 (2010)

Y.H. Qu, Y.Z. Zhang, J.D. Wang, A dynamic Bayesian network data fusion algorithm for estimating leaf area index using time-series data from in situ measurement to remote sensing observations. Int. J. Remote Sens. **33**, 1106–1125 (2012)

Y.H. Qu, Y.Q. Zhu, W.C. Han, J.D. Wang, M.G. Ma, Crop leaf area index observations with a wireless sensor network and its potential for validating remote sensing products. IEEE J. Sel. Top. Appl. Earth Obs. Remote Sen. **7**, 431–444 (2014)

M.R. Raupach, J.J. Finnigan, Scale issues in boundary-layer meteorology: surface energy balances in heterogeneous terrain. Hydrol. Process. **9**, 589–612 (1995)

D.D. Richter, M.L. Mobley, Monitoring Earth's critical zone. Science **326**, 1067–1068 (2009)

S.W. Running, S.T. Gower, A general model of forest ecosystem processes for regional applications.2. Dynamic carbon allocation and nitrogen budgets. Tree Physiol. **9**, 147–160 (1991)

K. Scipal, M. Arcioni, J. Chave, J. Dall, F. Fois, T. Le Toan, C.C. Lin, K. Papathanassiou, S. Quegan, F. Rocca, S. Saatchi, H. Shugart, L. Ulander, M. Williams, The BIOMASS mission 2014; an ESA Earth explorer candidate to measure the BIOMASS of the earth's forests, in *2010 IEEE International Geoscience and Remote Sensing Symposium* (2010), pp. 52–55

P.J. Sellers, Modeling the exchanges of energy, water, and carbon between continents and the atmosphere. Science **275**, 502–509 (1997)

P.J. Sellers, F.G. Hall, G. Asrar, D.E. Strebel, R.E. Murphy, The First ISLSCP field experiment (FIFE). Bull. Am. Meteorol. Soc. **69**, 22–27 (1987)

J.C. Shi, X.L. Dong, T.J. Zhao, J. Y. Du, L. M. Jiang, Y. Du, H. Liu, Z. Wang, D. B. Ji, C. Xiong, WCOM: the science scenario and objectives of a global water cycle observation mission, in *2014 IEEE Geoscience and Remote Sensing Symposium* (2014), pp. 3646–3649

G. Skofronick-Jackson, W.A. Petersen, W. Berg, C. Kidd, E.F. Stocker, D.B. Kirschbaum, R. Kakar, S.A. Braun, G.J. Huffman, T. Iguchi, P.E. Kirstetter, C. Kummerow, R. Meneghini, R. Oki, W.S. Olson, Y.N. Takayabu, K. Furukawa, T. Wilheit, The global precipitation measurement (GPM) mission for science and society. Bull. Am. Meteorol. Soc. **98**, 1679–1695 (2017)

Z. Su, The surface energy balance system (SEBS) for estimation of turbulent heat fluxes. Hydrol. Earth Syst. Sci. **6**, 85–99 (2002)

T. Le Toan, S. Quegan, M.W.J. Davidson, H. Balzter, P. Paillou, K. Papathanassiou, S. Plummer, F. Rocca, S. Saatchi, H. Shugart, L. Ulander, The BIOMASS mission: mapping global forest biomass to better understand the terrestrial carbon cycle. Remote Sens. Environ. **115**, 2850–2860 (2011)

T.L. van Zyl, I. Simonis, G. McFerren, The sensor web: systems of sensor systems. Int. J. Digital Earth **2**, 16–30 (2009)

H. Vereecken, T. Kamai, T. Harter, R. Kasteel, J. Hopmans, J. Vanderborght, Explaining soil moisture variability as a function of mean soil moisture: a stochastic unsaturated flow perspective. Geophys. Res. Lett. **34**, 315–324 (2007)

H. Vereecken, J.A. Huisman, H. Bogena, J. Vanderborght, J.A. Vrugt, J.W. Hopmans, On the value of soil moisture measurements in vadose zone hydrology: a review. Water Resour. Res. **44**, 253–270 (2008)

K. Wang, R.E. Dickinson, A review of global terrestrial evapotranspiration: observation, modeling, climatology, and climatic variability. Rev. Geophys. **50**(2), RG2005 (2012). https://doi.org/10.1029/2011RG000373

E. Ward, W. Buytaert, L. Peaver, H. Wheater, Evaluation of precipitation products over complex mountainous terrain: a water resources perspective. Adv. Water Resour. **34**, 1222–1231 (2011)

J.A. Wiens, Spatial scaling in ecology. Funct. Ecol. **3**, 385–397 (1989)

Z.W. Xu, S.M. Liu, X. Li, S.J. Shi, J.M. Wang, Z.L. Zhu, T.R. Xu, W.Z. Wang, M.G. Ma, Intercomparison of surface energy flux measurement systems used during the HiWATER-MUSOEXE. J. Geophys. Res. Atmos. **118**, 13,140–13,157 (2013)

K. Yang, J. Qin, L. Zhao, Y.Y. Chen, W.J. Tang, M.L. Han, Z. Lazhu, Q. Chen, N. Lv, B.H. Ding, H. Wu, C.G. Lin, A multi-scale soil moisture and freeze-thaw monitoring network on the Third Pole. Bull. Am. Meteorol. Soc. **94**, 1907–1916 (2013)

S. Zacharias, P. Dietrich, H.J. Vogel, R. Seppelt, D. Borchardt, S. Klotz, F. Messner, G. Teutsch, TERENO – methodical approach for the implementation of a terrestrial observatory for Environmental Research in Central Germany, *Consoil 2008*: *Theme B – Functions and Values of Soil-Water Systems; Understanding of Processes* (2008), pp. 85–89

G.F. Zhu, X. Li, K. Zhang, Z.Y. Ding, T. Han, J.Z. Ma, C.L. Huang, J.H. He, T. Ma, Multi-model ensemble prediction of terrestrial evapotranspiration across North China using Bayesian model averaging. Hydrol. Process. **30**, 2861–2879 (2016)

M. Zreda, W.J. Shuttleworth, X. Zeng, C. Zweck, D. Desilets, T.E. Franz, R. Rosolem, COSMOS: the COsmic-ray soil moisture observing system. Hydrol. Earth Syst. Sci. **16**, 4079–4099 (2012)

Ground-Based Soil Moisture Determination

François Jonard, Heye Bogena, David Caterina, Sarah Garré, Anja Klotzsche, Alessandra Monerris, Mike Schwank, and Christian von Hebel

Contents

F. Jonard (✉)
Institute of Bio- and Geosciences, Forschungszentrum Jülich GmbH, Jülich, Germany

Earth and Life Institute, Université catholique de Louvain, Louvain-la-Neuve, Belgium
e-mail: f.jonard@fz-juelich.de; francois.jonard@uclouvain.be

H. Bogena
Agrosphere Institute (IBG-3), Forschungszentrum Jülich GmbH, Jülich, Germany
e-mail: h.bogena@fz-juelich.de

D. Caterina · A. Klotzsche · C. von Hebel
Institute of Bio- and Geosciences, Forschungszentrum Jülich GmbH, Jülich, Germany
e-mail: d.caterina@fz-juelich.de; a.klotzsche@fz-juelich.de; c.von.hebel@fz-juelich.de

S. Garré
Gembloux Agro-Bio Tech, UR TERRA, Université de Liège, Gembloux, Belgium
e-mail: sarah.garre@ulg.ac.be

A. Monerris
Department of Infrastructure Engineering, University of Melbourne, Melbourne, Australia
e-mail: alessandra.monerris@unimelb.edu.au

M. Schwank
Mountain Hydrology and Torrents, Swiss Federal Institute for Forest, Snow and Landscape Research WS, Birmensdorf, Switzerland

GAMMA Remote Sensing AG, Gümligen, Switzerland
e-mail: mike.schwank@wsl.chv

X. Li, H. Vereecken (eds.), *Observation and Measurement of Ecohydrological Processes*, Ecohydrology, https://doi.org/10.1007/978-3-662-48297-1_2

Abstract

Soil water content is a key variable for understanding and modeling ecohydrological processes. In this chapter, we review the state of the art of ground-based methods to characterize the spatiotemporal dynamic of soil water content, from point to field scale. First, point measurements methods are briefly discussed. Then, field-scale hydrogeophysical approaches such as ground-penetrating radar, ground-based L-band radiometry, electromagnetic induction, electrical resistivity tomography, cosmic-ray neutron probes, global navigation satellite system reflectometry, and nuclear magnetic resonance are described in more details. The basic principles of the different techniques, the spatial and temporal characteristics of their measurements, their advantages and limitations, as well as the recent developments in the data processing are presented.

Keywords

Soil moisture · Spatial variability · Temporal variability · Modeling · Hydrogeophysical methods

Introduction

Soil water content (SWC) is a soil physical state variable which is defined as the water contained in the unsaturated soil zone or vadose zone. Knowledge of soil water content is essential, as it represents a key variable in many hydrological, climatological, environmental, and ecohydrological processes. In hydrology, SWC plays a major role in the water cycle by partitioning rainfall into runoff and infiltration and by controlling hydrological fluxes such as groundwater recharge. Soil water content is also a key variable of the climate system, as it governs the energy fluxes between the land surface and the atmosphere through its impact on evapotranspiration (Seneviratne et al. 2006, 2010). In addition, soil water availability influences plant transpiration and photosynthesis and, therefore, has an important effect on the biogeochemical cycles (Jonard et al. 2011a).

Determining the temporal and spatial variability of SWC is hence essential for many scientific issues and applications (Famiglietti et al. 2008). In that respect, a large number of SWC sensing techniques have been developed and used in the last 50 years (e.g., Robinson et al. 2008; Vereecken et al. 2008). The standard reference method to determine SWC is the gravimetric technique, which consists of extracting soil samples from the field. The samples are then weighted before and after drying in an oven at 105 °C during 24 h to derive their water content. The amount of water in the soil is typically expressed as volumetric [m^3 of water per m^3 of soil] or gravimetric [g of water per g of soil] water content. The gravimetric method is the only direct measurement technique, but many other techniques are available to estimate SWC. Two categories of soil moisture measurement techniques are often distinguished: contact-based (or invasive) and contact-free (or proximal/remote sensing) methods, depending on whether the methods require or not direct contact with the soil. The measurement techniques can also be classified, according to their spatial extent, as point measurement or field-scale measurement methods. Hereafter, the classification based on the spatial extent will be used.

Point Soil Water Content Measurement Methods

The most common point SWC measurement methods are the electromagnetic (EM) methods, which include time- or frequency-domain reflectometry (TDR or FDR), time- or frequency-domain transmissometry (TDT or FDT), as well as capacitance and impedance methods. The EM techniques are based on the dependency of the soil dielectric permittivity on the SWC. As the dielectric permittivity of liquid water dominates the dielectric permittivity of other soil components, water is the principal factor governing EM wave propagation in the soil.

The TDR measurement principle is based on the propagation velocity of guided EM waves emitted by a pulse generator and propagated along the waveguides of the TDR probe into the soil. The propagation velocity is determined from the measured travel time along the TDR probe (with a known length) which is dependent on the soil electromagnetic properties (Robinson et al. 2003). Similar to TDR, TDT sensors measure the propagation velocity of EM waves, but, in this case, the EM waves propagate along a closed transmission line. The measurement principle of the capacitance sensor is to incorporate soil medium that surrounds the sensor prong as part of the dielectric of the sensor capacitor. The permittivity of the soil is then determined by measuring the charge time from a starting voltage to a voltage with an applied capacitor voltage. Typically, capacitance and impedance sensors operate at a frequency between 50 and 150 MHz, while TDR/FDR and TDT/FDT sensors operate at higher frequencies. TDR/FDR and TDT/FDT sensors are then considered to be more accurate as their soil measurements are expected to be less influenced by the electrical conductivity and the imaginary dielectric permittivity of the soil.

All these instruments can perform continuous nondestructive measurements of SWC over a wide range of soils and with a very high temporal resolution. However, they are all invasive methods, which are restricted to local observation areas (< 1 m^2), and may not be representative of the soil moisture variability within the field.

Other point SWC measurement methods are also available, such as neutron probes, which are based on the estimation of the number of hydrogen nuclei in soils; heat pulse sensors, which are based on the estimation of soil thermal properties; and fiber optic sensors, relying on the attenuation or reflection of a light signal in the soil or the characteristics of hydrophilic polymers.

Wireless Sensor Network

More recently, an emerging technology, the wireless sensor network using cluster of point SWC measurement sensors, appeared as a promising approach to monitor SWC over large areas and with a high temporal resolution, which is particularly useful for observing ecohydrological processes (Bogena et al. 2010; Jin et al. 2014b). However, given the small support scale of the sensors, i.e., the area or volume integrated by an individual measurement, and the relatively large spacing between them, these networks do not allow to reveal the local-scale SWC patterns.

Because of the multitude of SWC measurements within the sensor network, the interpretation of the sensor signal should be straightforward and unambiguous as possible. Also, the cost of the SWC sensors should be reasonably low in order to maximize the number of sensor nodes. Capacitance and FDR sensors are relatively inexpensive and easy to operate and have been found to be a promising choice for SWC measurements with wireless sensor networks. A multitude of wireless communication technologies have been used for the development of wireless sensor networks, e.g., ZigBee, Bluetooth, Wibree, and Wi-Fi (Aqeel-ur-Rehman et al. 2014). The ZigBee wireless communication technology (IEEE 802.15.4) is mostly used for wireless sensor networks due to its low-cost and low-power consumption property. Typically, three different components are present in a wireless network application: a coordinator that initiates the wireless links within the network, router devices that pass data within a network, and the sensor nodes, which can host one or more sensors.

Recent applications of wireless sensor networks include monitoring of soil moisture combined with salinity in irrigated fields to optimize irrigation management (Yu et al. 2013), spatiotemporal observation of SWC in forested sites (Rosenbaum et al. 2012), and validation of remote sensing data (Bircher et al. 2012).

Field-Scale Hydrogeophysical Methods

Hydrogeophysical methods such as ground-penetrating radar (GPR), ground-based microwave (L-band) radiometry (MR), electromagnetic induction (EMI), and electrical resistivity tomography (ERT) are increasingly used to characterize the subsurface at the field scale.

Recently, novel promising geophysical technologies such as the use of cosmic-ray neutron probes (CRNP), global navigation satellite system reflectometry (GNSS-R), and nuclear magnetic resonance (NMR) have also been investigated to characterize the SWC variability at the field scale.

Ground-Penetrating Radar

Ground-penetrating radar (GPR) is a geophysical method that uses the propagation of EM waves in the subsurface. The propagation can mathematically be described by Maxwell's equations. In common practice, GPR uses a transmitter antenna (Tx) which emits electromagnetic waves. These waves are interacting with the surrounding medium and scattering; reflections and refractions of the wave can occur (Fig. 1a). The resulted signal can be sensed via a receiving antenna (Rx). The material properties affect the velocity and the attenuation of the EM wave. The electromagnetic wave velocity (v) depends on the relative dielectric permittivity (ε_r), whereas the attenuation (α) of the wave is related to the electrical conductivity (σ) of the medium through which the wave travels. The magnetic permeability (μ) is often simplified to the value in free space ($\mu_0 \approx 4\pi\ 10^{-7}$ H m^{-1}). For high-frequency ($f > 10^5$ Hz), low-loss, and non-magnetic materials, the permittivity and the electrical conductivity can be expressed by

$$\varepsilon_r = \left(\frac{c}{v}\right)^2 \tag{1}$$

and

$$\alpha = \frac{\sigma}{2}\sqrt{\frac{\mu}{\varepsilon}} \tag{2}$$

where c is the speed of light in free space ($\approx 3\ 10^8$ m s^{-1}). For conveniencc, ε_r is often used, whereas it is described by $\varepsilon_r = \varepsilon/\varepsilon_0$ with ε the absolute dielectric permittivity of the medium (F m^{-1}) and ε_0 the free space dielectric permittivity ($\varepsilon_0 \approx 1/36\pi\ 10^{-9}$ F m^{-1}). The geophysical responses caused by changes in permittivity and/or electrical conductivity can be linked to hydrogeologically relevant variables and soil properties such as SWC, porosity, water salinity, permeability, fluid content, pore structure, clay content, soil texture, and lithological variations. Due to the large disparity of the relative dielectric permittivity of air ($\varepsilon_r = 1$), soil minerals ($\varepsilon_r = 3$–5), and pure water ($\varepsilon_r = 80$ at 20 °C), permittivity can be used to determine the water content or porosity (in the case of a saturated medium) in the vadose zone or an aquifer. For example, wet sand has a permittivity range of 20–30, whereas completely dry sand has a permittivity range of 3–5. In contrast, the attenuation of the EM wave depends strongly on the electrical conductivity of the medium, which can give indications about clay content or the pore water salinity (Davis and Annan 1989). Therefore, GPR is well in providing soil water content and furthermore well suited to monitor infiltrations and recharge processes of aquifer systems and the critical zone.

Another important aspect of GPR is the frequency of the used antenna. Most available GPR systems are using a center frequency (f_c) between 10 MHz and 3.6 GHz. The frequency corresponds to the dominant wavelength (λ_c) of the signal and can be described by

$$\lambda_c = \frac{v}{f_c}. \tag{3}$$

The attenuation of the wave is combination of the electrical and the scattering losses and increases with increasing frequencies (Jol 2009). Lower frequencies have a larger wavelength and can penetrate deeper (depending of the attenuation), whereas higher frequencies have a smaller wavelength and less penetration but a higher resolution.

For measuring with GPR, several different configurations can be applied: surface, crosshole, and off-ground GPR. In the last decades, all of these methods have become more and more popular and showed a high potential to derive SWC at different scales. Before we discuss these different configurations, the petrophysical relationships that can be used to derive SWC from GPR data are described.

Petrophysical Relationships for GPR and Other Electromagnetic Methods

To obtain porosity or SWC from the permittivity and electrical conductivity distribution, soil petrophysical relationships are necessary. There are numerous empirical petrophysical relationships that can be used to obtain SWC from permittivity data. One of the most common ones relating permittivity to SWC is the Topp equation (Topp et al. 1980) given by

$$\theta_v = -5.3\ 10^{-2} + 2.92\ 10^{-2}\varepsilon_r - 5.5\ 10^{-4}{\varepsilon_r}^2 + 4.3\ 10^{-6}{\varepsilon_r}^3 \tag{4}$$

where θ_v is the volumetric water content in $m^3\ m^{-3}$. This equation uses a third-order polynomial function fitting observed laboratory permittivity responds of sand and loam soils under different soil water contents using TDRs. There are several other relationships available based on the Topp equation (overview by Steelman and Endres 2011). Nevertheless, this model is often inaccurate for clay- and organic-rich materials. A more theoretical approach to obtain soil water content from permittivity is the dielectric mixing model. Thereby, the permittivity of the bulk material $(\widetilde{\varepsilon})$ is related to the volume fraction and the permittivity of each of the soil components in the system (e.g., Steelman and Endres 2011). The general form is given by

$$(\widetilde{\varepsilon})^{\alpha} = \sum_{i=1}^{n} \chi_i(\varepsilon_i)^{\alpha} \tag{5}$$

where n is the number of dielectric components in the medium, α is the geometrical fitting parameter, and χ_i is the volume fraction of the component i. Typically near-surface soils can be described by a three-phase system using air, water, and soil components under consideration of the soil porosity Φ

$$\frac{\theta_V = \widetilde{\varepsilon}^{\alpha} - (1-\Phi)\varepsilon_s^{\alpha} - \Phi\varepsilon_a^{\alpha}}{\varepsilon_w^{\alpha} - \varepsilon_a^{\alpha}} \tag{6}$$

where ε_a, ε_w, and ε_s represent the relative dielectric permittivity of the air, water, and soil contributions, respectively.

This model can be extended to any number of known fractions in the soil, for example, splitting the soil component into clay and sand contributions. Fully saturated media can be described with a two-phase system using water and soil contributions. The geometrical fitting parameter α can vary between -1 and 1. By using $\alpha = 0.5$, the complex refraction index model (CRIM) is derived (Birchak et al. 1974), which showed adequate results in many studies (e.g., Roth et al. 1990):

$$\tilde{\varepsilon} = \left[\theta_{\mathrm{V}}\sqrt{\varepsilon_{\mathrm{w}}} + (1-\Phi)\sqrt{\varepsilon_{\mathrm{s}}} + (\Phi-\theta_{\mathrm{V}})\sqrt{\varepsilon_{\mathrm{a}}}\right]^2. \tag{7}$$

Surface GPR

For surface applications of GPR, we discriminate between two acquisition types. The common-offset (CO) reflection profiling survey, which is most widely used, has a fixed spacing between transmitter and receiver antennas at each measurement location (Fig. 1b). This technique allows fast mapping over large-scale structures. Thereby, the reflected signal is measured, indicating changes of the geophysical properties. Changes in reflection time and amplitude indicate variations in velocity (permittivity), reflection coefficient, and attenuation (electrical conductivity). This method is normally applied at straight lines or on rectangular grids allowing a 3-D view. CO surveys provide radargrams displaying time versus distance. To convert the time into depth, the velocity of EM waves in the near surface needs to be known.

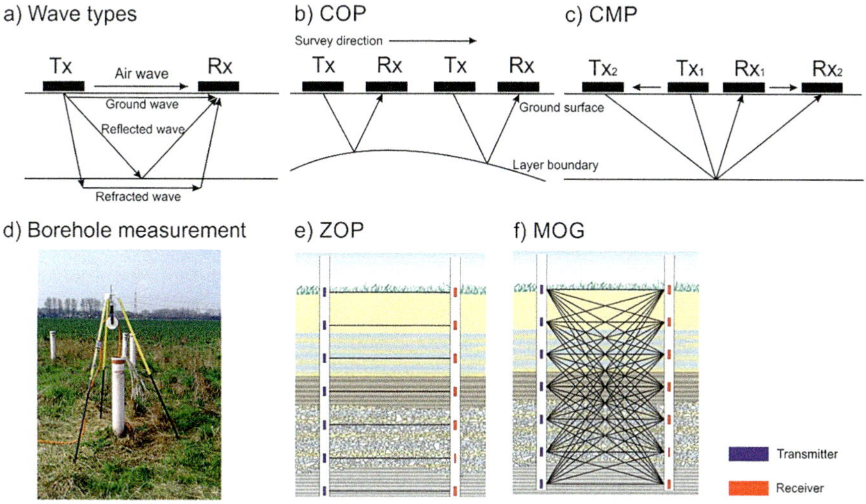

Fig. 1 Schematic GPR measuring setups for surface and crosshole measurements. (**a**) Illustrates the different wave types that can occur in a GPR survey. Surface measurement techniques are displayed in (**b**) common-offset profiling and (**c**) common-midpoint profiling. (**d**) Shows crosshole GPR measurements in the field, whereas (**e**) and (**f**) illustrate the zero-offset profiling and multi-offset gathers measuring technique, respectively

Common-midpoint (CMP) or wide-angle reflection and refraction (WARR) measurements can provide velocity information of the subsurface by varying the transmitter and receiver spacing (Fig. 1c). For CMP measurements, the transmitter and receiver antennas are moved with a fixed step size away from each other, while for WARR, the transmitter antenna is fixed at a certain point, and the receiver antenna is moved away or toward the antenna with a constant step size. Velocity semblance analysis can be applied to convert the observed reflection travel time to velocity-depth profile as well as to exploit the dielectric permittivity (Jol 2009). A definition of the structural organization of soil and soil layering can just be carried out when the contrasts in the physical properties is significant and the layer thickness is not too small compared to the wavelength of the signal. The direct ground wave that travels in the upper centimeters of the subsurface can be linked to water content using petrophysical parameters (Fig. 1c, e.g., Huisman et al. 2003). For studying soil water content over a large area, CO are suitable, whereas CMP or WARR acquisitions can provide depth information and permittivity changes at specific points with a higher resolution.

Recently, full-waveform developments are progressively done to obtain from CMP or WARR permittivity, electrical conductivity and layer depth information (Busch et al. 2012). The application of a full-waveform inversion (FWI) scheme can reliably estimate the permittivity and the conductivity for the same sensing volume, by analyzing reflected waves presented in surface WARR GPR. Busch et al. (2012) proposed a full-waveform inversion based on 3-D frequency-domain solution of Maxwell's equations assuming a layered model of the subsurface and reliably obtained permittivities, electrical conductivities, and layer depths. By obtaining two physical parameters at the same time, an improved characterization of the subsurface soil properties is possible.

Crosshole GPR

Crosshole GPR is minimal invasively and cannot be applied at large scale easily because of the need of boreholes. Nevertheless it can provide a higher resolution and insight in the medium than other techniques. For measuring bistatic cross-borehole GPR, two different measuring techniques can be applied. Zero-offset profiling (ZOP) is performed by systematically simultaneously lowering or rising the transmitter and the receiver antennas stepwise to the same depth in two different boreholes (Fig. 1d–e). Thereby, a 1-D velocity profile of the medium in between two boreholes can be achieved by assuming horizontally traveling ways and slowly changing properties. In heterogeneous soils, not only a direct wave occurs but also scattered waves caused by reflections and refractions at layer boundaries and changes in physical contrast.

An improved, but more time consuming, measuring method is multi-offset gather (MOG) shown in Fig. 1f. Thereby, the transmitter antenna is fixed at a certain position in one borehole, while the receiver antenna is moved constantly to different locations in another borehole. This is repeated for several positions of the transmitter and guarantees that the medium between the boreholes is sampled by a large number of rays with a larger number of angles. To improve the resolution, reciprocal measurements can be collected. MOGs are commonly applied for tomographic inversion that is used to derive physical parameters in the subsurface from geophysical data. Standard ray-based inversion methods applied to crosshole GPR data use the first arrival times and the first cycle amplitude information of the measured traces. These methods use only a small amount of the measured data, and therefore their resolution is limited. In the presence of small-scale structures, for example, related to high porosity zones, these methods often cannot provide sufficient good results. To resolve such small-scale structures, a more sophisticated method is necessary. In recent years, the full-waveform inversion of crosshole GPR data has proven a high potential to resolve decimeter-scale structures that could not be detected by standard ray-based methods (Klotzsche et al. 2012). The full-waveform inversion is based on solving Maxwell's equation and uses all the information contained in the measured traces, including reflection and refraction events (Ernst et al. 2007; Meles et al. 2010). This method has been applied to several different aquifers in Germany (Güting et al. 2015), Switzerland (Klotzsche et al. 2013), and the USA (Klotzsche et al. 2014) and showed high-resolution images of small-scale structures. Comparison with porosity logging data confirmed the presence of the resolved structures.

Off-Ground GPR

Similar to surface GPR, off-ground GPR is noninvasive and can be easily and fast applied to measure field-scale SWC variations. Typically, the off-ground GPR system is based on international standard vector network analyzer (VNA) technology, thereby operating in the frequency domain, and an accurate 3-D modeling of GPR wave propagation in the antenna-soil system (Lambot et al. 2004). The VNA can be connected to an ultrawideband and highly directive horn antenna acting simultaneously as transmitter and receiver (i.e., monostatic configuration). In general, the penetration depth is less than for surface GPR. The method is also sensitive to soil surface roughness as the EM wave reflection on the soil is dependent on the surface roughness with respect to the wavelength of the EM wave. A distinction can be made between smooth and rough surfaces based on the Rayleigh criterion ($hc = \lambda/8\ \cos(\gamma)$, where hc is the critical height of the surface protuberances, γ is the incidence angle, and λ is the wavelength). However, soil surface roughness can be accounted for in the inversion of off-ground GPR data for SWC retrieval by combining a roughness model to the GPR model as proposed by Jonard et al. (2012). This proximal sensing GPR method proved to be particularly appropriate for field-scaleSWC mapping and monitoring at a high spatial resolution, due its rapidity and to the air-launched configuration of the antenna (Minet et al. 2012; Jonard et al. 2013). Recently, off-ground GPR has also been successfully applied to characterize forest litter (André et al. 2016).

Summarizing, GPR is able to provide SWC and detailed information about soil properties. Thereby, different measurement techniques are possible, and new and sophisticated inversion approaches are able to derive high-resolution images of the subsurface and allow an improved understanding of the spatiotemporal variability of SWC at different scales.

Ground-Based L-Band Radiometry

Almost 35 years ago, it was suggested that soil water content could be retrieved from remotely sensed thermal radiance received with an L-band (1–2 GHz corresponding to vacuum wavelengths of 30–15 cm) radiometer (Schmugge 1985). The radiance T_B^p (also referred to as brightness temperature) emitted from a terrestrial surface at horizontal ($p = \mathrm{H}$) or vertical ($p = \mathrm{V}$) polarization depends on the effective temperature T_S of the soil and on the reflectivity R^p of the observed scene via $T_B^p \approx T_S\,(1 - R^p)$. Sensitivity of T_B^p with respect to volumetric SWC [$\mathrm{m}^3\ \mathrm{m}^{-3}$] is established through R^p, being dependent on the soil effective permittivity ε_S. The latter is a strong function of SWC due to the marked contrast between the permittivity of free water ($\varepsilon_r \approx 80$ for frequencies significantly smaller than the relaxation frequency ≈ 10 GHz) and dry soil ($\varepsilon_r \approx 3$ to 5) (see also section "Ground Penetrating Radar"). The just-mentioned relationship allows the soil surface water content to be estimated from T_B^p measured with an L-band radiometer by applying dielectric mixing (e.g., Dobson et al. 1985; Mironov et al. 2009) and radiative transfer models (e.g., Mo et al. 1982; Wigneron et al. 2007). Typically, L-band brightness temperatures T_B^p of a very dry bare soil can be 150 K higher than T_B^p for the same soil in its saturated moisture state. However, in many cases, a number of more complex radiative transfer processes complicate the retrieval of SWC from remote measurements of L-band brightness temperatures. For example, parameterization of ground roughness is critical for the use of microwave radiometry (MR) to achieve quantitative information on SWC, because soil roughness heavily impacts L-band emission T_B^p and thus affects SWC retrieved from it.

However, L-band radiometry is the most adequate remote sensing technique to monitor SWC since (i) passive L-band measurements T_B^p exhibit relatively large sensitive volumes as a consequence of moderate absorption and scattering in natural media such as soil, snow, and vegetation; (ii) impacts of soil surface roughness are less distinct compared with passive measurements at higher frequencies and also active measurements (radar) even at the same frequency; (iii) measurements T_B^p can be performed at almost any time because the atmosphere is largely transparent at the L-band and the T_B^p do not depend on sunlight; and (iv) the frequency range 1,400–1,427 GHz within the microwave L-band (1–2 GHz) is protected, which means that distortions of measured T_B^p due to man-made radio-frequency interferences (RFI) are minimized.

At the plot scale, SWC can be monitored with ground-based L-band radiometers mounted, e.g., on towers (de Rosnay et al. 2006; Guglielmetti et al. 2008; Schwank et al. 2012; Jonard et al. 2015) or mobile platforms (Jonard et al. 2011b; Temimi et al. 2014). As examples, Fig. 2 shows the ETH L-band radiometer II (ELBARA II) (Schwank et al. 2010) operated on a tower at the Mediterranean Ecosystem L-band Characterization Experiment III (MELBEX III) field site in Spain (Schwank et al. 2012), and Fig. 3 shows the ELBARA II operated on an arc (at ≈ 4 m height) in a controlled setup consisting of a sand box surrounded by a wire grid at the Terrestrial Environmental Observatories (TERENO) field test site in Selhausen, Germany (Jonard et al. 2015).

Fig. 2 L-band radiometer ELBARA II mounted on a tower at the MELBEX III site in Spain used for the calibration and validation of Soil Moisture and Ocean Salinity (SMOS) satellite observations

The technical specifications of the ELBARA II instrument can be considered as typical for ground-based L-band radiometers in application to the detection of SWC. As any L-band radiometer, it is a highly sensitive receiver for microwaves within the frequency band ranging from 1400 to 1427 GHz. This frequency band has become a protected radio astronomy allocation worldwide, in which it is forbidden to transmit any kind of electromagnetic radiance. An RFI-free environment is mandatory to measure brightness temperatures T_B^p emitted from terrestrial surfaces. However, the restriction of the receiver sensitivity to the bandwidth $B = 27$ MHz of the protected band implies that the received power level $P = k\ T\ B \approx 0.11\ 10^{-12}$ W (≈ -99.5 dBm) is very low (example for the observation of a black body at temperature $T = 300$ K; $k = 1.380658\ 10^{-23}$ J K^{-1} is the Boltzmann constant). To reliably detect radiance of such extremely low power, the radiometer must be very well temperature stabilized, and its residual noise (mostly caused by transmission losses) must be kept as low as possible. Beyond that, the gain of the radiometer microwave assembly must be very high ($\approx$80 dB), linear, and as stable as possible.

Fig. 3 Controlled setup for active and passive microwave remote sensing studies at the TERENO test site in Selhausen, Germany. The setup consists of an ELBARA II radiometer mounted on an arc and a sand box in the center of a wire grid.

Furthermore, the radiometer must be equipped with at least two internal calibration sources with known reference noise temperatures. This is needed to convert the raw data at the output of the power detector into calibrated brightness temperatures T_B^p.

A further highly critical component of any L-band radiometer is the antenna attached to the microwave receiver. In most cases, it features horizontal ($p = \mathrm{H}$) and vertical ($p = \mathrm{V}$) polarization, with the highest possible insulation between the H and V ports and lowest possible return loss. To allocate measurements T_B^p with a well-defined footprint area, the antenna requires a high spatial directivity. This requirement is driven by the fundamental resolution limit defined by Abbe's law of diffraction, implying that the diameter d of the antenna aperture must be significantly larger than the observation wavelength $\lambda \approx 21$ cm. For example, to achieve a beamwidth (at -10 dB sensitivity with respect to its sensitivity along the main direction) of $\Theta \approx \pm\, 12°$ around the antenna main direction, the required diameter of the aperture is $d \approx 1.4$ m. To comply with these requirements, a suitable antenna is necessarily rather bulky as can be seen in the example shown in Fig. 2. Of course, spatial resolution of a ground-based passive L-band observation is not only given by the antenna beamwidth Θ. The spatial extent of the footprint area at the ground results from the projection of the sensitive cone of the antenna (with aperture angle Θ) and the measurement configuration defined by the observation angle φ relative to nadir, and the radiometer installation-height H above ground. For the example shown in Fig. 2, the resulting footprint is an ellipse with long and short half-axes of $\approx$12 m $\times$ 7 m corresponding to an area of $\approx$264 m^2.

Temporal resolution of ground-based passive L-band measurements T_B^p can be rather high (order of seconds). The limiting factors with respect to the temporal resolution of SWC retrievals derived from $T_B^p(\varphi)$ are given by the requirement of multiangular measurements (e.g., $20° \leq \varphi \leq 60°$ in steps of 10°). Hence, the time used to direct the antenna to the different elevation angles φ and the integration time (e.g., 2 s) of measurements are the limiting factors for the temporal resolution of retrieved SWC.

The use of multiangular brightness temperatures $T_B^p(\varphi)$ measured at horizontal ($p = \mathrm{H}$) and vertical ($p = \mathrm{V}$) polarization is of advantage to estimate SWC especially for vegetated areas. It allows to disentangle radiative contributions associated with SWC and vegetation properties such as optical depth τ and single-scattering albedo ω. This is because of the qualitatively different impacts of SWC and vegetation on the angular pattern of $T_B^p(\varphi)$. In most cases, dense vegetation with high microwave attenuation (corresponding to high values of optical depth τ) and possibly significant volume scattering (corresponding to high values of single-scattering albedo ω) diminishes polarization difference, while polarization difference increases with increasing SWC as the result of increasing Fresnel-like emission of the soil surface. However, beyond the advanced physically based retrieval methodologies applied to estimate SWC from $T_B^p(\varphi)$ at L-band, a number of new approaches aiming to estimate other relevant land surface parameters from $T_B^p(\varphi)$ measured at L-band are currently under development. For example, this includes the retrieval approaches applicable to the (i) detection of annual soil freeze/thaw cycles across mid to high latitudes (Rautiainen et al. 2014), (ii) estimation of snow mass-density (Lemmetyinen et al. 2016), (iii) monitoring of drought and flooding events, (iv) estimation of soil hydraulic properties (Jonard et al. 2015), and (v) estimation of vegetation water content linked with retrieved vegetation optical depth τ (Lawrence et al. 2014).

Electromagnetic Induction

Electromagnetic induction (EMI) instruments measure contactless the bulk soil electrical properties and are particularly working well in electrical conductive environments, where the SWC, soil texture, salinization, fertilization, organic matter, and/or the residual pore water contribute to the recorded apparent electrical conductivity (ECa). To estimate the SWC at the catchment scale with a high lateral and vertical resolution, EMI instruments show a particular large potential because of their mobile use that allows to measure relatively large areas in comparable short time.

The EMI method was successfully used to measure and predict the temporal and spatial SWC changes, where a linear relationship between ECa and SWC was derived using data measured simultaneously over a period of 16 months along an approximately 2-km-long transect (Sheets and Hendrickx 1995). The soil water dynamics at a 4-ha-large deltaic zone were obtained by using repeated EMI measurements over several months to discriminate the time-invariant soil properties (e.g., clay content) and the dynamic SWC changes to ECa. In addition, measurements performed before and after a heavy rainfall event identified zones of water depletion and accumulation (Robinson et al. 2009). In a subsequent EMI time-lapse study performed in asemi-arid oak savanna catchment of 4 ha size, Robinson et al. (2012) estimated the relative SWC changes by subtracting the areal ECa values of the driest day from data collected during wetter phases. Consequently, (time-lapse) EMI measurements provide important insights into the soil water distribution and dynamical changes (Calamita et al. 2015) such that EMI surveys can preferably be used over classical SWC measurements (e.g., gravimetric soil sampling and time-domain reflectometry) because these methods are laborious when covering large areas and deliver limited depth and sparse spatial point-scale information.

However, soil is inherently complex and many factors beside the soil water content determine the geoelectrical properties such that a one-to-one relation of ECa to a single soil constituent does not exist. For example, Zhu et al. (2010) investigated a 19.5-ha-large agricultural test site and found that terrain attributes, depth to bedrock, and management practices mask the effect of SWC on ECa especially during dry states compared to wetter periods and at wetter locations. Similarly, the spatiotemporal variability of SWC was found to be less significant on ECa patterns compared to the stable soil properties (soil texture) as well as compared to the electrical conductivity of the soil solution in clay-poor soils (Martini et al. 2016).

The often site-specific correlations between ECa and some soil properties open broad perspectives of EMI usages. In agricultural studies, EMI surveys were applied to precision agriculture and to better understand soil-water-plant interactions (Corwin and Lesch 2003). In soil science, EMI data were successfully used to identify soil patterns andrelatively high correlations between ECa and clay content were obtained, e.g., across the north-central USA (Sudduth et al. 2005). EMI data were further used to predict soil constituents, e.g., organic matter (Altdorff et al. 2016) as well as to infer the water holding capacity of a watershed based on soil textural prediction using ECa values (Abdu et al. 2008).

EMI instruments transmit a low-frequency ($f < 10^5$ Hz) primary magnetic field (H_P) generated by an alternating current passing the transmitter coil (Tx). Due to induction phenomena, H_P induces eddy currents in an electrical conductive subsurface, which in turn generate secondary magnetic fields (H_S), in electrical conductive media (see Fig. 4a). The ratio of the superimposed H_S over H_P, i.e., H_S/H_P, is measured at the receiver coil (Rx) and related to the ground electrical properties.

The recorded ECa, i.e., the output value of common EMI instruments, reflects a weighted average value over the coil configuration-specific sensing depth or investigated soil volume. Recently developed fixed-boom multi-coil EMI instruments as indicated in Fig. 4b use one transmitter and multiple receiver coils that are oriented either horizontal coplanar (HCP), vertical coplanar (VCP), or perpendicular (PRP) with coil separations s ranging between 0.32 and 4 m to investigate depths of approximately 1.5 s, 0.75 s, and 0.5 s, respectively. For example, the CMD-

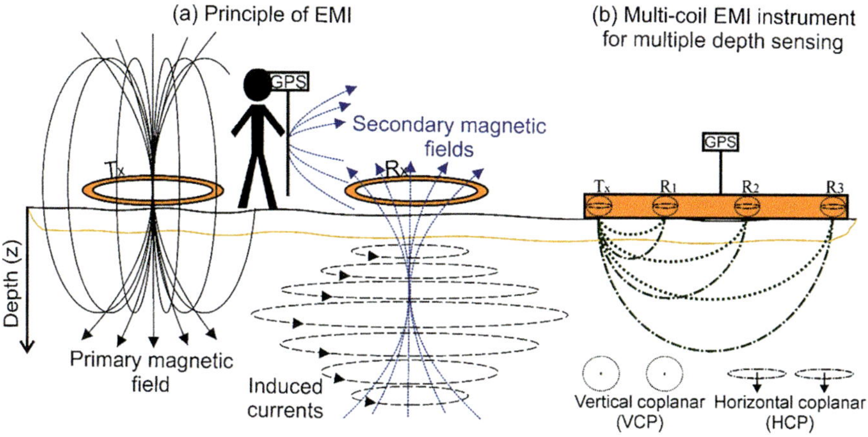

Fig. 4 (**a**) Principle of electromagnetic induction (EMI), where the transmitter Tx generates a primary magnetic field that induces eddy currents in the subsurface, which in turn generates secondary magnetic fields measured superimposed at the receiver coil Rx. Note that surrounding media may generate additional secondary magnetic fields. (**b**) Shows the principle of multi-coil EMI instruments that sense different overlapping depth intervals that depend on the coil configuration

MiniExplorer (GF-Instruments, Brno, Czech Republic) or the DualEM-421 (DualEM, Milton, Canada) carry three and six receiver coils, respectively. The EM38 (Geonics Ltd., Ontario, Canada) houses a single Tx-Rx coil pair with $s = 1$ m to investigate up to 1.5 and 0.75 m depth in HCP and VCP mode, respectively, while also multifrequency EMI devices, e.g., GEM-2 (Geophex Ltd., Raleigh, USA), can be used to explore different depths.

Using fixed-boom multi-coil EMI instruments, researchers attempt to obtain layered subsurface electrical conductivity models by using inverse-modeling approaches. Reliable results are obtained when inverting quantitative values. However, EMI measurements mostly record qualitative data because the induction phenomena occur in all electrical conductive media such as the operator, cables, or GPS systems surrounding the instrument, as illustrated in Fig. 4a. Quantitative EMI-ECa values were recently obtained by introducing a post-calibration procedure based on inverted electrical resistivity tomography (ERT) data (Lavoué et al. 2010). To calibrate the recorded EMI-ECa values for external influences close to the instrument, linear regressions between measured and predicted ECa values are performed, and the obtained multiplicative and additive regression factors can be applied to large-scale EMI-ECa (von Hebel et al. 2014). The approach uses collocated EMI and ERT measurements performed along relatively short calibration lines. The EMI instruments carried small (< 1.7 m) separated Tx-Rx coil pairs sensing relatively small subsurface volumes that approximately match the ERT information. The ERT delivers the subsurface electrical conductivity distributions, which are inserted into an EMI forward model to predict ECa values along the calibration line.

The post-calibrated EMI data can be inverted quantitatively. The modeling in EMI inversions are often performed using either cumulative response functions (McNeill 1980) or a Maxwell-based exact electromagnetic forward model (EM-FM) (Wait 1951), which can be implemented due to the available computational power, to compute the response of a stratified earth to an EMI instrument. In a combined global-local search, where the global search was performed with the cumulative response functions and the local search used the exact EM-FM, post-calibrated EMI data of multiple instruments were inverted to resolve a two-layered earth along a 120-m-long transect (Mester et al. 2011). In addition, they showed that the inversion of uncalibrated data returns unreliable results. This work was adapted by von Hebel et al. (2014), who used the exact EM-FM in a parallelized three-layer inversion scheme to obtain a quasi-3-D layered electrical conductivity model of 1.1-ha-large test site. Their approach basically turned the EMI usage from a proxy indicator toward a tool that quantitatively characterizes the shallow subsurface at the catchment scale.

Large-scale EMI measurements were linked with satellite derived leaf-area index (LAI) maps (Rudolph et al. 2015) showing that the deeper subsoil is mainly responsible for plant performance especially under drought conditions. The ECa maps indicated buried paleo-river channels; however only quantitative fixed-boom multi-coil EMI data inversions reveal the depth of these channels. Here, we use the inverse-modeling scheme of von Hebel et al. (2014) to invert the post-calibrated CMD-MiniExplorer data of an approximately 3-ha-large agricultural field showing prominent paleo-river channel structures. One structure runs approximately north-south (N-S) through the middle of the field; the second prominent structure is a westward running channel that seems to be connected to the N-S structure in the northern part of the field. Two depth slices through the large-scale quasi-3-D EMI inversion results are shown in Fig. 5.

In Fig. 5a, the depth slice at 0.1 m cuts the plowing layer, where a relatively homogeneous soil and no paleo-river channels were present. The deeper slice at 1.5 m depth presented in Fig. 5b shows the prominent paleo-river channels with larger electrical conductivities due to clay-rich material compared to the low electrical conductive gravelly surrounding. In drought conditions, the paleo-river channels still supply water due to higher water holding capacity of the finer-textured soil compared to the generally sand- and gravel-dominated sediments. The inversion results additionally provided new insight into the soil-plant interaction. Whereas linear regression between LAI and ECa of VCP coils with $s = 0.32$ m obtained a coefficient of determination (R^2) of 0.63, the R^2 of LAI and the inverted electrical conductivity (σ) of the upper layer was $R^2 = 0.15$. These results indicate, from the EMI perspective, that the signal is more influenced by material of deeper depths than estimated by the cumulative response definition (that states that 70% of the signal originate in the coil specific sensing depth (McNeill 1980)), i.e., 0.25 m or approximately the plowing depth for the VCP coils with $s = 0.32$ m. For ecohydrological modeling purposes, inverted electrical conductivities should be used instead of ECa to accurately characterize subsurface processes.

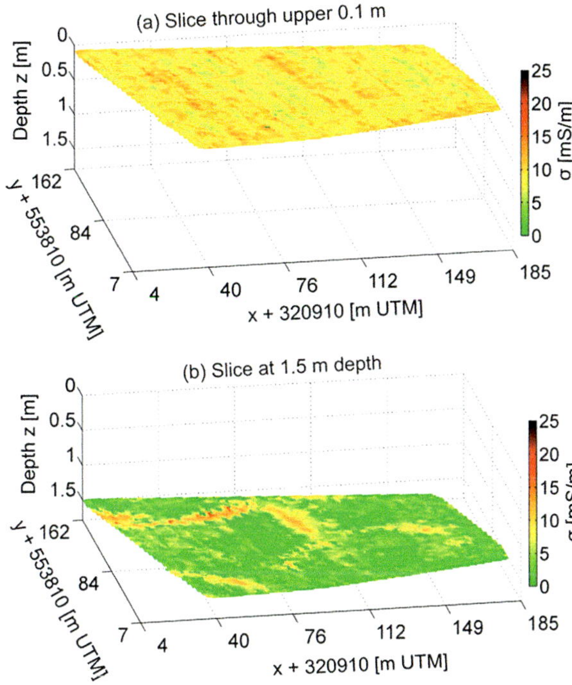

Fig. 5 Quasi-3-D inverted EMI data, where x and y given in UTM coordinates, zone 32 U, and z is the depth. (**a**) Depth slice through upper 0.1 m roughly showed homogenous soil. (**b**) Slice at 1.5 m depth, where buried paleo-river channel structures with larger electrical conductivity values compared to the surrounding material were visible. Whereas the soil is generally sand and gravel dominated, the paleo-river channels are characterized by finer-textured, clay-rich, soil having inherently a higher water holding capacity to supply plants especially under drought conditions

Electrical Resistivity Tomography

Whereas electrical resistivity tomography (ERT) was initially especially used for exploration and at larger scales (geology, groundwater), it has become increasingly popular for vadose zone research in the last decades (e.g., Samouelian et al. 2005) since the commercial equipment has become more powerful and adapted to monitoring, the potential of inversion codes has improved, and computing power has increased. Electrical monitoring has been used to monitor water fluxes and solute transport under agricultural crops (e.g., Banton et al. 1997; Michot et al. 2003), show interaction for water between different species (e.g., Garré et al. 2013), or measure water depletion by trees (Cassiani et al. 2015). In addition to soil monitoring, it has been used to measure water fluxes in tree stems (al Hagrey 2006). Nevertheless, some difficulties still need to be resolved when ERT is applied at decimeter resolution in the soil-plant continuum: (i) lack of complementary measurement methods and physically based models to take into account spatially variable bio-pedo-physical relationships, (ii) difficulties to resolve sharp contrasts due to smoothness-constraint inversion (see below), (iii) lack of standardized methods to take into account measurement and model errors and their temporal variation for monitoring studies, and (iv) practical difficulties with electrode contact arising when measuring under very dry conditions.

Electrical resistivity tomography is a technique which measures the bulk electrical resistivity of the soil between electrodes. The bulk electrical resistivity corresponds to the combined resistivity of soil particles, pore water, and air. A basic measuring system consists of four electrodes (A, B, M, N), a transmitter, and a receiver. The transmitter applies a quasi-DC current (I) (i.e., rectangular pulses or an AC current at low frequency to avoid polarization) to the ground using the electrodes A and B, whereas the receiver measures the voltage (V) between the electrodes M and N (see Fig. 6).

Figure 7 shows some possible four-electrode arrays. Resistivity meters then give a resistance (R) value for each combination of four electrodes, based on Ohm's law $R = V/I$.

For a tomography, more than four electrodes are used. During the measurements, many different combinations of current and potential electrodes are used for injection and measurement. The electrodes can be inserted in the soil surface or at several depths as borehole electrodes or a combination of both.

From the current, the voltage and a geometric configuration factor (k), the apparent electrical resistivity (ρ_a) is calculated. The geometric factor depends on the spatial arrangement of the four electrodes A, B, M, and N. Resistance is an extrinsic property (i.e., depends on the way it is measured); resistivity is an intrinsic property (i.e., depends only on the material in which the measurement is performed). We call the obtained resistivity "apparent," because it represents the resistivity of a hypothetical, homogeneous medium, which will give the same resistance value for the same electrode arrangement. The measured, apparent resistivity is a weighted average of the resistivities of the various materials that the current encounters. The closer the electrode spacing, the

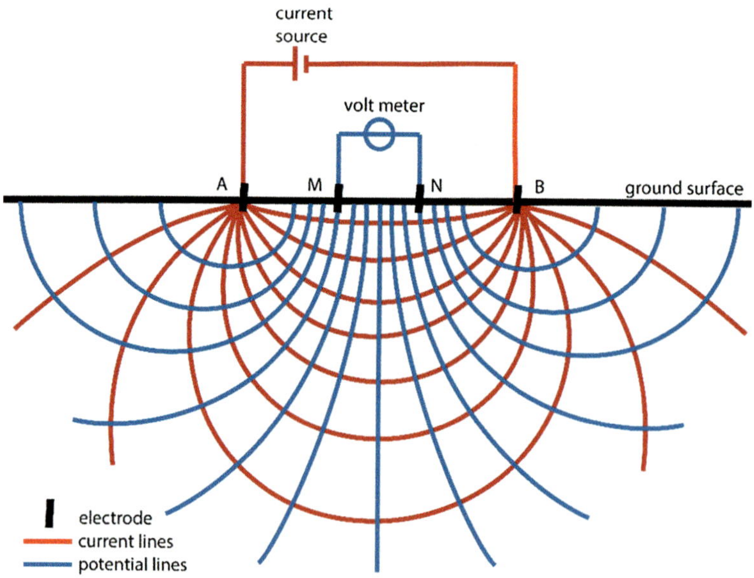

Fig. 6 Basic concept of electrical resistivity measurement of the subsurface

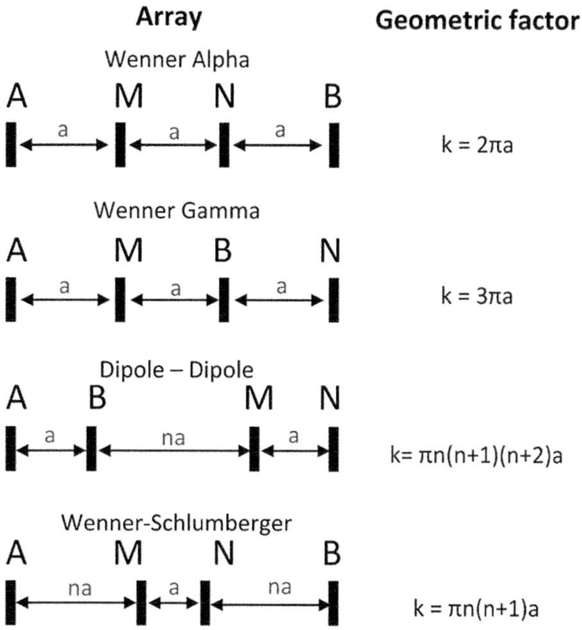

Fig. 7 Four exemplary standard electrical resistivity electrode arrays with corresponding geometric factor

more of the current will remain close to the surface, and the more the apparent resistivity value will be influenced by the properties and state of the material close to the surface.

The vector form of Ohm's law is the basis for resistivity studies:

$$\mathbf{J} = \sigma\mathbf{E} = \frac{1}{\rho}\mathbf{E} = -\frac{1}{\rho}\nabla V. \tag{8}$$

where $\mathbf{J}$ is the current density vector [A m^{-2}], $\mathbf{E}$ is the electric field vector [V m^{-1}], V is the electric potential (V), σ is the conductivity [Ω m^{-1}], and ρ is the resistivity [Ω m]. The electrodes depicted in Fig. 6 are treated as point sources/sinks of current flow. For surface electrodes, the total current (I) flows across the surface of a half sphere with area ½($4\pi r^2$), with r the radius of the sphere, and thus Ohm's law for one electrode becomes

$$J = \frac{I}{\frac{1}{2}\left(4\pi r^2\right)} = -\frac{1}{\rho}\frac{\mathrm{d}V}{\mathrm{d}r}. \tag{9}$$

Integrating this for constant resistivity yields the following equation for the potential at a distance r from the electrode:

$$V(r) = \frac{\rho I}{2\pi r}. \tag{10}$$

In real field measurements, the resistivity is generally not constant, so this expression actually corresponds to the earlier defined "apparent resistivity." If we now use this to express the potential at measurement electrodes M and N, we have to superpose the potential of the two source electrodes A and B:

$$V_{\mathrm{M}} = \frac{\rho I}{2\pi}\left(\frac{1}{AM} - \frac{1}{MB}\right). \tag{11}$$

where AM and MB represent the distances between the electrodes. The total potential difference between the electrodes M and N is therefore

$$V_{\mathrm{MN}} = V_{\mathrm{M}} - V_{\mathrm{N}} = \frac{\rho I}{2\pi}\left(\left(\frac{1}{AM} - \frac{1}{MB}\right) - \left(\frac{1}{AN} - \frac{1}{NB}\right)\right) = \rho I\frac{1}{k} \tag{12}$$

with k the abovementioned geometric configuration factor, which will yield a specific value for a given electrode spacing (see Fig. 7). Note that this is true for a flat surface. If an undulated topography is present, the configuration factor is unknown and can only be assessed by numerical modeling using homogeneous resistivity (Günther 2004).

Relation to State Variables

As stated before, the bulk electrical resistivity (and its reciprocal electrical conductivity EC) or "effective electrical resistivity" is the resistivity of the solid-water-air system, which is the (un)saturated soil in ecohydrological studies. The apparent electrical resistivity (ρ_{a}) in soil depends on the water content, pore water electrical conductivity, soil porosity, surface conductivity of the solid particles as well as the soil temperature. Hence, to derive soil water content from electrical resistivity measurements, one must rely on a "pedo-electrical" function. Several of these models have been elaborated with more or less complex approaches. Without aiming at being exhaustive, we list some of these models here. A well-known EC model is the empirical Waxman and Smits (1968) model based on Archie's law (Archie 1942). In 1998 Revil and Glover proposed a more physically based model also based on Archie's law. Derived more in the context of agricultural soils, the conceptual Rhoades et al. (1989) model relies on the assumption of two separate parallel electrical pathways: a continuous pathway through large water-filled pores and a series of coupled solid-liquid pathways.

The effect of temperature on ρ_{a} is often treated before applying a pedophysical relationship to normalize the data to a reference temperature. There are two distinct effects of temperature on soil bulk electrical conductivity: (i) the mobility of the ions in the soil solution and (ii) the quantity of total dissolved solids (TDS) in the pore water, which is an irreversible process. Most of the current models for temperature correction of electrical resistivity correct only for this first factor. Ma et al. (2011) gives an overview of temperature correction models. A temperature correction model (to convert to a reference temperature T of 25 °C) which is often used in hydrogeophysical investigation is the equation developed by Campbell (1949): $\sigma_{a\text{-}25} = \sigma_{a\text{-}T}/(1 + \alpha\ (T-25))$, with $\alpha = 0.02$ and σ_a the apparent electrical conductivity.

Nevertheless, the available theoretical models often fail when used without site-specific field calibration at field scale, and therefore, several researchers still prefer empirical, site-specific approaches to quantify the pedophysical relationship for their application. In those cases, SWC, electrical resistivity, and temperature are measured simultaneously in the field over a wide range of conditions and in the different soil horizons and used to establish a simple, empirical relationship (e.g., Michot et al. 2003; Garré et al. 2011).

Data Processing

The apparent resistivities of a 2-D survey are commonly visualized in the form of a pseudosection. In this diagram, the horizontal location of the point is placed at the midpoint of the set of electrodes used to make that measurement. The vertical location of the plotting point is placed at a distance, which is proportional to the separation between the electrodes. Following Edwards (1977), the vertical position is placed at the median estimated investigation depth, or pseudodepth, of the quadrupole used. This pseudodepth value is based on the sensitivity values or Frechet derivative for a homogeneous half space. The color of the point represents the apparent resistivity. A pseudosection gives a good idea of the true subsurface resistivity distribution, but not of its exact spatial organization since the contours depend on the type of array used and the true resistivity of the subsurface.

To determine the true resistivity of the subsurface, a mathematical procedure, called "inversion," must be carried out. The inversion procedure combines a forward modeling routine able to simulate the electrical field in any spatial model of resistivity distributions with a mathematical procedure to compare measured apparent resistivity values with simulated apparent resistivity values and move toward a model which suits the data well. In ERT studies, for the same measured data set, there is wide range of models resulting in the same calculated apparent resistivity values. To narrow down the range of possible models, some assumptions are made concerning the nature of the subsurface that can be incorporated into inversion subroutine. One common assumption is that smooth changes are more probable than sudden changes, a principle that is applied in the conventional smoothness-constrained inversion (de Groot-Hedlin and Constable 1990). Some commonly used inversion codes are RES2DINV/RES3DINV (Loke and Barker 1996), R2/R3T (Binley 2013), BERT (Günther 2004), and CRTOMO (Kemna 2000).

Spatial Extent, Resolution, and Sensitivity

In order to speak of the spatial extent and precision of ERT measurements, we first need to define a few concepts: depth of investigation, resolution, sensitivity, and data coverage.

Oldenburg and Li (1999) defined the depth of investigation (*DOI*) as the depth below which data no longer constrain earth structures. In their method, they require two successive inversions with different reference models to visualize regions of parameters only related to the choice of the reference model. The index tells you how much the model is constrained by the data and how much by the regularization, and Marescot et al. (2003) and Robert et al. (2011) followed similar approaches. Since its value depends on the way it is computed, it should be considered as a qualitative index indicating whether an anomaly is more likely to be data-related or constraint-related. The *DOI* should not be confused with the pseudodepth, which is described above.

Sensitivity is related to data coverage. The sensitivity matrix (S) shows how the data set is actually influenced by the respective resistivity of the model cells, i.e., how areas of the imaging region are "covered" by the data (Kemna 2000). The Jacobian or sensitivity matrix contains the partial derivatives of the model responses with respect to the model parameters. A poorly covered region is likely to be less well resolved, but it must be emphasized that good coverage does not necessarily imply high resolution. Sensitivity and resolution are correlated but do not give the same information.

Resolution is like the filter through which the inversion process sees the subsurface, so that the inverted model equals the resolution matrix (R) times the true resistivity distribution. Ideally R is defined by an identity matrix so that the inverted model equals the true distribution. However, this is never possible for a continuous inverse problem with incomplete data (Friedel 2003). A complete description of how R is computed can be found in Günther (2004).

These different quantities can either be obtained a priori using model simulations of the object under investigations yielding probable resistivity distributions on which different possible electrode arrays can be tested for their expected performance (e.g., Garré et al. 2012). As such, an informed choice can be made in the trade-off between measurement time, resolution, and measurement volume. A posteriori, these measures can be used to assess the reliability of the inverted parameters, as pointed out by Caterina et al. (2013).

Perspectives

Recently, several authors have pointed out the potential but also the complexity of using electrical measurements to characterize the root zone moisture and/or salinity dynamics. Some authors claim to be able to use electrical measurements to localize the root system. Many different types of pedophysical relationships have been used to isolate one of these variables. However, in addition to inherent soil heterogeneity, the presence of plant roots makes the soil a highly heterogeneous medium in space (root architecture) and time (root growth, maturation, and decay) in which the distribution of the electrical field and the bio-pedo-physical properties of the medium are uncertain. On the one hand, researchers are digging into the potential of using the distinct electrical signature of root tissue in order to map their presence noninvasively. Historically, botanists have explored electrical capacitance measurements. Recently, the potential of induced polarization to map roots is being explored. On the other hand, the spatial heterogeneity of the bio-pedo-physical relationship needs to be taken into account. Typically, soil moisture or salinity values are obtained applying one single pedophysical relationship for a given soil horizon, whereas it is clear that this relationship is heterogeneous even within one and the same horizon. The use of (geo)statistical techniques taking into account spatial (co-)variance might be a possibility. Nevertheless, it is not straightforward to obtain training images to characterize the spatial structure of this relationship.

The classical inversion approach suffers from spatially and temporally varying resolution and sometimes yields unrealistic solutions without uncertainty quantification, making their utilization for hydrogeological calibration less consistent. The inverse problem for electrical resistivity tomography is ill posed, and its solution is therefore nonunique. A regularization is used to reduce the amount of mathematical solutions to more "plausible" models. The most common regularization is the smoothness constraint, which selects a "smooth" distribution of resistivities above one with a lot of contrasts. However, the inversion can also be regularized using prior information on the system under consideration in the inversion. Recently, the addition of structural constraints in Occam's inversion (smoothness-constrained inversion) or the incorporation of geostatistical constraints has been implemented successfully. Another option is to avoid the geophysical inversion and use process models, e.g., a hydrodynamic physically based model representing the processes governing the system under consideration, to invert directly for soil hydraulic or root architectural parameters. This option is called a coupled inversion (e.g., Hinnell et al. 2010). However, the technique is difficult to apply in complex field cases and it remains computationally demanding to estimate uncertainty. Prediction-focused approaches (Hermans et al. 2016) might offer new perspectives circumventing the necessity to invert the data by seeking a direct relationship between the data and the subsurface variables we want to predict.

Cosmic-Ray Neutron Probes

Theoretical Background

Estimating and monitoring soil moisture at the appropriate spatial and temporal scale has proven to be a difficult task (Bogena et al. 2015a). One promising geophysical technique to help fill this need is the cosmic-ray neutron probe (CRNP) (Desilets et al. 2010; Zreda et al. 2008), which measures the ambient amount of low-energy secondary neutrons in the lower atmosphere. A detailed description of the CRNP technique can be found in Zreda et al. (2012). Here, only the basic principles are presented. A cascade of secondary neutrons with varying energy levels are created in the earth's atmosphere when incoming high-energy primary particles produced within supernovae interact with atmospheric nuclei (Zreda et al. 2012; Köhli et al. 2015). The secondary high-energy neutrons continue to lose energy during numerous collisions with nuclei in the atmosphere. Due to its higher density, soil effectively slows neutrons further down. In the final near-surface neutron energy spectrum, three different types of neutron energies are prominent (Köhli et al. 2015): highly energetic neutrons around 100 MeV, evaporation neutrons around 1 MeV, and low-energy neutrons which are in thermal equilibrium with the environment (< 0.5 eV). Epithermal fast neutrons (in the following referred as "fast neutrons") with energies between 0.5 eV and 100 eV are particularly sensitive to energy loss by elastic collisions with light atoms like hydrogen. Since soil moisture is one of the largest sources of hydrogen present in terrestrial systems, it largely controls the presence of fast neutrons in the lower atmosphere (Zreda et al. 2012). Thus, relative changes in the intensity of fast neutrons are strongly correlated to soil moisture changes.

Design and Calibration of CRNP

The fast neutron intensity in the lower atmosphere can be measured by CRNP's, which are neutron detectors with a tube filled with helium-3 or boron-10 (enriched to 96%) trifluoride ($^{10}BF_3$) proportional gases at high pressure (Zreda et al. 2012). A polyethylene shielding around the tube moderates fast neutrons to thermal neutrons before they enter the detector tube, in order to increase the probability of them being captured by the detector. Within the detector tube, fast neutrons that collide with helium-3 or boron-10 nuclei produce electrons that induce pulses of electrical current and which are counted by the detector. The size of the tube determines the probability of collisions and thus the sensitivity of the CRNP to measure soil moisture at higher temporal resolution (Bogena et al. 2013). The energies measured by the bare tube comprise a continuous distribution which is heavily weighted toward thermal neutrons (< 0.5 eV), with a small proportion of fast neutrons also being detected (< 10%) (Andreasen et al. 2016). The moderated detector is more sensitive to higher neutron energies (> 0.5 eV). Although the polyethylene shielding effectively attenuates the influx of thermal neutrons, a large proportion of the counts (approximately 40% of the thermal neutrons detected by the bare detector) still originates from below 0.5 eV (Andreasen et al. 2016). Since neutron counts follow Poissonian statistics, the measurement uncertainty of a given neutron intensity (N) decreases with increasing neutron intensity according to $N^{0.5}$.

Desilets et al. (2010) proposed a simple calibration function to relate fast neutron intensity measurements to volumetric soil moisture:

$$\theta_{\mathrm{v}} = \rho_{\mathrm{bd}}\left(a_0\left(\frac{N_{\mathrm{corr}}}{N_0} - a_1\right)^{-1} - a_2\right) \tag{13}$$

where θ_{v} is the volumetric soil water content [$m^3\ m^{-3}$]. N_{corr} refers to the measured fast neutron counts corrected for influences of atmospheric pressure and humidity and variations in incoming cosmic radiation (see Zreda et al. 2012 for a detailed discussion). The parameters a_0 and a_2 are divided by the dry soil bulk density ρ_{bd} [$g\ cm^{-3}$]. Using simulations of neutron transport for generic silica soils, Desilets et al. (2010) derived $a_0 = 0.0808$, $a_1 = 0.372$, and $a_2 = 0.115$ for values of $\theta > 0.02\ g\ g^{-1}$. Thus, only N_0, representing the count rate over dry soil conditions, needs to be calibrated.

This calibration function is not universal, because it depends on local soil and vegetation characteristics reflecting the variation of background hydrogen levels across landscapes (Zreda et al. 2012). To account for these influences, it needs to be fitted using soil samples taken within the footprint of the CRNP, which typically has a radius between 150 m and 250 m and depths between 0.1 m and 0.7 m (Köhli et al. 2015). Given this large footprint area, the CRNP method is an ideal complement to long-term surface energy balance monitoring with the eddy covariance technique (Jana et al. 2016). Recent neutron transport modeling has further refined the footprint area to be a function of atmospheric water vapor, sensor elevation, surface heterogeneity, vegetation, and soil water content (Köhli et al. 2015).

The application of the CRNP method is hampered by its susceptibility to additional sources of hydrogen (e.g., above- and belowground biomass, humidity of the lower atmosphere, lattice water of the soil minerals, organic matter and water in the litter layer, intercepted water in the canopy, and soil organic matter), e.g., Bogena et al. (2013), and Heidbüchel et al. (2016). In case that these hydrogen sources are temporally stable, the influence can be amended by estimating the contribution of each source and subtracting its contribution during the transformation of the neutron counts into SWC (Zreda et al. 2012). In the following chapter, methods for correcting for biomass effects on CRNP measurements are discussed in more detail.

Accounting for Biomass Effects on CRNP Measurements

Recently, Baatz et al. (2015) developed a simple empirical approach to correct for biomass effects based on long-term fast neutron intensity measurements from a network of 10 CRNPs located in the Rur catchment, Germany. In addition, they gathered fast neutron intensity measurements for shorter periods (between 24 and 405 h) at 13 locations in the forested research catchment Wuestebach (Germany), which was partly deforested. These locations were selected in such a way that the CRNP footprint contained distinctly different amounts of aboveground biomass. Using this extensive data set, they found that fast neutron intensity was reduced by 0.9% per kg dry aboveground biomass per m^2. Baatz et al. (2015) further developed an equation for correcting N_0 for biomass effects:

$$N_{0,\,\mathrm{corr}} = -r\,ABG_{\mathrm{dry}} + N_{0,\,ABG=0} \tag{14}$$

where $N_{0,\mathrm{corr}}$ is the biomass corrected N_0, $N_{0,AGB\,=\,0}$ is the reference N_0 for a site without standing biomass, ABG_{dry} is the dry aboveground biomass [kg/m^2], and r represents the reduction factor for N_0. It was found that r has a value of 11.19 per kg dry aboveground biomass per m^2. This regression explained 87% of the variation between biomass and neutron count rate. It has to be noted that this correction was derived for temporally stable biomass situations (e.g., forest sites). In addition, forests typically have a litter layer, whose water content can change rapidly adding additional temporal variability to the CRNP signal. Therefore, Bogena et al. (2013) recommended considering the water dynamics in the litter layer explicitly in the calibration of the CRNP. Furthermore, temporally changing above- and belowground biomass of growing vegetation (e.g., crops) and intercepted water in the canopy affect fast neutron count rates and calibration parameters in a complicated way. Baroni and Oswald (2015) proposed that the influence of aboveground biomass could be straightforwardly incorporated into a weighting approach. On the other hand, Coopersmith et al. (2014) found that soil moisture is overestimated for high-biomass situations, but it is underestimated when biomass is relatively low. In order to elucidate the biomass effect of growing vegetation on CRNP measurements in more detail, the field test site in Selhausen (Germany) of the TERENO project (Bogena et al. 2012) was instrumented with 7 CRNPs and a wireless soil moisture sensor network (Fuchs 2016). In order to track the biomass changes of the growing winter wheat, roots and plants were sampled approximately every 4 weeks. As expected, an increasing discrepancy between cosmic-ray-derived and in situ measured SWC during the growing season and a sharp decrease in discrepancy after the harvest were found (Fig. 8).

Figure 8 shows a good agreement between SWC derived from in situ measurements and cosmic-ray neutron data for the time period with low-standing biomass (<2 kg m^{-2}). As winter wheat begins to grow faster, SWC derived from the CRNP starts to deviate from the in situ measurements. In order to correct for the temporally varying effect of biomass on fast neutron intensity, Fuchs (2016) derived relationships between the calibration parameter N_0 and total fresh biomass. Using these relationships for the correction of fast neutron intensity reduced the discrepancy between cosmic-ray-derived and in situ measured SWC considerably (from 0.39 m^3 m^{-3} to 0.08 m^3 m^{-3}). The remaining uncertainty is due to the short-term fluctuations in the CRNP data that coincide with rainfall events, indicating effects of interception storage and sharp soil water gradients due to infiltration of rainfall water into the soil. A major further step in this direction would be the development of a method that enables the inference of biomass changes from cosmic-ray neutron intensity measurements without further measurements. Recently, it was shown that the temporal dynamics of the thermal-to-fast neutron ratio might be a potential

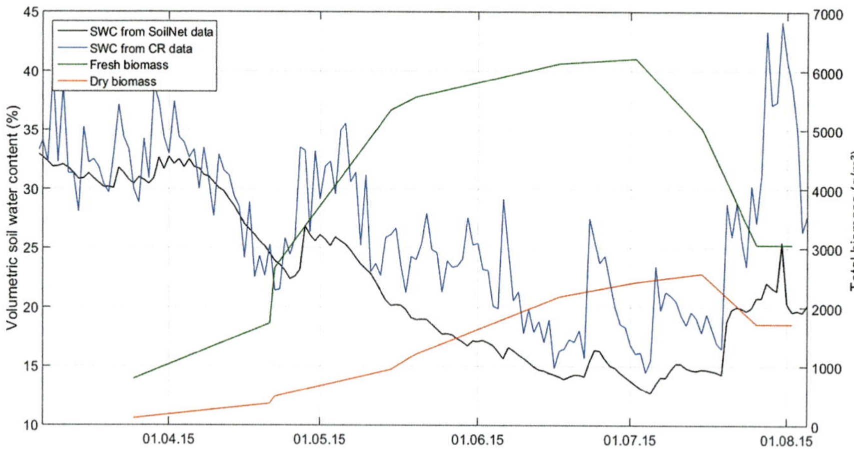

Fig. 8 Time series of soil water content derived from in situ measurements (SoilNet data) and cosmic-ray neutron data (CR), as well as measurements of fresh and dry biomass between March and September 2015 at the TERENO test site in Selhausen, Germany

predictor for canopy interception and biomass changes (Andreasen et al. 2016). However, to this date, no generic method exists that enables the direct dry biomass estimation without additional information on other sources of hydrogen in the footprint of the CRNP (e.g., water content of vegetation and soil).

Increasing the Spatial Scale with the COSMOS Rover

Although the CRNP footprint is already large (>10 ha), a single probe cannot capture soil moisture patterns at the catchment scale. Recently, a roving version of the CRNP (the COSMOS rover) has been developed to measure soil water content at larger scales (e.g., Chrisman and Zreda 2013). The main difference to the standard CRNP is the much larger size of the detector tubes used for the COSMOS rover, enabling measurements with much higher temporal resolutions (e.g., 1 min). The COSMOS rover thus enables the assessment of catchment-scale wetness conditions (up to hundreds of square kilometers in a single day under ideal situations). This makes the COSMOS rover a very promising method to close the critical scale gap in SWC monitoring toward the scale of a catchment. First feasibility experiments have been conducted with the COSMOS rover. For instance, Chrisman and Zreda (2013) attempted to compare road-effected rover data with SMOS satellite products. Later, Dong et al. (2014) measured and validated spatial SWC surveys with independent in situ SWC measurements. They concluded that the COSMOS rover is able to determine SWC with an accuracy of about 0.03 $m^3\ m^{-3}$. Recently, Franz et al. (2015) combined SWC measurements from roving and fixed cosmic-ray neutron probes with the aim to establish a real-time monitoring system for irrigation management. One of the major challenges is the calibration of the sensor under mobile conditions, since spatial variations in vegetation biomass and soil properties strongly affect the neutron counts. Therefore, in order to achieve a reliable estimation of spatial SWC, a new concept for signal processing is required that incorporates information of biomass distributions and soil properties in the calibration process.

Global Navigation Satellite System Reflectometry

The concept of Global Navigation Satellite System Reflectometry (GNSS-R) was first introduced 20 years ago, although it was years later when its use for remote sensing of environmental variables was finally considered. GNSS-R has several advantages over other technologies: all-weather and worldwide availability of GNSS signals; it is inexpensive and consists of lightweight, low-power consumption sensors; there are an increasing number of GNSS satellites in orbit; and GNSS operate in L-band (with wavelengths $\approx$ 20 cm) which is one of the optimal frequency bands for surface soil moisture estimation. Altogether, this highlights the potential of GNSS-R for near real-time monitoring of land properties, such as surface SWC, vegetation water content (VWC), and snow depth. Over land, GNSS-R takes advantage of the multiple GNSS satellites visible at any time and at various elevation angles, and also of the high sensitivity of the reflected signal to changes in the dielectric constant (and therefore SWC) of the ground. Although the development of a robust SWC retrieval algorithm from GNSS-R observations is a work in progress, a few algorithms have been derived for bare and vegetated surfaces using electromagnetic forward models and widely used dielectric constant models (Jin et al. 2014a; Zavorotny et al. 2014; Larson 2016). Some studies have found that constraining the use of satellites with incidence angles between 10° and 50° leads to more sensitive retrieval results (Small et al. 2016). One of the main limitations of GNSS-R, however, is that the likelihood of accurate SWC retrievals from GNSS-R observations decreases with increasing soil surface roughness. The footprint size of this technology, or first Fresnel zone, is in the order of tenths to hundreds of squared meters, depending on the elevation angle of the GNSS satellite, and the distance above the ground and the radiation pattern of the receiving antenna. The majority of GNSS-R alternatives in the literature have been assessed for the US Global Positioning System (GPS) constellation. However, its use could be extended to other GNSS, such as Galileo (Europe), GLONASS (Russia), and BeiDou (China).

A cost-effective GNSS-R approach is the GPS interferometric reflectometry (GPS-IR), which uses standard, commercial geodetic GPS antennas and receivers deployed in several networks around the globe. The number of units in these networks is rapidly increasing, and at present approximately 10,000 have their data publicly available in near real-time, several of them with long-term data records. For a typical antenna height of about 2 m, the footprint has a radius between 20 and 30 m. Soil moisture and vegetation water content are derived from the temporal changes in the signal-to-noise ratio (SNR) interferogram between the direct and reflected GPS signals. The phase of the SNR interferogram varies linearly with SWC, although additionally, the vegetation contribution to phase changes was found to be in the same order of magnitude at sites with VWC > 1 kg m^{-2}. Changes in the amplitude of the reflected GPS signal, however, are largely related to temporal changes in the vegetation, with the amplitude decreasing with increasing VWC. Therefore, some of the GPS-IR retrieval algorithms estimate first the vegetation conditions from the amplitude of the SNR interferogram, and then derive the SWC from the corrected phase. The accuracy of the GPS-IR surface SWC retrieval is ≈ 0.04 m^3 m^{-3} after taking into account the vegetation effects (Small et al. 2016; Larson 2016). This accuracy meets the target for satellite missions and is within the tolerance of in situ probes such as time-domain reflectometry and capacitance sensors.

Another ground-based alternative is to use sensors that have been specifically designed to measure the GPS reflections on the land. Over a 6-month period, Egido et al. (2012) evaluated the sensitivity of GNSS-R to SWC and VWC over a bare and vegetated agricultural area using a system at 25 m height, with three circular polarization antennas, one RHCP (right-hand circular polarization) uplooking and two (LHCP – left-hand circular polarization – and RHCP) downlooking, acquiring data sequentially. A sensitivity of 0.3 dB/SWC (%) and $R = 0.76$ was found if the LHCP downlooking antenna was used, whereas little sensitivity was found in the case of the RHCP antenna. Rodriguez-Alvarez et al. (2011) designed and implemented the SMIGOL (Soil Moisture Interference Pattern GNSS Observations at L-band) reflectometer, which measures the power of the interference between the direct and reflected GPS signals. This approach, known as interference pattern technique (IPT), uses one V-pol antenna pointing at the horizon and is able to estimate the SWC within a ≈20 m radius at an accuracy <0.05 m^3 m^{-3} and at a spatial resolution of tenths of centimeters. The concept was later extended to dual-polarization, H- and V-pol (PSMIGOL; Alonso-Arroyo et al. 2014), and has also been successfully used for vegetation height, topography, water level, and snow thickness applications. The main limitation of these and any other stationary sensors is, however, their non-capability to provide spatially distributed SWC information over larger areas, if required. To circumvent this limitation, some GNSS-R instruments have been designed for airborne applications over land (e.g., Alonso-Arroyo et al. 2016; Motte et al. 2016).

An example which illustrates the potential of roving GNSS-R systems for agriculture is shown hereafter. The advantages of such a system over airborne platforms, which have been the alternative to stationary antennas for farm-scale applications, are evident: lower deployment cost, on-demand availability, and higher spatial resolution. Figure 9 (left) shows an overview of the tractor setup used during the field experiments conducted to evaluate this proof of concept, which was a first of its kind (Australian Grains Research and Development Corporation (GRDC), through funding to the project UMO-004 "On-the-go soil moisture monitoring using GPS signals during routine farming practices"). The experiment site was a rain fed winter barley paddock in North-West Victoria, Australia. Due to the large size of the farm, a focus area of 400 m (E-W) × 300 m (N-S) was selected. The GNSS-R sensor used is an improved version of the LARGO (Light Airborne Reflectometer for GNSS-R Observations), for which the sensitivity to SWC changes had been successfully tested in an airborne (Alonso-Arroyo et al. 2016) and a ground-based configuration during the 2013–2014 GELOz experiments (GNSS-R experiments over land in Australia). The LARGO II sensor in Fig. 9 (left) consists of one

Fig. 9 (Left) Overview of the GNSS-R tractor setup used during the surveys conducted in North-West of Victoria, Australia. (Right) Maps showing the *in situ* and estimated volumetric SWC during one of the tractor surveys

RHCP uplooking and one LHCP downlooking, highly directive customized antennas at GPS L1 frequency. The antennas were installed at a 30° tilt from the horizontal and 2.75 m above the ground level. The GPS direct and reflected signals measured for all the satellites in view are stored internally every second. As the tractor surveys were conducted at a speed of 5 km h^{-1}, this translated into one acquisition every 1.4 m. The footprint size varied between 1 and 5 m^2, depending on the elevation of the GPS satellite. A series of auxiliary data sets were collected to gain knowledge of the in situ conditions at the farm and to be used for the validation and interpretation of the soil moisture retrievals. Optical and thermal sensors were mounted on the same frame as the LARGO II to derive crop properties, such as normalized difference vegetation index (NDVI) and temperature. Volumetric surface SWC was sampled every ≈ 15 m using EM sensors (Stevens Hydraprobe). Geo-located values of soil temperature, dielectric constant, and bulk electrical conductivity, and user-input crop data such as height and the presence of dew, were recorded as well.

The in situ and estimated SWC during one of the experiments is shown in Fig. 9 (right). The GNSS-R observations were filtered, and only GPS satellites with incidence angles between 10° and 50° were used in the retrieval algorithm. A simple forward model was used, in which the vegetation effect was modeled as an offset added to the bare soil contribution. Although the LARGO II instrument used has a 1 s measurement interval, the estimated SWC points were averaged at a coarser spatial resolution to reduce the noise inherent to the GNSS-R technique. A grid size of 10 m was found to be adequate, but it will depend on the speed of the tractor, and the position of the GPS satellites during the surveys. As there were no observations in some areas of the paddock, due to the geometry of the problem and the filtering of the data, the GNSS-R grid estimates were interpolated using the kriging method to obtain a full map of the focus area. No significant correlation between in situ and estimated SWC was found, due to the limited range of SWC conditions observed and the noise in the GNSS-R measurements. The spatial variation in SWC was, nevertheless, captured, and the probability distribution functions of both SWC data sets were similar, resulting in root-mean-square errors ≈0.04 m^3 m^{-3}. Although further experimental data is required to implement a robust retrieval algorithm, the use of this roving GNSS-R system has already proven to be an asset for farm-scale applications.

Nuclear Magnetic Resonance

Nuclear magnetic resonance (NMR) is a physical phenomenon that can yield molecular properties of matter by irradiating atomic nuclei, in a magnetic field, with electromagnetic radio waves (Blümich et al. 2014). NMR is particularly sensitive to the presence of protons ^1H and is therefore suited to investigate processes that involve water. When applied to the study of soil, NMR can be used to infer SWC, total porosity, and pore-size distribution and to quantify bound and producible fluid fractions.

Protons ^1H possess an intrinsic magnetic moment and angular momentum and consequently a nonzero spin (spin = ½). Under an external magnetic field of strength

B_0, they will thus behave like small bar magnets that may orient parallel or antiparallel to the direction of $\mathbf{B_0}$, defining two energy levels (low E^- and high E^+, respectively) as illustrated in Fig. 10. The relative number (N) of nuclei in each spin state is described by the Boltzmann distribution:

$$\frac{N^+}{N^-} = e^{\left(-\frac{(E^+ - E^-)}{k_b T}\right)} \tag{15}$$

where k_b is the Boltzmann constant and T is the temperature in K.

In thermal equilibrium, the parallel state, more energetically favorable, is slightly more populated than the antiparallel state.

The rotational axis of the spinning nucleus is not oriented exactly parallel or antiparallel to the direction of $\mathbf{B_0}$. It exhibits a certain angle which causes precession, similarly to a spinning-top, around $\mathbf{B_0}$ at a frequency, called the Larmor or resonance frequency. The latter can be calculated with the following expression:

$$\nu_0 = \frac{(E^+ - E^-)}{h} = \frac{\gamma B_0}{2\pi} \tag{16}$$

where h is Planck's constant (expressed in J s), γ is the gyromagnetic ratio which is specific for each atomic nucleus (expressed in rad s^{-1} T^{-1}), and B_0 is the strength of the magnetic field (expressed in T). In a 1 Tesla field, the Larmor frequency of protons ^1H is 42.576 MHz.

The macroscopic magnetization, $\mathbf{M_0}$, of a sample is given by the sum of each spin vectors, as illustrated in Fig. 10. This magnetization can be used to infer the proton density in the sample submitted to $\mathbf{B_0}$. Assuming that water is the main source of ^1H protons in the investigated volume, proton density can be directly related to the water content.

The quantification of $\mathbf{M_0}$ is not possible as long as it points in the same direction as $\mathbf{B_0}$. In order to measure it, it is necessary to tip the magnetization away from its alignment with the external magnetic field. This is achieved by

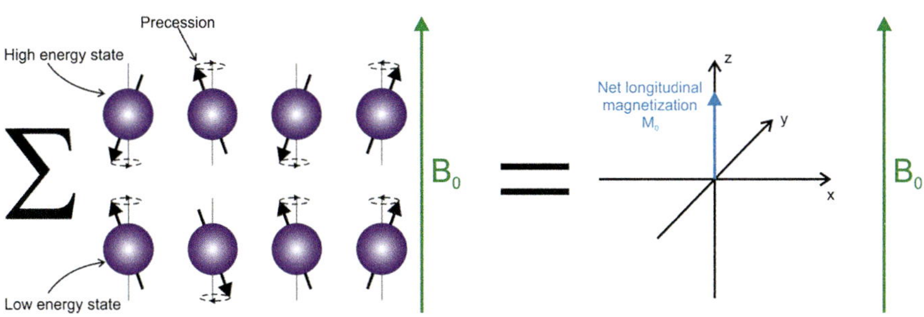

Fig. 10 Due to the slight difference of spin population in the two energy levels at thermal equilibrium, the sum of all the magnetic moments produces a macroscopic magnetization in the direction of $\mathbf{B_0}$, also known as the net longitudinal magnetization

applying a radio-frequency (rf) magnetic impulse via a coil, at the Larmor frequency, that generates an oscillating magnetic field, $\mathbf{B}_1$, in the perpendicular plane to $\mathbf{B}_0$. The resulting tipped magnetization now oscillates at the Larmor frequency in the transverse plane to $\mathbf{B}_0$ generating an electrical field that can be detected by an antenna (generally the coil used to produce the excitation rf impulse). For a single pulse, the maximum intensity of the signal is recorded when the flip angle is 90° so that all the magnetization $\mathbf{M}_0$ is transferred into the orthogonal plane to $\mathbf{B}_0$.

At the end of the pulse, the nuclei relax and return to their equilibrium positions and the recorded signal decays. Under a homogeneous magnetic field $\mathbf{B}_0$, the response signal decays with a time constant T_2, called the transverse relaxation time, whereas in an inhomogeneous magnetic field, it decays with a generally smaller time constant T_2^*. In order to compensate for the effect of inhomogeneous magnetic field, it is possible to apply a series of 180° pulses after the excitation pulse in order to refocus spins that would have encountered a dephasing. This measurement sequence is known as Carr-Purcell-Meiboom-Gill (CPMG) echo train and is the standard way to estimate T_2 in inhomogeneous magnetic field environment. The time required to rebuild the magnetization along $\mathbf{B}_0$ is called the longitudinal relaxation time T_1 (with $T_1 > T_2 \geq T_2^*$).

The principle of NMR signal acquisition is illustrated in Fig. 11.

NMR Parameters

Two types of NMR parameters, particularly interesting in soil physics, can be determined from the impulse response signal: (1) the spin density and (2) the relaxation times.

Spin density (or proton density when focusing on ^{1}H atomic nuclei) represents the total number of spins present in the investigated volume and is directly proportional to the initial amplitude of the impulse response (=$S(0)$ in Fig. 11). By comparing the initial amplitude of the signal with the initial amplitude of a reference sample, soil water content or total porosity (in fully saturated conditions) can be estimated. Spin density is probably the most basic NMR parameter and can generally be measured with any NMR instrument (Blümich et al. 2014).

Relaxation times T_1 and T_2 are important parameters in NMR for material characterization and discrimination. In complex media, such as in soil, there is generally not a single T_1 or T_2 value characteristic of the whole sample, but rather distributions of T_1 and T_2 values. In porous media, during the acquisition of the signal, water molecules may encounter two different environments where relaxation of the signal may occur: the pore surface and the bulk fluid volume. Three NMR relaxation mechanisms influence T_1 or T_2 relaxation times: (1) bulk fluid relaxation, (2) relaxation by molecular diffusion in internal and external magnetic field gradients, and (3) grain surface relaxation. Bulk fluid relaxation is the intrinsic relaxation property of the fluid. It can generally be neglected in porous media (because much slower than the two others relaxation mechanisms). The motion of water molecules by diffusion in magnetic field gradients during the measurement, meaning that water

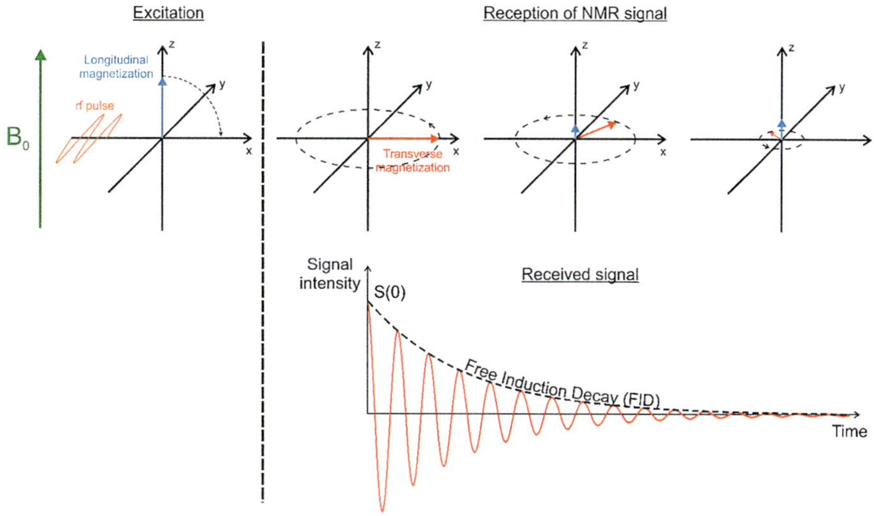

Fig. 11 During the excitation phase, the magnetization is tipped away from the direction of $\mathbf{B_0}$ by a rf pulse applied at the resonance frequency. If the amplitude and duration of the pulse is correctly adjusted, all the magnetization can be rotated in the *x*-*y* plane. The resulting transverse magnetization precesses now in the *x*-*y* plane at the Larmor frequency and generates a signal that can be detected by an antenna. The decay of the signal is due to two types of relaxation: (1) the realignment of spins along $\mathbf{B_0}$, also called longitudinal or T_1 relaxation (regrow of blue arrow), and (2) the loss of synchronization of the spins due to spin-spin interactions and magnetic field inhomogeneity, also called T_2^* relaxation (decrease of red arrow)

molecules may experience regions with different Larmor frequencies, induces a dephasing of spins which is responsible of T_2 relaxation. Grain surface relaxation is generally the dominant relaxation mechanism in porous media. It occurs when water molecules diffuse close to paramagnetic impurities at the pore wall or grain surface. Water molecules in large pores collide less frequently with pore walls and consequently relax slower than water molecules in small pores so that, in saturated media and under certain conditions (not detailed here), T_1 or T_2 distributions may mimic pore-size distribution (Kenyon et al. 1995).

In unsaturated media, T_2 distribution appears to be rather a measure of the distribution of the water within the pore space at a specific degree of saturation than a measure of the pore-size distribution as shown by Costabel and Yaramanci (2011).

Available in Situ NMR Sensors

Currently, three types of NMR sensors can be used in situ to measure SWC. Their implementation on the field is schematized in Fig. 12:

1. Mobile low-field NMR sensors (single-sided NMR) such as the NMR-MOUSE (Magritek Ltd.) that allows to investigate water content of topsoil down to a maximum depth of 2.5 cm (Blümich et al. 2014). The device is based on the

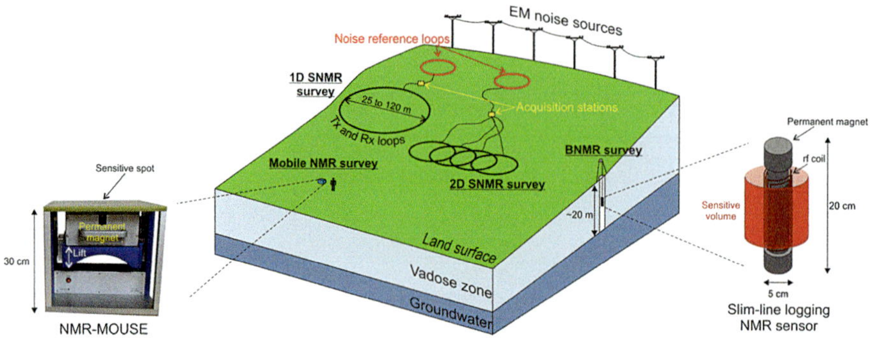

Fig. 12 Possible in situ applications of NMR for SWC measurements. Three methods are illustrated, each having its own advantages. Mobile NMR sensors such as the NMR-MOUSE developed by Magritek Ltd. offer a way to probe the very first centimeters of soil with a high vertical resolution. BNMR sensors, like the slim-line logging sensor schematized here (modified from Perlo et al. 2013) and developed by the RWTH Aachen, are particularly suited to quantify SWC in the vadose zone with a vertical resolution of 10 cm (Behroozmand et al. 2015). SNMR allows to investigate water content from shallow horizons to more than 100 m depth without requiring the drilling of boreholes (scheme modified from Vista Clara Inc 2016). All methods presented are very sensitive to ambient EM noise and particular precautions (such as the deployment of noise reference loops in SNMR) must be taken to reduce its effect on the data quality

principle of "inside-out" NMR and is composed of a permanent U-shape magnet generating a magnetic gradient perpendicular to the surface (Blümich et al. 1998). A rf coil, mounted above the magnet, is used to excite the spins and receive the NMR signal coming from a sensitive volume whose thickness (i.e., resolution) ranges from 5 to 100 μm depending on the device model. The magnet is mounted on a lift so that it is possible to acquire one-dimensional profiles.

2. Borehole NMR (BNMR) sensors were initially developed for the characterization of petroleum resources. Unfortunately, these sensors are not suited for shallow measurements of SWC due to their dimension that largely exceeds the classical size of wells in this depth range. To overcome that, slim-line logging tools were developed (e.g., Perlo et al. 2013; Walsh et al. 2013). The principle of measurement is similar to the NMR-MOUSE except that the sensitive zone is, in this case, a thin cylindrical region around the outside of the borehole.
3. Surface NMR (SNMR). The main difference of SNMR compared to other NMR methods is the use of the earth magnetic field instead of the large magnetic field created by magnets. The rf signal used to excite the spins is generated by a wire loop laid down on the ground (Legchenko et al. 2002). The same loop is also used to receive the NMR signal after the excitation. The depth of investigation of this method depends on the loop size and site-specific conditions, such as the presence of electrically conductive layers below the setup, but usually ranges

from 40 to 120 m (Legchenko et al. 2011). SNMR was initially designed for 1-D sounding (in that case, the term magnetic resonance sounding (MRS) is generally employed), but 2-D and even 3-D applications of the method are possible (e.g., Hertrich et al. 2007).

We refer the interested reader to the review paper of Behroozmand et al. (2015) for more information about the introduced methods.

Soil water content, after the collection of a sample, can also be measured in laboratory using various NMR tools. Among them, magnetic resonance imaging (MRI), a technique commonly used in medicine, holds particular interest because it allows the visualization in 2-D or in 3-D of SWC with a very high resolution (e. g., Pohlmeier et al. 2008; Merz et al. 2016).

NMR techniques have demonstrated for years their usefulness to quantify water content at great depths (e.g., BNMR logging tools in petroleum industry). However, the development of NMR sensors to investigate SWC at shallower depth with high spatial resolution is relatively recent and improvements are still required in order for these sensors to be routinely used in that scope. Among possible optimizations, one can point out the reduction of the dead time (= time between excitation and signal acquisition) and/or echo time in order to capture fast-decaying signals (possibly caused by fine pore materials or the presence of paramagnetic impurities) and the development of new processing tools to reduce the effect of EM noise and increase the signal-to-noise ratio.

Summary and Outlook

Soil water content is a key variable for understanding and modeling ecohydrological processes. In this chapter, we reviewed the state of the art with respect to the spatiotemporal characterization of SWC from point to field scale using ground-based techniques. The number of SWC measurement methods has significantly increased over the last years, in particular with the emergence of novel techniques such as cosmic-ray neutron probes, navigation satellite system reflectometry, and nuclear magnetic resonance. Large improvements have also been made in order to fully exploit the information contained in soil moisture data by using data assimilation techniques, more sophisticated data processing approaches, and inversion schemes (e.g., coupled hydrogeophysical inversion, full-waveform analysis for GPR). More accurate instruments (e.g., better signal-to-noise ratio, smaller instrument drift for EMI) have also been developed during the last decade. There is still a need for soil moisture data with higher temporal and spatial resolution, which is particularly useful for observing ecohydrological processes. Finally, combination of different measurement techniques (such as ERT and EMI or active and passive microwave systems) with complementary characteristics might be an interesting avenue to further explore.

References

H. Abdu, D.A. Robinson, M. Seyfried, S.B. Jones, Geophysical imaging of watershed subsurface patterns and prediction of soil texture and water holding capacity. Water Resour. Res. **44**(4), W00D18 (2008)

S.A. al Hagrey, Electrical resistivity imaging of tree trunks. Near Surf. Geophys. **4**, 179–187 (2006)

A. Alonso-Arroyo, A. Camps, A. Monerris, C. Rüdiger, J.P. Walker, A. Aguasca, G. Forte, H. Park, D. Pascual, R. Onrubia, Dual-polarization GNSS-R interference pattern technique for soil moisture mapping. IEEE J. Sel. Top. Appl. Earth Obs. Remote Sens. **7**(5), 1533–1544 (2014)

A. Alonso-Arroyo, A. Camps, A. Monerris, C. Rüdiger, J.P. Walker, R. Onrubia, J. Querol, H. Park, D. Pascual, On the correlation between GNSS reflectometry and L-band radiometry. IEEE J. Sel. Top. Appl. Earth Obs. Remote Sens. **9**(12), 5862–5879 (2016)

D. Altdorff, M. Bechtold, J. van der Kruk, H. Vereecken, J.A. Huisman, Mapping peat layer properties with multi-coil offset electromagnetic induction and laser scanning elevation data. Geoderma **261**, 178–189 (2016)

F. André, F. Jonard, M. Jonard, S. Lambot, In situ characterization of forest litter using ground-penetrating radar. J. Geophys. Res. Biogeo. **121**(3), 879–894 (2016)

M. Andreasen, K.H. Jensen, M. Zreda, D. Desilets, H. Bogena, M.C. Looms, Modeling cosmic ray neutron field measurements. Water Resour. Res. **52**(8), 6451–6471 (2016)

U. Aqeel-ur-Rehman, A.Z. Abbasi, N. Islam, Z.A. Shaikh, A review of wireless sensors and networks' applications in agriculture. Comput. Stand. Interfaces **36**(2), 263–270 (2014)

G.E. Archie, The electrical resistivity log as an aid in determining some reservoir characteristics. Trans. AIME **146**, 54–61 (1942)

R. Baatz, H. Bogena, H.-J. Hendricks Franssen, J.A. Huisman, C. Montzka, H. Vereecken, An empirical vegetation correction for soil water content quantification using cosmic ray probes. Water Resour. Res. **51**(4), 2030–2046 (2015)

O. Banton, M.K. Seguin, M.A. Cimon, Mapping field-scale physical properties of soil with electrical resistivity. Soil Sci. Soc. Am. J. **61**, 1010–1017 (1997)

G. Baroni, S. Oswald, A scaling approach for the assessment of biomass changes and rainfall interception using cosmic-ray neutron sensing. J. Hydrol. **525**, 264–276 (2015)

A.A. Behroozmand, K. Keating, E. Auken, A review of the principles and applications of the NMR technique for near-surface characterization. Surv. Geophys. **36**, 27–85 (2015)

A. Binley, R3t (version 1.8), Lancaster University (2013) [Logiciel]. Repéré à http://www.es.lancs.ac.uk/people/amb/Freeware/R3t/R3t.htm

J.R. Birchak, C.G. Gardner, J.E. Hipp, J.M. Victor, High dielectric constant microwave probes for sensing soil moisture. Proc. IEEE **62**, 93–98 (1974)

S. Bircher, N. Skou, K. Jensen, J. Walker, L. Rasmussen, A soil moisture and temperature network for SMOS validation in Western Denmark. Hydrol. Earth Syst. Sci. **16**, 1445–1463 (2012)

B. Blümich, P. Blümler, G. Eidmann, A. Guthausen, R. Haken, U. Schmitz, K. Saito, G. Zimmer, The NMR-mouse: Construction, excitation, and applications. Magn. Reson. Imaging **16**, 479–484 (1998)

B. Blümich, S. Haber-Pohlmeier, W. Zia, *Compact NMR* (Walter de Gruyter GmbH, Berlin., 276 pp, 2014)

H. Bogena, M. Herbst, J.A. Huisman, U. Rosenbaum, A. Weuthen, H. Vereecken, Potential of wireless sensor networks for measuring soil water content variability. Vadose Zone J. **9**(4), 1002–1013 (2010)

H. Bogena, R. Kunkel, E. Krüger, S. Zacharias, T. Pütz, M. Schwank, O. Bens, E. Borg, A. Brauer, P. Dietrich, I. Hajnsek, H. Kunstmann, J.-C. Munch, H. Papen, E. Priesack, H.P. Schmid, U. Wollschläger, H. Vereecken, TERENO – Long-term monitoring network for terrestrial research. Hydrol. Wasserbewirtsch. **56**(3), 138–143 (2012)

H. Bogena, J.A. Huisman, R. Baatz, H.-J. Hendricks Franssen, H. Vereecken, Accuracy of the cosmic-ray soil water content probe in humid Forest ecosystems: The worst case scenario. Water Resour. Res. **49**(9), 5778–5791 (2013)

H. Bogena, J.A. Huisman, A. Güntner, C. Hübner, J. Kusche, F. Jonard, S. Vey, H. Vereecken, Emerging methods for non-invasive sensing of soil moisture dynamics from field to catchment scale: A review. WIREs Water **2**(6), 635–647 (2015a)

S. Busch, J. van der Kruk, J. Bikowski, H. Vereecken, Quantitative conductivity and permittivity estimation using full-waveform inversion of onground GPR data. Geophysics **77**(6), H79–H91 (2012)

G. Calamita, A. Perrone, L. Brocca, B. Onorati, S. Manfreda, Field test of a multi-frequency electromagnetic induction sensor for soil moisture monitoring in southern Italy test sites. J. Hydrol. **529**, 316–329 (2015)

R.B. Campbell, Change of electrical conductivity with temperature and the relation of osmotic pressure to electrical conductivity and ion concentration for soil extracts. Soil Sci. Soc. Am. J. **13**, 66–69 (1949)

G. Cassiani, J. Boaga, D. Vanella, M.T. Perri, S. Consoli, Monitoring and modelling of soil-plant interactions: The joint use of ERT, sap flow and eddy covariance data to characterize the volume of an orange tree root zone. Hydrol. Earth Syst. Sci. **19**(5), 2213–2225 (2015)

D. Caterina, J. Beaujean, T. Robert, F. Nguyen, A comparison study of different image appraisal tools for electrical resistivity tomography. Near Surf. Geophys. **11**, 639–657 (2013)

B.B. Chrisman, M. Zreda, Quantifying mesoscale soil moisture with the cosmic-ray rover. Hydrol. Earth Syst. Sci. **17**(12), 5097–5108 (2013)

E.J. Coopersmith, M.H. Cosh, C.S.T. Daughtry, Field-scale moisture estimates using COSMOS sensors: A validation study with temporary networks and leaf-area indices. J. Hydrol. **519**(Part A), 637–643 (2014)

D.L. Corwin, S.M. Lesch, Application of soil electrical conductivity to precision agriculture: Theory, principles, and guidelines. Agron. J. **95**(3), 455–471 (2003)

S. Costabel, U. Yaramanci, Relative hydraulic conductivity and effective saturation from earth's field nuclear magnetic resonance-a method for assessing the vadose zone. Near Surf. Geophys. **9**, 155–167 (2011)

J. Davis, A. Annan, Ground-penetrating radar for high-resolution mapping of soil and rock stratigraphy. Geophys. Prospect. **37**, 531–551 (1989)

C. de Groot-Hedlin, S. Constable, Occam's inversion to generate smooth, two dimensional models form magnetotelluric data. Geophysics **55**, 1613–1624 (1990)

P. de Rosnay, J.-C. Calvet, Y. Kerr, J.-P. Wigneron, F. Lemaître, M.-J. Escorihuela, J. Muñoz Sabater, K. Saleh, J. Barrié, G. Bouhours, L. Coret, G. Cherel, G. Dedieu, R. Durbe, F.N.E. Dine, F. Froissard, J. Hoedjes, A. Kruszewski, F. Lavenu, D. Suquia, P. Waldteufel, SMOSREX: A long term field campaign experiment for soil moisture and land surface processes remote sensing. Remote Sens. Environ. **102**, 377 (2006)

D. Desilets, M. Zreda, T.P.A. Ferre, Nature's neutron probe: Land surface hydrology at an elusive scale with cosmic-rays. Water Resour. Res. **46**, W11505 (2010)

M.C. Dobson, F.T. Ulaby, M.T. Hallikainen, M.A. El-Rayes, Microwave dielectric behavior of wet soil-part II: Dielectric mixing models. IEEE Trans. Geosci. Remote Sens. **23**(1), 35–46 (1985)

J.N. Dong, T.E. Ochsner, M. Zreda, M.H. Cosh, C.B. Zou, Calibration and validation of the COSMOS rover for surface soil moisture measurement. Vadose Zone J. **13**(4) (2014)

L.S. Edwards, A modified pseudosection for resistivity and IP. Geophysics **42**(5), 1020–1036 (1977)

A. Egido, M. Caparrini, G. Ruffini, S. Paloscia, E. Santi, L. Guerriero, N. Pierdicca, N. Floury, Global navigation satellite systems reflectometry as a remote sensing tool for agriculture. Remote Sens. **4**, 2356–2372 (2012)

J.R. Ernst, H. Maurer, A.G. Green, K. Holliger, Full-waveform inversion of crosshole radar data based on 2-D finite-difference time-domain solutions of Maxwell's equations. IEEE Trans. Geosci. Remote Sens. **45**(9), 2807–2828 (2007)

J.S. Famiglietti, D.R. Ryu, A.A. Berg, M. Rodell, T.J. Jackson, Field observations of soil moisture variability across scales. Water Resour. Res. **44**(1), W01423, 1–16 (2008)

T.E. Franz, T. Wang, W. Avery, C. Finkenbiner, L. Brocca, Combined analysis of soil moisture measurements from roving and fixed cosmic ray neutron probes for multiscale real-time monitoring. Geophys. Res. Lett. **42**, 3389–3396 (2015)

S. Friedel, Resolution, stability and efficiency of resistivity tomography estimated from a generalized inverse approach. Geophys. J. Int. **153**, 305–316 (2003)

H. Fuchs, Effects of biomass on soil moisture measurements using cosmic-ray neutron probes. Master Thesis, Radboud University and University of Duisburg-Essen, 59 pp. (2016)

S. Garré, M. Javaux, J. Vanderborght, L. Pages, H. Vereecken, Three-dimensional electrical resistivity tomography to monitor root zone water dynamics. Vadose Zone J. **10**(1), 412–424 (2011)

S. Garré, T. Günther, J. Diels, J. Vanderborght, Evaluating experimental design of ERT for soil moisture monitoring in contour hedgerow intercropping systems. Vadose Zone J. **11**(4), 1–14 (2012)

S. Garré, I. Coteur, C. Wongleecharoen, T. Kongkaew, J. Diels, J. Vanderborght, Non-invasive monitoring of soil water dynamics in mixed cropping systems: A case-study in Ratchaburi province, Thailand. Vadose Zone J. **12**(2), 1–12 (2013)

M. Guglielmetti, M. Schwank, C. Mätzler, C. Oberdörster, J. Vanderborght, H. Flühler, FOSMEX: Forest soil moisture experiments with microwave radiometry. IEEE Trans. Geosci. Remote Sens. **46**(3), 727 (2008)

T. Günther, Inversion Methods and Resolution Analysis for the 2D/3D Reconstruction of Resistivity Structures from DC Measurements, PhD Thesis, University of Mining and Technology, Freiberg (2004)

N. Güting, A. Klotzsche, J. van der Kruk, J. Vanderborght, H. Vereecken, A. Englert, Imaging and characterization of facies heterogeneity in an alluvial aquifer using GPR full-waveform inversion and cone penetration tests. J. Hydrol. **524**, 680–695 (2015)

I. Heidbüchel, A. Güntner, T. Blume, Use of cosmic-ray neutron sensors for soil moisture monitoring in forests. Hydrol. Earth Syst. Sci. **20**(3), 1269–1288 (2016)

T. Hermans, E. Oware, J. Caers, Direct prediction of spatially and temporally varying physical properties from time-lapse electrical resistance data. Water Resour. Res. **52**, 7262–7283 (2016)

M. Hertrich, M. Braun, T. Gunther, A.G. Green, U. Yaramanci, Surface nuclear magnetic resonance tomography. IEEE Trans. Geosci. Remote Sens. **45**, 3752–3759 (2007)

A.C. Hinnell, T.P.A. Ferré, J.A. Vrugt, J.A. Huisman, S. Moysey, J. Rings, M.B. Kowalsky, Improved extraction of hydrologic information from geophysical data through coupled hydrogeophysical inversion. Water Resour. Res. **46**, W00D40 (2010)

J.A. Huisman, S.S. Hubbard, J.D. Redman, A.P. Annan, Measuring soil water content with ground penetrating radar, A review. Vadose Zone J. **2**(4), 476–491 (2003)

R.B. Jana, A. Ershadi, M.F. McCabe, Examining the relationship between intermediate scale soil moisture and terrestrial evaporation within a semi-arid grassland. Hydrol. Earth Syst. Sci. **20**(9), 3987–4004 (2016)

R. Jin, X. Li, B. Yan, X. Li, W. Luo, M. Ma, J. Guo, J. Kang, Z. Zhu, S. Zhao, A nested ecohydrological wireless sensor network for capturing the surface heterogeneity in the midstream areas of the Heihe River Basin, China. IEEE Geosci. Remote Sens. Lett. **11**(11), 2015–2019 (2014)

S.G. Jin, E. Cardellach, F. Xie, *GNSS Remote Sensing: Theory, Methods and Applications* (Springer, Heidelberg., 276 pp, 2014)

H. M. Jol (ed.), *Ground Penetrating Radar: Theory and Applications* (Elsevier Science, Amsterdam., 524 pp, 2009)

F. Jonard, F. André, Q. Ponette, C. Vincke, M. Jonard, Sap flux density and stomatal conductance of European beech and common oak trees in pure and mixed stands during the summer drought of 2003. J. Hydrol. **409**, 371–381 (2011a)

F. Jonard, L. Weihermüller, K.Z. Jadoon, M. Schwank, H. Vereecken, S. Lambot, Mapping field-scale soil moisture with L-band radiometer and ground-penetrating radar over bare soil. IEEE Trans. Geosci. Remote Sens. **49**(8), 2863–2875 (2011b)

F. Jonard, L. Weihermüller, H. Vereecken, S. Lambot, Accounting for soil surface roughness in the inversion of ultrawideband off-ground GPR signal for soil moisture retrieval. Geophysics **77**(1), H1–H7 (2012)

F. Jonard, M. Mahmoudzadeh, C. Roisin, L. Weihermüller, F. André, J. Minet, H. Vereecken, S. Lambot, Characterization of tillage effects on the spatial variation of soil properties using ground-penetrating radar and electromagnetic induction. Geoderma **207–208**, 310–322 (2013)

F. Jonard, L. Weihermüller, M. Schwank, K.Z. Jadoon, H. Vereecken, S. Lambot, Estimation of hydraulic properties of a sandy soil using ground-based active and passive microwave remote sensing. IEEE Trans. Geosci. Remote Sens. **53**(6), 3095–3109 (2015)

A. Kemna, *Tomographic Inversion of Complex Resistivity-Theory and Application* (Ruhr-Universität, Bochum, 2000)

B. Kenyon, R. Kleinberg, C. Straley, G. Gubelin, C. Morriss, Nuclear magnetic resonance imaging-technology for the 21st century. Oilfield Rev **7**, 19–33 (1995)

A. Klotzsche, J. van der Kruk, G. Meles, H. Vereecken, Crosshole GPR full-waveform inversion of waveguides acting as preferential flow paths within aquifer systems. Geophysics **77**(4), H57–H62 (2012)

A. Klotzsche, J. van der Kruk, N. Linde, J. Doetsch, H. Vereecken, 3-D characterization of high-permeability zones in a gravel aquifer using 2-D crosshole GPR full-waveform inversion and waveguide detection. Geophys. J. Int. **195**(2), 932–944 (2013)

A. Klotzsche, J. van der Kruk, J. Bradford, H. Vereecken, Detection of spatially limited high porous layers using crosshole GPR signal analysis and full-waveform inversion. Water Resour. Res. **50**(8), 6966–6985 (2014)

M. Köhli, M. Schrön, M. Zreda, U. Schmidt, P. Dietrich, S. Zacharias, Footprint characteristics revised for field-scale soil moisture monitoring with cosmic-ray neutrons. Water Resour. Res. **51**, 5772–5790 (2015)

S. Lambot, E.C. Slob, I. van den Bosch, B. Stockbroeckx, M. Vanclooster, Modeling of ground-penetrating radar for accurate characterization of subsurface electric properties. IEEE Trans. Geosci. Remote Sens. **42**(11), 2555–2568 (2004)

K.M. Larson, GPS interferometric reflectometry: Applications to surface soil moisture, snow depth, and vegetation water content in the western United States. WIREs Water **3**(6), 775 (2016)

F. Lavoué, J. van der Kruk, J. Rings, F. André, D. Moghadas, J.A. Huisman, S. Lambot, L. Weihermüller, J. Vanderborght, H. Vereecken, Electromagnetic induction calibration using apparent electrical conductivity modelling based on electrical resistivity tomography. Near Surf. Geophys. **8**(6), 553–561 (2010)

H. Lawrence, J.-P. Wigneron, P. Richaume, N. Novello, J. Grant, A. Mialon, A. Al Bitar, O. Merlin, D. Guyon, D. Leroux, S. Bircher, Y. Kerr, Comparison between SMOS vegetation optical depth products and MODIS vegetation indices over crop zones of the USA. Remote Sens. Environ. **140**, 396 (2014)

A. Legchenko, J.-M. Baltassat, A. Beauce, J. Bernard, Nuclear magnetic resonance as a geophysical tool for hydrogeologists. J. Appl. Geophys. **50**(1–2), 21–46 (2002)

A. Legchenko, M. Descloitres, C. Vincent, H. Guyard, S. Garambois, K. Chalikakis, M. Ezersky, Three-dimensional magnetic resonance imaging for groundwater. New J. Phys. **13**, 025022 (2011)

J. Lemmetyinen, M. Schwank, K. Rautiainen, A. Kontu, T. Parkkinen, C. Mätzler, A. Wiesmann, U. Wegmüller, C. Derksen, P. Toose, A. Roy, J. Pulliainen, Snow density and ground permittivity retrieved from L-band radiometry: Application to experimental data. Remote Sens. Environ. **180**, 377 (2016)

M.H. Loke, R.D. Barker, Rapid least-squares inversion of apparent resistivity pseudosections by a quasi-Newton method. Geophys. Prospect. **44**, 131–152 (1996)

R. Ma, A. McBratney, B. Whelan, B. Minasny, M. Short, Comparing temperature correction models for soil electrical conductivity measurement. Precis. Agric. **12**, 55–66 (2011)

L. Marescot, M.H. Loke, D. Chapellier, R. Delaloye, C. Lambiel, E. Reynard, Assessing reliability of 2D resistivity imaging in mountain permafrost studies using the depth of investigation index method. Near Surf. Geophys. **1**(2), 57–67 (2003)

E. Martini, U. Werban, S. Zacharias, M. Pohle, P. Dietrich, U. Wollschläger, Repeated electromagnetic induction measurements for mapping soil moisture at the field scale: Validation with data from a wireless soil moisture monitoring network. Hydrol. Earth Syst. Sci. **21**(1), 495–513 (2017)

J.D. McNeill, *Electromagnetic Terrain Conductivity Measurement at Low Induction Numbers, Technical Note TN-6* (Geonics Ltd., Mississauga, 1980)
G.A. Meles, J. van der Kruk, S.A. Greenhalgh, J.R. Ernst, H. Maurer, A.G. Green, A. New Vector, Waveform inversion algorithm for simultaneous updating of conductivity and permittivity parameters from combination Crosshole/borehole-to-surface GPR data. IEEE Trans. Geosci. Remote Sens. **48**(12), 4329–4329 (2010)
S. Merz, A. Pohlmeier, B.J. Balcom, R. Enjilela, H. Vereecken, Drying of a natural soil under evaporative conditions: A comparison of different magnetic resonance methods. Appl. Magn. Reson. **47**, 121–138 (2016)
A. Mester, J. van der Kruk, E. Zimmermann, H. Vereecken, Quantitative two-layer conductivity inversion of multi-configuration electromagnetic induction measurements. Vadose Zone J. **10**(4), 1319–1330 (2011)
D. Michot, Y. Benderitter, A. Dorigny, B. Nicoullaud, D. King, A. Tabbagh, Spatial and temporal monitoring of soil water content with an irrigated corn crop cover using surface electrical resistivity tomography. Water Resour. Res. **39**, SBH 14 (2003)
J. Minet, P. Bogaert, M. Vanclooster, S. Lambot, Validation of ground penetrating radar full-waveform inversion for field scale soil moisture mapping. J. Hydrol. **424–425**, 112–123 (2012)
V.L. Mironov, L.G. Kosolapova, S.V. Fomin, Physically and Mineralogically based spectroscopic dielectric model for moist soils. IEEE Trans. Geosci. Remote Sens. **47**(7), 2059 (2009)
T. Mo, B.J. Choudhury, T.J. Schmugge, J.R. Wang, T.J. Jackson, A model for microwave emission from vegetation-covered fields. J. Geophys. Res. **87**, 11229 (1982)
E. Motte, M. Zribi, P. Fanise, A. Egido, J. Darrozes, A. Al-Yaari, N. Baghdadi, F. Baup, S. Dayau, R. Fieuzal, P.-L. Frison, D. Guyon, J.-P. Wigneron, GLORI: A GNSS-R dual polarization airborne instrument for land surface monitoring. Sensors **16**(5), 732 (2016)
D.W. Oldenburg, Y. Li, Estimating depth of investigation in DC resistivity and IP surveys. Geophysics **64**, 403–416 (1999)
J. Perlo, E. Danieli, J. Perlo, B. Blümich, F. Casanova, Optimized slim-line logging NMR tool to measure soil moisture in situ. J. Magn. Reson. **233**, 74–79 (2013)
A. Pohlmeier, A. Oros-Peusquens, M. Javaux, M. Menzel, J. Vanderborght, J. Kaffanke, S. Romanzetti, J. Lindenmair, H. Vereecken, N. Shah, Changes in soil water content resulting from root uptake monitored by magnetic resonance imaging. Vadose Zone J. **7**, 1010–1017 (2008)
K. Rautiainen, J. Lemmetyinen, M. Schwank, A. Kontu, C.B. Ménard, C. Mätzler, M. Drusch, A. Wiesmann, J. Ikonen, J. Pulliainen, Detection of soil freezing from L-band passive microwave observations. Remote Sens. Environ. **147**, 206 (2014)
A. Revil, P.W.J. Glover, Nature of surface electrical conductivity in natural sands, sandstones, and clays. Geophys. Res. Lett. **25**, 691–694 (1998)
J.D. Rhoades, N.A. Manteghi, P.J. Shouse, W.J. Alves, Soil electrical conductivity and soil salinity: New formulations and calibrations. Soil Sci. Soc. Am. J. **53**, 433–439 (1989)
T. Robert, A. Dassargues, S. Brouyère, O. Kaufmann, V. Hallet, F. Nguyen, Assessing the contribution of electrical resistivity tomography (ERT) and self-potential (SP) methods for a water well drilling program in fractured/karstified limestones. J. Appl. Geophys. **75**(1), 42–53 (2011)
D.A. Robinson, S.B. Jones, J.M. Wraith, D. Or, S.P. Friedman, A review of advances in dielectric and electrical conductivity measurement in soils using time domain reflectometry. Vadose Zone J. **2**, 444–475 (2003)
D.A. Robinson, C.S. Campbell, J.W. Hopmans, B.K. Hornbuckle, S.B. Jones, R. Knight, F. Ogden, J. Selker, O. Wendroth, Soil moisture measurement for ecological and hydrological watershed-scale observatories: A review. Vadose Zone J. **7**(1), 358–389 (2008)
D.A. Robinson, I. Lebron, B. Kocar, K. Phan, M. Sampson, N. Crook, S. Fendorf, Time-lapse geophysical imaging of soil moisture dynamics in tropical deltaic soils: An aid to interpreting hydrological and geochemical processes. Water Resour. Res. **45**, W00D32 (2009)
D.A. Robinson, H. Abdu, I. Lebron, S.B. Jones, Imaging of hill-slope soil moisture wetting patterns in a semi-arid oak savanna catchment using time-lapse electromagnetic induction. J. Hydrol. **416**, 39–49 (2012)

N. Rodriguez-Alvarez, X. Bosch-Lluis, A. Camps, A. Aguasca, M. Vall-llossera, E. Valencia, I. Ramos-Perez, H. Park, Review of crop growth and soil moisture monitoring from a ground-based instrument implementing the interference pattern GNSS-R technique. Radio Sci. **46**, RS0C03 (2011)

U. Rosenbaum, H. Bogena, M. Herbst, J.A. Huisman, T.J. Peterson, A. Weuthen, A. Western, H. Vereecken, Seasonal and event dynamics of spatial soil moisture patterns at the small catchment scale. Water Resour. Res. **48**(10), W10544 (2012)

K. Roth, R. Schulin, H. Flühler, W. Attinger, Calibration of time-domain reflectometry for water-content measurement using a composite dielectric approach. Water Resour. Res. **26**(10), 2267–2273 (1990)

S. Rudolph, J. van der Kruk, C.v. Hebel, M. Ali, M. Herbst, C. Montzka, S. Patzold, D.A. Robinson, H. Vereecken, L. Weihermuller, Linking satellite derived LAI patterns with subsoil heterogeneity using large-scale ground-based electromagnetic induction measurements. Geoderma **241**, 262–271 (2015)

A. Samouelian, I. Cousin, A. Tabbagh, A. Bruand, G. Richard, Electrical resistivity survey in soil science: A review. Soil Tillage Res. **83**, 173–193 (2005)

T. Schmugge, Remote sensing of soil moisture. in *Encyclopedia of Hydrological Forecasting*, ed. by M.G. Anderson, T. Burt (Wiley., Chichester/London, 101–124 1985)

M. Schwank, A. Wiesmann, C. Werner, C. Mätzler, D. Weber, A. Murk, I. Völksch, U. Wegmüller, ELBARA II, an L-band radiometer system for soil moisture research. Sensors MDPI **10**, 584 (2010)

M. Schwank, J.-P. Wigneron, E. Lopez-Baeza, I. Völksch, C. Mätzler, Y. Kerr, L-band radiative properties of vine vegetation at the SMOS Cal/Val site MELBEX III. IEEE Trans. Geosci. Remote Sens. **50**(5), 1587 (2012)

S.I. Seneviratne, D. Luthi, M. Litschi, C. Schar, Land-atmosphere coupling and climate change in Europe. Nature **443**(7108), 205–209 (2006)

S.I. Seneviratne, T. Corti, E.L. Davin, M. Hirschi, E.B. Jaeger, I. Lehner, B. Orlowsky, A.J. Teuling, Investigating soil moisture-climate interactions in a changing climate: A review. Earth Sci. Rev. **99**(3–4), 125–161 (2010)

K.R. Sheets, J.M.H. Hendrickx, Noninvasive soil water content measurement using electromagnetic induction. Water Resour. Res. **31**(10), 2401–2409 (1995)

E.E. Small, K.M. Larson, C.C. Chew, J. Dong, T.E. Ochsner, Validation of GPS-IR soil moisture retrievals: Comparison of different algorithms to remove vegetation effects. IEEE J. Sel. Top. Appl. Earth Obs. Remote Sens. **9**(10), 4759–4770 (2016)

C.M. Steelman, A.L. Endres, Comparison of petrophysical relationships for soil moisture estimation using GPR ground waves. Vadose Zone J. **10**, 270–285 (2011)

K.A. Sudduth, N.R. Kitchen, W.J. Wiebold, W.D. Batchelor, G.A. Bollero, D.G. Bullock, D.E. Clay, H.L. Palm, F.J. Pierce, R.T. Schuler, K.D. Thelen, Relating apparent electrical conductivity to soil properties across the north-central USA. Comput. Electron. Agric. **46**(1–3), 263–283 (2005)

M. Temimi, T. Lakhankar, X. Zhan, M.H. Cosh, N. Krakauer, A. Fares, V. Kelly, R. Khanbilvardi, L. Kumassi, Soil moisture retrieval using ground-based L-band passive microwave observations in northeastern USA. Vadose Zone J. **13**(3), 1–10 (2014)

G.C. Topp, J.L. Davis, A.P. Annan, Electromagnetic determination of soil-water content – Measurements in coaxial transmission-lines. Water Resour. Res. **16**(3), 574–582 (1980)

H. Vereecken, J.A. Huisman, H. Bogena, J. Vanderborght, J.A. Vrugt, J.W. Hopmans, On the value of soil moisture measurements in vadose zone hydrology: A review. Water Resour. Res. **44**(21), W00D06 (2008)

Vista Clara Inc. GMR (2016), http://www.vista-clara.com/instruments/gmr/

C.von Hebel, S. Rudolph, A. Mester, J.A. Huisman, P. Kumbhar, H. Vereecken, J. van der Kruk, Three-dimensional imaging of subsurface structural patterns using quantitative large-scale multiconfiguration electromagnetic induction data. Water Resour. Res. **50**, 2732–2748 (2014)

J.R. Wait, The magnetic diploe over the horizontally stratified earth. Can. J. Phys. **29**(6), 577–592 (1951)

D. Walsh, P. Turner, E. Grunewald, H. Zhang, J.J. Butler, E. Reboulet, S. Knobbe, T. Christy, J.W. Lane, C.D. Johnson, A small-diameter NMR logging tool for groundwater investigations. Groundwater **51**, 914–926 (2013)

M.M. Waxman, L.J.M. Smits, Electrical conductivity in oil-bearing shaly sand. Soc. Pet. Eng. J. **8**, 107–122 (1968)

J.-P. Wigneron, Y. Kerr, P. Waldteufel, K. Saleh, P. Richaume, P. Ferrazzoli, M.-J. Escorihuela, J.P. Grant, B. Hornbuckle, P. de Rosnay, J.-C. Calvet, T. Pellarin, R. Gurney, C. Mätzler, L-band microwave emission of the biosphere (L-MEB) model: Description and calibration against experimental data sets over crop fields. Remote Sens. Environ. **107**(4), 639 (2007)

X.Q. Yu, P.T. Wu, W.T. Han, Z.L. Zhang, A survey on wireless sensor network infrastructure for agriculture. Comput. Stand. Interfaces **35**(1), 59–64 (2013)

V. Zavorotny, S. Gleason, E. Cardellach, A. Camps, Tutorial on remote sensing using GNSS bistatic radar of opportunity. Geosci. Remote Sens. Mag. **2**(4), 8–45 (2014)

Q. Zhu, H. Lin, J. Doolittle, Repeated electromagnetic induction surveys for determining subsurface hydrologic dynamics in an agricultural landscape. Soil Sci. Soc. Am. J. **74**(5), 1750–1762 (2010)

M. Zreda, D. Desilets, T.P.A. Ferre, R.L. Scott, Measuring soil moisture content non-invasively at intermediate spatial scale using cosmic-ray neutrons. Geophys. Res. Lett. **35**(21), L21402 (2008)

M. Zreda, W.J. Shuttleworth, X. Zeng, C. Zweck, D. Desilets, T. Franz, R. Rosolem, COSMOS: The COsmic-ray soil moisture observing system. Hydrol. Earth Syst. Sci. **16**(11), 4079–4099 (2012)

Airborne and Spaceborne Passive Microwave Measurements of Soil Moisture

Jiancheng Shi, Tianjie Zhao, Qian Cui, and Panpan Yao

Contents

Abstract

Soil moisture is the key variable that controls the surface water movements including infiltration, evapotranspiration, and groundwater recharge. It is one of the most important surface conditions of the exchange processes between land and atmosphere. Microwave remote sensing provides an efficient way to map surface soil moisture at a large scale from space and has achieved rapid development especially in large aperture systems at L-band. This chapter describes the basic theory and methodology for retrieving surface soil moisture including the soil roughness and vegetation effects. Special cases of soil moisture estimates

J. Shi (✉) · T. Zhao · Q. Cui · P. Yao
State Key Laboratory of Remote Sensing Science, Institute of Remote Sensing and Digital Earth, Chinese Academy of Science, Beijing, China
e-mail: shijc@radi.ac.cn; zhaotj@radi.ac.cn

X. Li, H. Vereecken (eds.), *Observation and Measurement of Ecohydrological Processes*, Ecohydrology, https://doi.org/10.1007/978-3-662-48297-1_3

from airborne and spaceborne measurements are presented. Results demonstrated that multichannel (multi-angle or multifrequency) microwave observations can be combined to enhance the retrieval accuracy and spatial resolution of remote sensed soil moisture.

Keywords
Soil moisture · Passive Microwave · Airborne measurements · Satellite remote sensing

Introduction

Soil moisture is one of the most active elements of the global water cycle and is an indispensable variable in ecohydrology study, as the plant physiology is directly related to the supply of water. It is especially important in the event of surface water evaporation and seepage. In semiarid climate areas, the vegetation types and its distribution are highly related to the amount of water that plants can obtain from soil. When the soil moisture is insufficient, soil water stress would inhibit the effects of transpiration and photosynthesis. In addition, timely soil moisture observations can be directly used for flood and drought monitoring. And long-term soil moisture data can help improve the performances of land surface and hydrological models and further improve the accuracy of various numerical models in weather forecasting and climate prediction. Therefore, soil moisture has been endorsed by the Global Climate Observing System as an Essential Climate Variable (ECV).

How to effectively carry out measurement of soil moisture in large spatial scale and over a long-term period is challenging. Due to the regional climate complexity, diverse vegetation conditions, topography, and also the soil type variation, the distribution of soil moisture shows strong spatial heterogeneity and dynamic variability, which brings a lot of inconvenience to the observation of soil moisture. The traditional local measurements cannot meet the various application requirements. Fortunately, satellite remote sensing technology, which emerged and developed rapidly in the 1960s, provided the possibility of effective observation of soil moisture on a certain spatial-temporal scale. These attempts include thermal, optical, active, and high-frequency microwave measurements. Among them, the microwave remote sensing provides a measure of electromagnetic properties of surface objects in all-weather conditions. By using the huge differences in the dielectric properties between dry soil and liquid water, microwave remote sensing has an unparalleled advantage in the monitoring of soil moisture. As the passive microwave remote sensing is more sensitive to the change of soil moisture and is less sensitive to surface structure characteristics, it has gradually become the key tool in the observation of large-scale soil moisture.

In recent years, there has been a great interest in L-band radiometers. The European Space Agency Soil Moisture and Ocean Salinity (SMOS) mission was launched on November 2, 2009 (Kerr et al. 2010). It provides global microwave brightness temperature observations at L-band with dual polarization and a wide range of incidence angles (5–60° approximately). The Soil Moisture Active Passive (SMAP) mission was launched on January 31, 2015, (Entekhabi et al. 2010) by the National Aeronautics and Space Administration (NASA). It makes global measurements of both L-band brightness temperature and backscatter at a single incidence angle of 40°. Aquarius/SAC-D has a combined passive/active L-band microwave instrument (Le Vine et al. 2007) that also provides measurements of land surface soil moisture (Bindlish et al. 2015). China is planning a Water Cycle Observation Mission (WCOM) to combine active and passive microwave measurements on the key elements of land, sea, and atmosphere (Shi et al. 2014), which also will include L-band. Because of their potential to measure through vegetation canopies, these missions can provide valuable information for mapping soil moisture at a global scale.

In this chapter, basic physics rationale for remote sensing of soil moisture at L-band is introduced in section "Rationale for Passive Remote Sensing of Soil Moisture." It is followed by special study cases of soil moisture retrieval with L-band airborne and spaceborne measurements as presented in sections "Soil Moisture from Airborne Measurements" and "Soil Moisture from Spaceborne Measurements." Section "Summary" gives the summary of this chapter.

Rationale for Passive Remote Sensing of Soil Moisture

Accurate modeling of microwave emission from land surface is essential for improved estimation of soil moisture from microwave radiometers. Despite the soil moisture, soil surface roughness has a significant effect on the microwave polarized emission (Escorihuela et al. 2007; Schwank et al. 2010; Mialon et al. 2012; Lawrence et al. 2013). Roughness generally increases the emissivity and decreases the sensitivity of brightness temperatures to soil moisture. Furthermore, vegetation's contribution is more complex that it not only attenuates the emission of microwave energy but also attenuates the energy emitted by itself, which is reflected upward again by the soil surface. Generally, vegetation coverage would result in an increase in microwave brightness temperature observed from space and present a depolarization effect. The situation can vary depending on the vegetation water content and vegetation structure. Therefore, accurate soil moisture retrieval generally relies on sufficient corrections of both roughness and vegetation effects (Jackson 1993).

Soil Roughness Effects

Soil roughness is defined as the randomly varying height of the soil surface with regard to a reference surface (Ulaby et al. 1982a). This reference surface can be the unperturbed surface of a periodic pattern or can be the mean surface if only random variations exist (Verhoest et al. 2008). Currently, most studies in remote sensing are based on the single-scale roughness that uses root mean squared height (RMS) s and correlation length l to describe the vertical and horizontal conditions in surface roughness. These roughness parameters are obtained from the statistical analysis of observed height variations along transects. Assuming that a one-dimensional surface profile contains N points with surface height z_i, the RMS height s is calculated as:

$$\mathrm{s} = \sqrt{\frac{1}{N}\left[\left(\sum_{i=1}^{N} z_i^2\right) - N \cdot \bar{z}^2\right]} \tag{1}$$

where

$$\bar{z} = \frac{1}{N}\sum_{i=1}^{N} z_i \tag{2}$$

The correlation length l is defined as the horizontal distance (ξ) over which the surface profile is autocorrelated with a value larger than $1/e$. The autocorrelation function (ACF) is given by:

$$\rho(\xi) = \frac{\sum_{i=1}^{N-j} z_i z_{i+j}}{\sum_{i=1}^{N} z_i} \tag{3}$$

The ACF describes the similarity between height values at two different points as a function of the distance lag between them. The autocorrelation decreases as the distance between sample points increases, and they are considered to be uncorrelated when the distance between them is larger than the correlation length l. In theoretical models, the ACF of soil profile is needed to characterize the surface roughness condition. Often two types of autocorrelation functions (Fung et al. 1994) are used: the Gaussian and the exponential function. The Gaussian ACF is given by:

$$\mathrm{ACF}(\xi) = e^{-\xi^2/l^2} \tag{4}$$

The exponential function with the power coefficient equal 1 is defined as:

$$\mathrm{ACF}(\xi) = e^{-|\xi|/l} \tag{5}$$

Normally, the larger RMS heights or smaller correlation lengths are indicative of a rougher surface. This is illustrated by randomly rough surfaces with different roughness conditions and correlation functions in Fig. 1. This figure shows that the Gaussian function describes a smoother surface than an exponential one. The exponential function characterizes smaller correlations at small lags, with a better description of the micro-roughness in the soil profile than the Gaussian one.

Theoretically, the reflectivity (R_p^e) of a rough surface consists of two components: coherent (R_p^{coh}, specular) and noncoherent (R_p^{non}, diffuse). The noncoherent component could be computed by integrating over the upper hemisphere the bistatic scattering coefficient (σ_{pp}, σ_{pq}), which describes the radiation from the incident direction (θ, φ) scattered into the outgoing direction (θ_s, φ_s). The two components can then be written as:

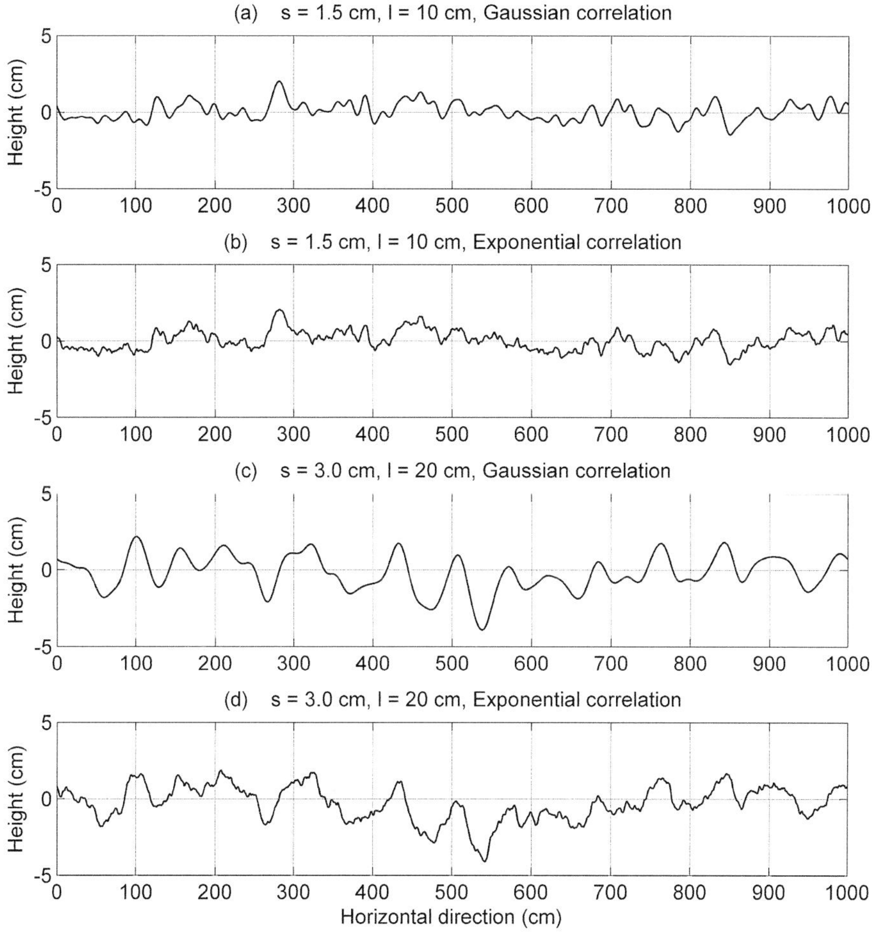

Fig. 1 Profiles of soil surface side view with different roughness parameters and shapes of the autocorrelation function (Gaussian and exponential)

$$\begin{aligned} R_p^e &= R_p^{\mathrm{coh}} + R_p^{\mathrm{non}} \\ &= r_p \cdot \exp\left[-(2ks\cdot\cos\theta)^2\right] + \frac{1}{4\pi\cdot\cos\theta}\int_0^{2\pi}\int_0^{\pi/2}\left(\sigma_{pp}+\sigma_{pq}\right)\cdot\sin\theta_s\cdot d\theta_s \\ &\quad \cdot d\varphi_s \end{aligned} \tag{6}$$

where k is the wave number. The specular reflectivity r_p for the flat surface is calculated by the incidence angle θ and the soil dielectric permittivity, which could be simply computed from a dielectric mixing model (i.e., Wang and Schmugge 1980; Dobson et al. 1985; Mironov et al. 2009). The coherent component only depends on the vertical variations of the surface height, while the noncoherent component depends on RMS height, horizontal correlation length, and the type of the correlation function.

According to the Kirchhoff's law, the polarized soil microwave emissivity (E_p, $p = v$ or h for the vertical and horizontal polarization respectively) can be expressed by applying the effective soil reflectivity (R_p):

$$E_p^s = 1 - R_p^e \tag{7}$$

Theoretical models can provide accurate simulation but require significant computational effort. It is extremely difficult to incorporate it into soil moisture retrieval algorithms due to its complexity. As an alternative to theoretical models for accounting for the surface roughness, the most commonly used semiempirical model is the so-called Q/H model proposed by Wang and Choudhury (1981). This model computes the rough reflectivity based on three parameters Q, h, and N. Q is the polarization mixing factor due to the surface roughness effects, h is a measurement of the intensity of the roughness effects to decrease the surface reflectivity, and N describes the angular dependence. The model is written as:

$$R_p^e = \left[(1 - Q) \cdot r_p + Q \cdot r_q\right] \cdot \exp\left(-h \cdot cos^N(\theta)\right) \tag{8}$$

Based on most of previous studies, the parameter Q is generally lower than 0.2. Furthermore, Q is commonly considered to be 0 at L-band based on a large data set (Mo and Schmugge 1987). Some studies (Wigneron et al. 2001) show that the roughness effects are related to soil moisture conditions. It is found the roughness effects could be stronger as the soil moisture decrease, as the microwave-observed roughness might be not equal to the local measured roughness. Despite this, we use a simple roughness parameter H_p hereafter to represent the roughness effects for different incidence angles, and as a result the rough surface reflectivity can be simply written as:

$$R_p^e = r_p \cdot H_p \tag{9}$$

Parameterization of the H_p can be conducted based on experimental or simulation data. To achieve a wide representation with fast computation ability, parametric models based on theoretical simulations are good compromise. The advanced integral equation method (AIEM) model (Chen et al. 2003), which is often used with radar systems, was selected in this study for calculating the passive microwave emissivity through Eq. 6. Its validity has been checked using three-dimensional Monte Carlo simulations and field experimental data (Shi et al. 2005; Chen et al. 2010; Guo et al. 2013). AIEM has been proven to be accurate over a wide range of soil moisture, surface roughness conditions, and incidence angles and has been widely applied in microwave remote sensing. By using the soil permittivity calculated by the dielectric model, a microwave emissivity/reflectivity simulation database at L-band was generated by the AIEM model to investigate the roughness effects. This database covers a wide range of soil moisture, RMS height, correlation length, and incidence angle. It was found that the roughness effects (H_p) are a function of incidence angle and "geometric roughness" in terms of spatial variations in the soil surface height. Chen et al. (2010) found that the best "geometric roughness" variable for a Gaussian correlated surface is the ratio s/l. For an exponentially correlated surface, the best variable, which is determined by comparing the fitting results, was found to be the slope which was defined as (Zhao et al. 2015a):

$$\mathrm{m} = s^2/l \tag{10}$$

To obtain the best results for all the incidence angles, the effective roughness parameter can be expressed as an exponential function of the slope:

$$H_p = A_p \cdot \exp\left(B_p \cdot m^2 + C_p \cdot m\right) \tag{11}$$

The reason that we use an exponential function is that it is equal to 1 when the roughness slope is 0, which corresponds to no roughness effects on the surface emission. The coefficients of A_p, B_p, and C_p were found to be dependent only on the incidence angle. This angular dependence of the coefficients can be described by a quadratic equation:

$$A_p, B_p, C_p = \mathrm{a} \cdot \theta^2 + b \cdot \theta + \mathrm{c} \tag{12}$$

The values of a, b, and c for both V-pol and H-pol are listed in Table 1, which is obtained from the AIEM model simulation database that covers a wide range of soil moisture and roughness conditions.

The relationship between the roughness parameter and the slope is shown in Fig. 2. The best-fit formulations for different incidence angles are also illustrated in this figure. Smaller values of the roughness parameter represent larger roughness effects on the surface emission. It can be seen that the roughness parameter can decrease when the slope parameter increases and increase with increasing incidence angle. For most cases, the roughness parameter is less than 1. This suggests that the roughness would actually reduce the reflectivity or increase the emission in comparison with that from a flat surface. And this phenomenon is more significant at H-pol than that at V-pol. However, the effects of roughness on V-pol would change in both magnitude and direction at larger incidence angles. For V-pol at 55°, the

Table 1 Coefficients for calculating A_p, B_p, and C_p

	a	b	c
A_v	0.296412	−0.169075	0.993309
B_v	−0.049491	−0.012277	0.048694
C_v	0.55279	0.086331	−0.636948
A_h	0.068502	−0.058486	0.976321
B_h	−0.051377	0.014978	0.045456
C_h	0.601618	−0.151848	−0.607679

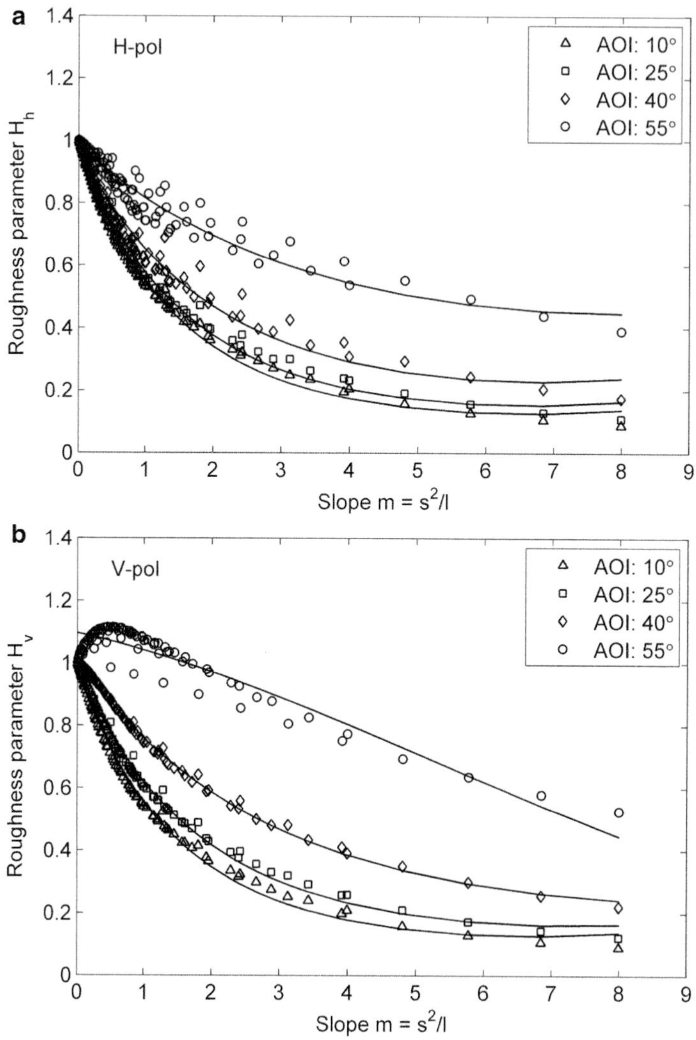

Fig. 2 Retrieved values of the roughness parameter H_p versus the slope parameter at different angles of incidence (AOI)

roughness parameter may be larger than 1, which means that the microwave emission would be decreased due to the roughness effects. These results indicate that the L-band soil emission characteristics are greatly affected by the surface roughness effects and differ depending on incidence angles and polarizations. This should be considered in development of soil moisture inversion algorithm to form a strong physical basis.

Generally, the soil effective reflectivity from the Gaussian simulations is higher than that from the exponential simulations. This feature can be explained by the difference in surface roughness characteristics. The exponential correlated surface is characterized by short-range, small-amplitude fluctuations, whereas the Gaussian surface appears "smooth" on this length scale (Ogilvy and Foster 1989). Therefore, the soil roughness effects on soil emissivity vary with the correlation function used. For the natural surface, an autocorrelation function was found with a shape between the exponential and Gaussian function (power coefficient ranges between 1 and 2) (Shi et al. 1997). To cover a wide range of surface conditions, roughness effects are further examined and parameterized based on three autocorrelation functions: Gaussian, 1.5-power, and exponential. Here we only focus on H polarization.

By analyzing the AIEM simulations, we present $S_r = (k \cdot s)^{2-N} \left(\frac{s}{l}\right)^N$ (k is wave number the unit of cm^{-1}, and N is the power coefficient) as the surface effective roughness parameter, which is similar to Zribi's (Zribi et al. 2014) findings in modeling radar backscattering. A parameterized surface reflectivity model at H-pol can be expressed as:

$$R_h^e = r_h \cdot A \cdot \exp\left(B \cdot S_r^D + C \cdot S_r^{2D}\right) \tag{13}$$

where coefficients A, B, C, and D can be determined by regression analysis using AIEM-simulated database and are dependent on incidence angle. The dependence on incidence angle of these coefficients can be described as:

$$A,B,C,D(\theta) = e + f \cdot \theta + g \cdot \theta^2 + h \cdot \theta^3 \tag{14}$$

Table 2 gives the coefficients of e, f, g, and h. Generally, the parametric model agrees well with AIEM simulations (Cui et al. 2016), and its accuracy is acceptable for passive microwave remote sensing. These indicate that the parametric model can accurately simulate soil effective reflectivity over a wide range of soil conditions.

Figure 3 shows the relationships of the soil effective emissivity simulated by the AIEM at two incidence angles with a difference of 10° under the three correlation functions for V (top) and H polarizations (bottom), respectively. We note that the soil effective emissivity at two incidence angles are highly correlated and could be characterized as a linear function. These relationships are neither dependent on the soil dielectric property nor on the roughness properties. It is also found that it presents a stronger correlation with the H-pol than with V-pol.

The linear function of the bare soil effective emissivity at two incidence angles can be described as:

Table 2 Coefficients for calculating A, B, C, and D

Coefficients	e	f	g	h
A	1.0225	0.0303	−0.1550	0.1217
B	−2.2110	0.4824	0.1569	1.3219
C	0.6285	−0.6073	1.0226	−1.3638
D	0.9214	−0.3925	0.7734	−0.9355

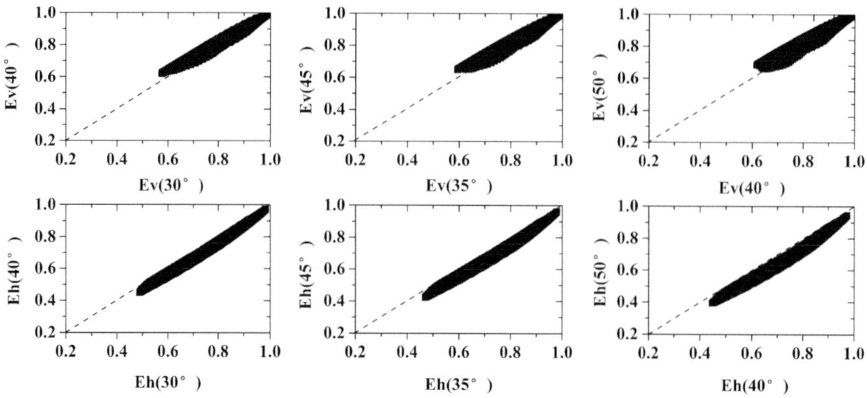

Fig. 3 Relationships of soil effective emissivity at two incidence angles with a 10° difference at V and H polarizations as simulated by AIEM

$$E_p^s(\theta_2) = \alpha_p(\theta_1, \theta_2) + \beta_p(\theta_1, \theta_2) \cdot E_p^s(\theta_1) \quad (15)$$

where α and β are only dependent on the polarization and the pair of incidence angles used. These parameters can be easily determined by regression analysis using the AIEM simulated database. The linear relationship of the bare soil effective emissivity at two incidence angles described above is essential for minimizing the effects of soil emission signals from multiangular radiometer measurements, which is presented in "Vegetation Effects" below.

Vegetation Effects

When vegetation is present, it emits microwaves and also attenuates the radiation from the soil surface. Vegetation can be the most significant influence during the soil moisture inversion. It requires a good understanding of the involved complex electromagnetic interactions to model the emission and scattering from vegetation canopies, and many models have been proposed in the literature. The discrete random media models describe a vegetation canopy as a continuous random media with discrete scatters. These scatters have varying dielectric constant, and their scattering features can be obtained through averaging over the distribution of size and orientation. A typical example is the Michigan Microwave Canopy Scattering Model (MIMICS), which consists of canopy, trunk, and ground layers' components.

In the case with passive microwave remote sensing, when the multiple scattering in the vegetation layer is neglected (Njoku and Chan 2006), vegetation effects are usually described using the $\omega-\tau$ model (Ulaby et al. 1982; Shi et al. 2008), which is a zeroth-order approximation that can be expressed as a four-component model. The first term is the upward radiation from vegetation, and the second term is

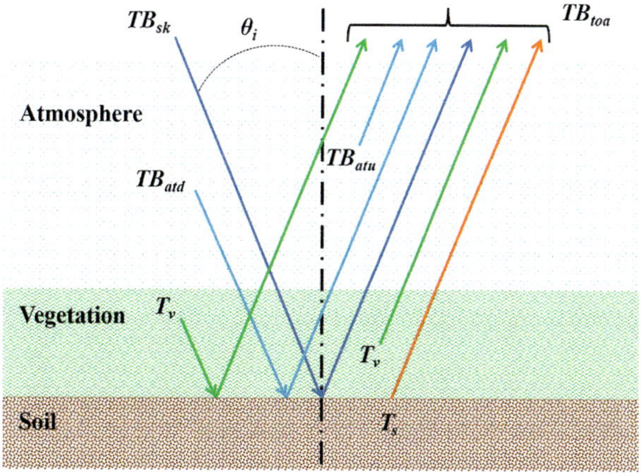

Fig. 4 Radiative transfer from vegetated surface at L-band (from SMOS ATBD)

the downward vegetation radiation reflected by the soil and then attenuated by the vegetation again. The third term is the soil radiation attenuated by the vegetation, and the last term is the direct soil emission. This model is normally implemented in the current soil moisture retrieval algorithm (Fig. 4).

$$T_{Bp} = F_v \cdot E_p^v \cdot T_v + F_v \cdot E_p^v \cdot \left(1 - E_p^s\right) \cdot L_p \cdot T_v + F_v \cdot E_p^s \cdot L_p \cdot T_s + (1 - F_v) \cdot E_p^s \cdot T_s \tag{16}$$

In the above model, F_v is the fraction of vegetation within the footprint or the remote sensing pixel. The values T_v and T_s are the physical temperatures of the vegetation and soil, respectively. The value for E_p^v is the emissivity of the vegetation canopy that is given by the equation $E_p^v = (1 - \omega) \cdot \left(1 - L_p\right)$. The ω is the single-scattering albedo of the canopy. L_p is the one-way attenuation factor for the vegetation, which is also known as the transmissivity. It is usually defined in terms of the nadir optical depth τ and the sensor viewing angle θ as:

$$L_p = \exp(-\tau_{\mathrm{NAD}} \cdot \sec\theta) \tag{17}$$

The two most important parameters in the model are the single-scattering albedo and the optical depth, which is related to vegetation water content, vegetation structure, and vegetation types. Normally, these parameters are empirically set based on ground-based experiments. The use of this model requires some assumptions: the single-scattering albedo is less than 0.2; there is no obvious air-vegetation interface that the vegetation reflectivity can be set to 0; the vegetation layer has a higher refraction than air; and the air-soil reflectivity can be used instead of vegetation-soil reflectivity. It should be noted that the coherence effect is only important in the active microwave observations with smaller field of view, and it is relatively weak in large-scale passive microwave observations.

Several studies have found that the vegetation optical depth is related to the total vegetation water content W (kg/m^2) using an empirical parameter b (Jackson and Schmugge 1991; Van de Griend and Wigneron 2004) as:

$$\tau_{\mathrm{NAD}} = b \cdot W \tag{18}$$

where b is a coefficient that is believed to be dependent on frequency and polarization. It also changes with the type and growing period of canopy. Vegetable water content is the total amount of water contained in all vegetation per unit area and is usually expressed in kg/m^2. Traditional measurements of vegetation water content are destructive. All the vegetation above the ground should be harvested and measured timely for wet weight. If necessary, the vegetation can be divided into branches, stems, and leaves. Vegetation should be dried as soon as possible for dry weight measurement and then used for calculation.

Wigneron et al. studied the relationship between the optical depth of the vegetation at different polarizations (Wigneron et al. 2007) and its dependence on the incidence angle. It is argued that the optical depth for different polarizations could be corrected by:

$$\tau_v = \tau_{\mathrm{NAD}}\left(\cos^2\theta_i + C_{\mathrm{pol.}}\sin^2\theta_i\right) \tag{19}$$

And the nadir optical depth is expressed as a function of the leaf area index (LAI):

$$\tau_{\mathrm{NAD}} = b' \cdot \mathrm{LAI} + b'' \tag{20}$$

The model in Eq. 16 could be rearranged as a two-component model:

$$T_{Bp} = V_e + V_t \cdot E_p^s \tag{21}$$

The intercept of this two-component model is denoted as vegetation emission component, and the slope is described as the vegetation attenuation component:

$$V_e = \left[F_v \cdot E_p^v \cdot (1 + L_p)\right] \cdot T_v \tag{22}$$

and

$$V_t = \left[1 - F_v + F_v \cdot L_p\right] \cdot T_s - \left[F_v \cdot E_p^v \cdot L_p\right] \cdot T_v \tag{23}$$

As described above, through analysis of numerical simulations from AIEM, a strong linear correlation was found between the bare soil emissivity at two adjacent incidence angles or frequencies. And the soil emissivity can be rearranged as:

$$E_p^s = \frac{T_{Bp} - V_e}{V_t} \tag{24}$$

By considering observations at two adjacent channels (frequencies or incidence angles), we could have:

$$\frac{T_B(C_2) - V_e(C_2)}{V_t(C_2)} = \alpha(C_1, C_2) + \beta(C_1, C_2) \cdot \frac{\mathrm{T}_{Bh}(C_1) - V_e(C_1)}{V_t(C_1)} \tag{25}$$

Rearranging this equation, we obtain:

$$T_B(C_2) = \alpha(C_1, C_2) \cdot V_t(C_2) + V_e(C_2) - \beta(C_1, C_2) \cdot \frac{V_t(C_2)}{V_t(C_1)} \cdot V_e(C_1) + \beta(C_1, C_2) \cdot \frac{V_t(C_2)}{V_t(C_1)} \cdot T_B(C_1) \tag{26}$$

Equation 26 indicates that the brightness temperature at two adjacent channels could be approximated by a linear function. We define the slope and intercept as microwave vegetation indices (MVIs):

$$\mathrm{MVI}_A(C_1, C_2) = \alpha(C_1, C_2) \cdot V_t(C_2) + V_e(C_2) - \mathrm{MVI}_B(C_1, C_2) \cdot V_e(C_1) \tag{27}$$

$$\mathrm{MVI}_B(C_1, C_2) = \beta(C_1, C_2) \cdot \frac{V_t(C_2)}{V_t(C_1)} \tag{28}$$

MVIs are not affected by soil signals and are only dependent on vegetation properties and temperature. Therefore, vegetation parameters can be directly retrieved from radiometer observations without the influence of soil emission signals.

Additionally, we study the pure vegetation surface ($F_v = 1$) and assume that the attenuation of vegetation from different channels of observation could be expressed as:

$$V_t(C_2) = a \cdot V_t(C_1)^b \tag{29}$$

$$\mathrm{MVI}_B = a \cdot V_t(C_1)^{b-1} \cdot \beta(C_1, C_2) \tag{30}$$

$$V_t(C_1) = \sqrt[b-1]{\frac{\mathrm{MVI}_B}{a \cdot \beta(C_1, C_2)}} \tag{31}$$

From this equation, the B parameter of MVIs is directly related to the vegetation attenuation component. Under the configuration of SMOS or AMSR-E, it can be expressed as:

$$V_t(\theta_1) = \sqrt[\frac{\sec\theta_2}{\sec\theta_1} - 1]{\frac{\mathrm{MVI}_B}{a \cdot \beta(\theta_1, \theta_2)}} \tag{32}$$

$$V_t(f_1) = {}^{\frac{f_2}{f_1}-1}\sqrt{\frac{\mathrm{MVI}_B}{a \cdot \beta(f_1, f_2)}} \tag{33}$$

Thus, by theoretical derivation, we have obtained a clear mathematical relationship between the B parameter of MVIs and vegetation attenuation properties. This provides an excellent way to eliminate the vegetation effects in soil moisture inversion.

Soil Moisture from Airborne Measurements

Experimental Data

Passive data from SMEX02 passive and active L/S-band sensor (PALS) are used in this study. PALS has radiometers operating at 1.41 GHz (L) and 2.69 GHz (S) with multiple polarization, both with an incidence angle of 45°. The PALS instrument was flown over the SMEX02 region on June 25 and 27 and July 1, 2, 5, 6, 7, and 8, 2002. We delete the data on July 1 and 2 which has only several flight lines because of experimental instrument.

SMEX03 Two-Dimensional Synthetic Aperture Radiometer (2DSTAR) and the Polarimetric Scanning Radiometer (PSR) data are used in this study. 2DSTAR is a radiometer operating at 1.4 GHz (L) with an incidence angle of 40°, and PSR is a radiometer operating at C/X-band with an incidence angle of 55°. We selected the overlap data (6 days) of 2DSTAR and PSR over the SMEX03 region on July 2, 10, 12, 13, 14, and 15, 2003.

Soil Moisture Retrieval from L- and S-Bands

The bare surface emission signals can be described as a function of the surface roughness and the Fresnel reflectivity (flat surface) that calculated by dielectric constant and incidence angle, as shown in Eq. 9. So:

$$R_p^e \cdot V_p^t = r_p \cdot H_p \cdot V_p^t = r_p \cdot \mathrm{CRV}_p \tag{34}$$

where the surface roughness function H_p differs for different polarizations p. CRV_p is a combined correction parameter for roughness and vegetation effects at polarization p.

The surface type is very complex, but Fresnel reflex law which describes the plane surface is accurate. It can be found:

$$r_h = {r_v}^a \tag{35}$$

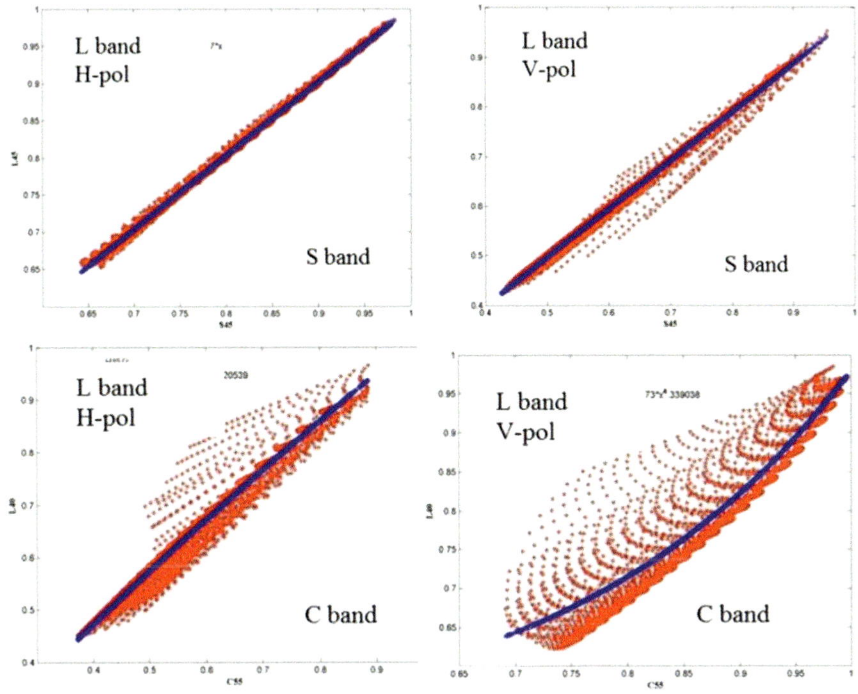

Fig. 5 Numerical simulated surface emissivities for H on the left and V on the right by AIEM with L-band as y-axis and S-band as x-axis (upper panels), with L-band as y-axis and C-band as x-axis (lower panels)

$$\frac{\left(R_v^e \cdot V_v^t\right)^a}{R_h^e \cdot V_h^t} = \frac{\mathrm{CRV}_v{}^a}{\mathrm{CRV}_h} \tag{36}$$

From Fig. 5 in next section "Brightness Temperature Downscaling with L/S Bands," the reflectivity of L- and S-band can be approximately equal, so it leads to:

$$\frac{R_p^e(S) \cdot V_p^t(S)}{R_p^e(L) \cdot V_p^t(L)} = \frac{\mathrm{CRV}_p(S)}{\mathrm{CRV}_p(L)} \approx \frac{V_p^t(S)}{V_p^t(L)} \tag{37}$$

Furthermore, through analyses of our simulated database, we found the good relationships between vegetation components V_p^e and V_p^t at each frequency. We can describe the relationships as:

$$V_p^e = \begin{cases} 1 - V_p^t \\ \left(1 - V_p^t\right)^b \end{cases} \tag{38}$$

From above analyses, we have developed the relationships for the combined roughness and vegetation correction parameters and the vegetation components at different frequencies and polarizations. These relationships can significantly reduce the number of unknown in retrieval process. We have reduced the total unknowns to four. The equations can be solved numerically by using four dual-frequency and dual polarization measurements with no assumption on the polarization independence of the vegetation effects.

PALS data obtained in Soil Moisture Field Experiment for 2002 (SMEX02) with intensive field soil moisture measurements are used for testing the algorithm. We compared the soil moistures of the field measurements and the estimated soil moistures from PALS dual frequency radiometer data. The RMSE is 0.048 m^3/m^3 with L-band, and RMSE is improved to 0.035 m^3/m^3 with L- and S-bands.

Brightness Temperature Downscaling with L/S-Bands

In order to characterize the bare surface emission signals of L/S-bands in 45° (corresponding to the SMEX02 PALS L/S data) and L/C-bands in 40° and 55° (corresponding to the SMEX03 2DSTAR L-band data and PSR C-band data), we first generated a microwave surface emissivity/reflectivity database using AIEM model. This database covers a wide range of soil roughness (RMS height and correlation length) and soil moisture. Soil RMS height varies from 0.25 cm to 3.5 cm with a 0.25 cm step, and the correlation length varies from 2.5 cm to 30 cm with a 2.5 cm step. The soil moisture varies from 2% to 50% with a 2% step.

From Fig. 5 we can notice that the bare surface emissivities at L- and S-band are highly correlated and can be approximated as an equal, namely, $E_p^s(L) \approx E_p^s(S)$. The bare surface emissivities at L- and C-band are highly correlated and can be described as an exponential relationship.

The principle of this downscaling algorithm is to downscale L-band TBs (coarse resolution) using S-band TBs (medium resolution) based on the linear relationship between the L- and S-band TBs and get the soil moisture products with medium resolution.

Based on $\omega - \tau$ model in Eq. 16, and the two components in Eqs. 22 and 23, we got

$$\frac{E_p^t(L) - V_p^e(L)}{V_p^t(L)} = E_p^s(L) \approx E_p^s(S) = \frac{E_p^t(S) - V_p^e(S)}{V_p^t(S)} \tag{39}$$

$$E_p^t(L) = \left[V_p^e(L) - \frac{V_p^t(L)}{V_p^t(S)} \cdot V_p^e(S)\right] + \frac{V_p^t(L)}{V_p^t(S)} \cdot E_p^t(S) \tag{40}$$

Finally, we deduced the following linear relationship between L-band TBs and S-band TBs:

$$\mathrm{TB}_p^t(L) = \left[V_p^e(L) - \frac{V_p^t(L)}{V_p^t(S)} \cdot V_p^e(S) \right] \cdot T_v + \frac{V_p^t(L)}{V_p^t(S)} \cdot \mathrm{TB}_p^t(S) \tag{41}$$

Namely,

$$\mathrm{TB}_p^t(L) = a + b * \mathrm{TB}_p^t(S) \tag{42}$$

For a particular research area and in a short time period, the slope b and the intercept a mainly rely on vegetation.

The vegetation changes with the passage of time. In order to find a more accurate description of the relationship between L-band and S-band TBs, we adopt the microwave vegetation index MVI (Shi et al. 2008), which is sensitive to vegetation and surface features, to correct the linear equation. We tested some different equations forms:

$$\mathrm{TB}_p^t(L) = \left(a + c^* \frac{\mathrm{MVI}^t}{\overline{\mathrm{MVI}}^t} \right) + \left(b + d^* \frac{\mathrm{MVI}^t}{\overline{\mathrm{MVI}}^t} \right) * \mathrm{TB}_p^t(S) \tag{43}$$

$$\mathrm{TB}_p^t(L) = a^* \frac{\mathrm{MVI}^t}{\overline{\mathrm{MVI}}^t} + \left(b + d^* \frac{\mathrm{MVI}^t}{\overline{\mathrm{MVI}}^t} \right) * \mathrm{TB}_p^t(S) \tag{44}$$

$$\mathrm{TB}_p^t(L) = a^* \frac{\mathrm{MVI}^t}{\overline{\mathrm{MVI}}^t} + b * \mathrm{TB}_p^t(S) \tag{45}$$

$$\mathrm{TB}_p^t(L) = \left(a + c^* \frac{\mathrm{MVI}^t}{\overline{\mathrm{MVI}}^t} \right) + b * \mathrm{TB}_p^t(S) \tag{46}$$

Finally, we carried out the downscaling work with best results using the modified relationship of TB_s in Eq. 43. At coarse resolution, we regressed and obtained the coefficient a, b through the time sequence observations on each coarse grid. At medium resolution, we calculated the downscaled TBs of L-band using the equation with medium-resolution TBs of S-band.

There is no satellite radiometer data at S-band; available L- and S-band radiometer data are PALS data from SGP99, SMEX02 CLASIC, and SMAPVEX08 field experiments. We selected L- and S-band TBs data of PALS sensor in SMEX02 experiment to validate the downscaling method. After the preprocessing, H/V polarizations L-band and S-band TB data are gridded to 0.8 km (medium resolution) and 4 km (coarse resolution) spatial resolution of the 8 days during the SMEX02 experiment.

The results are shown in Figs. 6 and 7. Figure 6 is the downscaling results of L/S-band TB, and Fig. 7 is the scatterplot of downscaling results and original observations. In Fig. 6, the downscaled L-band TB at 800 m in the right column can basically present the spatial distribution condition of original L-band TB in the medium column, and it has more detail heterogeneity in coarse grid than that of the observed L-band TB at 4 km in the left column. But there are obvious plaque

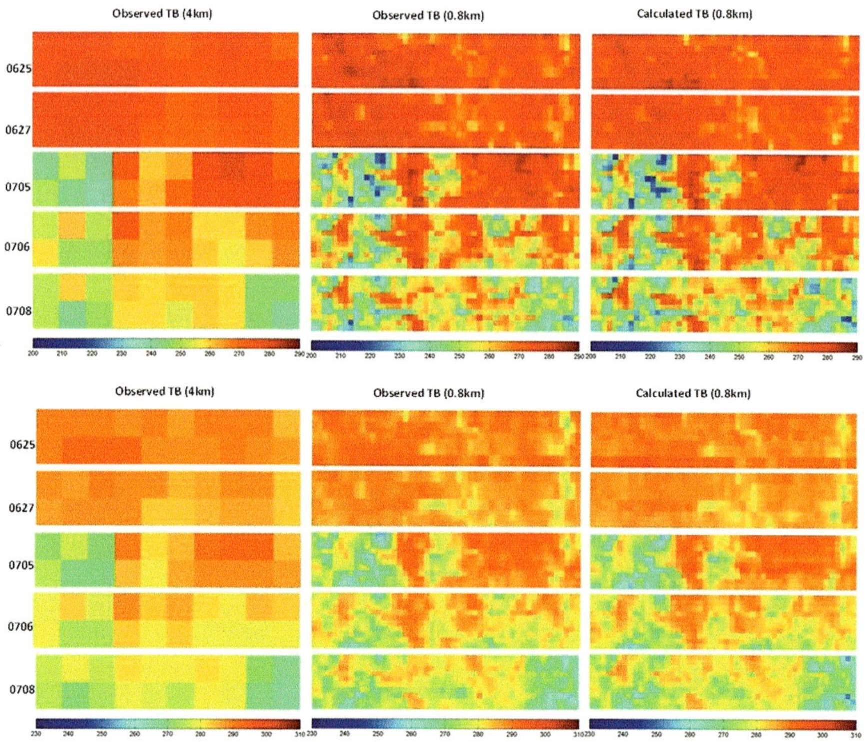

Fig. 6 Downscaling results of L-band TBs using S-band TBs (**a**) H polarization (**b**) V polarization (from left to right, 4 km original observations, 0.8 km original observations, 0.8 km downscaling results, respectively)

phenomenon and unsmooth space transition in the downscaled L-band TB of the right column. From the precision displayed in scatterplots in Fig. 7, the RMSE are 2.86 K and 2.19 K for H/V polarization, respectively.

Brightness Temperature Downscaling with L/C-Bands

Similar to section "Brightness Temperature Downscaling with L/S Bands," we deduced the following exponential relationship between L-band TBs and C-band TBs, namely:

$$\mathrm{TB}_p^t(L) = a + b * \mathrm{TB}_p^t(C)^c \tag{47}$$

As the method we used in section "Brightness Temperature Downscaling with L/S Bands," we adopted the vegetation index MVI to correct the exponential equation:

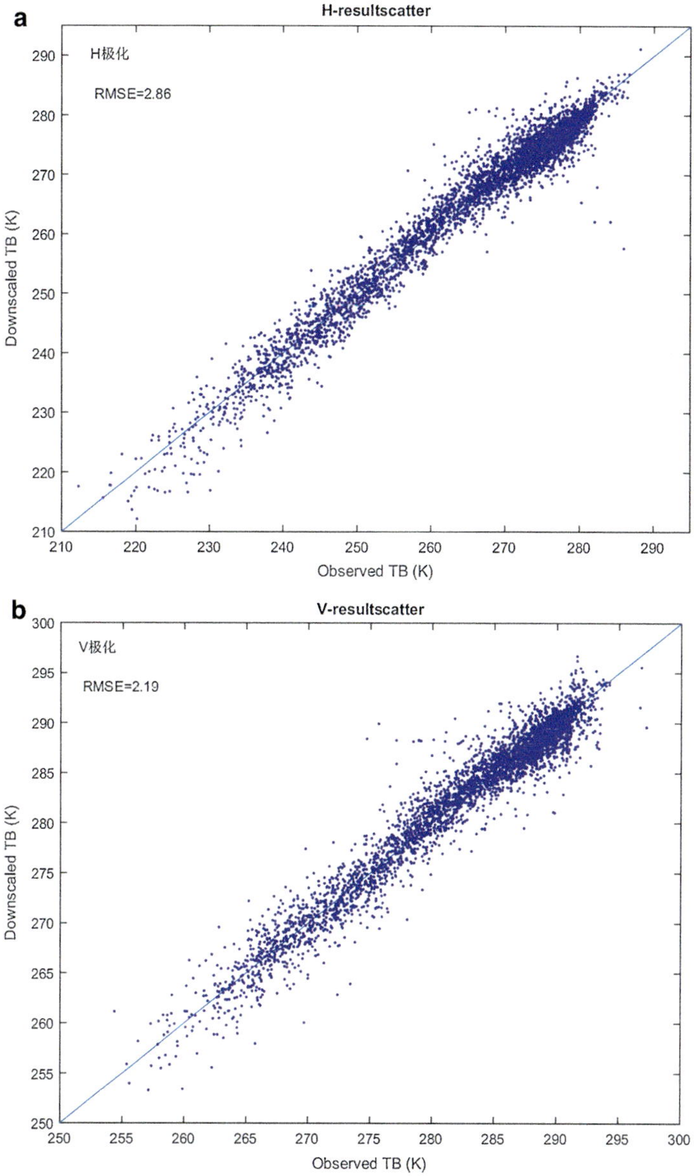

Fig. 7 The scatter plot of L-band TBs downscaling results and original observations: (**a**) H polarization and (**b**) V polarization

$$\mathrm{TB}_p^t(L) = \left(a + d^* \frac{\mathrm{MVI}^t}{\overline{\mathrm{MVI}}^t}\right) + \left(b + e^* \frac{\mathrm{MVI}^t}{\overline{\mathrm{MVI}}^t}\right) * \mathrm{TB}_p^t(C)^c \tag{48}$$

We carried out the downscaling work using the relationship of TBs in Eq. 48. At coarse resolution, we regressed and obtained the coefficients *a*, *b*, *c*, *d*, and *e* through the time sequence observations on each coarse grid. At medium resolution, we calculated the downscaled TBs of L-band using the equation with medium TBs of C band.

There is no appropriate satellite radiometer data at L-band and C-band. We selected the 2DSTAR L band data and PSR C band data in SMEX03 experiment to validate the downscaling method. After the preprocessing, H/V polarizations L band and C band TB data are gridded to 0.8 km (medium resolution) and 4 km (coarse resolution) spatial resolution of the overlapped 6 days during the SMEX03 experiment.

The results are shown in Figs. 8 and 9. Figure 8 is the downscaling results of L/C band TB, and Fig. 9 is the scatterplot of downscaling results and original observations. In Fig. 8, the downscaled L band TB at 800 m in the bottom row can basically present the spatial distribution condition of original L band TB in the medium row, and it has more detail heterogeneity in coarse grid than that of the observed L band TB at 4 km in the top row. But there are obvious plaque phenomenon and unsmooth space transition in the downscaled L band TB of the bottom row. From the precision displayed in scatterplots in Fig. 9, the RMSE are 3.36 K and 2.27 K for H/V polarizations, respectively.

Soil Moisture from Spaceborne Measurements

Satellite Data

The SMOS Level 1c (L1c) (V5.04) brightness temperature product is used in this study. The L1c product is provided over the ISEA-4H9 (Icosahedral Snyder Equal Area Earth fixed) grid, which is referred to as the discrete global grid (DGG) with nodes that are equally spaced 15 km apart (Kerr et al. 2012). Separate products are provided for the ascending (6:00 AM) and descending (6:00 PM) half-orbits. However, the well-known radio frequency interference (RFI) significantly affects the quality of observations in some regions of the world. In addition, multiangular observations are not obtained at fixed observation angles and are not suitable for our retrieval algorithm, which requires observations at specific observation angles. To address these issues, a two-step regression approach is used to refine the SMOS L1c multiangular brightness temperatures and fit them to the expected observation angles. Details on this method can be found in Zhao et al. (2015b). The results showed in this study are only from SMOS ascending pass data.

When the approach (Cui et al. 2015) described in this study is applied to refined global SMOS L1c brightness temperature data, the grids with land cover types of water, permanent wetlands, and urban and built-up and barren or sparsely vegetated land based on MODIS land cover type data (MCD12C1) classified by the International Geosphere Biosphere Programme (IGBP) global vegetation scheme are masked and not retrieved. To avoid the effects of snow and frozen soil conditions,

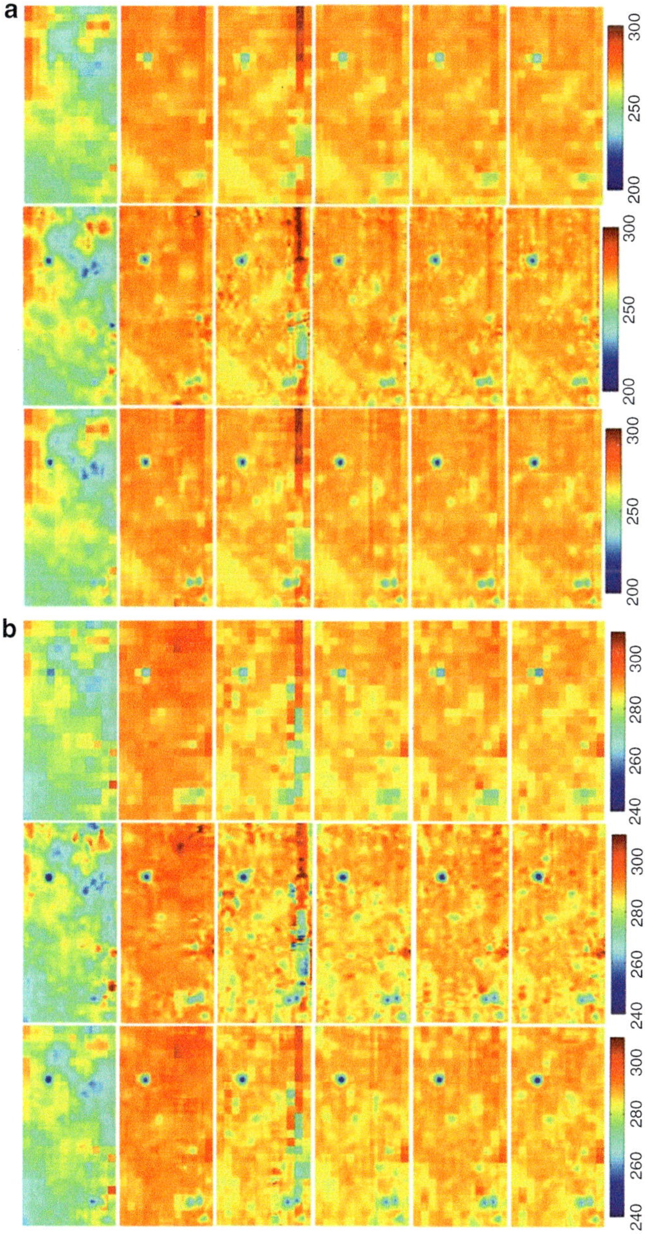

Fig. 8 Downscaling results of L band TBs using C band TB (**a**) H polarization (**b**) V polarization (from top to bottom, 4 km original observations, 0.8 km original observations, 0.8 km downscaling results, respectively)

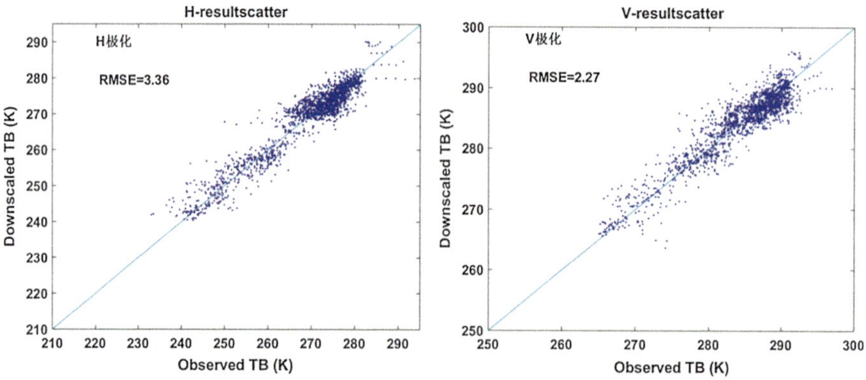

Fig. 9 The scatter plot of L band TBs downscaling results and original observations (H polarization on the left and V polarization on the right)

the retrieval algorithm operates only when the temperature is greater than 0 °C. The temperature data are from ancillary soil temperature data (0–7 cm) from the European Center for Medium-Range Weather Forecasting (ECMWF).

Vegetation Optical Depth (VOD) Retrieval from SMOS

To explore the H-polarized MVIs responses with respect to the vegetation parameters, the normalized MVI_A(40°, 50°) (MVI_A(40°, 50°)/TS) and MVI_B(40, 50) are computed from Eqs. 22, 23, 27, and 28. As seen in Fig. 10, both the normalized MVI_A and MVI_B are highly sensitive to the VOD. The normalized MVI_A increases with VOD, while MVI_B decreases with VOD. However, the MVIs shown in Fig. 11 demonstrate a much lower sensitivity to the single-scattering albedo. The MVIs are nearly independent of the single-scattering albedo, particularly when the VOD is lower than 1.0. Therefore, retrieving the VOD is more effective than retrieving the albedo using the H-polarized MVIs.

For low vegetation types (grass, crops, shrubs, and savannas), ω was found to be rather low with values lower than 0.05. In this study, we assume a default ω value of 0.05. In forests, the scattering process within forests is complex and significant because tree trunks and branches are large relative to the microwave length of the L-band. However, the multiple-scattering effect within the vegetation canopy is not considered in the $\omega - \tau$ model. For the application of surface parameters retrieval, the $\omega - \tau$ model could still be applied to forests, provided that the VOD and albedo are defined as equivalent or effective parameters (Ferrazzoli et al. 2002). The forests' effective albedo retrieved from the zero-order radiative transfer model here is different from the theoretical definition of the single-scattering albedo. The effective albedo is a global parameter that depends on all the processes taking place within the canopy at the large scale of the SMOS view, including multiple-scattering and canopy ground interactions (Kurum et al. 2012). Therefore, based on published

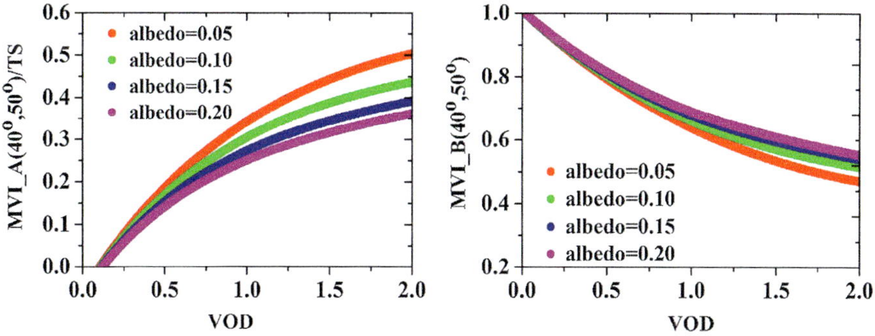

Fig. 10 H-polarized MVIs from the incidence angle pair (40°, 50°) plotted versus the VOD ranging from 0.01 to 2.0, with vegetation single-scattering albedo of 0.05, 0.10, 0.15, and 0.20

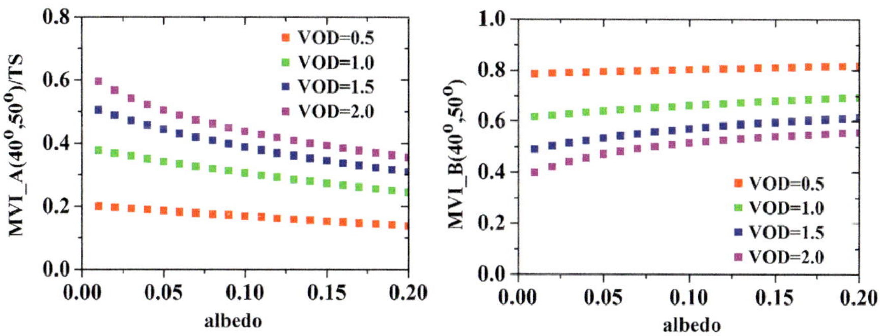

Fig. 11 H-polarized MVIs from the incidence angle pair (40°, 50°) plotted versus vegetation single-scattering albedo ranging from 0.01 to 0.20, with VOD of 0.5, 1.0, 1.5, and 2.0

information, we set the ω value for forests to a representative value of 0.10 (Della Vecchia et al. 2006; Grant et al. 2008; Kerr et al. 2012).

Given soil temperature, ω, and parameters a and b, only one vegetation parameter is unknown in Eq. 26 (i.e., the VOD). Table 3 provides three pairs of observation angles (θ_1, θ_2) and the corresponding parameters of α and β in Eq. 15. The correlation coefficient (R^2) and RMSE are also shown in Table 3. Using three pairs of brightness temperatures from different observation angles (described in Table 3), the VOD can be best fitted based on three equations like 26.

The validation of the VOD using ground measured vegetation parameters, such as vegetation water contents, is very difficult due to SMOS's relatively coarse spatial resolution. Meanwhile, many soil moisture networks can provide long-term series of ground measurements for satellite calibration/validation activities. Therefore, we attempted here to indirectly validate the VOD retrievals. Using the approach described in this study, the VOD can be retrieved from multiangular H-polarized brightness temperature data. Meanwhile, in situ soil moisture measurements are used to calculate smooth surface reflectivity based on Fresnel equations and soil dielectric

Table 3 Parameters a and b for different pairs of incidence angles in Eq. 15 at H polarization

(θ_1, θ_2)	α	β	R^2	RMSE
(30°, 40°)	−0.0514	1.0208	0.9958	0.0082
(40°, 50°)	−0.0632	1.0069	0.9925	0.011
(35°, 45°)	−0.0577	1.0163	0.9944	0.0096

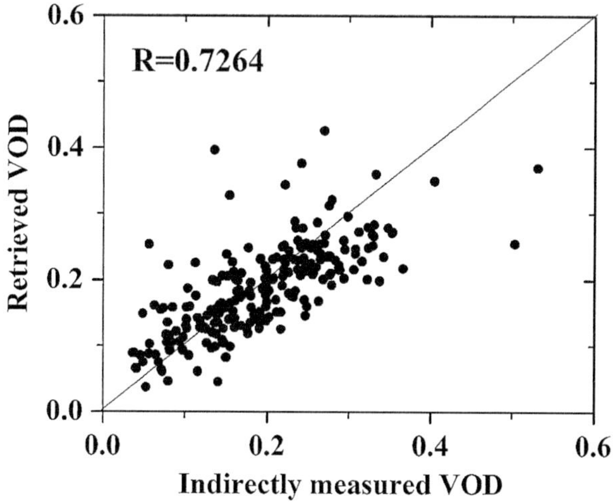

Fig. 12 Scatter plots of the retrieved VOD versus the indirectly measured VOD in the Little Washita Watershed

model (Wang and Schmugge 1980). Considering the surface roughness effects, the surface effective emissivity at H polarization $\varepsilon_h^s(\theta)$ can be estimated from a simple model (Escorihuela et al. 2007). Given the vegetation single-scattering albedo and vegetation canopy temperature, the indirectly measured VOD can be calculated using the H-polarized brightness temperature at 40° based on the $\omega - \tau$ model.

The in situ soil moisture data over 2010–2011 was obtained from the USDA ARS Little Washita Watershed network. The Little Washita Watershed is located in Southwest Oklahoma in the Great Plains region of the USA; it covers an area of approximately 610 km^2. The land cover is dominated by rangeland and pasture, but it also includes winter wheat and croplands. Within this watershed, a total of 20 ARS Micronet stations are available that have been used to validate AMSR-E and SMOS soil moisture products (Jackson et al. 2010, 2012). Note that the brightness temperature data are only from the DGG grids containing one or more in situ stations and the retrieved and indirectly measured VOD for every grid are averaged, respectively, to represent the entire experiment region. Figure 12 is a scatter plot of the regional retrieved VOD versus indirectly measured ones. The retrieved VOD agrees well with the indirectly measured ones with a high correlation coefficient $R = 0.7264$, which demonstrates the feasibility of the retrieval method in large scale as SMOS's view.

The VOD is distinct from the optical vegetation index in that it is sensitive to both leaf and woody biomass. Here we show an interesting comparison between VOD, NDVI, and aboveground biomass. The comparison uses an aboveground standing biomass (AGB) density map (in megagrams of mass per unit area) across three continental regions (Latin America, Africa, and Southeast Asia) at 1 km resolution (Saatchi et al. 2011). It should be noted that global scale AGB data are very limited and the AGB map we used only represent aboveground biomass circa the year of 2000. However, these data are still valuable for comparing microwave-derived VOD and optical vegetation indices such as NDVI in quantitative sense. A complete description of the map can be found in Saatchi et al. (2011). Here, the aboveground biomass (AGB) map was resampled to 0.25° resolution.

The 0.25° AGB distribution map is shown in Fig. 13a. The VOD and NDVI are the 3-month averages (December of 2010, January and February of 2011) and shown in Fig. 13b, c, respectively; both were resampled to 0.25° resolution. As expected, the distributions of the AGB and VOD generally show high values for forests and low values for low vegetation regions. For example, both AGB and VOD show the highest values in the forest areas in northern and eastern Amazonia along the Guiana Shield, southwestern Peru in Latin America, and Congo Basin in Africa. However, along the Amazon River, the AGB shows high values, while the values of VOD are rather low. The presence of open water may lead to the underestimates of VOD. Differences can also be detected in deciduous forests in Latin America, where the VOD and NDVI exhibit moderate-high values, while the biomass is rather low. In South Asian region including Vietnam, northwestern Myanmar, and southern China, the difference between the VOD and the biomass or NDVI is greater. These areas are covered with evergreen forests and exhibit high AGB and NDVI values; however, the corresponding VOD values are relatively low. The most likely cause for this is assigned to the strong RFI still existing in the refined brightness temperature data; thus, the approach for the refinement of SMOS L1c brightness temperature data may fail (Zhao et al. 2015b).

Figure 14 shows the global mean VOD over 2010–2011 at 0.25° resolution. The VOD retrievals are only available when the surface temperature is above 0 °C, so the VOD values are the averages of warm days only. The summary of global mean VOD and NDVI for different land covers is shown in Table 4. As expected, the VOD shows highest values in forests and lowest values in grasslands. However, the VOD values are rather low in the areas of South China where are covered by forests, most likely because the RFI effect is not minimized by the refined method. In addition, the forests in North America exhibit relatively low VOD values, possibly because the extensive open water causes an underestimation of the VOD. It should be noted that in the retrieval of the VOD, the temperature of vegetation canopy is assumed to equal that of the soil. With an overpass of dawn, when the vegetation begins to conducting plant photosynthesis and transpiration, the temperature of vegetation can be expected to close to that of soil. However, the satellite passes do not occur in the early morning everywhere, particularly in the high-latitude region; therefore, temperature differences exist between the vegetation canopy and the soil. The retrieved VOD for the high-latitude regions may be affected by this temperature differences.

The retrieved global daily VOD over 2 years (2010–2011) was processed into 8-day mean datasets at 0.25° resolution. Global pixel-wise linear correlations between the VOD and NDVI datasets over 2 years (2010–2011) are shown in Fig. 15. As seen

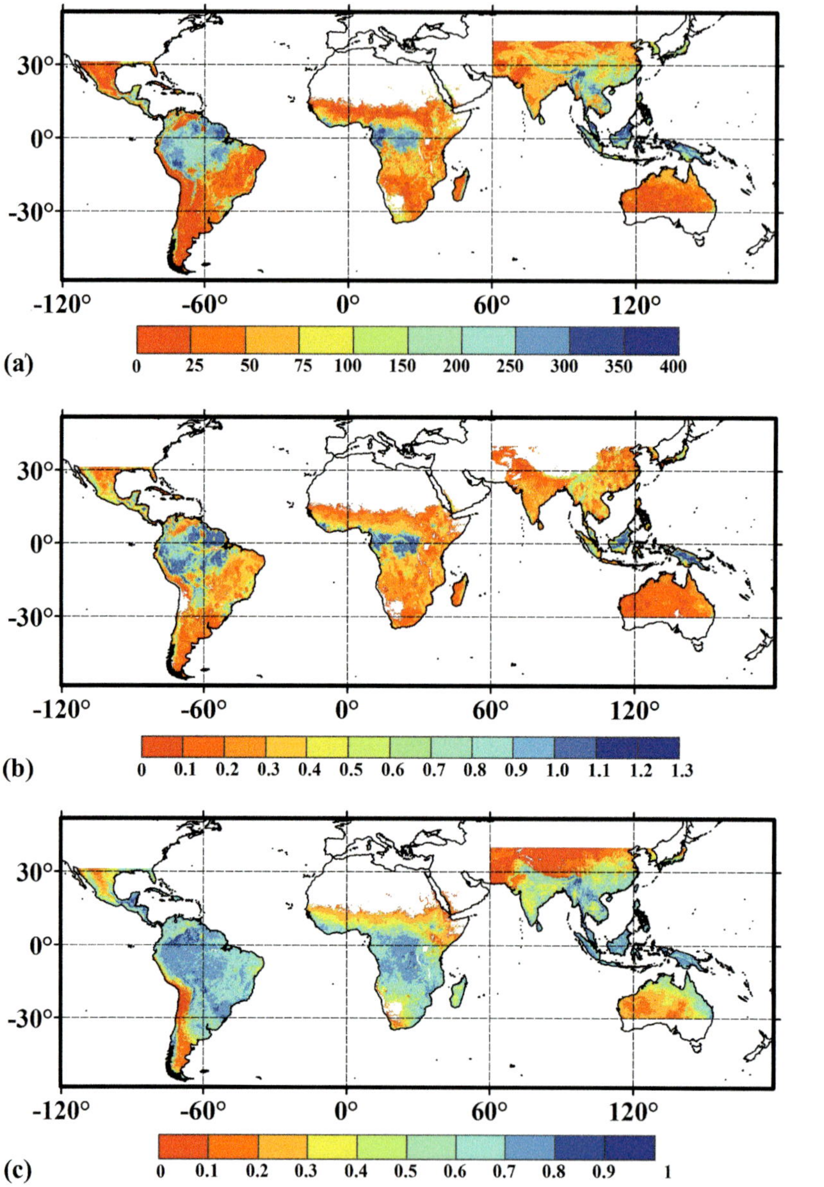

Fig. 13 Distribution maps of aboveground biomass (AGB, units, Mg/ha) (**a**), SMOS-derived VOD (**b**) and MODIS-derived NDVI (**c**) at 0.25° resolution

from Fig. 15, the croplands in North America savannas in Southern America and Africa and shrublands and forests in North America and North Asia show moderate correlations ($0.5 < R < 0.75$). Low correlations ($-0.25 < R < 0.25$) are mainly present in the Amazon region, the central region of Africa and southeastern of Asia,

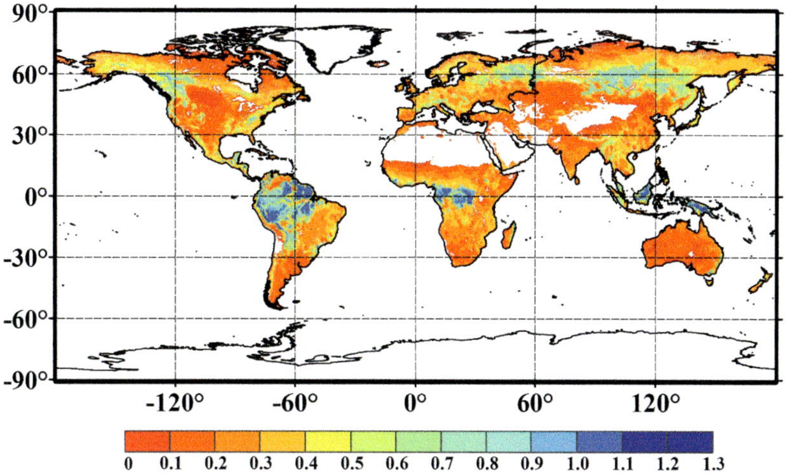

Fig. 14 Global mean SMOS-derived VOD over 2010–2011

Table 4 Summary of global mean values of VOD and NDVI for different land cover types over 2010–2011

Land cover type	VOD	NDVI
Evergreen needleleaf forest	0.521	0.432
Evergreen broadleaf forest	0.710	0.735
Deciduous needleleaf forest	0.773	0.334
Deciduous broadleaf forest	0.575	0.573
Mixed forests	0.515	0.512
Closed shrubland	0.216	0.327
Open shrubland	0.194	0.198
Woody savannas	0.430	0.506
Savannas	0.261	0.504
Grasslands	0.179	0.263
Croplands	0.219	0.423
Cropland/natural vegetation	0.253	0.551

which are mainly covered by evergreen broadleaf forests. No significant seasonal changes in the time series of the VOD and NDVI may occur. Unexpectedly, the entirety of China and India shows low or even negative correlations. Croplands are widely distributed in these regions. The strong RFI influence is still in the refined brightness temperature data, and the approach for the refinement of SMOS L1c brightness temperature data may fail (Zhao et al. 2015b); thus, the retrieved results may be unreliable. On the global scale, the correlations between the VOD and NDVI in this study are generally lower than those for the AMSR-E-derived VOD. AMSR-E has a high-frequency (18.7 GHz) configuration, which results in its more related to top layer of vegetation canopy. Because of the longer wavelength that SMOS

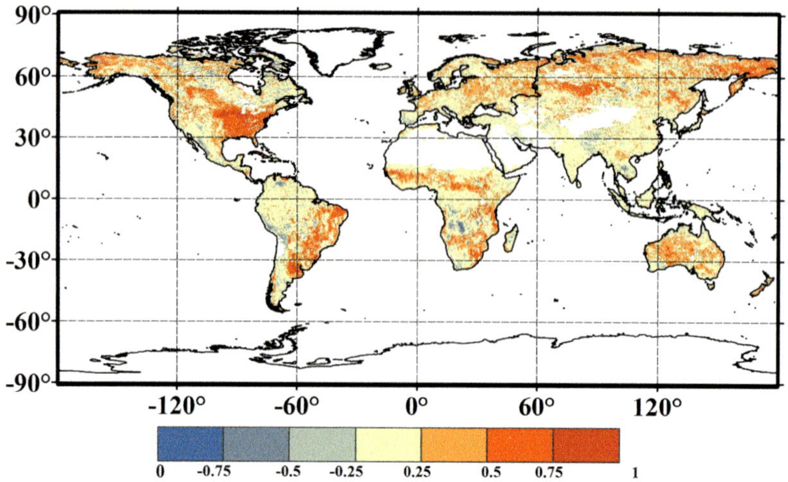

Fig. 15 Global pixel-wise correlations between the SMOS-derived VOD and MODIS-derived NDVI over 2010–2011

operates at, SMOS is more sensitive to the woody parts of the canopy than AMSR-E. The NDVI is sensitive to the greenness of a thin layer of vegetation leaves. Therefore, the correlation between the AMSR-E-derived VOD and NDVI is higher than that between the SMOS-derived VOD and NDVI.

Soil Moisture Retrieval from SMOS

After estimating vegetation parameters of VOD, H-polarized multiangular surface reflectivity $R_h^s\left(R_h^s = 1 - \varepsilon_h^s\right)$ can be calculated from the $\omega - \tau$ model. Soil dielectric constant k and effective surface roughness parameter S_r can be simultaneously retrieved by minimizing a cost function (CF) through iteration. Then, soil moisture is estimated by the soil dielectric model. The cost function is the squared difference between the calculated multiangular soil reflectivity (R_h^s) from the $\omega - \tau$ model and the parameterized ones $(R_{h\_par}^s)$ by Eq. 13 using high incidence angles ($30° \le \theta \le 50°$ were considered in this study). The cost function was computed as:

$$\mathrm{CF} = \sqrt{\sum_{i=1}^{n} \left(R_h^s(\theta_i) - R_{h\_par}^s(\theta_i)\right)^2} \tag{49}$$

The soil moisture retrieval algorithm above was applied to refined SMOS L1c brightness temperature data. It should be noted that some areas over Europe and Asia are influenced by extremely strong RFI. For these areas, the brightness temperature refined approach may fail, or the fitted results might not be recommended for use. Therefore, the soil moisture retrieval algorithm did not operate in these areas. For vegetation-covered areas, ω was given a constant value of 0.05 and 0.10 for short-vegetation-covered region and forests, respectively. This study does not challenge to retrieve soil moisture over dense forests. Thus, soil moisture is not retrieved in the areas where VOD is greater than 0.6. Retrieved soil moisture is then resampled to 0.25° resolution. Figure 16 shows the Africa soil moisture in January, April, July, and October of 2011, respectively. SMOS soil moisture from L2SMUDP (v501) is also shown as a reference. It can be seen from Fig. 16 that the Sahara Desert is very arid with soil moisture values less than 0.05 m^3/m^3 and generally does not show seasonal variations. From January to April, south of the equator, less areas exhibit a high soil moisture value (>0.25 m^3/m^3). In July, soil moisture in the grass-covered areas located between the Sahara Desert and tropical rainforest increases to the range of 0.15–0.40 m^3/m^3 and soil moisture in the regions south of the equator decreases to the range of 0.0–0.15 m^3/m^3. In October, soil moisture at the south of equator increases. The changes of soil moisture match the moving of the rain belt. Although the newly retrieved and SMOS soil moisture shares a similar pattern, some differences exist between these two soil moistures. The arid area is drier, and the moist area is wetter in SMOS L2 product. SMOS soil moisture exhibits a larger dynamic range than the new result. Quality of brightness temperature and methods of vegetation and surface roughness correction are all possible causes for these differences.

In situ soil moisture over 2010–2011 from two study areas were used to validate this retrieval algorithm. One study area is the Little Washita watershed, which is introduced in section "Vegetation Optical Depth (VOD) Retrieval from SMOS." The other study area is the Yanco region located east of the Murrumbidgee River catchment, Australia. It covers about 3600 km^2 in area. In this region, the land use is mainly crops and native pasture. In situ soil moisture of 13 stations (not including the cluster sites) in Yanco was used. Detailed information on these two study areas and ground-measured data can be found in Jackson et al. (2010) and Smith et al. (2012), respectively. All of the SMOS footprints within the study area were processed using the retrieval algorithm. The average of the retrieved soil moisture within a study area was compared to the in situ soil moisture derived by averaging all points in the study area at the time closest to the overpass time.

Time series of retrieved soil moisture by this algorithm along with the in situ soil moisture for both study areas are showed in Fig. 17. The in situ soil moisture were measured from a depth of 0–5 cm. Daily precipitation data are also illustrated for comparison. It shows that, except some overestimated soil moisture, the variation of retrieved results matches quite well with the in situ measurements in both areas. The RMSE values are 0.031 and 0.045 m^3/m^3 for LW watershed and Yanco, respectively. Figure 18 demonstrates the comparison between retrieved soil moisture and in situ measurements. The correlation coefficient (R) values is 0.829 for both areas. The SMOS L2 soil moisture product is also included as a reference. Good agreements exist between in situ and both retrieved soil moisture. Generally, the accuracy is close to or better than the mission target accuracy of 0.04 m^3/m^3, which demonstrates that this new algorithm can accurately retrieve soil moisture over these two

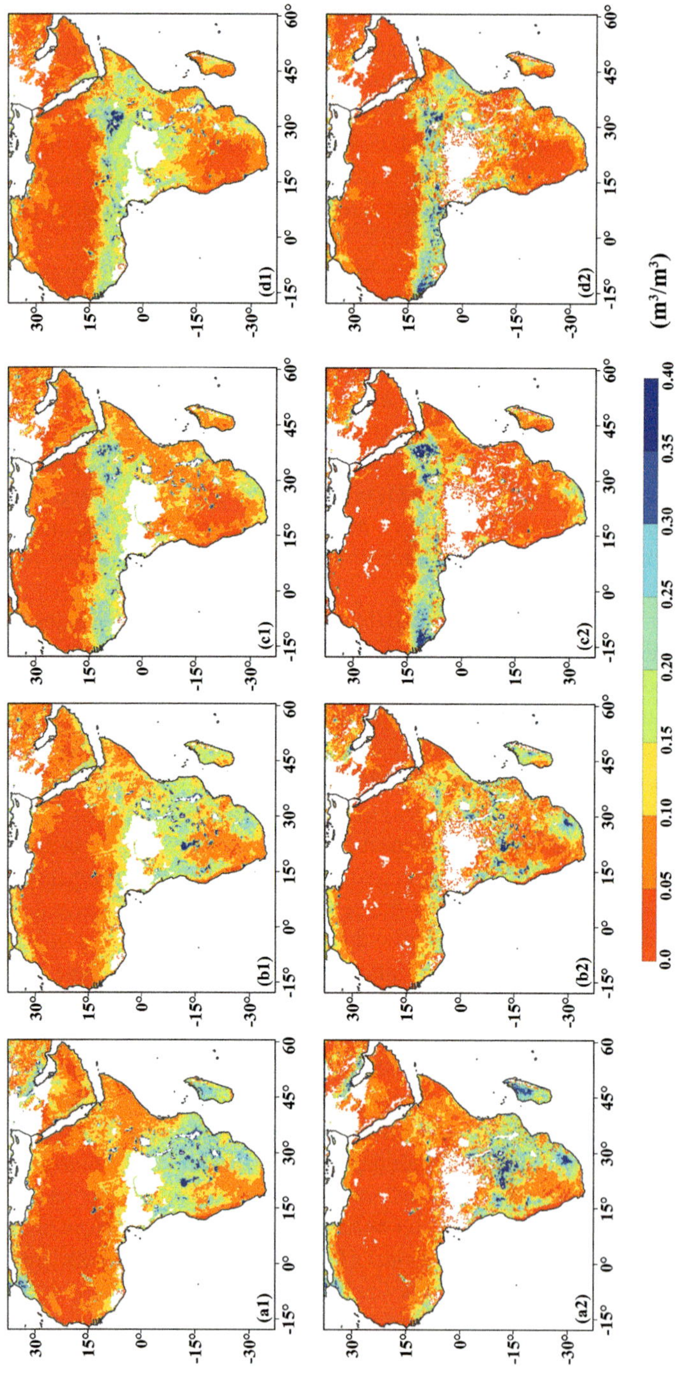

Fig. 16 Africa soil moisture of new retrievals (top row) and SMOS L2 product (bottom row) at 0.25° resolution for (a) January, (b) April, (c) July, and (d) October 2011

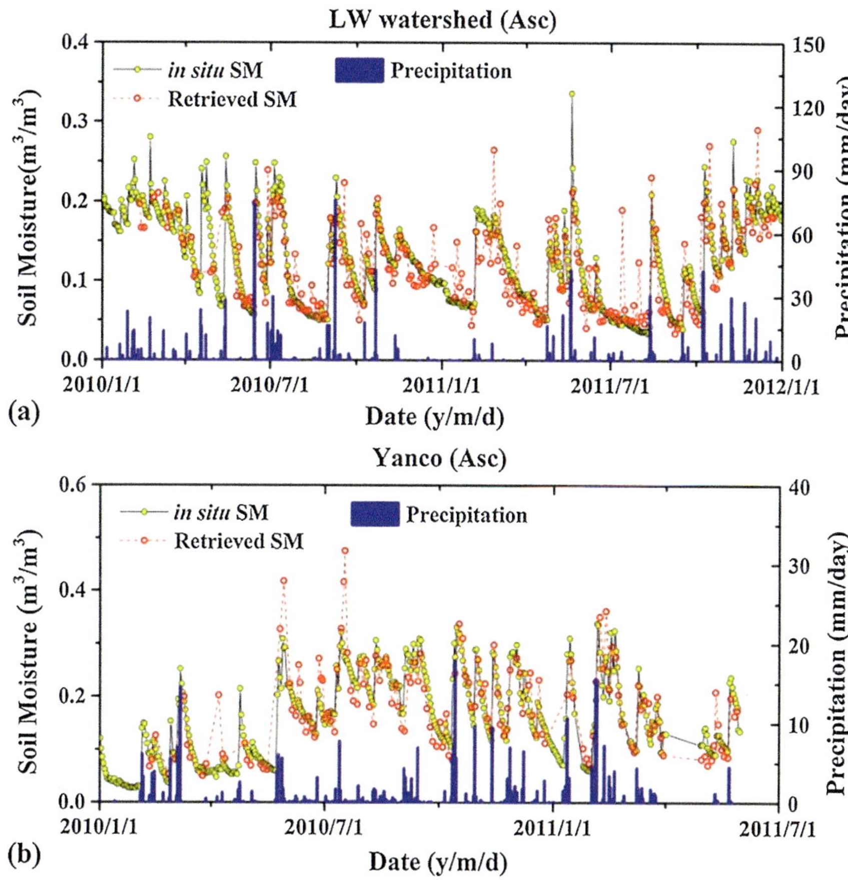

Fig. 17 Time series of in situ and retrieved soil moisture by the new algorithm for (**a**) LW watershed and (**b**) Yanco

study areas. The average values of retrieved effective surface roughness parameter S_r are 0.085 and 0.078 for LW watershed and Yanco, respectively.

Summary

Surface roughness and vegetation effects are the two critical factors in the microwave remote sensing of soil moisture, and this chapter describes the related model and algorithm development toward soil moisture retrieval practice. One solution to solve the roughness issue is the parametric model based on theoretical model simulations; it is found that the best geometric roughness variable for an exponential correlated surface is the slope ratio of s^2/l and s/l for the Gaussian correlated surface. Through optimization, the parametric model could achieve similar

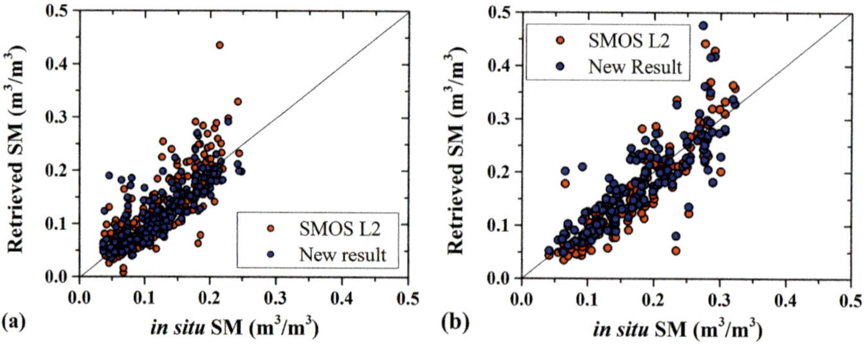

Fig. 18 Scatterplots of retrieved and in situ soil moisture for (**a**) LW watershed and (**b**) Yanco

performance as the theoretical model and can be useful in soil moisture retrieval. The vegetation effects are more complex that researchers always utilize the simple zero-order model to retrieve soil moisture. Microwave vegetation indices (MVIs) are developed based on the assumption that emissions at two adjacent channels (frequencies or incidence angels) are linearly correlated. The B parameter of MVI is theoretically correlated to attenuation features of vegetation. The MVI is enabling us to separate the soil and vegetation information from observations from remote sensing.

Soil moisture retrieval with SMEX airborne data demonstrated that the combination use of L- and S-bands could increase the retrieval accuracy due to better correction of the vegetation effects. In addition, S-band and C-band observations have higher spatial resolution compared with L-band, so that they are implemented for the downscaling of L-band observations. The disaggregation method is based on the linear relationship between L- and S-bands, or the exponential relationship between L- and C-bands. These strategies achieve a general RMSE of about 2–3 K and are under consideration for the future WCOM satellite. Based on the SMOS multiangular MVI, iterative retrieval algorithm is developed to estimate both the vegetation optical depth and soil moisture. Validation over specific sites indicates the accuracy is within the scientific requirements of RMSE $= 0.04\ m^3/m^3$.

Soil moisture is one of the most important and tractable variable of global water, and it is the key to understand the water cycle evolution under global climate changes. L-band measurements from airborne and spaceborne sensors provide the optimal information on soil moisture fields, and hopefully it can be continued with WCOM after the SMOS and SMAP missions to have high science impacts and broad societal benefits.

References

R. Bindlish, T. Jackson, M. Cosh, T. Zhao, P. O'Neill, Global soil moisture from the aquarius/SAC-D satellite: description and initial assessment. IEEE Geosci. Remote Sens. Lett. **12**, 923–927 (2015)

K.S. Chen, T.D. Wu, L. Tsang, Q. Li, Emission of rough surfaces calculated by the integral equation method with comparison to three-dimensional moment method simulations. IEEE Trans. Geosci. Remote Sens. **41**, 90–101 (2003)

L. Chen, J. Shi, J.-P. Wigneron, K.-S. Chen, A parameterized surface emission model at L-band for soil moisture retrieval. IEEE Geosci. Remote Sens. Lett. **7**, 127–130 (2010)

Q. Cui, J. Shi, J. Du, T. Zhao, C. Xiong, An approach for monitoring global vegetation based on multiangular observations From SMOS. IEEE J. Sel. Top. Appl. Earth Obs. Remote Sens. **8**, 604–616 (2015)

Q. Cui, X. Dong, J. Shi, T. Zhao, C. Xiong, An algorithm for retrieving soil moisture using L-band H-polarized multiangular brightness temperature data. IEEE Geosci. Remote Sens. Lett. **13**, 1295–1299 (2016)

A. Della Vecchia, P. Ferrazzoli, F. Giorgio, L. Guerriero, A large scale approach to estimate L band emission from forest covered surfaces, in *Proceedings of 2nd Recent Advances in Quantitative Remote Sensing* (2006), pp. 925–930

M.C. Dobson, F.T. Ulaby, M.T. Hallikainen, M.A. Elrayes, Microwave dielectric behavior of wet soil. 2. Dielectric mixing models. IEEE Trans. Geosci. Remote Sens. **23**, 35–46 (1985)

D. Entekhabi, E.G. Njoku, P.E. O'Neill, K.H. Kellogg, W.T. Crow, W.N. Edelstein, J.K. Entin, S.D. Goodman, T.J. Jackson, J. Johnson, J. Kimball, J.R. Piepmeier, R.D. Koster, N. Martin, K.C. McDonald, M. Moghaddam, S. Moran, R. Reichle, J.C. Shi, M.W. Spencer, S.W. Thurman, L. Tsang, J. Van Zyl, The soil moisture active passive (SMAP) mission. Proc. IEEE **98**, 704–716 (2010)

M.J. Escorihuela, Y.H. Kerr, P. De Rosnay, J.-P. Wigneron, J.-C. Calvet, F. Lemaitre, A simple model of the bare soil microwave emission at L-band. IEEE Trans. Geosci. Remote Sens. **45**, 1978–1987 (2007)

P. Ferrazzoli, L. Guerriero, J.P. Wigneron, Simulating L-band emission of forests in view of future satellite applications. IEEE Trans. Geosci. Remote Sens. **40**, 2700–2708 (2002)

A.K. Fung, M.R. Shah, S. Tjuatja, Numerical-simulation of scattering from 3-dimensional randomly rough surfaces. IEEE Trans. Geosci. Remote Sens. **32**, 986–994 (1994)

J.P. Grant, K. Saleh-Contell, J.-P. Wigneron, M. Guglielmetti, Y.H. Kerr, M. Schwank, N. Skou, A.A. Van de Griend, Calibration of the L-MEB model over a coniferous and a deciduous forest. IEEE Trans. Geosci. Remote Sens. **46**, 808–818 (2008)

P. Guo, J. Shi, Q. Liu, J. Du, A new algorithm for soil moisture retrieval with L-band radiometer. IEEE J. Sel. Top. Appl. Earth Obs. Remote Sens. **6**, 1147–1155 (2013)

T.J. Jackson, Measuring surface soil-moisture using passive microwave remote-sensing. 3. Hydrol. Process. **7**, 139–152 (1993)

T.J. Jackson, T.J. Schmugge, Vegetation effects on the microwave emission of soils. Remote Sens. Environ. **36**, 203–212 (1991)

T.J. Jackson, M.H. Cosh, R. Bindlish, P.J. Starks, D.D. Bosch, M. Seyfried, D.C. Goodrich, M.S. Moran, J. Du, Validation of advanced microwave scanning radiometer soil moisture products. IEEE Trans. Geosci. Remote Sens. **48**, 4256–4272 (2010)

T.J. Jackson, R. Bindlish, M.H. Cosh, T. Zhao, P.J. Starks, D.D. Bosch, M. Seyfried, M.S. Moran, D.C. Goodrich, Y.H. Kerr, D. Leroux, Validation of soil moisture and ocean salinity (SMOS) soil moisture over watershed networks in the U.S. IEEE Trans. Geosci. Remote Sens. **50**, 1530–1543 (2012)

Y.H. Kerr, P. Waldteufel, J.-P. Wigneron, S. Delwart, F. Cabot, J. Boutin, M.-J. Escorihuela, J. Font, N. Reul, C. Gruhier, S.E. Juglea, M.R. Drinkwater, A. Hahne, M. Martín-Neira, S. Mecklenburg, The SMOS mission: new tool for monitoring key elements of the global water cycle. Proc. IEEE **98**, 666–687 (2010)

Y.H. Kerr, P. Waldteufel, P. Richaume, J.P. Wigneron, P. Ferrazzoli, A. Mahmoodi, A. Al Bitar, F. Cabot, C. Gruhier, S.E. Juglea, D. Leroux, A. Mialon, S. Delwart, The SMOS soil moisture retrieval algorithm. IEEE Trans. Geosci. Remote Sens. **50**, 1384–1403 (2012)

M. Kurum, P.E. O'Neill, R.H. Lang, A.T. Joseph, M.H. Cosh, T.J. Jackson, Effective tree scattering and opacity at L-band. Remote Sens. Environ. **118**, 1–9 (2012)

H. Lawrence, J.-P. Wigneron, F. Demontoux, A. Mialon, Y.H. Kerr, Evaluating the semiempirical H-Q model used to calculate the L-band emissivity of a rough bare soil. IEEE Trans. Geosci. Remote Sens. **51**, 4075–4084 (2013)

D.M. Le Vine, G.S.E. Lagerloef, F.R. Colomb, S.H. Yueh, F.A. Pellerano, Aquarius: an instrument to monitor sea surface salinity from space. IEEE Trans. Geosci. Remote Sens. **45**, 2040–2050 (2007)

A. Mialon, J.-P. Wigneron, P. de Rosnay, M. Jose Escorihuela, Y.H. Kerr, Evaluating the L-MEB model from long-term microwave measurements over a rough field, SMOSREX 2006. IEEE Trans. Geosci. Remote Sens. **50**, 1458–1467 (2012)

V.L. Mironov, L.G. Kosolapova, S.V. Fomin, Physically and mineralogically based spectroscopic dielectric model for moist soils. IEEE Trans. Geosci. Remote Sens. **47**, 2059–2070 (2009)

T. Mo, T.J. Schmugge, A parameterization of the effect of surface-roughness on microwave emission. IEEE Trans. Geosci. Remote Sens. **25**, 481–486 (1987)

E.G. Njoku, S.K. Chan, Vegetation and surface roughness effects on AMSR-E land observations. Remote Sens. Environ. **100**, 190–199 (2006)

J.A. Ogilvy, J.R. Foster, Rough surfaces-gaussian or exponential statistics. J. Phys. D Appl. Phys. **22**, 1243–1251 (1989)

S.S. Saatchi, N.L. Harris, S. Brown, M. Lefsky, E.T. Mitchard, W. Salas, B.R. Zutta, W. Buermann, S.L. Lewis, S. Hagen, S. Petrova, L. White, M. Silman, A. Morel, Benchmark map of forest carbon stocks in tropical regions across three continents. Proceedings of the National Academy of Sciences of the United States of America. **108**, 9899–9904 (2011)

M. Schwank, I. Voelksch, J.-P. Wigneron, Y.H. Kerr, A. Mialon, P. de Rosnay, C. Maetzler, Comparison of two bare-soil reflectivity models and validation with L-band radiometer measurements. IEEE Trans. Geosci. Remote Sens. **48**, 325–337 (2010)

J. Shi, J. Wang, A.Y. Hsu, P.E. O'Neill, E.T. Engman, Estimation of bare surface soil moisture and surface roughness parameter using L-band SAR image data. IEEE Trans. Geosci. Remote Sens. **35**, 1254–1266 (1997)

J.C. Shi, L.M. Jiang, L.X. Zhang, K.S. Chen, J.P. Wigneron, A. Chanzy, A parameterized multifrequency-polarization surface emission model. IEEE Trans. Geosci. Remote Sens. **43**, 2831–2841 (2005)

J. Shi, T. Jackson, J. Tao, J. Du, R. Bindlish, L. Lu, K.S. Chen, Microwave vegetation indices for short vegetation covers from satellite passive microwave sensor AMSR-E. Remote Sens. Environ. **112**, 4285–4300 (2008)

J. Shi, X. Dong, T. Zhao, J. Du, L. Jiang, Y. Du, H. Liu, Z. Wang, D. Ji, C. Xiong, WCOM: the science scenario and objectives of a global water cycle observation mission, in *2014 IEEE International Geoscience and Remote Sensing Symposium (IGARSS), IEEE International Symposium on Geoscience and Remote Sensing IGARSS* (IEEE, New York, 2014), pp. 3646–3649

A.B.. Smith, J.P. Walker, A.W. Western, R.I. Young, K.M. Ellett, R.C. Pipunic, R.B. Grayson, L. Siriwardena, F.H.S. Chiew, H. Richter, The Murrumbidgee soil moisture monitoring network data set. Water Resour. Res. **48**, 7701 (2012)

F.T. Ulaby, F. Kouyate, A.K. Fung, A.J. Sieber, A backscatter model for a randomly perturbed periodic surface. IEEE Trans. Geosci. Remote Sens. **20**, 518–528 (1982a)

F.T. Ulaby, A. Aslam, M.C. Dobson, Effects of vegetation cover on the radar sensitivity to soil-moisture. IEEE Trans. Geosci. Remote Sens. **20**, 476–481 (1982b)

A.A. Van de Griend, J.P. Wigneron, The b-factor as a function of frequency and canopy type at H-polarization. IEEE Trans. Geosci. Remote Sens. **42**, 786–794 (2004)

N.E.C. Verhoest, H. Lievens, W. Wagner, J. Alvarez-Mozos, M.S. Moran, F. Mattia, On the soil roughness parameterization problem in soil moisture retrieval of bare surfaces from synthetic aperture radar. Sensors **8**, 4213–4248 (2008)

J.R. Wang, B.J. Choudhury, Remote-sensing of soil-moisture content over bare field at 1.4 GHz frequency. J. Geophys. Res. Atmos. **86**, 5277–5282 (1981)

J.R. Wang, T.J. Schmugge, An empirical-model for the complex dielectric permittivity of soils as a funcion of water-content. IEEE Trans. Geosci. Remote Sens. **18**, 288–295 (1980)

J.P. Wigneron, L. Laguerre, Y.H. Kerr, A simple parameterization of the L-band microwave emission from rough agricultural soils. IEEE Trans. Geosci. Remote Sens. **39**, 1697–1707 (2001)

J.P. Wigneron, Y. Kerr, P. Waldteufel, K. Saleh, M.J. Escorihuela, P. Richaume, P. Ferrazzoli, P. de Rosnay, R. Gurney, J.C. Calvet, J.P. Grant, M. Guglielmetti, B. Hornbuckle, C. Mätzler, T. Pellarin, M. Schwank, L-band microwave emission of the biosphere (L-MEB) model: description and calibration against experimental data sets over crop fields. Remote Sens. Environ. **107**, 639–655 (2007)

T. Zhao, J. Shi, R. Bindlish, T.J. Jackson, M. Cosh, L. Jiang, Z. Zhang, H. Lan, Parametric exponentially correlated surface emission model for L-band passive microwave soil moisture retrieval. Phys. Chem. Earth **83–84**, 65–74 (2015a)

T.J. Zhao, J.C. Shi, R. Bindlish, T.J. Jackson, Y.H. Kerr, M.H. Cosh, Q. Cui, Y.Q. Li, C. Xiong, T. Che, Refinement of SMOS multi-angular brightness temperature toward soil moisture retrieval and its analysis over reference targets. IEEE J. Sel. Top. Appl. Earth Obs. Remote Sens. **8**, 589–603 (2015b)

M. Zribi, A. Gorrab, N. Baghdadi, A new soil roughness parameter for the modelling of radar backscattering over bare soil. Remote Sens. Environ. **152**, 62–73 (2014)

Remote Sensing Precipitation: Sensors, Retrievals, Validations, and Applications

Yang Hong, Guoqiang Tang, Yingzhao Ma, Qi Huang, Zhongying Han, Ziyue Zeng, Yuan Yang, Cunguang Wang, and Xiaolin Guo

Contents

Abstract

Precipitation is one of the most important water cycle components. The chapter reviews modern instruments and techniques for global precipitation retrieval,

Y. Hong (✉)
State Key Laboratory of Hydroscience and Engineering, Department of Hydraulic Engineering, Tsinghua University, Beijing, China

School of Civil Engineering and Environmental Science, University of Oklahoma, Norman, OK, USA
e-mail: yanghong@ou.edu

G. Tang · Y. Ma · Q. Huang · Z. Han · Z. Zeng · Y. Yang · C. Wang · X. Guo
State Key Laboratory of Hydroscience and Engineering, Department of Hydraulic Engineering, Tsinghua University, Beijing, China

X. Li, H. Vereecken (eds.), *Observation and Measurement of Ecohydrological Processes*, Ecohydrology, https://doi.org/10.1007/978-3-662-48297-1_4

including weather radars and satellites. Some of the most popular global multi-satellite precipitation products are introduced, including PERSIANN-CCS, TMPA, and IMERG. In addition, we extend to the typical regional and global studies about the assessment of various products and their application in flood detection and prediction.

Keywords
Remote sensing · Precipitation · Sensors · Retrievals · Validations · Applications

Introduction

Precipitation plays a vital role in understanding the mechanism and interaction of global water and energy balance and is the main input of hydrometeorological models and climate studies (Hong et al. 2007; Kidd and Huffman 2011). Researchers can hardly conduct simulations of the water cycle and eco-hydrological process without accurate precipitation inputs. Traditionally, people can measure rainfall and snowfall using various types of ground-based rain gauges. However, rain gauge-based techniques for precipitation observations have nonnegligible limitations in eco-hydrology research due to large spatial nonuniformity and temporal availability in rainfall fields inherently. And unfortunately, the number of ground rain gauges continues to decline over the world (Stokstad 1999; Shiklomanov et al. 2002). In addition, rain gauges have trouble observing precipitation in the oceans, polar areas, and mountainous regions which attracts more and more attention in the context of climate change (Ma et al. 2016).

Fortunately, the development of remote sensing techniques promotes the capability of humans in observing global precipitation. Weather radars and space-based meteorological satellites are two major remote sensing techniques in precipitation observation. After World War II, weather radars have dramatically improved our ability to measure precipitation at high temporal and spatial resolutions (Farbry et al. 1994; Morin et al. 1995; Harris et al. 2001; Berne et al. 2004). For example, the United States has built a dense radar network, namely, Next-Generation Weather Radar (NEXRAD) consisting of around 160 high-resolution S-band Doppler weather radars (Heiss et al. 1990; Vivekanandan et al. 1990). Regional-scale studies can be performed across the United States, Western Europe, and a few other regions across the globe. In China, more than 100 S-band and C-band weather radars have also been installed and been applied in meteorological and hydrological research (Fang et al. 2002). However, weather radar systems are still inadequate for obtaining global precipitation and suffer from similar problems with gauges such as limited spatial coverage, particularly in oceanic, remote, or undeveloped regions.

Currently, the only practical way to achieve global precipitation observation is space-based meteorological satellites (Hong et al. 2012; Tang et al. 2016a). Satellite-based precipitation estimates have great advantages in terms of spatial coverage compared with rain gauges and weather radars. Moreover, satellite measurements are continuous and uniform and can avoid the high cost of ground observation networks (Hong et al. 2016). Global satellite-based rainfall products are currently mainly based on passive microwave (PMW), calibrated infrared (IR), and PMW plus IR observations. IR sensors are on geostationary Earth orbit (GEO) satellites which provide precipitation estimates at high temporal resolutions, but the accuracy of IR-based rainfall estimates is generally not good due to the indirect linkage between IR signals and precipitation. PMW sensors are onboard low Earth orbiting (LEO) satellites and can provide precipitation estimation with a more direct link to rain and ice particles. Currently, some countries, such as the United States and China, have been trying to put PMW sensors onto GEO satellites which, if realized, can promote the capability of observing rainfall from space significantly. Active microwave sensors can provide the most accurate estimation of rainfall and snowfall, i.e., the Ku-band precipitation radar (PR) onboard the TRMM satellite (Simpson et al. 1988), the W-band Cloud Profiling Radar (CPR) onboard the CloudSat and the Ku-/Ka-band Dual-frequency Precipitation Radar (DPR) onboard the GPM Core Observatory (Hou et al. 2014). However, those spaceborne precipitation radars also suffer from long revisiting periods.

There have been great numbers of multi-satellite precipitation products since the launch of TRMM in late 1997, such as Multi-satellite Precipitation Analysis (TMPA) (Huffman et al. 2007), Climate Prediction Center (CPC) MORPHing technique (CMORPH) (Joyce et al. 2004), Precipitation Estimation from Remotely Sensed Information using Artificial Neural Networks (PERSIANN) (Sorooshian et al. 2000), PERSIANN Cloud Classification System (PERSIANN-CCS) (Hong et al. 2004), and Global Satellite Mapping of Precipitation (GSMaP) (Kubota et al. 2007). Those products are generally based on IR, MW, and rain gauge data which have promoted our knowledge of global precipitation and the development of hydrological and eco-hydrological models.

This chapter first provides an overview of remotely sensed rainfall, including satellite sensors and weather radars, precipitation retrieval algorithms, and global precipitation products. Then we briefly introduce the validation and application of some typical remotely sensed precipitation products in hydrology, meteorology, and hazard monitoring and forecast. Finally, we summarize current achievement and discuss the prospects of global satellite precipitation products.

Spaceborne and Ground-Based Sensors and Precipitation Retrievals

Review of Precipitation Sensors

Satellite sensors and ground weather radars are two main sources of remotely sensed rainfall. Satellite sensors include multichannels, such as visible and infrared (VIS/IR), passive microwave, and active microwave. Infrared sensors are generally onboard GEO which can provide precipitation estimates with relatively high temporal and spatial resolutions. Infrared data from the Geostationary Operational

Environmental Satellites (GOES) E/W; the Meteosat 5,7,8; and the Multifunctional Transport Satellites (MTSAT) have been widely used in various satellite precipitation products. FY-2 is the first generation of meteorological satellites in China, six geostationary satellites from FY-2A to FY-2F. Currently, FY-2 satellites have provided large amounts of data of precipitation, land surface, and sea surface temperature (http://fy3.satellite.cma.gov.cn/portalsite/default.aspx). Microwave data can be applied for more accurate rainfall estimation due to its more direct linkage with precipitation. The Advanced Microwave Sounding Unit (AMSU) onboard NOAA 10–12 and Special Sensor Microwave Imager (SSM/I) onboard Defense Meteorological Satellite Program (DMSP) F-13/14/15/16 are all popular passive microwave data sources.

Satellite-borne sensors are of great importance in estimating precipitation over the vast majority of the Earth's surface which lacks adequate in situ observing systems. Precipitation radars, such as TRMM PR, CloudSat CPR, and GPM DPR, can estimate precipitation best from space, while the orbit width of the three radars is very small (from 1.4 km for CPR to ~250 km for PR/DPR). Weather radars can provide accurate rainfall estimate on the ground. Combination of ground radars and rain gauges can generate precipitation products of high quality. All those satellite sensors and ground radars will be introduced detailed below.

Infrared Sensors

The Earth's surface, clouds, and atmosphere absorb solar energy and emit part of it into the outer space. The infrared sensors can sense these energies, which are often carried by GEO satellites, providing IR imagery 24 h a day with a high resolution (~ 4 × 4 km). There is an international constellation of GEO satellites covering the globe (Table 1), among which the Geostationary Operational Environmental Satellite system (GOES) operated by the National Oceanic and Atmospheric Administration (NOAA) is the primary one.

In the field of meteorology, radiation information collected by the IR sensors always relates to brightness temperature (T_b) of land surface, water surface, and

Table 1 Some geostationary satellites and sensors

Satellite	Agency	Sensor	Channels	Resolution (km)
GOES E/W	National Oceanic and Atmospheric Administration (NOAA)	GOES I-M Imager	5	1–4
Meteosat	European Organisation for the Exploitation of Meteorological Satellites (EUMETSAT)	MVIRI and SEVIRI	3–12	1–4
MTSAT/ Himawari	Japan Meteorological Agency (JMA)	Imager	5	1–5

cloud top. Most of IR-based rainfall estimation methods try to establish relationship model between surface rainfall rates and cloud-top brightness temperatures (Arkin 1979; Kummerow et al. 1998). According to the method used in processing the T_b values for different sensors, IR-based rainfall retrieval algorithms are divided into three main types, including pixel-based, window-based, and patch-based. The pixel-based algorithm is one of the most basic methods in satellite rainfall retrieval, and GOES Precipitation Index (GPI) is the most well-known one of them (Arkin and Meisner 1987). To improve the accuracy, the initial result is accumulated over a longer time scale (usually weekly or monthly) (Bastiaanssen et al. 1998; Liang et al. 2001).

The autoestimator (AE) algorithm utilizes a power law function to fit the Rainfall-T_b relationship. Further, researchers developed hydroestimator (HE) algorithm which takes more feature variables into account, such as relative humidity and precipitation water (Scofield and Kuligowski 2003). Besides the IR4 and IR5 channels, all other three channels of GOES Imager are used to modify the identification of raining clouds (Ba and Gruber 2001). Then, the rainfall rate of each raining cloud is assigned with the reference of its T_b. The cloud window-based algorithm is an extension of the pixel-based one. Hsu et al. (1997) developed the Precipitation Estimation from Remotely Sensed Information Using Artificial Neural Networks (PERSIANN) system, which relates pixel rainfall rate to a range of the neighborhood pixel coverage like a window (Hsu et al. 1997). The cloud window-based algorithm can involve in cloud local texture information based on T_b, making the retrieval more reasonable than the former two methods.

Cloud-patch-based algorithms extract feature information which may be related to rainfall rate from the entire cloud coverage. Various cloud segmentation approaches accompanied by pixel rainfall rate assignment tricks were developed. The Griffith-Woodley Technique (Griffith et al. 1978) assumes a specific T_b threshold under 253 K to separate cloud from IR T_b imagery. The Convective-Stratiform Technique (Adler and Negri 1988) firstly screens convective patches according to the local minimum value of T_b and assigns different rainfall amounts to convective clouds and stratiform clouds separately. Pixel rainfall rates are proportionally distributed according to the T_b inside cloud patches. Xu et al. (1999) used SSM/I microwave rainfall estimation to separate rain and no-rain pixels in a cloud patch, which made T_b thresholds variable in different situations (Xu et al. 1999). Hong et al. (2005) developed an automated neural network model named self-organizing nonlinear output (SONO) for patch-based rainfall estimation on the basis of PERSIANN system (Hong et al. 2005). The SONO model establishes various nonlinear T_b-Rainfall relations by classifying varied cloud patch into different clusters based on their extracted features which can be divided into three categories, i.e., coldness, geometry, and texture. The first category is directly associated with cloud brightness temperature (i.e., minimum and mean temperature of a cloud patch); the second one is derived from the geometric properties of cloud patches (i.e., cloud-path size and shape index); and the third category is the texture variation of cloud brightness temperature (i.e., standard deviation, mean value of local standard deviation, standard deviation of local standard deviation, and gradient).

Feature extraction is another important part of cloud-patch-based approaches. It's theoretically ideal to include as many features associated with land surface rainfall as possible. In PERSIANN-CCS system (Hong et al. 2004), nine key feature indices including the eight indices in Hong et al. (2005) which have been introduced above and gray image texture were extracted from cloud patches.

Microwave Sensors

Microwave sensors carried by LEO satellites have evolved steadily from the early Electronically Scanning Microwave Radiometer in the 1970s to the current SSM/I on DMSP satellites, the AMSU on the NOAA satellite series, the TRMM TMI, the Aqua Advanced Microwave Scanning Radiometer-Earth Observing System (AMSR-E) sensors, and the Global Precipitation Measurement (GPM) Microwave Imager (GMI). However, due to low temporal sampling resolution of passive microwave (PMW) data, significant coverage gaps exist even merging all of the available PMW datasets within every 3-h time window.

Rainfall from microwave sensors is estimated based on the principle that rainfall at the surface is related to microwave emission from raindrops at low-frequency channels and microwave scattering from ice at high-frequency channels (Kummerow et al. 1998). The satellite MW-based rainfall estimation algorithms can be roughly categorized into three classes: (1) the "emission"-type algorithms (Wilheit et al. 1991), which detect the increased radiances caused by rain over radiometrically cold oceans using low-frequency channels; (2) the "scattering" algorithms (Ferraro and Marks 1995), which correlate rainfall to radiance depressions caused by ice scattering present in many precipitating clouds; and (3) the "multichannel inversion"-type algorithms (Kummerow et al. 2001), which aim at inverting the entire radiance vector simultaneously.

The TMI as well as the AMSR-E uses the algorithms developed by Wilheit et al. (1991) and Kummerow et al. (2001), while the SSM/I uses the algorithms developed by Wilheit et al. (1991) and Ferraro and Marks (1995) in the GPCP over ocean and land, respectively. Precipitation estimation derived from the Advanced Microwave Sounding Unit-B (AMSU-B) at the National Environmental Satellite Data and Information Service (NESDIS) is supported by the algorithm developed by Weng et al. (2003). Each algorithm appears to have strengths and weaknesses related to specific applications, because they are optimized for the corresponding satellite sensor.

NASA GPM Integrated Multi-satellite Retrievals for GPM (IMERG), an advancement of TRMM TMPA, is an up-to-date product of global precipitation estimation. The algorithm used in GPM is transparent, parametric, and unified which ensures uniform rainfall products across all MW sensors from all satellite platforms. Huffman et al. (2015) suggests that the algorithm used in IMERG performs careful intercalibration of microwave estimates and provides finer time and space scales. Also, the robust algorithm including a generalized parametric framework which avoids the impasse of cross-evaluation of previous MW algorithms can attribute the differences between sensors to physical differences between observed scenes rather than artifacts of the algorithm.

Ground-Based Weather Radars

In order to address the growing interest in and demand for ground-based weather radars, this section serves as a guidance toward the understanding of this advanced instrument and relevant algorithms that constantly provides various kinds of meteorological data. The development of weather radars originated from World War II when radars were invented mainly for the battlefield (Buderi 1996; Brown 1999; Geiger 2008). With the advent of weather radars, our ability to detect high-resolution precipitation temporally and spatially has encountered unprecedented improvement, thus facilitating such application as flash flood forecasting and precipitation estimation (Doswell et al. 1996; Davis 2001; Krajewski and Smith 2002; Delrieu et al. 2005; Smith et al. 2007). For instance, with the radar systems available, not only regional-scale studies but also global-scale researches can be carried out. Recent decades have witnessed the invention of ground-based weather radars, enabling us to perform research and operations much more easily with radars.

Weather radars typically consist of the basic components: a klystron transmitter, a waveguide, an antenna, the pedestal, a feedhorn, a radome, and a receiver (Rinehart 1991; Doviak 1993). Weather Surveillance Radar-1988 Doppler (WSR-88D) that constitutes the NEXRAD network in operation across the United States is a widely used ground-based weather radar (Klazura and Imy 1993). The NEXRAD system is a network of 160 high-resolution Doppler weather radars operated by the National Weather Service, an agency of the National Oceanic and Atmospheric Administration (NOAA) within the US Department of Commerce. NEXRAD is capable of detecting precipitation and atmospheric movement or wind. When processed, the data returning can be displayed in a mosaic map which showcases patterns of precipitation and its movement. The greatest benefit of weather radars for hydrology lies in its potential to estimate rainfall rates at relatively high spatiotemporal resolution (i.e., 1 km/5 min), in real time, within a radius of approximately 250 km of the radar. Weather radar has enabled the transformation from a system of manual reporting and reacting to weather to one of the automated observations and anticipating weather impacts.

Steps for computing quantitative precipitation estimation (QPE) using conventional, single-polarization radar usually include the following. The first is radar calibration. The calibration of radar has a major influence on the accuracy of rainfall rates. A miscalibration of only 1 dB can result in bias in rainfall rates of 15%. Several methods for calibrating radar are provided in Atlas (2002). The second is quality control. Now that the Z data (radar reflectivity factor) have been bias-corrected for radar miscalibration, which requires large samples of comparisons over hours or even days of precipitation, every single bin of radar data must be carefully scrutinized to remove deleterious effects from non-meteorological scatters on the ground, biota in the atmosphere, planes, chaff, etc. The third is precipitation rate estimation. Calibrated reflectivity values illustrating the size, shape, state, and concentration of the hydrometeors within the radar sampling volume are used subsequently to compute precipitation rate. In regions with complex terrain, bins are selected at multiple elevation angles to construct the hybrid scan for QPE. Reflectivity from the hybrid scan has made it possible to compute two-dimensional fields of precipitation rate in spherical coordinates (range, azimuth). The general form of reflectivity-to-rainfall relationships, or *Z-R* equations, is a power law as

$$Z = aR^b$$

where a is the prefactor and b is the exponent. The two most common *Z-R* relations are the NEXRAD default for convection ($a = 300$, $b = 1.4$) and the Marshall-Palmer relation ($a = 200$, $b = 1.6$) generally applied to stratiform rain (Marshall and Palmer 1948).

Global Multi-satellite Precipitation Products

Overview

A single sensor or single satellite can hardly provide global precipitation products with satisfying quality. Therefore, researchers have increasingly moved toward combining data from GEO VIS/IR and LEO MW sensors (Huffman et al. 2015; Wanders et al. 2015; Ashouri et al. 2015). A lot of such multi-satellite precipitation products have been developed and released to the public which are characterized by free access, quasi-global coverage, fine spatiotemporal resolutions, and continuous self-renewal, which, in turn, promote their development and applications (Tang et al. 2016b). The most commonly used satellite global rain products are summarized in Table 2. Most products provide both near-real-time and post-real-time datasets to satisfy different timeliness and quality requirement in climatology, meteorology,

Table 2 Summary of global satellite rainfall products

Product name	Agency/country	Spatial resolution	Temporal resolution	Period
GPCP	NASA/USA	2.5°	Monthly	1979–present
CMAP	NOAA/UAS	2.5°	5-day	1979–present
TMPA	NASA GSFC/USA	0.25°	3-hourly	1998–present
CMORPH	NOAA CPC/USA	0.25°/8 km	3-hourly/30-min	1998–present
PERSIANN	University of Arizona/USA	0.25°	3-hourly	2000–present
PERSIANN-CCS	University of California Irvine/USA	4 km	30-min	2006–present
PERSIANN-CDR	NOAA/USA	0.25°	Daily	1983–present
GSMaP	JAXA/Japan	0.1°	30-min	2014–present
IMERG	NASA/USA	0.1°	30-min	2014–present

hydrology, hazard forecast, etc. Compared with the information listed in Hong et al. (2012), the resolutions, periods, and accuracy of those products have been increased significantly in just a few years indicating the rapid advance of satellite remotely sensed precipitation. For example, the most recent IMERG products of the GPM mission are built on previous algorithms from PERSIANN-CCS, TMPA, and CMORPH which can provide three products at 30-min and 0.1° resolution between 60°N-S. Here we briefly introduce the algorithms and products of the three typical mainstream satellite precipitation products of PERSIANN-CCS, TMPA, and IMERG due to their widespread applications.

PERSIANN-CCS Algorithm

Through extracting local and regional cloud features from infrared (10.7 μm) geostationary satellite imagery, with microwave and ground radar rainfall data blending or training, the PERSIANN-CCS algorithm can provide fine-scale (4 km and 30-min) rainfall distribution. As shown in Fig. 1, the PERSIANN-CCS algorithm processes satellite cloud images into pixel rain rates by following four steps:

1. Segmentation of Satellite Infrared Cloud Images
   The incremental temperature threshold (ITT) is proposed to effectively segment cloud images which contain several convective cells into a large number of separated cloud patches by gradually increasing threshold temperatures. The ITT algorithm contains two main steps. First, the algorithm locates the local minimum temperature and initializes seeds. Second, the threshold temperature is set higher, and seeded regions are expanded from the seeded points until the border of other seeded regions or cloud-free regions is reached.
2. Extraction of Cloud-Patch Features
   A feature extraction scheme that retrieves both local pixel temperature textures and regional cloud-patch features instead of only local pixel features as in PERSIANN. Coldness, geometry, and texture are used to discriminate cloud types. Specifically, the coldness features of cloud patch include the minimum and mean temperature of a cloud patch; the geometric features include cloud-patch area and cloud-path shape index; and the texture features include standard deviation of cloud-patch temperature, mean value of local standard deviation of cloud temperature, standard deviation of local standard deviation of cloud, gradient of cloud-top brightness temperature, and gray image texture.
3. Classification of Cloud Patches
   The clouds are classified into different groups based on the cloud-patch features extracted from different temperature levels by a clustering algorithm, the self-organizing feature map (SOFM) (Hsu et al. 1997). SOFM projects the high-dimensional classification space of many input variables into various clusters arranged in a two-dimensional coordinate. Two main steps are involved. First, the distance between patch features and SOFM cluster center is calculated. Second,

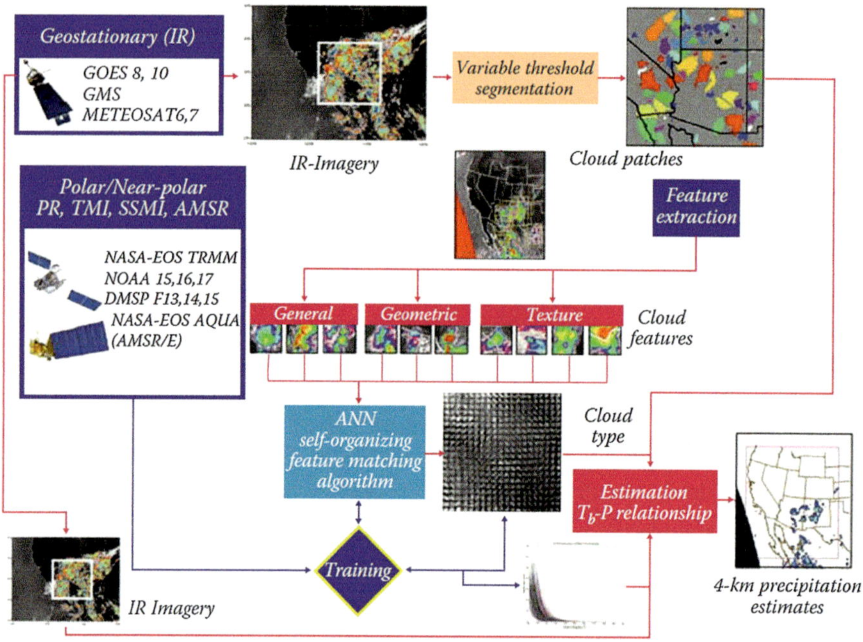

Fig. 1 Satellite cloud image segmentation, feature extraction, classification, multisensor blending, and rainfall estimation of the PERSIANN-CCS algorithm. (From Hong et al. (2012))

the best matching SOFM cluster center from the minimum distance between the input vector and the SORM connection weights is found.

4. Estimation of Patch and Pixel Rainfall
   Instead of calibrating only one T_b-R function for all clouds in PERSIANN, the PERSIANN-CCS searches a nonlinear T_b-R relationship for the full spectrum of cloud-rainfall conditions. Therefore, PERSIANN-CCS overcomes the limitation of a single statistical T_b-R function and can generate various rain rates at a given brightness temperature and variable rain/no-rain IR thresholds for different cloud types. In each classified cloud-patch group, the probability matching method is used to redistribute the T_b-R pixel pairs so that the proportion of the R distribution above a given rain rate is equal to the proportion of the T_b distribution below the associated T_b threshold value (Atlas et al. 1990). Then a nonlinear exponential function is fitted to the redistributed pixels, whose parameters are calibrated by numerous GOES infrared images and their collocated gauge-corrected radar rainfall.

The PERSIANN-CCS algorithm provides pixel rainfall distribution as well as cloud-rainfall system analysis basing on real-time GOES cloud images (Hong et al. 2004). Afterward, an automated neural network for cloud-patch-based rainfall estimation, entitled SONO (Hong et al. 2005), was developed to adjust the T_b-R mapping function by using passive microwave precipitation estimates from low Earth orbiting satellite platforms, such as TRMM and Defense Meteorological Satellite Program. Evaluated at various temporal and spatial scales, PERSIANN-CCS shows good performance of estimation accuracy both in rain intensity and in the detection of rain/no-rain pixels (Hong et al. 2004, 2005, 2007). Real-time data from the current version of PERSIANN-CCS are available online both at regional and global scales (http://chrsdata.eng.uci.edu/).

TRMM-Based Multi-satellite Precipitation Analysis Algorithm

TMPA provides a calibration-based sequential scheme for combining precipitation estimates from multiple satellites, as well as gauge analyses where feasible, at $0.25° \times 0.25°$ and 3-hourly resolutions (Huffman et al. 2007). The TMPA provides a real-time product (3B42RT) and a gauge-adjusted post-real-time research product (3B42V7). The TMPA estimates are produced in four stages: (1) the MW precipitation estimates are calibrated and combined; (2) the IR precipitation estimates are created using the calibrated MW precipitation; (3) the MW and IR estimates are combined; and (4) the rain gauge data are incorporated.

Sources of passive microwave satellite precipitation estimates include TRMM TMI, SSM/I, Special Sensor Microwave Imager/Sounder (SSMIS) (3B42V7 only), AMSR-E, AMSU-B, and Microwave Humidity Sounder (MHS). The real-time product, 3B42RT, uses TMI estimates as the initial RT calibrator to calibrate precipitation estimates derived from available LEO MW radiometers and then merges all of the estimates at 3-h intervals. Gaps in the analyses are filled with GEO IR data regionally calibrated to the merged MW product. The post-real-time product, 3B42V7, adjusts the monthly accumulations of the 3-hourly fields from 3B42RT based on a monthly gauge analysis, including the new Global Precipitation Climatology Centre (GPCC) "full" gauge analysis whenever available and the GPCC "monitoring" gauge analysis since 2010. However, the previous monitoring product encompasses periods up to April 2005, and the Climate Assessment and Monitoring System (CAMS) analysis is used thereafter (Huffman et al. 2010). The final step of creating the research product is to introduce monthly rain gauge data. The monthly ratio of the satellite-only and satellite-gauge combination is used to rescale the individual 3-hourly estimates.

The TMPA intended to provide the "best" estimate of quasi-global precipitation. The most successful use of the TMPA data is that the analysis takes advantage of the fine-scale data to create time/space averages appropriate to the user's application. However, TMPA has some aspects that need to be improved which include improved error estimation and extension to higher latitudes (Li et al. 2009; Yong et al. 2013, 2014; Tang et al. 2016a, 2016b). After about 17 years of productive data gathering, the instruments on TRMM were turned off on April 8, 2015.

Integrated Multi-satellite Retrievals for GPM

The GPM mission has an international constellation of satellites, including one Core Observatory satellite and approximately ten partner satellites. The GPM Core Observatory was deployed on February 28, 2014, by a joint effort of NASA and the Japan Aerospace Exploration Agency (JAXA), marking a transition from the TRMM era to the GPM era. The GPM Core Observatory carries the DPR (the Ku-band at 13.6 GHz and Ka-band at 35.5 GHz) and GMI (frequencies range between 10 and 183 GHz). GPM extends the sensor package compared to TRMM instruments with the single-frequency PR (the Ku-band at 13.8 GHz) and TMI (frequencies range between 10 and 85.5 GHz). IMERG is a unified US algorithm, which is intended to intercalibrate, merge, and interpolate all microwave estimates of the GPM constellation, IR estimates, gauge observations, and other data from potential sensors at $0.1^\circ \times 0.1^\circ$ and half-hour temporal resolutions (Huffman et al. 2015). IMERG provides three kinds of products, including the near-real-time "Early" and "Late" run products, and the post-real-time "Final" run product. IMERG combines intermittent precipitation estimates from all constellation microwave sensors, IR-based observations from geosynchronous satellites, and monthly gauge precipitation data (Hou et al. 2014). IMERG employs the 2014 version of the Goddard Profiling Algorithm (GPROF2014) to compute precipitation estimates from all passive microwave (PMW) sensors onboard GPM satellites, which is an improvement compared with TMPA (GPROF2010) (Huffman et al. 2015). The IMERG "Final" run combines the GPCC Monitoring Product (currently Version 4) in the product, whose data source is limited to the Global Telecommunications System (GTS) with only about 7000 stations over the globe. The Full Data Reanalysis (currently Version 6) involves much more stations than the Monitoring Product but only covers the period 1901–2010. It is anticipated that the IMERG datasets would be reprocessed using the Full Data Reanalysis when it is updated to a longer period (Huffman et al. 2015). The IMERG data can be downloaded from the PMM website (http://pmm.nasa.gov/data-access/downloads/gpm). The IMERG product has been proved to be better than the previous TMPA products by some latest studies which will be introduced in detail in following sections.

Validation and Applications of Remote Sensing Precipitation

Regional and Global Assessment

The performance of Day-1 Integrated Multi-satellitE Retrievals for GPM IMERG and its predecessor, TRMM 3B42V7, was cross-evaluated using data from the best-available hourly gauge network over the Tibetan Plateau (TP) (Ma et al. 2016). IMERG Final run was used. Analyses of 3-hourly rainfall estimates in the warm season of 2014 reveal that IMERG shows appreciably better correlations and lower errors than 3B42V7, though with very similar spatial patterns for all assessment indicators. IMERG also appears to detect light rainfall better than 3B42V7.

However, IMERG shows the slightly lower probability of detection (POD) than 3B42V7 for elevations above 4200 m. Both IMERG and 3B42V7 successfully capture the northward dynamic life cycle of the Indian monsoon reasonably well over the TP. In particular, the relatively light rain from early and end Indian monsoon moisture surge events often fails to be captured by the sparsely distributed gauges. In spite of limited snowfall field observations, IMERG shows the potential of detecting solid precipitation, which cannot be retrieved from the 3B42V7 products (Fig. 2).

In addition, we also compare IMERG and 3B42V7 products over Mainland China using hourly rain data from around 2400 rain gauges (Tang et al. 2016b). The point gauge data are interpolated to areal precipitation field using the inverse distance weight (IDW) method. Both products agree well with the gauge data over East and South Mainland China. However, the correlation coefficient (CC) was a little lower over North China and even under 0.2 over West China. Several factors could contribute to relatively low CC of IMERG and 3B42V7 products over such areas: (1) the topography and climate over West China are complex, posing a great challenge for accurate satellite precipitation estimation; (2) the IMERG Final run was corrected using GPCC, whereas few gauges are used in the production of GPCC monthly gauge analysis in the TP, and thus the quality of IMERG products is potentially degraded; and (3) the interpolated precipitation could deviate far from the real values due to the sparse gauge networks, which would reduce reliability of metrics calculated against such ground reference. Figure 3 shows spatial distributions of CC, bias ratio (BIAS), and critical success index (CSI) computed from IMERG and 3B42V7 products against IDW interpolated precipitation at 3-hourly and $0.1^\circ \times 0.1^\circ$ resolutions over Mainland China. The distribution of CC seemed better for 3B42V7 than IMERG at the 3-hourly resolution (Fig. 3a–b). But at the daily resolution, IMERG performs better than 3B42V7 in North China.

The Coupled Routing and Excess Storage (CREST) model (Wang et al. 2011) is developed by the University of Oklahoma (http://hydro.ou.edu) and the NASA SERVIR Project Team (www.servir.net). The CREST model has been implemented successfully in a variety of multi-scale meteorological and hydrological studies (Tang et al. 2016c; Kan et al. 2017; Li et al. 2017). The CREST V2.1 (Shen et al. 2016) is used to evaluate the quality of IMERG as well as its hydrological continuity compared with TRMM era products (TMPA 3B42V7 and 3B42RT) in the Ganjiang River basin, which is the seventh largest sub-catchment of the Yangtze River. The China Gauge-based Daily Precipitation Analysis (CGDPA) product is used as the ground truth (Shen and Xiong 2016).

In the validation period (May 1, 2014–September 30, 2014), the CREST model was forced by CGDPA, 3B42V7, 3B42RT, and IMERG precipitation data based on parameter calibrated by CGDPA from 2003 to 2009. CGDPA has the best skill scores in terms of streamflow simulation as expected, closely followed by IMERG (IMERG/CGDPA: NSCE = 0.77/0.86, CC = 0.91/0.94, BIAS = −14.09/−8.76%, and RMSE = 1080.87/822.73m^3/s). The hydrograph of IMERG is remarkably similar to that of CGDPA (Fig. 4a and b). Compared with 3B42V7 and 3B42RT, IMERG performed the best and matched well with the observed streamflow,

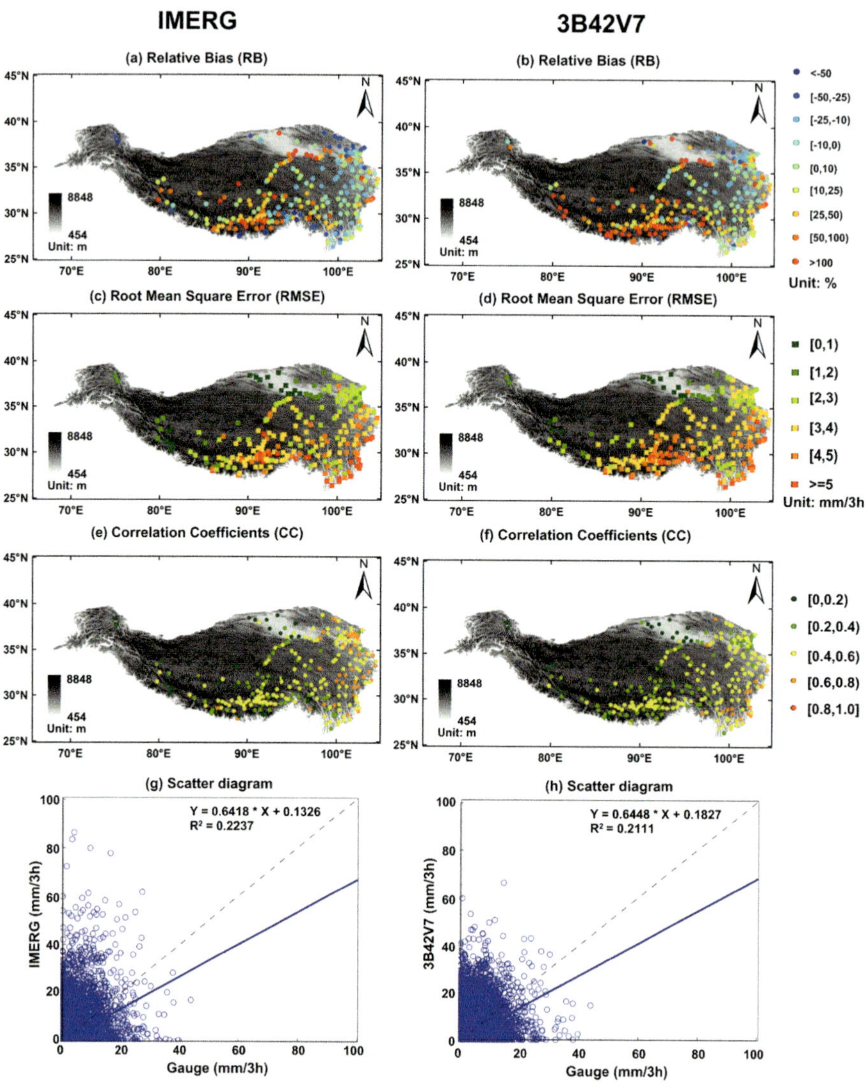

Fig. 2 Maps of (**a**–**b**) relative bias (RB), (**c**–**d**) root mean square error (RMSE), (**e**–**f**) correlation coefficients (CC), and (**g**–**h**) scatter diagram between satellite- and gauge-based 3-hourly rainfall estimates in the warm season (April–September) of 2014 over the TP. Notes: the left panel, i.e., (**a**, **c**, **e**, **g**), stands for IMERG rainfall estimates, and the right panel, i.e., (**b**, **d**, **f**, **h**) stands for 3B42V7 products. (From Ma et al. 2016)

especially for the second flow peak compared with 3B42V7 and 3B42RT (Fig. 4b–d). The 3B42RT product came at the bottom as its NSCE declined to 0.46 and RMSE increased to 1637.53 m^3/s.

The accuracy of satellite precipitation products varies with regions. Therefore, although regional validation studies can help reveal the error characteristics of some

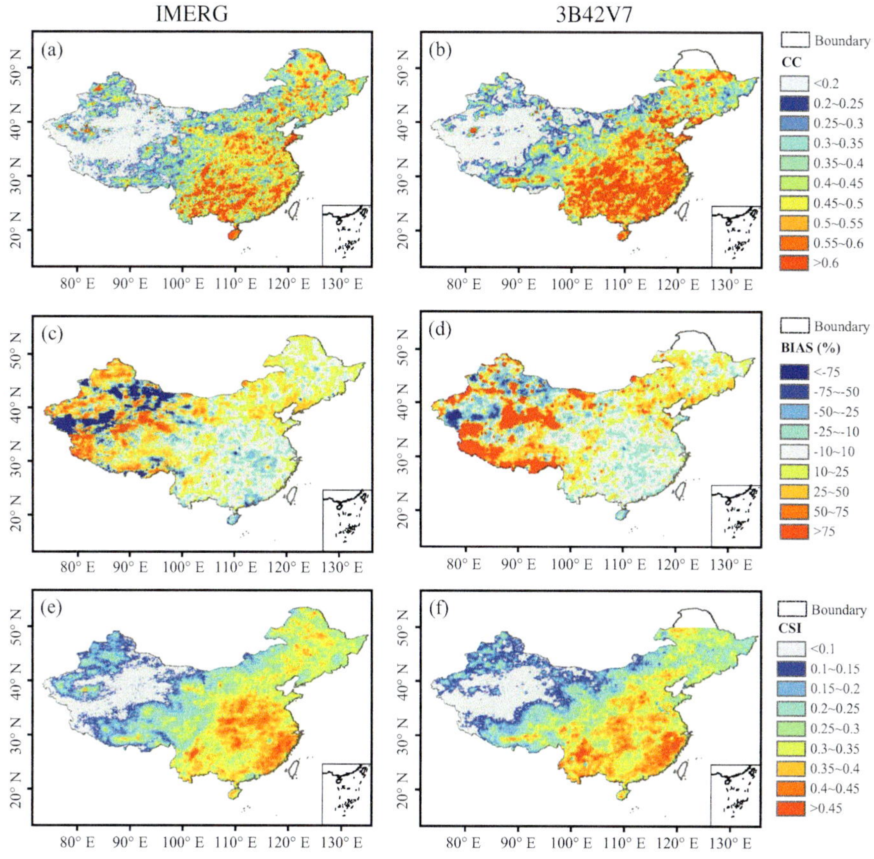

Fig. 3 Spatial distributions of assessment metrics for IMERG (the left panel) and 3B42V7 (the right panel) precipitation at 3-hourly and $0.1^\circ \times 0.1^\circ$ resolution over Mainland China: (**a**–**b**) CC, (**c**–**d**) BIAS, and (**e**–**f**) CSI. (From Tang et al. 2016a)

satellite products, global evaluation and comparison are necessary. Beck et al. (2017) evaluated the performance of 23 precipitation products in the global scale using 76,086 gauges worldwide. A conceptual hydrological model HBV is employed to evaluate the ten gauge-corrected datasets in 9053 small- to medium-sized basins to avoid the independence problem caused by overlapped gauges. However, it is hard for researchers to obtain worldwide ground observations. Massari et al. (2017) uses the triple collocation (TC) method to characterize uncertainties of satellite precipitation products in the globe. The TC method requires the input of three independent datasets and can output the error statistics of input datasets. In addition, some studies try to evaluate satellite precipitation products by using data from spaceborne radars as the benchmark due to their high accuracy than passive microwave and infrared sensors. For example, Behrangi et al. (2014) evaluated the capability of spaceborne sensors in detecting light precipitation and snowfall using CloudSat Cloud Profiling

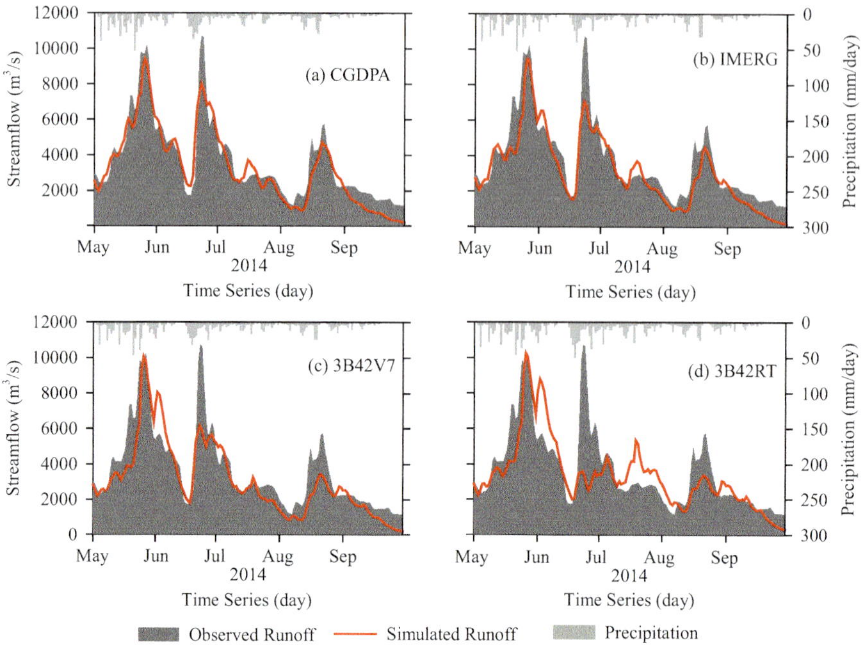

Fig. 4 Comparison of CREST simulated streamflow with gauge-calibrated parameters and observed streamflow validation period 2 (May 1, 2014–September 30, 2014). (**a**) Daily data from CGDPA; (**b**) daily data from IMERG; (**c**) daily data from 3B42V7; and (**d**) daily data from 3B42RT. (From Tang et al. 2016c)

Radar (CPR). Tang et al. (2017) intercompared the rainfall and snowfall performance of the three existing spaceborne precipitation radars, i.e., TRMM PR, GPM DPR, and CloudSat CPR.

Application in Flood Detection and Prediction

Satellite precipitation products can be applied in hydrologic studies to predict flood events (Khan et al. 2011). Wang et al. (2011) established a real-time Global Hydrological Prediction System (GHPS) with the CREST model to investigate the detectability and predictability of flooding (Fig. 5). GHPS is forced by the NASA Tropical Rainfall Measuring Mission (TRMM) Multi-satellite Precipitation Analysis at near-real-time and by the deterministic and ensemble precipitation forecast products from NOAA Global Forecast System (GFS) at several lead times (Zhang et al. 2015). The CREST model is currently running within the near-real-time global hydrological simulation and flood monitoring demonstration system (http://eos.ou.edu) at the University of Oklahoma. Presently, it is driven by the TRMM 3B42RT. Another global flood system is Global Flood Monitoring System (GFMS; http://flood.umd.edu/) (Wu et al. 2014) which is supported by TMPA and GPM products.

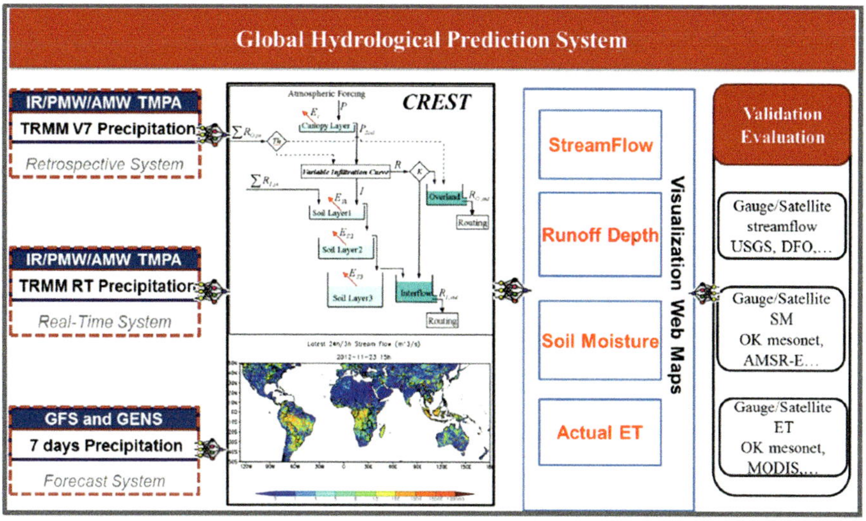

Fig. 5 Structure of the Global Hydrological Prediction System. (From Zhang et al. 2015)

GFMS has played an important role in flood monitoring and can be utilized for decision support (Kirschbaum et al. 2017).

Floods usually lead to huge casualties and damages to urban areas due to its high population density and intensively developed infrastructure. Moreover, the impervious surface in urban areas would also increase the runoff rate. Thus, it is an emergency to predict urban to reduce their impacts. For example, the flood event on July 21, 2012, in Beijing, the Capital City of China, is detected and predicted by the GHPS. The "7·21" Beijing storm lasted for around 16 h, and the rain rate reached as high as 215 mm/day in the urban areas, resulting in 79 fatalities and around 1.6 billion dollars of damages. We used rain gauge observations, TRMM RT, TRMM V7, and both GFS deterministic and ensemble precipitation forecasts at different initializations (with different lead times, as in Table 3) to simulate the hydrological predictions of surface runoff in urban areas and streamflow in the watersheds in Beijing. Before the simulation, soil states in the global CREST model are initialized by running the model using TRMM RT rainfall forcing from July 1, 2012, until the initial time of each experiment. Results indicate that the disastrous "7·21" storm was detectable by TRMM satellite precipitation estimates and predictable by deterministic GFS rainfall forecasts at least 4 days in advance.

In flash flood warning and forecasting, remotely sensed rainfall can provide useful information to determine when and how to make alerts and take evacuation actions. Based on the widely used Flash Flood Guidance (FFG) system (Georgakakos 1987), we developed a Cascading Flash Flood Guidance (CFFG) system, progressively from the Flash Flood Potential Index (FFPI), the Flash Flood Hazard Index (FFHI), and the Flash Flood Risk Index (FFRI) (Zeng et al. 2006). Together with the land cover, vegetation cover, SRTM slope, and soil

Table 3 Summary of characteristics of the precipitation products used in the detecting and predicting of the "7·21" Beijing event[a]

Rainfall products	Spatial resolution	Time interval	Lead time	Reference
Rain Gauge Observation	Interpolated onto 0.25°	Hourly; then accumulated to 3-hourly	N/A	Huffman et al. 2007
TRMM RT	0.25°	3-hourly	N/A	
TRMM RP	0.25°	3-hourly	N/A	
Deterministic GFS Precipitation	0.25°	3-hourly	180 h	Wang 2010; Wang et al. 2013
Ensemble GFS Precipitation	0.25°	3-hourly	168 h	

[a]From Zhang et al. (2015)

property data for the FFPI calculation, average daily amount in flood seasons and maximum 6 h and maximum 24 h amount from TRMM V7 are used as rainfall effect indicators to determine the relative hazard potential of flash flood, which is represented by the value of FFHI (ranging from 1 to 10, 1 means lowest potential and 10 the highest). The weights between indicators are determined by an integrated approach of the analytic hierarchy process and the information entropy theory. Further inclusion of GDP, population, and flood prevention measures as vulnerability factors for the FFRI enables the prediction of the flash flood risk. Until now, the CFFG system has been implemented in a fine resolution (1 km) in Yunnan Province, and the characteristics of China's flash floods are also mapped in a cascading manner.

Conclusive Remarks and Outlooks

The rapid development of remote sensing techniques has made precipitation estimation more accurate and cover broader regions compared with traditional rain gauges around the world. Satellite-based precipitation products supported by VIS/IR and passive/active MW sensors promote our knowledge of global water and energy cycle. Currently, the newest GPM based on the heritage of precedent multi-satellite algorithms such as TMPA, CMORPH, and PERSIANN-CCS has released its Level-1 to Level-3 data to the public for more than 2 years. The Level-4 products combining with models are also anticipated to be developed and released in the future, and the spatial coverage could be extended to near 90°N/S compared with 60°N/S of current products. However, satellite precipitation products still have some way to go in further improving its quality in high mountains (such as the TP), high latitude regions, and arid regions.

Active microwave sensors can estimate precipitation with comparable quality with ground gauges and radars. However, after the TRMM PR was decommissioned, only two precipitation radars operate in space. The future development of spaceborne radars has two complementary directions, multi-frequency radar instruments on single platform and constellation of many small to cube radar satellites, for improved global precipitation estimation. The employment of more spaceborne radar data is the future trend of achieving accurate enough estimation of global rainfall and snowfall only based on satellites. In addition, future radar design should consider including more bands on a single platform, such as Ka-, Ku-, and W-bands. Jet Propulsion Laboratory (JPL) has developed an architecture with a low-cost simplified spaceborne radar system, namely, RainCube (Peral et al. 2015). The inexpensive CubeSats can compose a constellation with a number of satellites monitoring the earth frequently. Together with the advanced but expensive big spaceborne radars, estimation of global rainfall and snowfall can be greatly improved.

Acknowledgment This study was financially supported by the National Natural Science Foundation of China (Grant No. 71461010701), National Key Research and Development Program of China (2016YFE0102400), and National Natural Science Foundation of China (Grant No. 91437214).

References

R.F. Adler, A.J. Negri, A satellite infrared technique to estimate tropical convective and stratiform rainfall. J. Appl. Meteorol. **27**, 30–51 (1988)

P.A. Arkin, The relationship between fractional coverage of high cloud and rainfall accumulations during GATE over the B-scale array. Mon. Weather Rev. **107**(10), 1382–1387 (1979)

P.A. Arkin, B.N. Meisner, The relationship between large-scale convective rainfall and cold cloud over the western hemisphere during 1982–84. Mon. Weather Rev. **115**, 51–74 (1987)

H. Ashouri, K.L. Hsu, S. Sorooshian, et al., PERSIANN-CDR: Daily precipitation climate data record from multisatellite observations for hydrological and climate studies. J. Bull. Am. Meteorol. Soc. **96**(1), 69–83 (2015)

D. Atlas, Radar calibration: Some simple approaches. Bull. Am. Meteorol. Soc. **83**(9), 1313 (2002)

D. Atlas, D. Rosenfeld, D.B. Wolff, Climatologically tuned reflectivity-rain rate relations and links to area-time integrals. J. Appl. Meteorol. **29**, 1120–1135 (1990)

M. Ba, A. Gruber, GOES multispectral rainfall algorithm (GMSRA). J. Appl. Meteorol. **40**, 1500–1514 (2001)

W.G.M. Bastiaanssen, H. Pelgrum, J. Wang, et al., A remote sensing surface energy balance algorithm for land (SEBAL): Part 2: Validation. J. Hydrol. **212**, 213–229 (1998)

H.E. Beck et al., Global-scale evaluation of 23 precipitation datasets using gauge observations and hydrological modeling. Hydrol. Earth Syst. Sci. Discuss., 1–23 (2017)

A. Behrangi et al., What does CloudSat reveal about global land precipitation detection by other spaceborne sensors? Water Resour. Res. **50**(6), 4893–4905 (2014)

A. Berne, G. Delrieu, J.D. Creutin, C. Obled, et al., Temporal and spatial resolution of rainfall measurements required for urban hydrology. J. Hydrol. **299**(3–4), 166–179 (2004)

L. Brown, A radar history of World War II. J. Am. Hist. Res. (1999)

R. Buderi, *The Invention that Changed the World: How a Small Group of Radar Pioneers Won the Second World War and Launched a Technological Revolution* (Simon and Schuster, New York, 1996)

R.S. Davis, Flash flood forecast and detection methods, in *Severe Convective Storms*, (American Meteorological Society, Boston, 2001), pp. 481–525

G. Delrieu, J. Nicol, E. Yates, et al., The catastrophic flash-flood event of 8–9 September 2002 in the Gard region, France: A first case study for the Cévennes–Vivarais Mediterranean Hydrometeorological Observatory. J. Hydrometeorol. **6**(1), 34–52 (2005)

C.A. Doswell, H.E. Brooks, R.A. Maddox, Flash flood forecasting: An ingredients-based methodology. J. Weather Forecast. **11**(4), 560–581 (1996)

R.J. Doviak, *Doppler Radar and Weather Observations* (Courier Corporation, Chelmsford, 1993)

L.H.Z.P.N. Fang, H.S.Z.X.G. Runsheng, X. Bao, China operational weather radar data processing system. J. Appl. Meteorlol. **6**, 014 (2002)

F. Farbry, A. Bellon, M.R. Duncan, G.L. Austin, et al., High resolution rainfall measurements by radar for very small basins: The sampling problem reexamined. J. Hydrol. **161**(1–4), 415–428 (1994)

R. Ferraro, G. Marks, The development of SSM/I rain-rate retrieval algorithms using ground-based radar measurements. J. Atmos. Ocean Technol. **12**(4), 755–770 (1995)

R.L. Geiger, *Research and Relevant Knowledge: American Research Universities Since World War II* (Transaction Publishers, New Brunswick, 2008)

K.P. Georgakakos, Real-time flash flood prediction. J. Geophys. Res. Atmos. **92**(D8), 9615–9629 (1987)

C.G. Griffith, W.L. Woodley, P.G. Grube, D.W. Martin, J. Stout, D.N. Sikdar, Rain estimation from geosynchronous satellite imagery-visible and infrared studies. Mon.Weather Rev. **106**, 1153–1171 (1978)

D. Harris, E. Foufoula-Georgiou, K. Droegemeier, et al., Multiscale statistical properties of a high-resolution precipitation forecast. J. Hydrometeorol. **2**(4), 406–418 (2001)

W.H. Heiss, D.L. McGrew, D. Sirmans, et al., NEXRAD: Next generation weather radar (WSR-88D). J. Microwave **33**(1), 79–89 (1990)

Y. Hong, K.L. Hsu, S. Sorooshian, et al., Precipitation estimation from remotely sensed imagery using an artificial neural network cloud classification system. J. Appl. Meteorol. **43**, 1834–1853 (2004)

Y. Hong, K. Hsu, S. Sorooshian, et al., Self-organizing nonlinear output (SONO): A neural network suitable for cloud patch–based rainfall estimation at small scales. Water Resour. Res. **41**, 477–490 (2005)

Y. Hong, D. Gochis, J. Cheng, et al., Evaluation of PERSIANN-CCS rainfall measurement using the NAME event rain gauge network. J. Hydrometeorol. **8**, 469–482 (2007)

Y. Hong, S. Chen, X. Xue, G. Hodges, Global precipitation estimation and applications, in *Multiscale Hydrologic Remote Sensing: Perspectives and Applications*, (CRC Press, Boca Raton, 2012), pp. 371–386

Y. Hong, Y. Zhang, S. Khan, *Hydrologic Remote Sensing: Capacity Building for Sustainability and Resilience* CRC Press, (2016)

A.Y. Hou, R.K. Kakar, S. Neeck, A.A. Azarbarzin, C.D. Kummerow, M. Kojima, . . . T. Iguchi, The global precipitation measurement mission. J. Am. Meteorol. Soc. **95**(5), 701–722 (2014)

K. Hsu, X. Gao, S. Sorooshian, H.V. Gupta, Precipitation estimation from remotely sensed information using artificial neural networks. J. Appl. Meteorol. **36**, 1176–1190 (1997)

G.J. Huffman, D.T. Bolvin, E.J. Nelkin, D.B. Wolff, R.F. Adler, G. Gu, . . . E.F. Stocker, The TRMM multisatellite precipitation analysis (TMPA): Quasi-global, multiyear, combined-sensor precipitation estimates at fine scales. J. Hydrometeorol. **8**, 38–55 (2007)

G.J. Huffman, R.F. Adler, D.T. Bolvin, E.J. Nelkin, The TRMM multi-satellite precipitation analysis (TMPA), in *Satellite Rainfall Applications for Surface Hydrology*, (Springer Netherlands, Dordrecht, 2010), pp. 3–22

G.J. Huffman, D.T. Bolvin, E.J. Nelkin, Integrated Multi-satellitE Retrievals for GPM (IMERG) Technical Documentation. NASA/GSFC Code 612, 47 pp. http://pmm.nasa.gov/sites/default/files/document_files/IMERG_doc.pdf (2015)

R.J. Joyce, J.E. Janowiak, P.A. Arkin, P. Xie, CMORPH: A method that produces global precipitation estimates from passive microwave and infrared data at high spatial and temporal resolution. J. Hydrometeorol. **5**(3), 487–503 (2004)

G. Kan, G. Tang, Y. Yang, Y. Hong, J. Li, L. Ding, et al., An improved coupled routing and excess storage (CREST) distributed hydrological model and its verification in Ganjiang River basin, China. Water **9**(11), 904 (2017)

I.S. Khan et al., Satellite remote sensing and hydrologic modeling for flood inundation mapping in Lake Victoria basin: Implications for hydrologic prediction in ungauged basins. IEEE Trans. Geosci. Remote Sens. **49**, 85–95 (2011)

C. Kidd, G. Huffman, Global precipitation measurement. Meteorol. Appl. **18**, 334–353 (2011)

D.B. Kirschbaum, G.J. Huffman, R.F. Adler, S. Braun, K. Garrett, E. Jones, et al., NASA's remotely sensed precipitation: A reservoir for applications users. Bull. Am. Meteorol. Soc. **98**(6), 1169–1184 (2017)

G.E. Klazura, D.A. Imy, A description of the initial set of analysis products available from the NEXRAD WSR-88D system. J. Am. Meteorol. Soc. **74**(7), 1293–1311 (1993)

W.F. Krajewski, J.A. Smith, Radar hydrology: Rainfall estimation. J. Adv. Water Resour. **25**(8–12), 1387–1394 (2002)

T. Kubota, S. Shige, H. Hashizume, K. Aonashi, N. Takahashi, S. Seto, ... K. Iwanami. Global precipitation map using satellite-borne microwave radiometers by the GSMaP project: Production and validation. IEEE T. Geosci. Remote. **45**(7), 2259–2275 (2007)

C. Kummerow, W. Barnes, T. Kozu, et al., The tropical rainfall measuring mission (TRMM) sensor package. J. Atmos. Ocean. Technol. **15**(3), 809–817 (1998)

C. Kummerow, Y. Hong, W. Olson, et al., The evolution of the Goddard Profiling Algorithm (GPROF) for rainfall estimation from passive microwave sensors. J. Appl. Meteorol. **40**(11), 1801–1820 (2001)

L. Li, Y. Hong, J. Wang, et al., Evaluation of the real-time TRMM-based multi-satellite precipitation analysis for an operational flood prediction system in Nzoia Basin, Lake Victoria, Africa. J. Nat. Hazards **50**(1), 109–123 (2009)

N. Li, G. Tang, P. Zhao, Y. Hong, Y. Gou, K. Yang, Statistical assessment and hydrological utility of the latest multi-satellite precipitation analysis IMERG in Ganjiang River basin. Atmos. Res. **183**, 212–223 (2017)

L. Liang, C. Liu, Y.Q. Xu, et al., Real-time texture synthesis by patch-based sampling. J. ACM Trans. Graphics (ToG) **20**(3), 127–150 (2001)

Y. Ma, G. Tang, D. Long, B. Yong, L. Zhong, W. Wan, Y. Hong, Similarity and error intercomparison of the GPM and its predecessor-TRMM multisatellite precipitation analysis using the best available hourly gauge network over the Tibetan plateau. Remote Sens. **8**(7), 569 (2016)

J.S. Marshall, W.M.K. Palmer, The distribution of raindrops with size. J. Meteor. **5**(4), 165–166 (1948)

C. Massari et al., An assessment of the performance of global rainfall estimates without ground-based observations. Hydrol. Earth Syst. Sci. **21**(9), 4347–4361 (2017)

J. Morin, D. Rosenfeld, E. Amitai, et al., Radar rain field evaluation and possible use of its high temporal and spatial resolution for hydrological purposes. J. Hydrol. **172**(1–4), 275–292 (1995)

E. Peral, S. Tanelli, Z. Haddad, et al., Raincube: A proposed constellation of precipitation profiling radars in CubeSat, in *C. Geoscience and Remote Sensing Symposium (IGARSS), 2015 I.E. International*, (IEEE, 2015), Milan, Italy, pp. 1261–1264

R.E. Rinehart, *Radar for Meteorologists* (University of North Dakota, Office of the President, 1991). Grand Forks, North Dakota

R.A. Scofield, R.J. Kuligowski, Status and outlook of operational satellite precipitation algorithms for extreme-precipitation events. Mon. Weather Rev. **18**, 1037–1051 (2003)

Y. Shen, A. Xiong, Validation and comparison of a new gauge-based precipitation analysis over mainland China. Int. J. Climatol. **36**(1), 252–265 (2016)

X. Shen, Y. Hong, K. Zhang, Z. Hao, Refining a distributed linear reservoir routing method to improve performance of the CREST model. J. Hydrol. Eng. **22**, 04016061 (2016)

A.I. Shiklomanov, R.B. Lammers, C.J. Vörösmarty, Widespread decline in hydrological monitoring threatens pan-arctic research. Eos, Trans. Am. Geophys. **83**(2), 13–17 (2002)

J. Simpson, R.F. Adler, G.R. North, A proposed tropical rainfall measuring mission (TRMM) satellite. J. Am. Meteorol. Soc. **69**(3), 278–295 (1988)

J.A. Smith, M.L. Baeck, K.L. Meierdiercks, et al., Radar rainfall estimation for flash flood forecasting in small urban watersheds. J. Adv. Water Resour. **30**(10), 2087–2097 (2007)

T.M. Smith, V. Lakshmanan, G.J. Stumpf, et al., Multi-radar multi-sensor (MRMS) severe weather and aviation products: Initial operating capabilities. J. Am. Meteorol. Soc. **97**(9), 1617–1630 (2016)

S. Sorooshian, K.L. Hsu, X. Gao, H.V. Gupta, B. Imam, D. Braithwaite, Evaluation of PERSIANN system satellite-based estimates of tropical rainfall. Bull. Am. Meteor. Soc. **81**(9), 2035–2046 (2000)

E. Stokstad, Scarcity of rain, stream gages threatens forecasts. Science **285**(5431), 1199–1200 (1999)

G. Tang, Y. Ma, D. Long, L. Zhong, Y. Hong, Evaluation of GPM Day-1 IMERG and TMPA Version-7 legacy products over mainland China at multiple spatiotemporal scales. J. Hydrol. **533**, 152–167 (2016a)

G. Tang, D. Long, Y. Hong, Systematic anomalies over inland water bodies of High Mountain Asia in TRMM precipitation estimates: No longer a problem for the GPM era? IEEE Geosci. Remote Sens. **13**, 1762 (2016b)

G. Tang, Z. Zeng, D. Long, X. Guo, B. Yong, W. Zhang, Y. Hong, Statistical and hydrological comparisons between TRMM and GPM Level-3 products over a Midlatitude Basin: Is Day-1 IMERG a good successor for TMPA 3B42V7? J. Hydrometeorol. **17**(1), 121–137 (2016c)

G. Tang et al., Similarities and differences between three coexisting spaceborne radars in global rainfall and snowfall estimation. Water Resour. Res. **53**(5), 3835–3853 (2017)

J. Vivekanandan, S.M. Ellis, R. Oye, et al., Cloud microphysics retrieval using S-band dual-polarization radar measurements. J. Am. Meteorol. Soc. **80**(3), 381–388 (1990)

N. Wanders, M. Pan, E.F. Wood, Correction of real-time satellite precipitation with multi-sensor satellite observations of land surface variables. J. Remote Sens. Environ. **160**, 206–221 (2015)

X. Wang, Incorporating ensemble covariance in the gridpoint statistical interpolation variational minimization: A mathematical framework. Mon. Weather Rev. **138**(7), 2990–2995 (2010)

J. Wang, Y. Hong, L. Li, et al., The coupled routing and excess storage (CREST) distributed hydrological model. Hydrol. Sci. J. **56**(1), 84–98 (2011)

X. Wang, D. Parrish, D. Kleist, et al., GSI 3DVar-based ensemble–variational hybrid data assimilation for NCEP global forecast system: Single-resolution experiment. Mon. Weather Rev. **141**(11), 4098–4117 (2013)

F. Weng, L. Zhao, R. Ferraro, et al., Advanced microwave sounding unit cloud and precipitation algorithms. Radio Sci. **38**(4), 8068 (2003)

T. Wilheit, A. Chang, L. Chiu, Retrieval of monthly rainfall indices from microwave radiometric measurements using probability distribution functions. J. Atmos. Ocean Technol. **8**(1), 118–136 (1991)

H. Wu, R.F. Adler, Y. Tian, G.J. Huffman, et al., Real-time global flood estimation using satellite-based precipitation and a coupled land surface and routing model. Water Resour. Res. **50**(3), 2693–2717 (2014)

L. Xu, X. Gao, S. Sorooshian, P.A. Arkin, A microwave infrared threshold technique to improve the GOES precipitation index. J. Appl. Meteorol. **38**, 569–579 (1999)

B. Yong, L. Ren, Y. Hong, et al., First evaluation of the climatological calibration algorithm in the real-time TMPA precipitation estimates over two basins at high and low latitudes. J. Water Resour. Res. **49**(5), 2461–2472 (2013)

B. Yong, B. Chen, J.J. Gourley, et al., Intercomparison of the Version-6 and Version-7 TMPA precipitation products over high and low latitudes basins with independent gauge networks: Is the newer version better in both real-time and post-real-time analysis for water resources and hydrologic extremes? J. Hydrol. **508**, 77–87 (2014)

Z. Zeng, G. Tang, D. Long, et al., A cascading flash flood guidance system: development and application in Yunnan Province, China. Natural Hazards, 84(3), 2071–2093

Y. Zhang, Y. Hong, X. Wang, et al., Hydrometeorological analysis and remote sensing of extremes: Was the July 2012 Beijing flood event detectable and predictable by global satellite observing and global weather modeling systems? J. Hydrol. **16**(1), 381–395 (2015)

Inhomogeneity in Winter Precipitation Measurements

Daqing Yang, Antonina Simonenko, and Xiaobo He

Contents

Abstract

Analyses of the long-term (1991–2010) intercomparison data quantify the consistency in winter precipitation observations by six identical Tretyakov gauges at the Valdai research station in Russia. Relative to the standard Tretyakov gauge, the mean catch ratios vary from 97% to 106% for dry snow, 94–104% for wet snow, 87–109% for blowing snow, 96–103% for mixed precipitation, to 98–101% for winter rain. The differences between the highest and lowest mean catches are about 10–11% for snow, 7% for mixed precipitation, and 3% for rain. On average, this difference is about 0.2 mm over the 12-h observation period. The catch difference for blowing

D. Yang (✉)
Watershed Hydrology and Ecology Research Division, Water Science and Technology, Environment and Climate Change Canada, Victoria, Canada
e-mail: daqing.yang@canada.ca

A. Simonenko
State Hydrologic Institute, St. Petersburg, Russia

X. He
State Key Laboratory of Cryosphere Sciences, Cold and Arid Regions Environment and Engineering Research Institute, Chinese Academy of Sciences, Lanzhou, Gansu, China

X. Li, H. Vereecken (eds.), *Observation and Measurement of Ecohydrological Processes*, Ecohydrology, https://doi.org/10.1007/978-3-662-48297-1_5

snow is much higher, up to 22%, or average of 0.6 mm per observation. Comparisons of 12-h observations show a better consistency in gauge performance for the low snowfall events, and a large variation in gauge catch for the high snowfall cases. The differences in 12-h snow catches are mostly less than 2 mm among the 6 gauges. The difference in the 12-h observations is less than 1% for rain and 4% for mixed precipitation. Close linear relationships exist between the 12-h gauge observations for all precipitation types. The maximum differences in gauge snow catches increase very weakly with the wind speed, and higher differences are associated with the warmer temperatures from –5 °C to 0 °C. There is, however, no significant relationship between the max catch difference and mean wind speed or temperature over the 12-h period.

Keywords
Tretyakov gauge · Valdai station · Precipitation · Measurement · Consistency

Background

Systematic and random errors exist in gauge observations of precipitation. Systematic errors include the wind-induced gauge undercatch, evaporation and wetting losses, and trace amount of precipitation (Goodison et al. 1998). Random errors are not easy to define, as they depend on many factors, such as human operation errors and instability of gauge installation and performance. Both random and systematic error affect the accuracy of precipitation measurements; they cause variability and difference in gauge observations and lead to uncertainties in regional precipitation datasets and products, thus affecting climate change analyses, water budget calculations, and calibrations of remote sensing algorithms and land surface models particularly over the cold regions. To quantify the systematic errors in precipitation measurements, intercomparison experiments have been carried out at national and international levels, such as the WMO Solid Precipitation Measurement Intercomparison study during 1986 to 1992 (Goodison et al. 1988, 1998). Many national standard gauges and instruments have been tested during the WMO experiment. The intercomparison data collected during the WMO project are very useful to advance precipitation science and research, such as the evaluation of reference systems for field experiments (Golubev 1989; Yang et al. 1993, 2000; Yang and Simonenko 2013), examination of biases in gauge precipitation measurements (Goodison et al. 1998; Yang et al. 1995, 1998a), development of bias-correction methods for the major national gauges (Goodison et al. 1998; Yang et al. 1995, 1998a, 1999a, b), quantification of wind shield effects on national gauge performance (Yang et al. 1999c), and documentation of incompatibility in national gauge observations (Yang et al. 2001). Applications of the WMO results have produced reliable precipitation data over many countries and large regions (Melcalfe et al. 1993; Yang 1999; Yang et al. 1998b, 1999b, 2005; Yang and Ohata 2001; Zhang et al. 2004; Ye et al. 2004; Adam and Lettenmaier 2003), and these datasets have significantly improved our understanding of cold region climate and hydrology, including regional climate change (Ding et al. 2007), basin water balance (Ye et al. 2012), and large-scale land surface model of the arctic hydrology system (Tian et al. 2007).

It is known that discontinuity exists across the national boundaries owing to the different instruments and observation methods used in the neighboring countries (Sevruk and Klemm 1989; Yang et al. 2001; Sanderson 1975; Nitu and Wong 2011). For instance, the NWS 8-inch gauge is used for precipitation measurements in the US, and the Nipher snow gauge is the standard instrument for snow observations in Canada. Different instruments have also been used at the observational networks within a county. The Type-B rain gauge and Nipher gauge are the standard instruments for rain and snow observations, respectively (Melcalfe and Goodison 1993; Mekis and Vincent 2011); and recently, the Geonor gauges have been installed at the synoptic stations across Canada. Instruments also change over time at most operational networks, resulting in significant breaks in data records. It has been realized that combination of regional precipitation records from different sources may result in inhomogeneous precipitation time series and can lead to incorrect spatial interpretations (Yang et al. 2005). Efforts have been reported to examine the discontinuity across national borders (Sanderson 1975; Yang et al. 2001). However, less is known regarding the inhomogeneity of precipitation records within a country (Groisman et al. 1999). Homogeneous data are essential for studies of climatic fluctuations and changes. At most stations with long-term records, instruments have been altered or relocated, and surrounding buildings and vegetation changed as well. For precipitation measurements, relocation of the gauge is the most frequent reason for inhomogeneities, although instrument changes can introduce abrupt change and discontinuity (break point) in the measurements. Inhomogeneity is an important issue, because most existing national precipitation data and products have been compiled and derived from the combination of various data sources, assuming these data and observations were compatible across the regions and among the observational networks.

Gauge intercomparison experiments mainly compare the various gauges against a reference instrument, so as to quantify the difference between a given gauge and the reference and derive the correction methods to reduce the biases in gauge observations. This chapter examines the repeatability in precipitation observations by the Tretyakov gauge. The Tretyakov gauge is the standard instrument for measuring both solid and liquid precipitation in the former USSR climatological and hydrological networks since the late 1940s (Groisman et al. 1991). The cross-sectional area of the gauge opening is 200 cm^2. At the Russian hydro-meteorological networks, the Tretyakov gauge is placed at 2 m above the ground with a wind shield to improve the catch efficiency (Groisman et al. 1991). Many studies on the Tretyakov gauge have been conducted since the 1960s. Bogdanova (1966) compared the Tretyakov gauge with the pit gauge at about 50 sites in the FUSSR and related the gauge catch of rainfall with storm mean wind speed at the gauge level and rainfall intensity. The Tretyakov gauge was tested during 1972–1976 in the International Rainfall Comparison of National Precipitation Gauges with a reference pit gauge (Sevruk and Hamon 1984). Golubev (1985, 1989) examined various designs of the double fence with this gauge for snowfall measurement at the Valdai Hydrological

Research Station against the so-called Valdai Control System (a shielded Tretyakov gauge located in the sheltered bush site at Valdai), and found that the double fence (gauge) system catches 92–96% of the "true" snowfall (i.e., the Valdai Control System). Based on the experimental data at the Valdai station, Golubev (1985) also developed a relation of the Tretyakov gauge catch of snowfall versus wind speed. Goodison (1981) investigated the Tretyakov gauge catch of snowfall versus snowboard measurements in a sheltered site in Canada and quantified the catch efficiency as a function of wind speed during snowfall period. During the WMO Solid Precipitation Measurement Intercomparison, the Tretyakov gauges were tested against the DFIR reference at 11 stations in seven countries (Goodison et al. 1998; Yang et al. 1995). The catch efficiency of the gauge vs. the DFIR was derived and tested using the WMO intercomparison data (Goodison et al. 1998; Yang et al. 1995). The result has been applied to the historical precipitation records collected by the Tretyakov gauges over large regions, including Siberia (Yang and Ohata 2001), Arctic Ocean (Yang 1999), Mongolia (Zhang et al. 2004), and the Arctic regions as a whole (Yang et al. 2005).

Reducing the known biases in gauge measurements of precipitation is a major challenge particularly for the cold regions. Efforts are ongoing to refine the bias-correction methods for the national standard gauges and to develop new approaches to better observe snowfall with automatic instruments and techniques. The purpose of this chapter is not to address the biases in gauge precipitation measurements (i.e., Yang et al. 1995) or to assess the reference for gauge intercomparison experiment (i.e., Yang and Simonenko 2013); rather it is to investigate the (in)consistency in precipitation observations by the same gauges. This is an important issue for any intercomparison experiments, because it is necessary to determine first how the same gauges will measure precipitation at the test sites. It is assumed that the same gauges will measure similar amounts of precipitation, and that the difference in same gauge observations is usually smaller than that between two different gauges. To test this assumption, determination of the consistency in gauge performance is necessary and useful to accurately evaluate various gauges tested in the intercomparison experiments. In other words, the knowledge of consistency in same gauge observations is critical to decide the acceptable degree of difference (or similarity) in precipitation observations by different gauges.

The objective of this chapter is to quantify and document the difference in winter precipitation observations among the six identical Tretyakov gauges at Valdai in Russia. Specifically, this analysis covers the data period during 1991–2010; it examines the relationship between the six Tretyakov gauges and investigates the major factors contributing to any significant differences among these gauges. It also compares the results with other relevant studies and discusses future needs for similar research and applications. The methods and results of this work will directly contribute to the design and data analysis of gauge intercomparison experiments, including the WMO Solid Precipitation Intercomparison Experiment (SPICE) project for the automatic gauges and instruments (http://www.wmo.int/pages/prog/www/IMOP/intercomparisons/SPICE/SPICE.html). They will also support climate and hydrology research, particularly precipitation changes, snow cover processes, and streamflow modeling over the cold regions.

Design of the Experiment

The Valdai Hydrological Research Station is situated in the northwestern part of the East European Plain. It is in the middle between Moscow and St. Petersburg. The station (57 59′N, 33 15′E, 194 m above the sea level) is located on the flat shore of Valdai lake in the center of the Valdai Hills. The Valdai Hills are characterized by a hilly moraine landscape where hills alternate with depressions. Predominant elevations are about 200 m above sea level. The area studied in this chapter is in the southern taiga zone mainly with spruce and deciduous forests and pine woods. The forests and bushes occupy 75% and 20% of the region, respectively. The climate near Valdai is mainly continental. Mean annual air temperature is about 3.4 °C, with July mean of 16.4 °C and January average of −9.7 °C. Total annual precipitation is about 800 mm, with annual precipitation days of 207. A stable snow cover is usually observed from late November to mid-April, with the peak accumulation (about 70 cm) in early March. Within the Valdai Hills, the southwestern winds prevail. Mean annual wind velocity is about 4 m/s. The maximum wind speeds (4.0–4.5 m/s) are observed during the winter months.

The Valdai Station has an open/meadow area and a bush site. The meadow site is about 100 × 100 m in area and 25 m from the latticed fence. This site is flat and about 1.5 m above the lake water level during the low-water season. The grass surface during the warm period is regularly mowed. The average vertical angle of obstacles is 1.4°. The instruments on the site are exposed to winds from all directions. The Valdai Station has a long history for testing various meteorological instruments. The site was modified in 1988 for the WMO Solid Precipitation Measurements Intercomparison. Eighteen gauges from six countries were installed during 1988–1995. Some instruments remained during the change in 1988 so as to ensure observation homogeneity and reliability; these instruments included the Valdai control system in the bush, and six Tretyakov gauges located at the corners and in the center of the meadow site (Fig. 1). The gauges in the center of the plot are about 20 m apart, and the four gauges at the corners were 90 m apart. Approximately 300 m northwest of the open site is the bush site, where 2–4 m high shrubs occupy a three-hectare area. There is a fenced area of 70 × 70 m, where the bushes are pruned (in autumn) at height of 2 m above the ground. The mean bush density is about 4 stems/m^2, and the mean diameter of the shrub tops (2 m above the ground) is approximately 25–50 cm. At the center of the site sit the bush gauges (Tretyakov gauge with a wind shield).

Precipitation measurements at the Valdai Station were conducted generally twice daily: at 9:00 and 21:00 of Moscow standard time. The contents of the gauges were both weighed and measured volumetrically with the resolution of 0.1 mm to determine precipitation amount, and over a period of time, an average wetting loss was determined by comparing the difference between the weighed and volumetric measurements. Since 1966, a correction for wetting loss of the Tretyakov gauge (i.e., 0.11 mm/event for rain and 0.06 mm/event for snow) has been added to every volumetric measurement, and therefore no additional correction for this systematic loss is required (Golubev 1989). Wind speed and direction were measured at 2 and

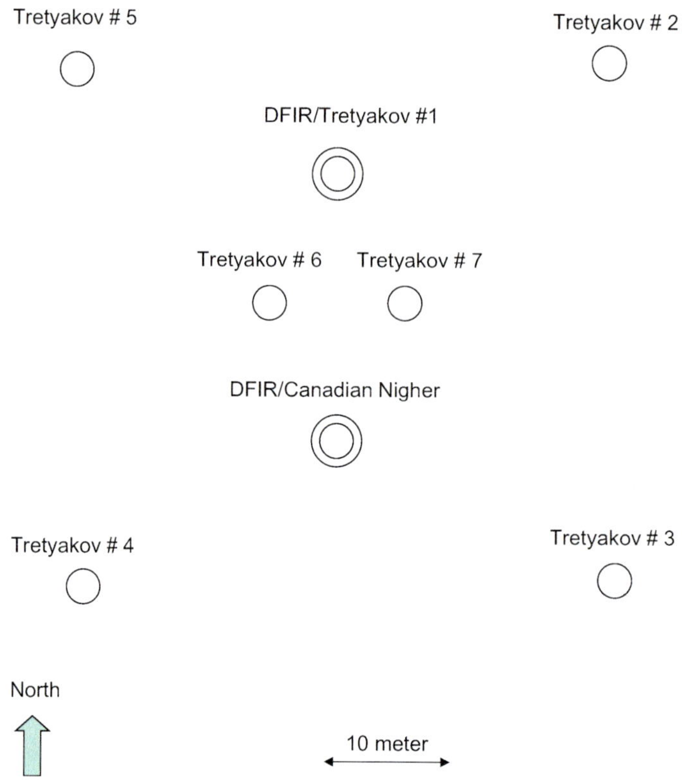

Fig. 1 Valdai meadow site layout and precipitation gauges

3 m heights in the center of the meadow site. Atmospheric pressure, air temperature, and humidity were also measured at the Valdai operational meteorological station located 0.9 km southwestward of the experimental site.

During the data collections, precipitation types were classified by the observers at the times of observations as dry snow, wet snow, mixed precipitation, and rain. Drifting or blowing snow events were also manually identified and reported by the observers. This chapter focuses on the analyses of winter (October to April) season precipitation data, i.e., snow, mixed precipitation, and winter rain. Statistical analyses of long-term 12-h precipitation data measured by the six Tretyakov gauges were carried out. The methods used include calculation of total precipitation amounts during the study period for various precipitation types, determinations of mean wind speed and air temperature on precipitation days, and regression and correlation analyses of precipitation, wind speed, and temperature data. The statistical tools used in this chapter have been recommended and tested in the previous WMO gauge intercomparison. The consistency in methodology is important as it ensures that the results from this work are comparable with those from the last WMO project and other relevant studies. This work also discusses and documents the magnitudes of systematic and random errors in precipitation measurements by other national standard gauges.

Gauge Catch Difference and Wind/Temperature Effect

Based on the data analyses, the results for the mean gauge catch, comparisons of 12-h gauge measurements, and the effects of wind speed and temperature on gauge catch differences are presented and discussed below.

Mean Difference Among the Gauge Measurements

To determine the mean differences among the gauge observations, the total precipitation amounts and mean temperature and wind speed over the long-term period during 1991–2010 were calculated. The gauge observations to the standard Tretyakov gauge (gauge #7 – the working reference at this station) for dry snow, wet snow, blowing snow, mixed precipitation, and rain were compared. The results below were presented by precipitation types (Table 1).

There were 640 events of dry snow reported for the study period. The total snowfall ranged from 992 to 1,019 mm among the six gauges; these values are systematically less than the DFIR and Bush gauge observations due to Tretyakov gauge undercatch of snowfall (Golubev 1988; Goodison et al. 1998; Yang et al. 1995). Mean temperature and wind speed at 2 m during these events were about –6 °C and 3.7 m/s. Relative to the standard gauge (#7), the other Tretyakov gauges recorded 97–106% of the total snowfall. The difference between the highest and lowest total snowfall (measured by gauges #3 and #5, respectively) is about 11%, or 128 mm for the 640 events, and average of 0.2 mm per observation. This difference is quite significant, since the six gauges were identical; they were located 10–50 m apart at an open site and were emptied regularly at the same time in the same way according to the Russian standard observation procedure (Groisman et al. 1991).

There were 506 wet snow events during the study period; the mean temperature and wind speed were –3 °C and 3.7 m/s, respectively. The six gauges recorded total snowfall from 685 to 768 mm. Relative to the gauge # 7, the mean catches of other Tretyakov gauges ranged from 94% to 104%. The difference between the lowest and highest totals (i.e., gauges # 4 and 5, respectively) is 10%, or 79 mm for all the events, average of 0.2 mm per observation. These results are similar to the dry snow data. It is important to note that wind conditions (i.e., long-term mean wind speeds) were similar for both wet and dry snow cases at this site. This may lead to some degree of consistency in gauge catch of snowfall in this region, although the mean temperature for the dry snow is lower than that for the wet snow.

In addition to the dry and wet snow data, there are 79 blowing snow cases reported at Valdai during the study period. The mean temperature and wind speed for these events were –5 °C and 5.2 m/s at 2 m height, respectively. The total accumulation ranged from 179 to 226 mm among the six gauges. Relative to

Table 1 Summary of all 12-h precipitation observations by various gauges at the Valdai experimental station during 1991 to 2010

	No. Event	Tmax (c)	Tmin (c)	Tmn (c)	Wdir (Dgr)	W3m (m/s)	W2m (m/s)	DFIR (mm)	Bush/Tret (mm)	Tret7 (mm) (%)	Tret2 (mm) (%)	Tret3 (mm) (%)	Tret4 (mm) (%)	Tret5 (mm) (%)	Tret6 (mm) (%)
Dry snow															
mean/total	640	−3.8	−8.1	−5.9	200.8	3.8	3.5	1378.4	1444.5	1019.9	1045.3	992.1	995.3	1080.5	1009.7
RCR										100%	102%	97%	98%	106%	99%
Wet snow															
mean/total	506	−1.1	−4.9	−3.0	226.9	3.9	3.7	952.3	1012.3	726.0	724.5	684.5	679.7	758.2	706.5
RCR										100%	100%	94%	94%	104%	97%
Mixed P															
mean/total	323	3.2	−0.8	1.2	218.8	4.4	4.1	1476.0	1554.1	1315.7	1325.5	1276.6	1263.6	1355.3	1290.6
RCR										100%	101%	97%	96%	103%	98%
Blowing snow															
mean/total	79	−2.6	−7.6	−5.1	216.7	5.7	5.2	338.2	379.6	207.3	205.6	182.4	179.4	225.8	196.0
RCR										100%	99%	88%	87%	109%	95%
Rain															
mean/total	500	9.1	4.7	7.5	215.3	4.0	3.6	1786.4	1872.7	1729.0	1730.1	1716.7	1694.2	1753.0	1722.4
RCR										100%	100%	99%	98%	101%	100%

Notes:

W3m wind speed at 3m height; *W2m* wind speed at 2m height; *bush/tret* Tretyakov gauge in bush; *RCR* relative catch ratio to Tret #7 the reference gauge #7, the mean catches of the gauges varied from 87% to 109%. The difference between the highest (gauge# 5) and lowest (gauge # 4) totals is about 47 mm, or 22%, averaging of 0.6 mm per observation. This difference is roughly twice as high as that for both dry and wet snow cases, clearly suggesting potential blowing snow impact to gauge performance at this site.

There were 323 mixed precipitation events, with the mean temperature and wind speed being 1.2 °C and 4.1 m/s at 2 m. Total precipitation for all the events ranged from 1,263 to 1,355 mm among the six gauges. Relative to the standard gauge #7, the mean catch was between 96% (gauge #4) and 103% (gauge# 5). The difference between the highest and lowest totals was 92 mm, or averaging of 0.3 mm per observation. This value is similar to the snow data, indicating some consistency in gauge catch of snow and mixed precipitation at this location.

Winter rainfall data were also collected at this site during October to April. In the winter season, the Tretyakov gauges at Valdai were used without a funnel; this is a standard configuration for snow observations at the Russian networks. In the other seasons, a funnel is installed in the gauge for rainfall collections. There are notes in the Valdai data sheets about the dates of installing/removing the funnel in spring and fall seasons. The wetting loss may be a bit higher for the gauge without a funnel. During the study period, 500 rainfall events were registered by the six gauges in the winter season. These events were between 0.1 and 30 mm for the 12-h period. The mean temperature and wind speed were 7.5 °C and 3.6 m/s, respectively. The total rainfall amounts ranged from 1,694 to 1,753 mm among the six gauges. Relative to the standard gauge (#7), the mean catches varied from 98% to 101%. As expected, the difference in gauge catch (about 3% between the highest and lowest catches) is much smaller than the snow cases, partly due to smaller wind effect and gauge undercatch of liquid precipitation (Goodison et al. 1998; Yang et al. 1995).

To determine the consistency in gauge catch and performance, the six gauges were ranked by total accumulation for various precipitation types, including blowing snow. A clear north-south gradient was found over the site for all precipitation types. This means the two gauges (#2 and #5) on the north site of the plot collected the highest precipitation, and the gauge pair (# 3 and #4) in the south side measured the lowest amount of precipitation. Previous data analysis for this site suggested an even distribution of yearly precipitation over the experiment area, although predominant winds were from the southwest (Golubev et al. 1989). Our results, however, demonstrate a noticeable difference in gauge measurements across the site. Site location and selection are important for instrument intercomparison and test. Strangeways (2009) used the Google Earth to assess several GCOS stations over the UK. Examination of a (May 2009) Google Earth image for the Valdai site and its surroundings reveals a patch of trees on the west side of plot for the most common wind direction. These trees may alter wind flows around the instruments and affect gauge catch of precipitation. The effect of the nearby lake to the site and instruments also needs attention. Future investigation is necessary to consider the influence of micrometeorology, particularly wind distribution and turbulence across this site, on gauge observations.

Comparison of 12-h Observations

Many studies show that wind speed and temperature affect gauge catch of precipitation (Yang et al. 1998a, 1999a; Goodison et al. 1998). Precipitation amount may also play a role in determining gauge performance, particularly for rainfall measurements (Nešpor and Sevruk 1999; Sugiura et al. 2006). To understand the variability in gauge catch, the gauge observations at the 12-h time interval were compared for various precipitation types.

Random errors exist in precipitation gauge observations. To reduce the impact of random errors, the mean precipitation amounts for the six gauges at the 12-h time interval were calculated, so as to identify the lowest and highest snow measurements by the gauges for 12 h. Figure 2 shows the mean, maximum, and minimum snowfall values for the six gauges. The data presented in Fig. 2 are not the time series of observations, but rearranged (or sorted) by the mean precipitation amount, i.e., from the lowest to the highest values for all the snowfall events. It is clear that most dry snow events were less than 2 mm for the 12-h period, and only about 80 cases (about 10%) were greater than 4 mm. The range of measurements by the gauges (i.e., the difference between highest and lowest values) was less than 1 mm for most low snowfall events, and it increased to 2–4 mm for the high snowfall events. This result suggests a better consistency of gauge performance for the low snowfall events, and a large variation in gauge catch for the high snowfall cases. Wet snow data generally demonstrate very similar result, although the difference in gauge catch was smaller relative to the dry snow data.

To reveal the difference in gauge catch, the 12-h data between the gauges were examined. Figure 3 shows the comparisons of dry snow data measured by the six Tretyakov gauges in three pairs. Gauges # 2 and 3 are located about 50 m apart at the east corners of the plot. The 12-h snow data are very similar between these two gauges, with the differences being less than 2 mm for most events, except for a few cases of higher snowfall up to 10 mm. There exists a close linear relationship between the 12-h observations. This relationship can be used as a transfer function between the gauge observations. On average, gauge #3 measures 7% less snow than gauge #2. The gauge pair of #4 and # 5 was located on the west corners of the plots (about 90 m apart); on average, gauge # 5 measured about 8% more snowfall than gauge #4. On the 12-h basis, gauge # 5 systematically collected more snowfall, particularly for the precipitation range from 2 to 8 mm. For a few cases, the differences were about 2–3 mm for the 12-h period. Gauges # 6 and #7 were installed in the center of the plot, about 20 m apart. Both gauges caught similar amount of snowfall. On average, they measure the same amount of snowfall for the 640 cases. But the difference for the individual observations between these two gauges is big, sometimes up to 4 mm for 12 h, which is higher than that for the other gauge pairs. This result is not expected, as these 2 gauges are located very close to each other.

Figure 4 compares the wet snow data collected by the three pairs of the Tretyakov gauges. For the gauge pair of #2 and #3, the 12-h snowfall ranged from trace amount to 14 mm. For most cases, both gauges measured similar amounts of wet snow, with the differences being less than 2 mm. Overall, gauge #3 collected 6% less snow than the gauge #2, although the linear correlation between the two gauges is very high.

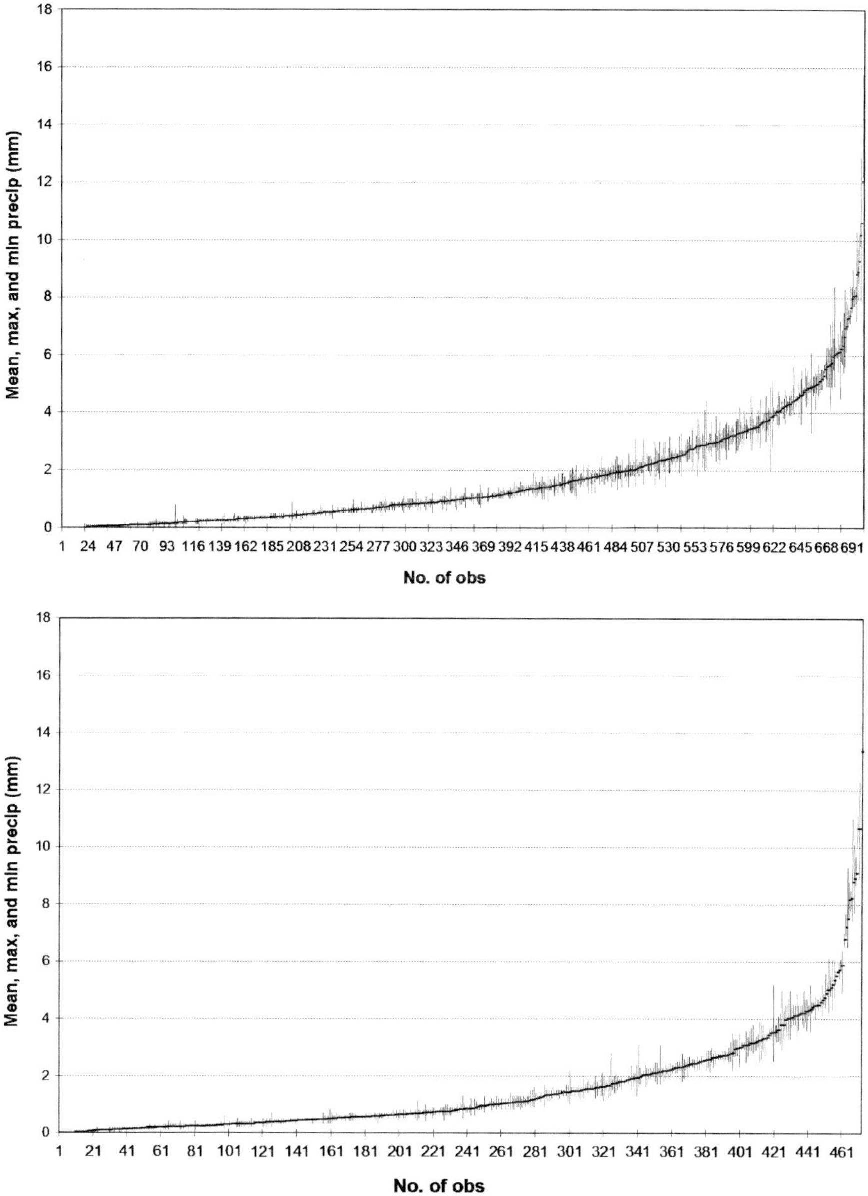

Fig. 2 Mean, maximum, and minimum snowfall measurements by the 6 Tretyakov gauges

This result is very similar to the dry snow case. For the gauge pair of #4 and #5, both gauges measured the similar amounts of snowfall in most cases, with the difference being less than 2 mm, except for a few outliers with the differences greater than 3 mm. On average, gauge #5 measured 8% more snow than gauge #4, although a

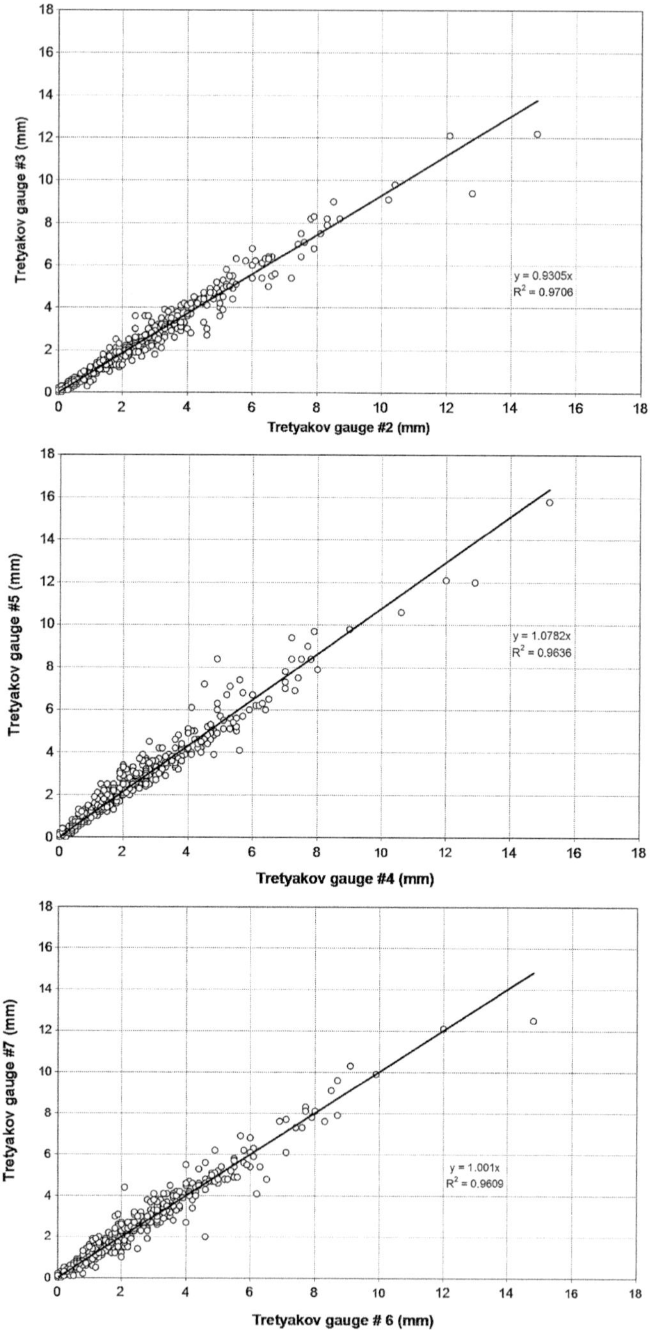

Fig. 3 Comparison of 12-h dry snow data among the 6 Tretyakov gauges

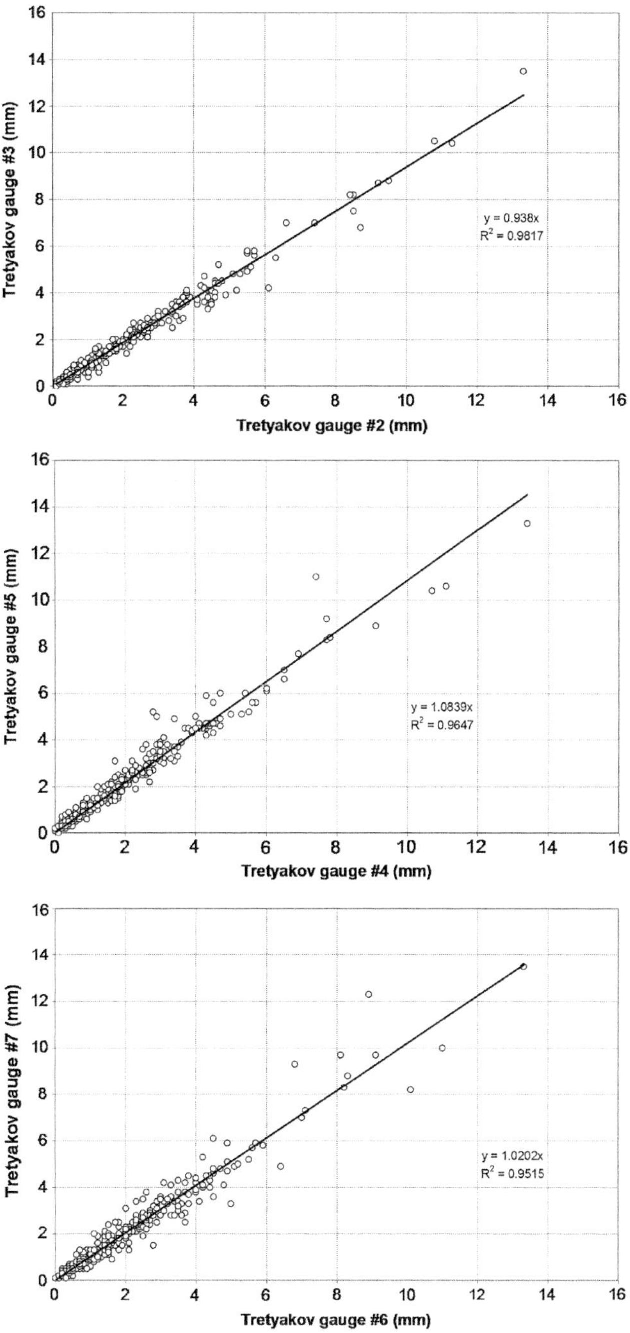

Fig. 4 Comparison of 12-h wet snow data among the 6 Tretyakov gauges

close linear relationship exists. This result is similar to the dry snow data. For the gauge pair #6 and #7, there were many events of less than 6 mm for the 12-h period. Both gauges reported similar amounts of snowfall. For snowfall cases ranging from 6 to 14 mm, the scatter is large with the highest difference of more than 3–4 mm for the 12-h period. Overall, gauge #7 measured 2% more snow than gauge #6. This difference is not very significant, although it is slightly different from that for the dry snow data, when the two gauges, on average, collected same amount of snowfall.

Figure 5 shows the 12-h blowing snow data collected by the six Tretyakov gauges. The data include both dry and wet snow events, but they are hardly distinguishable in terms of both catch difference and its scatter. There are systematic differences among the gauge catch. Gauge #3 measured less snow than gauge #2 for most blowing snow events (including wet and dry snow); on average, it reported 12% less snow than gauge #2. Gauge #5 caught more snow than gauge #4 by 26%, while gauge #7 measured more snow than gauge #6 by about 5% for all blowing snow cases. It is important to point out the consistency in gauge catch between blowing snow and non-blowing snow events. For instance, gauge #2 caught more snow than gauge #3 for both blowing snow and non-blowing snow cases, although the difference in gauge measurements for blowing snow events is much higher, i.e., 12% for the blowing snow vs. 6–7% for dry/wet snow. It is also interesting to note a very small catch difference between the gauges #6 and #7 for snow and blowing snow data, perhaps because these two gauges are situated about 10 m apart in the center of the plot and expose to similar wind and snow conditions.

Figure 6 displays the comparisons of the mixed and rain data during the study period. Mixed precipitation at Valdai ranged from trace amount to 26 mm over the 12-h period. For most measurable events, both gauges #2 and #3 reported similar amount of precipitation, with the differences being less than 1 mm. Overall, gauge #3 collected 4% less than gauge #2, although the correlation between the gauge measurements is very high. This result is very similar to the wet snow data. Rainfall data suggest that Tretyakov gauges #2 and #3 caught very similar amounts of precipitation for the 12-h period. There is a very close linear relationship between the gauge measurements, with the overall difference being less than 1%. The other gauge pairs have similar results to gauges #2 and #3 for both mixed precipitation and rain, respectively.

Maximum Catch Difference vs. Temperature and Wind Speed

The six Tretyakov gauges measure differently for most 12-h individual precipitation events. To understand the gauge performance for all precipitation measurements, the minimum and maximum values for each precipitation observations were identified, so as to calculate the difference between these high and low values and define the maximum catch difference among the six gauges. Examination of the maximum difference vs. temperature and wind speed during the 12-h observation was done. This analysis allows us to identify the factors controlling the variation in gauge observations of snowfall, including blowing snow conditions.

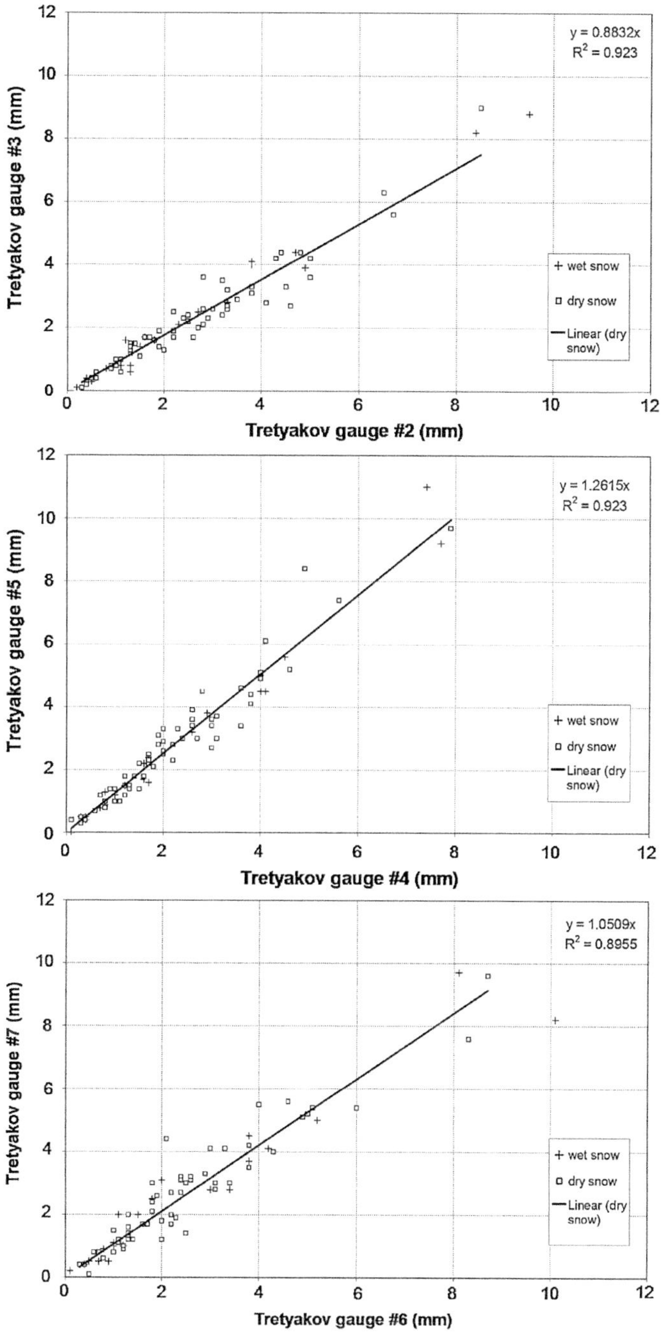

Fig. 5 Comparison of 12-h blowing snow data among the 6 Tretyakov gauges

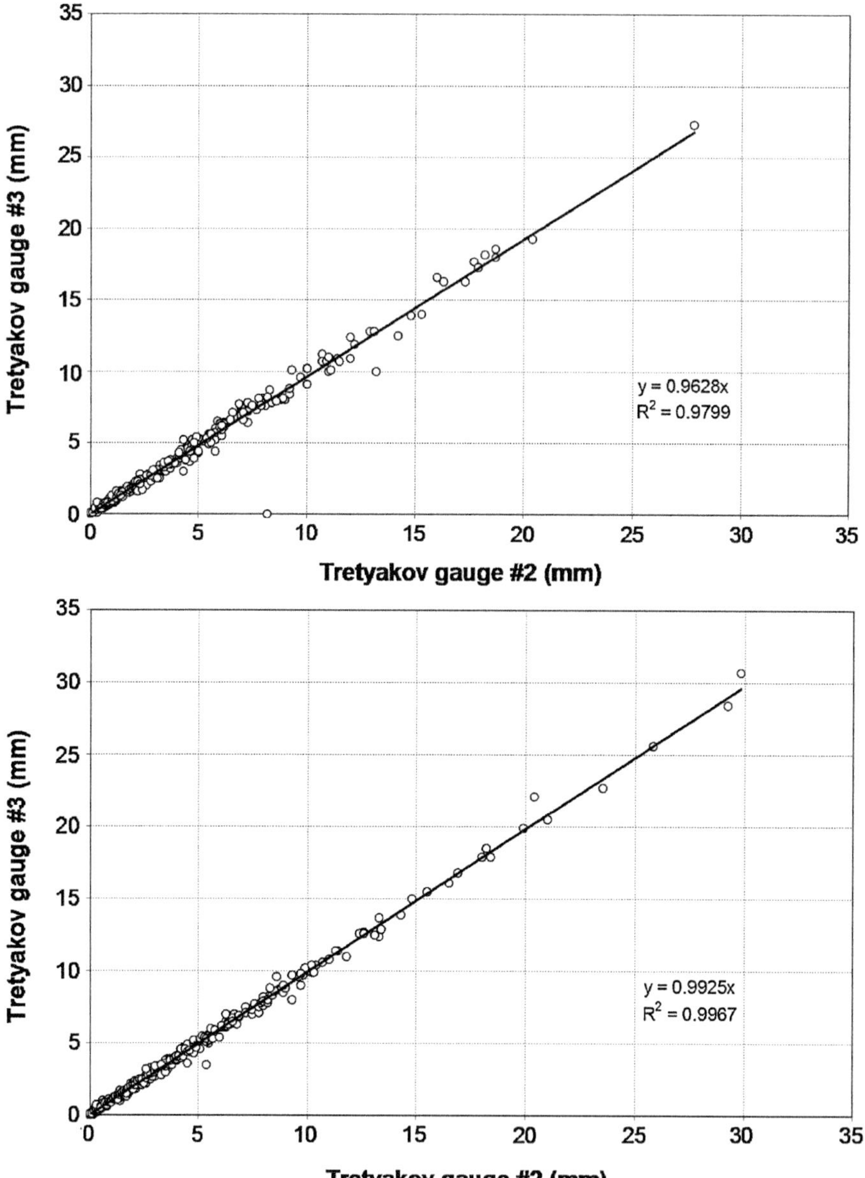

Fig. 6 Comparison of 12-h observations between 2 Tretyakov gauges for mixed precipitation (top) and winter rain (bottom)

Figure 7a shows the scatter plot of the maximum difference in gauge measurements vs. wind speed at 2 m. The range of the max difference varies from less than 1–3 mm for the 12-h period; it seems to rise very weakly with the wind speed. For instance, for higher wind speeds up to 5–8 m/s, there are more data points with higher differences among the six gauges, although the relationship between wind speed and max difference is not significant. This result is reasonable, since many studies show that wind speed is the most important factor to gauge catch efficiency of snowfall (Goodison et al. 1998; Yang et al. 1993, 1998a). The six Tretyakov gauges are installed over an area of 150x150 m; it is possible that they may experience different wind conditions during the snowfall periods. Figure 7b displays temperature vs. the max catch difference among the six gauges. For temperatures range from −20 °C to +5 °C, the max differences vary greatly. As such, there is almost no relationship between these two variables, although bigger differences seem to be associated with warmer temperatures from −5 °C to 0 °C. This association may be related with wet snow and snow sticking on the rimes of the gauges.

Wind speeds during the wet snow events were from 0.5 to 7.5 m/s. Most max differences were less than 1 mm, with some ranging from 1 to 2 mm, and only four cases between 2 and 3 mm (Fig. 8a). There is a weak tendency of larger catch difference associated with the higher wind speeds. This is very similar to the result for the dry snow data. Temperatures fluctuated from −16 °C to 4 °C for the 12-h events (Fig. 8b). The max difference did not change much with temperature, although the difference in gauge catch is higher (up to 3–4 mm) for temperatures between −5 °C and 0 °C. This result is very similar to the dry snow.

For the blowing snow events, wind speeds were from 3 to 8 m/s at the 2 m height and the corresponding max difference in gauge catch was less than 2.5 mm for most events, except one outlier of 2.8 mm with the highest wind speed of 8 m/s (Fig. 9a). It is interesting to note that the wind speed range for the blowing snow events is similar to that for the non-blowing snow cases. There is no clear relationship between the gauge catch difference and wind speed. Temperatures during blowing snow events ranged from −18 °C to 1 °C, the max catch difference varied widely, particularly for temperatures range from −10 °C to 0 °C. Similar to the snow data, there is a weak tendency of higher catch difference associated with the warmer temperatures during the blowing snow (Fig. 9b).

Uncertainties in Gauge Intercomparison

There are uncertainties in data collections and analyses for precipitation gauge intercomparison experiments. These include, for instance, determination of precipitation types, observations, and calculations of mean wind speed and temperature for a given time interval when precipitation was observed. At Valdai, the observers classified precipitation types at the time of the observations (2–4 times a day). Some misclassifications are likely particularly for the mixed precipitation and blowing snow events. Air temperature and humidity are useful to estimate precipitation types (Legates and Bolgart 2009). Yang et al. (1999, 2005) used daily air temperature to determine precipitation types when this information is not available for the northern regions. In this chapter, wind speed and air temperature are the 12-h means; they do not accurately represent the weather condition during the storm. The use of such mean wind speed may lead to uncertainties in gauge comparisons. Data collections

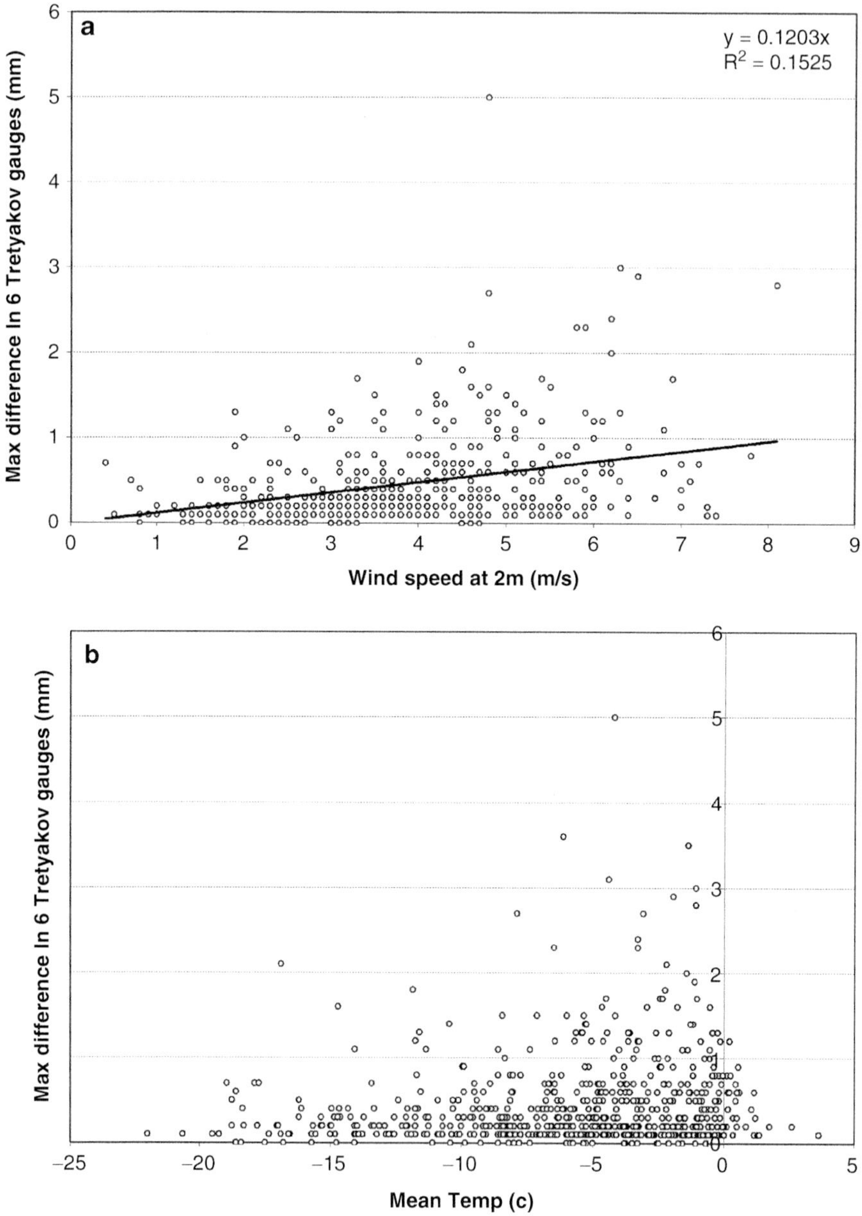

Fig. 7 Scatter plots of maximum catch difference vs. wind speed (**a**) and temperature (**b**) for dry snow

and analyses on shorter timescales, such as hourly or 6-hourly, is expected to produce more reliable results, since wind speeds may vary throughout the day and 12-hourly mean wind speeds may not be representative of wind conditions over the precipitation periods. Automatic sensors will also be important to detect precipitation types at operational and research networks.

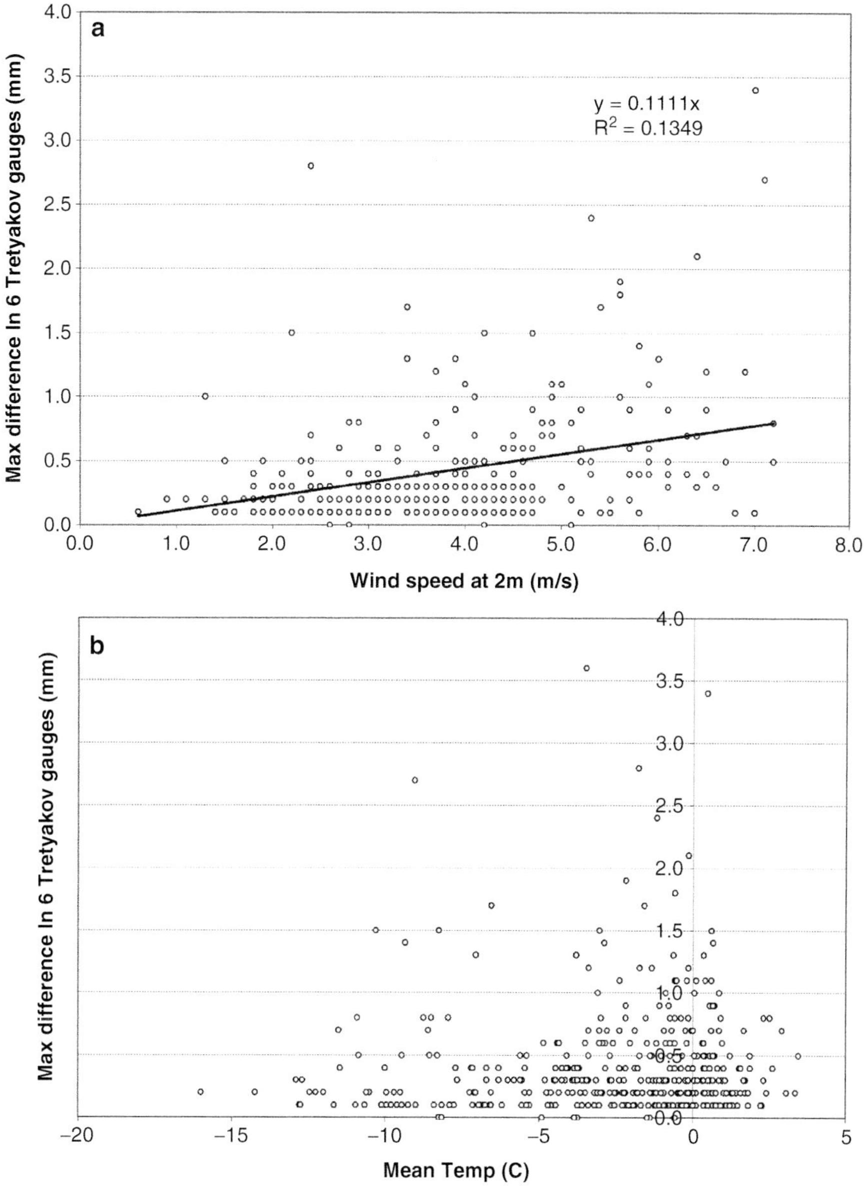

Fig. 8 Scatter plots of maximum catch difference vs. wind speed (**a**) and temperature (**b**) for wet snow

As recommended by the past WMO gauge intercomparison (Goodison et al. 1998), the identification and separation of blowing snow are necessary, because blowing snow conditions are a special case when assessing gauge performance. It is a challenge to quantify the effects of blowing snow on quality of snowfall measurements due to lack of necessary information. Blowing snow fluxes collected by precipitation gauges are called false precipitation (Golubev 1998). The amount of

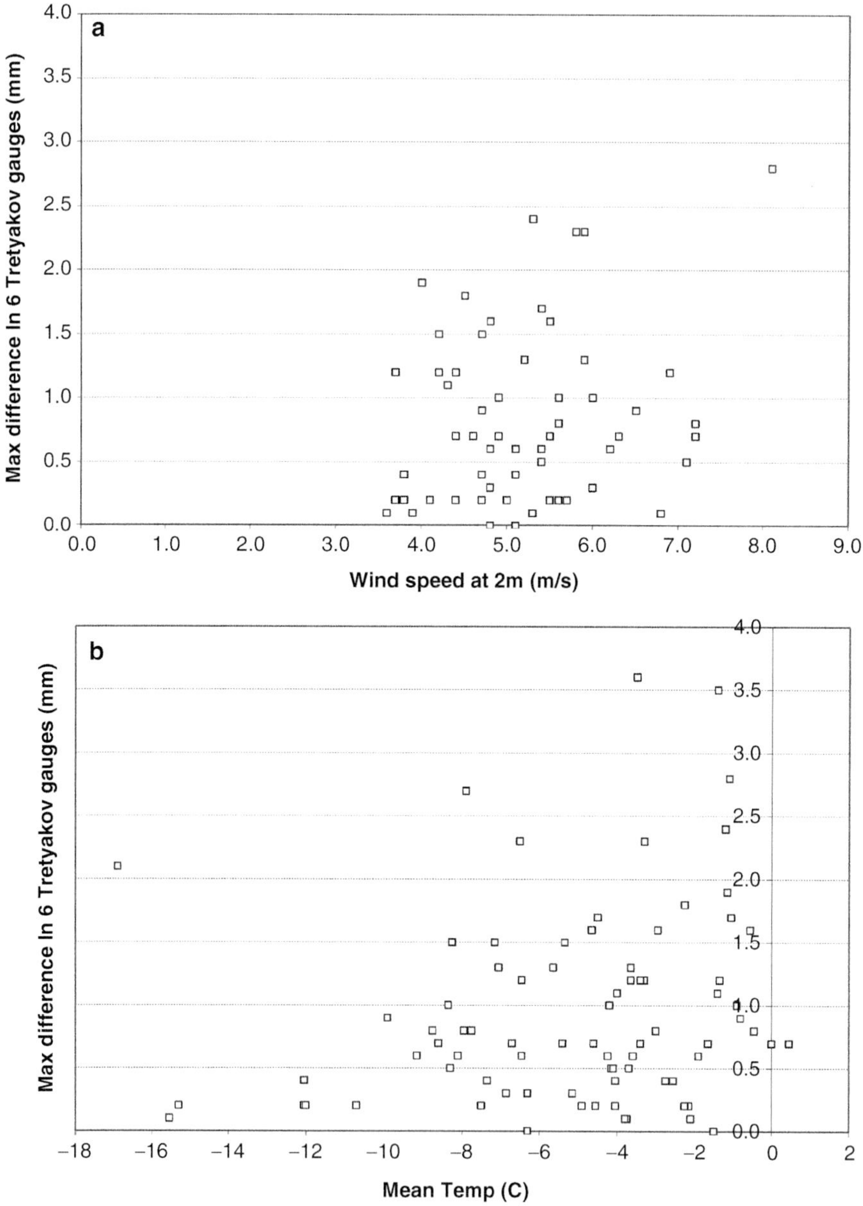

Fig. 9 Scatter plots of maximum catch difference vs. wind speed (**a**) and temperature (**b**) for blowing snow

false precipitation is proportional to the intensity of blowing snow and its duration. Based on field observations at a windy alpine location in the Colorado Front Range, Bardsley and Williams (1997) reported that blowing snow events often occur after the storms at high wind speeds over 20 m/s and may introduce 50% overcatch over a winter season. Yang and Ohata (2001) found an association of higher wind speeds with higher snow measurement by the Tretyakov gauges at the windy and cold Tiksi and Dekson stations in northern Siberia coast perhaps due to blowing snow into the gauges. Many blowing snow events were recorded during the intercomparison experiment at Valdai. The occurrence of blowing snow events was reported for an observation period at Valdai (i.e., 12 h). However, information of blowing snow duration and intensity are critical but mostly unavailable to determine blowing snow flux and its impact to gauge observations in the cold regions (Sugiura et al. 2006, 2009). Blowing snow generally occurs near Valdai station at mean wind speeds of 3–8 m/s at 2 m height (Fig. 5). During blowing snow events, the difference in the six gauge catches is generally higher than that for the non-blowing snow cases. Because of the uncertainty in gauge performance in high wind conditions, it is difficult to assess which Tretyakov gauges at Valdai would have measured snowfall better than the other counterparts. More data collection and analysis of snowfall in higher winds, including blowing snow events, are necessary with automatic instruments at this site and over other northern locations.

Automatic precipitation gauges have been used in the operational networks over many nations. The transformation from the manual to automatic observation systems will have a major impact to climate monitoring, including climate change investigations. It is thus useful to relate and compare the manual approach with automatic technique for precipitation observations (Groisman et al. 1999). For instance, the Belfort precipitation gauges have been widely used in many regions, and they have been tested at five sites during the past WMO gauge intercomparison project (Goodison et al. 1998; Yang et al. 2001). It is possible therefore to analyze the intercomparison data to study the consistency of the Belfort gauge observations of precipitation. Other ongoing efforts in USA, Canada, and through the WMO (Rasmussen et al. 2012) have also tested numerous automatic instruments for snowfall observations in various climate regimes. The data collections and analyses in these projects will improve our capability to better measure snowfall in the cold regions.

It is known that most national standard gauges, including the Russian Tretyakov, Canadian Nipher, and US 8-in. gauges, under measure precipitation especially for snowfall (Goodison 1981; Goodison et al. 1998; Yang et al. 1995, 1998a, 1999a). Compatibility analysis of precipitation measurements by various national gauges suggests little difference (less than 5%) for rainfall observations, but a significant discrepancy (up to 110%) for snowfall measurements (Yang et al. 2001). For instance, many national gauges have been tested at the Valdai station for many years; the experimental data demonstrated that the Canadian Nipher gauge caught, on average, more than 13% snowfall than the Tretyakov gauges (Fig. 10). The U.S. 8-inch gauge at Valdai systematically measured 30–50% less snow and mixed precipitation than the Canadian Nipher gauge (Yang et al. 2001). This difference in national gauge catch has introduced a significant discontinuity in precipitation records between the US and Canada borders particularly in windy and cold regions. It is clear that the catch difference among the national gauges is much higher than that among the six Tretyakov gauges (i.e., 5–10% for snow and 4–8% for mixed precipitation) at Valdai; this result suggests the systematic biases

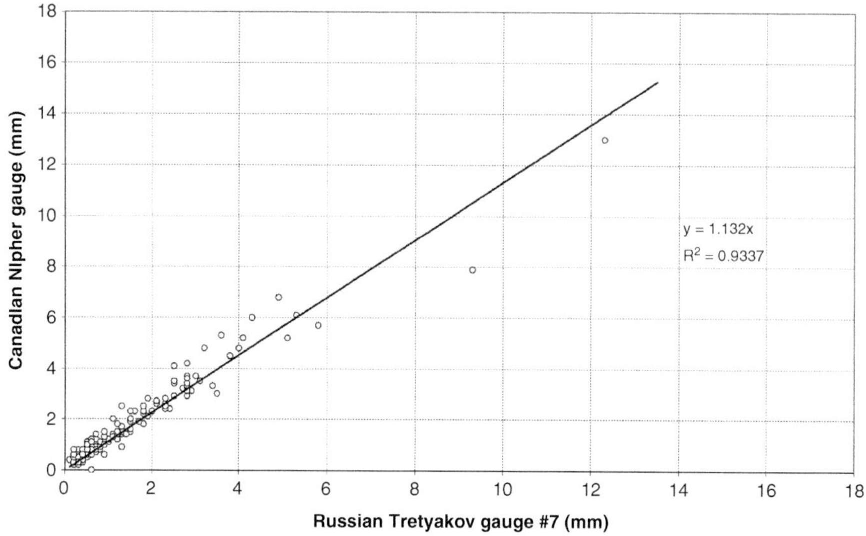

Fig. 10 Comparison of 12-h snow data between the Canadian Nipher gauge and Russian Tretyakov gauge at Valdai

and differences in gauge observations are quite high, and they deserve more research attention. However, documentation and quantification of the catch difference for a given national gauge are important and useful to determine the homogeneity of precipitation data collected by a standard gauge within the national and regional networks.

Summary

Analyses of the long-term intercomparison data define the mean catch of the six Tretyakov gauges for various precipitation types, including blowing snow. Relative to the standard Tretyakov gauge (#7), the mean gauge catch ratios vary from 97–106% for dry snow, 94–104% for wet snow, 87–109% for blowing snow, 96–103% for mixed precipitation, to 98–101% for winter rain. The differences between the highest and lowest mean gauge catches are about 10–11% for snow, 7% for mixed precipitation, and 3% for rain; On average, this difference is about 0.2 mm over the 12-h observation period. The catch difference for blowing snow is, however, much higher, up to 22%, or average of 0.6 mm per observation. It is likely that blowing snow impacts gauge catch and performance at Valdai. Calculations of total accumulation demonstrate a clear south to north gradient for all precipitation types, with the two gauges on the south (north) side of the plot collecting the lowest (highest) amounts of precipitation. Trees and the lake near the site are likely to affect wind regime and gauge catch difference. Our effort continues to examine these factors, mainly via the WMO SPICE experiment at this site, with better wind and snow data collections analyses.

Comparisons of 12-h observations show a better consistency in gauge performance for the low snowfall events and a large variation in gauge catch for the high snowfall cases. There are similarities in gauge catch among the three pairs of gauges and close linear relationships between the 12-h observations. For dry snow, the catch differences are mostly less than 2 mm between the gauges #2 and #3, 2–5 mm for gauges #4 and #5, and 3–4 mm for gauges #6 and #7. The differences in the 12-h observations are much higher for blowing snow (12% between gauges #2 and #3, and 26% for gauges #4 and #5), except for gauges #6 and #7 with 2% (similar to non-blowing snow). The six gauges generally report very similar amounts of 12-h precipitation for both mixed phase and rain, with the difference being less than 1 mm between gauges #2 and #3 for most cases, or less than 1% for rain and 4% for mixed precipitation, respectively. The maximum differences in 12-h gauge snow measurements increase very weakly with the wind speed, and higher differences seem to be associated with warmer temperatures from −5 °C to 0 °C. There is, however, almost no significant relationship between the maximum catch difference and wind speed or temperature over the 12-h period. The effects of wind speed and temperature to gauge catch difference for blowing snow event were weak, and more data analyses are needed to better understand gauge performance in high wind conditions.

A recent WMO survey indicates a large variety of automatic gauges currently used worldwide, including in the same country, for routine precipitation measurements at the national networks (Nitu and Wong 2010). These gauges differ in the measuring system, orifice area, capacity, sensitivity, and configuration. The variety in automatic gauges is much greater relative to the manual standard gauges (Severuk and Kemm 1989; Goodison et al. 1998). The extensive use of different instruments and configurations significantly impacts the accuracy and consistency of regional and global precipitation time series. It is very clear from this analysis and many other studies that field experiments are critical to address the issues of precipitation data accuracy and consistency; they are essential to evaluate national standard and automatic gauges, including quantifications of random and systematic errors in precipitation observations and their possible relationship with meteorological factors. This chapter is only possible thanks to the long-term data collection at the Valdai station. It is necessary to expand this work to other national standard gauges, such as the US NWS 8-in. gauge widely used in many nations and regions. It will be also useful to examine the automatic gauge data and their consistency at the shorter time periods, i.e., hourly or sub-hourly, as the dynamic and climatology of precipitation may differ at shorter time scales. The WMO SPICE project, currently collecting and analyzing gauge intercomparison data at many test sites around the globe, provide much needed new opportunities to improve snowfall precipitation observation and analysis techniques (Kochendorfer et al. 2017).

References

J. Adam, D.P. Lettenmaier, Adjustment of global gridded precipitation for systematic bias. J. Geophys. Res. **108**(D9), 4257 (2003). https://doi.org/10.1029/2002JD002499

T. Bardsley, M.W. Williams, Overcollection of sold precipitation by a standard precipitation gauge, Niwot Ridge, Colorado. in *Proceedings of Western Snow Conference,* Canada, Western Snow Conference 1997, pp. 354–362

E.G. Bogdanova, Investigation of precipitation measurement loss due to wind. Trans. Main. Geophys. Obs. **195**, 40–62 (1966)

Y. Ding, D. Yang, B. Ye, N. Wang, Effects of bias correction on precipitation trend over China. J. Geophys. Res. **112**(D13), D13116 (2007). https://doi.org/10.1029/2006JD007938

V.S. Golubev, On the problem of actual precipitation measurements at the observations site, in *Proceeding of the International Workshop on the Correction of Precipitation measurements* W MO/TD 104, 1985, pp. 61–64

V.S. Golubev, Assessment of accuracy characteristics of the reference precipitation gauge with a double-fence shelter. in *Final Report of the Fourth Session of the International Organizing Committee for the WMO Solid Precipitation Measurement Intercomparison* (St.Moritz, WMO, 1989), pp. 34–41

B.E. Goodison, Compatibility of Canadian snowfall and snowcover data. Water Resour. Res. **17**, 893–900 (1981)

B.E. Goodison, S. Klemm, B. Sevruk, in *WMO Solid Precipitation Measurement Intercomparison.* TECO-1988 WMO/TD-No. 222, Leipzig, 1988, pp. 255–262

B. Goodison, P.Y.T. Louie, D. Yang, WMO solid precipitation measurement intercomparison. Final Report, WMO/TD-No. 872, 1998, 212 pp

P.Y. Groisman, V.V. Koknaeva, T.A. Belokrylova, T.R. Karl, Overcoming biases of precipitation measurement: a history of the USSR experience. Bull. Am. Meteorol. Soc. **72**, 1725–1732 (1991)

P.Y. Groisman, E.L. Peck, R.G. Quayle, Intercomparison of recording and standard nonrecording U. S. gauges. J. Atmos. Ocean. Technol. **165**, 602–609 (1999)

J. Kochendorfer, R. Nitu, M. Wolff, E. Mekis, R. Rasmussen, B. Baker, M.E. Earle, A. Reverdin, K. Wong, C.D. Smith, D. Yang, Y.-A. Roulet, S. Buisan, T. Laine, G. Lee, J.L.C. Aceituno, J. Alastrué, K. Isaksen, T. Meyers, R. Brækkan, S. Landolt, A. Jachcik, A. Poikonen, Analysis of single-Alter-shielded and unshielded measurements of mixed and solid precipitation from WMO-SPICE. Hydrol. Earth Syst. Sci. **21**, 3525–3542 (2017). https://doi.org/10.5194/hess-21-3525-2017

D.R. Legates, T.A. Bogart, Estimating the proportion of monthly precipitation that falls in solid form. J. Hydrometeorol., 1209–1306 (2009). https://doi.org/10.1175/2009JH1086.1

É. Mekis, L.A. Vincent, An overview of the second generation adjusted daily precipitation dataset for trend analysis in Canada. Atmosphere-Ocean **49**(2), 163–177 (2011)

J.R. Metcalfe, B.E. Goodison, Correction Canadian winter precipitation data. in *Eighth Symposium on Meteorological Observations and Instrumentation*, Anaheim, 17–22 January 1993, AMS Boston, pp. 338–343

V. Nešpor, B. Sevruk, Estimation of wind-induced error of rainfall gauge measurements using a numerical simulation. J. Atmos. Ocean. Technol. **16**, 450–464 (1999)

R. Nitu, K. Wong, CIMO survey on national summaries of methods and instruments for solid precipitation measurement at automatic weather stations, WMO/TD-No. 1544, CIOM No. 102 2010, 57pp

R. Rasmussen, Coauthors, How well are we measuring snow: the NOAA/FAA/NCAR winter precipitation test bed. Bull. Amer. Meteor. Soc. **93**, 811–829. (2012). https://doi.org/10.1175/BAMS-D-11-00052.1

M. Sanderson, A comparison of Canadian and United States standard methods of measuring precipitation. J. Appl. Meteorol. **14**(10), 1197–1199 (1975)

B. Sevruk, W.R. Hamon, International comparison of national precipitation gauges with a reference pit gauge. WMO Instrument and Observing Methods Report 17, 1984, 111 pp

B. Sevruk, S. Klemm, Types of standard precipitation gauges, in *Proceedings of International Workshop on Precipitation Measurement*, (St.Moritz, WMO/IAHS/ETH, 1989), pp. 227–236
I. Strangeways, Using Google Earth to evaluate GCOS weather station sites. Weather **64**(1) (2009). 8pp
K. Sugiura, T. Ohata, D. Yang, Catch characteristics of precipitation in high-latitude regions with high winds. J. Hydrometeorol. **7**, 984–994 (2006)
K. Sugiura, T. Ohata, D. Yang, Application of a snow particle counter to solid precipitation measurements under the Arctic condition. Cold Reg. Sci. Technol. **58**(2009), 77–83 (2009)
X. Tian, A. Dai, D. Yang, Z. Xie, Effects of precipitation-bias corrections on surface hydrology over northern Latitudes. J. Geophys. Res. **112**(D14), D14101 (2007). https://doi.org/10.1029/2007JD008420
D. Yang, An improved precipitation climatology for the Arctic Ocean. Geophys. Res. Lett. **26**(11), 1625–1628 (1999)
D. Yang, T. Ohata, A bias-corrected Siberian regional precipitation climatology. J. Hydrometeorol. **2**, 122–139 (2001)
D. Yang, A. Simonenko, Comparison of winter precipitation measurements by Six Tretyakov gauges at the Valdai experimental site. Atmosphere-Ocean (2013). https://doi.org/10.1080/07055900.2013.865156
D. Yang, J.R. Metcalfe, B.E. Goodison, E. Mekis, An evaluation of double fence intercomparison reference (DFIR) gauge. in *Proceedings of Eastern Snow Conference*, 50th Meeting, Quebec City, 1993, pp. 105–111
D. Yang, B.E. Goodison, J.R. Metcalfe, V.S. Golubev, E. Elomaa, T. Gunther, R. Bates, T. Pangburn, C. Hanson, D. Emerson, V. Copaciu, J. Milkovic, Accuracy of Tretyakov precipitation gauge: result of WMO intercomparison. Hydrol. Process. **9**(8), 877–895 (1995)
D. Yang, B.E. Goodison, J.R. Metcalfe, V.S. Golubev, R. Bates, T. Pangburn, C.L. Hanson, Accuracy of NWS 8″ standard non-recording precipitation gauge: result and application of WMO intercomparison. J. Atmos. Ocean. Technol. **15**, 54–68 (1998a)
D. Yang, B.E. Goodison, C.S. Benson, S. Ishida, Adjustment of daily precipitation at 10 climate stations in Alaska: application of World Meteorological Organization intercomparison results. Water Resour. Res. **34**(2), 241–256 (1998b)
D. Yang, E. Elomaa, T. Gunther, V. Golubev, B.E. Goodison, B. Sevruk, H. Madsen, J. Milkovic, Wind-induced precipitation undercatch of the Hellmann gauges. Nord. Hydrol. **30**, 57–80 (1999a)
D. Yang, S. Ishida, B.E. Goodison, T. Gunther, Bias correction of precipitation data for Greenland. J. Geophys. Res.-Atmos. **105**(D6), 6171–6182 (1999b)
D. Yang, B.E. Goodison, J.R. Metcalfe, P. Louie, G. Leavesley, D. Emerson, V. Golubev, E. Elomaa, T. Gunther, C.L. Hanson, T. Pangburn, E. Kang, J. Milkovic, Quantification of precipitation measurement discontinuity induced by wind shields on national gauge. Water Resour. Res. **35**(2), 491–508 (1999c)
D. Yang, D.L. Kane, L.D. Hinzman, B.E. Goodison, J.R. Metcalfe, P.Y.T. Louie, G.H. Leavesley, D. G. Emerson, C.L. Hanson, An evaluation of the Wyoming gauge system for snowfall measurement. Water Resour. Res. **36**(9), 2665–2677 (2000)
D. Yang, B.E. Goodison, J.R. Metcalfe, P.Y.T. Louie, E. Elomaa, C.L. Hanson, V.S. Golubev, T. Gunther, J. Milkovic, M. Lapin, Compatibility evaluation of national precipitation gauge measurements. J. Geophys. Res.-Atmos. **16**(D2), 1481–1491 (2001)
D. Yang, D. Kane, Z. Zhang, D. Legates, B. Goodison, Bias corrections of long-term (1973–2004) daily precipitation data over the northern regions. Geophys. Res. Lett. **32**(19), L19501 (2005)
B. Ye, D. Yang, Y. Ding, T. Han, T. Koike, A bias-corrected precipitation climatology for China. J. Hydrometeorol. **5**(6), 1147–1160 (2004)
B. Ye, D. Yang, L. Ma, Effect of precipitation bias-correction on water budget calculation in Upper Yellow River, China. Environ. Res. Lett. (2012). https://doi.org/10.1088/1748-9326/7/2/025201
Y. Zhang, T. Ohata, D. Yang, G. Davaa, Bias correction of daily precipitation measurements for Mongolia. Hydrol. Process. **18**, 2991–3005 (2004)

Remotely Sensed Evapotranspiration

Yuting Yang

Contents

Abstract

Land surface evapotranspiration (E) is a key component of the water and energy balance over terrestrial ecosystems, quantification of which has long been an important topic in hydrological, meteorological, agricultural, and ecological studies. This chapter focuses on the quantification of E using remote sensing-based approaches, which provide a promising opportunity for spatially consistent and temporally continuous E mapping. An introduction of surface energy balance is firstly presented, followed by three typical methods of estimating E from remote sensing imageries (i.e., surface energy balance-based models, vegetation

Y. Yang (✉)
State Key Laboratory of Hydroscience and Engineering, Department of Hydraulic Engineering, Tsinghua University, Beijing, China
e-mail: yuting_yang@tsinghua.edu.cn

X. Li, H. Vereecken (eds.), *Observation and Measurement of Ecohydrological Processes*, Ecohydrology, https://doi.org/10.1007/978-3-662-48297-1_6

index-land surface temperature space models, and Penman-Monteith-based models) and the ETWatch model that combines the energy balance and Penman-Monteith models. A working example of comparing three remote sensing E models is then provided to better inform the model physics, as well as advantages and drawbacks. Additionally, temporal and spatial scaling methods are presented. Finally, existing terrestrial E products that have a global coverage and are publically accessible are introduced.

Keywords

Terrestrial evapotranspiration · Remote sensing · Surface energy balance · Temporal and spatial scaling

Introduction

Evaporation is the phase change from liquid to gas. Evaporation of water may occur from the Earth's surface (e.g., the soil or a water body), through plant leaves (termed transpiration), or from rainfall intercepted by objects aboveground (primarily vegetation in natural environments). The combined term evapotranspiration (E) covers these three components, though interception is not explicitly used in the compound word. Among different processes within the terrestrial hydrological cycle, land surface E plays an important role in determining the global water balance and linking the global water and energy cycles. On one hand, it moves a vast amount of water vapor upward into the atmosphere to support land surface precipitation. On the other hand, it controls land-atmosphere feedbacks via modulating the surface energy budget (Rosen 1999). From a water balance's perspective, land surface E is the second largest water flux (after precipitation) within the terrestrial hydrological cycle, which accounts for about 65% of land surface rainfall (Oki and Kanae 2006). This ratio can be as high as more than 90% in arid regions (Bonan 2002). In the northern part of China, agricultural water use accounts for about 70% of total water consumption in the area, with more than 90% of them consumed via E (Liu et al. 2002). This highlights the water resource crisis in the area, where it owns 65% of agricultural land but only 20% of the water resources of the national totals (Deng et al. 2006).

E, like precipitation, has the dimensions of depth per time and common units of mm day^{-1}. When spatially integrated over an area (m^2, such as a paddock, catchment, basin, or country), the dimensions of E become volume per time, and common units are m^2 day^{-1} or GL day^{-1}. In addition, because the phase change of water from liquid to gas is associated with the increase in internal energy, the latent heat of vaporization (λ; the heat that must be supplied to convert unit mass of water from liquid to vapor without a change in temperature) is usually used to convert the units of E into its energy equivalent (i.e., latent heat; λE). The value of λ (in MJ kg^{-1}) is a function temperature of the evaporating surface (T_s in °C):

$$\lambda = 2.501 - 0.002361T_s \tag{1}$$

Accurate quantification of E and its components can be achieved by using a combination of in situ measurements, such as the eddy covariance systems, Bowen ratio systems, weighing lysimeter, sap flow meters, and isotope sampling (see chapter ▶ "Micrometeorological Methods to Determine Evapotranspiration"). However, such field-based observations are often costly and can only provide measurements of E at the point scale. As a result, mathematical modeling in combination with more or less meteorological and/or surface property information becomes a powerful tool to quantify E over larger areas and longer periods. Quantification of E at a regional or even larger scale has been rapidly advancing since satellite remote sensing data became publically available in the 1970s. Satellite remote sensing has provided an unprecedented opportunity for capturing E across a variety of spatial and temporal scales that are not attainable by conventional techniques (e.g., Carlson 2007). In general, the remote sensing-based E models can be grouped into three types, i.e., surface energy balance models, vegetation index-land surface temperature space models, and Penman-Monteith-based models.

Remote Sensing Evapotranspiration Models

Surface Radiation Balance and Energy Budget

Radiation provides energy onto the Earth's surface and ultimately drives the E process. All surfaces receive shortwave radiation (typically with a wavelength of 0.1 μm and 4.0 μm) during daytime and exchange longwave radiation continuously (during both daytime and nighttime) with the above atmosphere. The net rate at which a surface receives shortwave and longwave radiations is called the net radiation, which is the energy input in most of E models (if not all). The surface net radiation can be written as:

$$R_n = R_{s,i} - R_{s,o} + R_{l,i} - R_{l,o} \tag{2}$$

where R_n is the net radiation, $R_{s,i}$ is the incoming shortwave radiation, $R_{s,o}$ is the outgoing shortwave radiation, $R_{l,i}$ is the incoming longwave radiation, and $R_{l,o}$ is the outgoing shortwave radiation. Common units used to express radiation received on a unit surface per unit time include MJ m^{-2} day^{-1}, W m^{-2}, and in its water equivalent mm day^{-1}; the conversion factors between those units can be found in Table 3 in Allen (1998).

Alternatively, the surface radiation balance (Eq. 2) is often written as:

$$R_n = (1 - \alpha)R_{s,i} + R_{l,i} - R_{l,o} \tag{3}$$

where α indicates the ability of a surface to reflect $R_{s,i}$ back into the atmosphere and is known as the surface reflectivity (or surface albedo; unitless). The surface reflectivity is mostly determined by surface material and color. For example, a typical dark soil surface has a reflectivity of 0.08–0.13, a green grass surface has a reflectivity of 0.24–0.26, and a fresh snow surface has a reflectivity of 0.75–0.95.

The surface energy (or heat) budget is expressed as:

$$R_n + M = H + \lambda E + G + S + A_d \tag{4}$$

where M is the net gain of heat from metabolism, H is the loss of sensible heat by convection into the atmosphere, λE is the loss of latent heat by E, and G is the loss of sensible heat by convection into the interior (e.g., soil) and is also known as ground heat flux. S is the heat stored within the system, and A_d is the loss of heat associated with horizontal air movement (e.g., advection).

In most conditions, the magnitude of M, S, and A_d is relatively much smaller than the other components in the surface energy budget equation (i.e., Eq. 3) and, thus, is often neglected. This suggests that the surface energy balance can be simplified as:

$$R_n = H + \lambda E + G \tag{5}$$

The above equation states that the net incoming energy flux by radiation is partitioned into three energy flux components at the Earth's surface, i.e., sensible heat flux, latent heat flux, and ground heat flux. In most of state-of-the-art remote sensing E models, R_n is either determined from a set of empirical/semiempirical equations (details can be found in Allen et al. 1998) or is taken from global/region reanalysis products, and G is considered as a relatively constant fraction of net radiation allocated to the soil surface and subjected to adjustments as empirical functions of surface temperature, vegetation cover, and surface reflectivity. The key is then to explicitly estimate either H or λE in order to close the surface energy budget equation based on classic heat and/or vapor transfer models in which some key variables/parameters are provided by remote sensing.

Surface Energy Balance Model

The surface energy balance (SEB) models explicitly deal with sensible heat flux and calculate latent heat flux (or E) as a residual in the surface energy budget equation. A common similarity in all SEB models lies in the use of remotely sensed land surface temperature as a key model input. Based on the difference of how each model parameterizes its resistance network (surface and aerodynamic resistances to heat and vapor transfer; Fig. 1), the SEB models can be broadly grouped into two categories, i.e., single-source model and multisource model.

The single-source model treats the land surface as a uniform layer or a "big leaf," which does not distinguish between evaporation from soil (E_s) and transpiration from vegetation canopy (E_c) (Fig. 1a). Examples of SEB model that employs a single-source resistance network include the Surface Energy Balance Algorithm for Land (SEBAL; Bastiaanssen et al. 1998), the Mapping EvapoTranspiration with

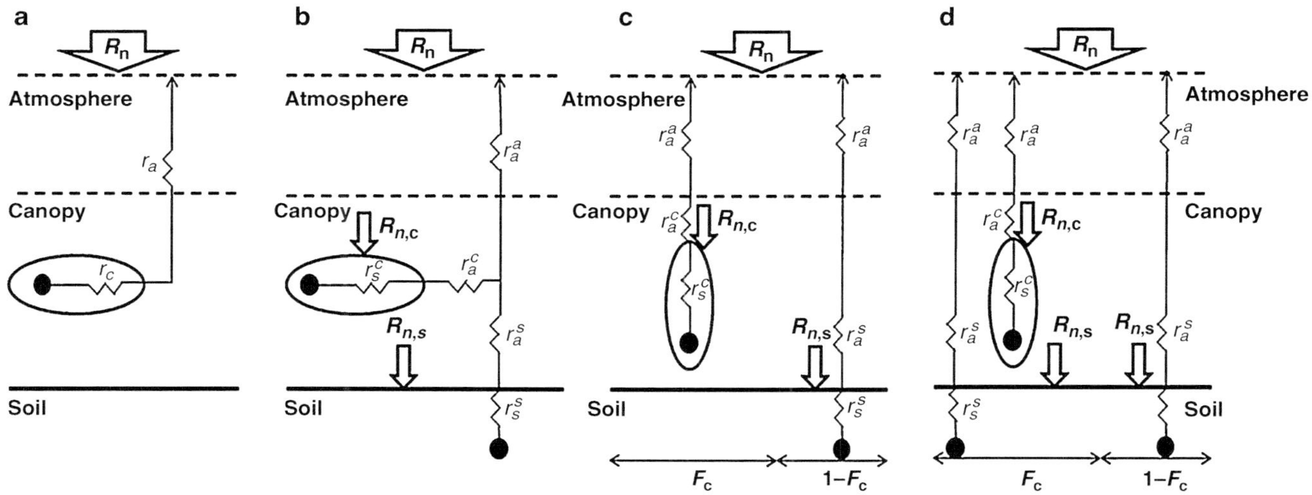

Fig. 1 Schematic diagrams of different parameterizations of resistance network: (**a**) single-source model, (**b**) two-layer model, (**c**) two-patch model, and (**d**) hybrid dual-source model. The nomenclature is given in section "Surface Energy Balance Model". "(Modified from Fig. 2.1 in Yang 2015)

high Resolution and Internalized Calibration (METRIC; Allen et al. 2007), the Surface Energy Balance System (SEBS; Su 2002), and the Simplified Surface Energy Balance Index algorithm (S-SEBI; Roerink et al. 2000).

The multisource model divides the land surface into multiple components at which energy fluxes are calculated separately. The most common multisource models are two-source models, which calculate sensible and latent flux separately from soil and vegetation (Fig. 1b–d). Three types of dual-source configurations of resistance network are used in the literature, i.e., the two-layer approach where each source of flux is superimposed and coupled (Fig. 1b), the two-patch approach where the flux from each source independently interacts with the above atmosphere (Fig. 1c), and the hybrid dual-source approach which combines the two-layer and two-patch approaches (Fig. 1d; Guan and Wilson 2009). The Two-Source Energy Balance (TSEB) model proposed by Norman et al. (1995) represents a classic SEB model that adopts a two-source resistance parameterization. Subsequent revisions/developments were made to the TSEB model, which yield many two-source SEB models including the Two-Source Time-Integrated model (TSTIM; Anderson et al. 1997, 2007), the Atmosphere-Land Exchange Inverse model (ALEXI; Anderson et al. 1997, 2007), the Disaggregated Atmosphere-Land Exchange Inverse model (DisALEXI; Norman et al. 2003), the Simple Remote Sensing EvapoTranspiration model (Sim-ReSET; Sun et al. 2009), etc.

Single-Source Model: SEBAL

The Surface Energy Balance Algorithm for Land (SEBAL) represents a single-source resistance transfer scheme based on remotely sensed data in combination with concurrent meteorological data for deriving components within the surface energy balance equation (Bastiaanssen et al. 1998). Once net radiation is estimated or prescribed, SEBAL calculates sensible heat flux (H in W m^{-2}) as:

$$H = \frac{\rho_a C_p dT}{r_{ah}} \tag{6}$$

where ρ_a is the air density ($\approx$1.292 kg m^{-3}), C_p is the specific heat of air at constant pressure ($\approx$1013 J kg^{-1} K^{-3}), dT is the temperature difference between the aerodynamic surface temperature (T_{Aero}) and the air temperature at reference height (T_a), and r_{ah} is the aerodynamic resistance to heat transfer (s m^{-1}).

One difficulty in direct applying Eq. 6 is that T_{Aero} is not equal to the radiometric surface temperature (T_s) retrievable from remote sensing data. To circumvent this difficulty, SEBAL assumes that dT is linearly correlated to T_s, given as:

$$dT = b + aT_s \tag{7}$$

where b and a are the fitted coefficients, which are determined by using two "anchor" points where a value for H can be reliably estimated based on a prior knowledge of the fluxes over dry land (hot pixel) and wet land (cold pixel). In the original version of SEBAL, it was prescribed that $H = 0$ for cold pixel (so that $\lambda E = R_n - G$) and $H = R_n - G$ for hot pixel (so that $\lambda E = 0$). However, $H = 0$ can rarely be found in the natural environment during the daytime, even over open water surfaces. Consequently, in a later version of SEBAL, for cold pixel, λE is prescribed as 1.05 times of reference crop E (E_0), i.e., $\lambda E = 1.05\lambda E_0$, where E_0 is given by:

$$E_0 = \frac{0.408\Delta(R_n - G) + \gamma \dfrac{900}{T_a + 273} u_2(e_s - e_a)}{\Delta + \gamma(1 + 0.34u_2)} \tag{8}$$

where R_n and G are given in MJ m^{-2} day^{-1}, e_s is the saturated vapor pressure (kPa), e_a is the actual vapor pressure (kPa), γ is the psychrometric constant (kPa °C^{-1}), T_a is the air temperature in °C at 2 m height, u_2 is the wind speed at 2 m height (m s^{-1}), and Δ is the slope of the e_s–T_a relationship (kPa °C^{-1}).

The aerodynamic resistance is determined by iteration of air stability corrections based on Monin-Obukhov similarity theory. For the initial estimation of H, the aerodynamic condition is assumed to be neutral, and r_{ah} is calculated from:

$$r_{ah} = \frac{\ln(z_2/z_1)}{u_* k} \tag{9}$$

where z_1 and z_2 are heights above the zero plane displacement of the vegetation where the endpoints of dT are defined (m), k is the von Karman's constant (=0.41), and u_* is the friction velocity (m s^{-1}) calculated by using the logarithmic wind law for neutral atmospheric condition for the first iteration as:

$$u_* = \frac{k u_{200}}{\ln(200/z_{om})} \tag{10}$$

where u_{200} is the wind speed at blending height, which is set to be 200 m in SEBAL. z_{om} is the momentum roughness length (m) and is calculated as a function of vegetation height (h), i.e., $z_{om} = 0.12\,h$. u_{200} is estimated from wind speed measured at a weather station by applying the logarithmic wind law, i.e.,

$$u_{200} = u_* \frac{\ln(200/z_{om})}{k} \tag{11}$$

During subsequent iterations, the Monin-Obukhov length (L) is firstly computed to examine the atmospheric stability:

$$L = -\frac{\rho_a C_p u_*^3 T_s}{kgH} \tag{12}$$

where g is the gravitational acceleration (=0.9807 m s^{-1}).

Then, corrections are made to u_* and r_{ah}, respectively, under non-neutral aerodynamic conditions for each successive iteration as:

$$u_* = \frac{ku_{200}}{\ln(200/z_{om}) - \psi_m(200)} \quad (13)$$

$$r_{ah} = \frac{\ln(z_2/z_1) - \psi_h(z_2) + \psi_h(z_1)}{u_* k} \quad (14)$$

where $\Psi_m(200)$ is the stability correction for momentum transport at 200 m height and $\Psi_h(z_2)$ and $\Psi_h(z_1)$ are stability corrections for heat transport at heights z_2 (the height at which meteorological variables are measured) and z_1 (the height immediately above the aerodynamic boundary layer of the evaporating surface), respectively. These correction terms can be calculated following Paulson (1970) and Webb (1970).

For unstable conditions (i.e., $L < 0$):

$$\psi_m(200) = 2\ln\left(\frac{1 + x(z)}{2}\right) + \ln\left(\frac{1 + x(z)^2}{2}\right) - 2ARCTAN(x(z)) + 0.5\pi \quad (15)$$

$$\psi_h(z) = 2\ln\left(\frac{1 + x(z)^2}{2}\right) \quad (16)$$

in which

$$x(z) = \left(1 - 16\frac{z}{L}\right)^{0.25}$$

For stable conditions ($L > 0$):

$$\psi_m(z) = -\frac{2}{5L} \quad (17)$$

$$\psi_h(z) = -\frac{z}{5L} \quad (18)$$

For neutral conditions ($L = 0$), $\Psi_m = \Psi_h = 0$.

The corrected values of u_* and r_{ah} are then used to recompute H until the value for r_{ah} stabilizes. A flowchart of iterative process for estimating H is shown in Fig. 2. Once H is quantified, λE is calculated as a residual in the surface energy budget equation as:

$$\lambda E = R_n - G - H \quad (19)$$

Dual-Source Model: TSEB

The Two-Source Energy Balance (TSEB) model is a two-source remote sensing-based E model, which estimates soil evaporation (E_s) and plant transpiration (E_c) separately

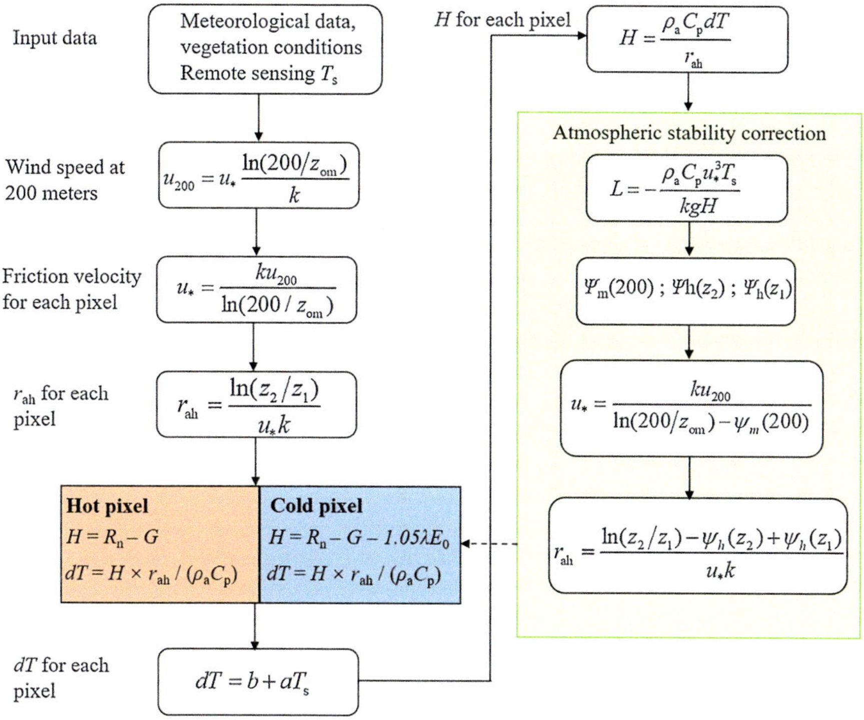

Fig. 2 Flowchart of the iterative process for estimating sensible heat in SEBAL. (Modified from Fig. 4 in Waters et al. 2002)

based on surface energy budget for each component (or source) (Norman et al. 1995). In TSEB, R_n is firstly partitioned and allocated to each component according to Beer's law:

$$R_{n,c} = R_n[1 - \exp(0.9\ln(1 - F_c))] \tag{20}$$

$$R_{n,s} = R_n \times \exp(0.9\ln(1 - F_c)) \tag{21}$$

where $R_{n,c,}$ and $R_{n,s}$ are net radiation allocated to vegetation and soil component, respectively. The constant of 0.9 is obtained with the assumptions of a random leaf positioning and a spherical leaf angle distribution and an extinction coefficient of radiation attenuation within the canopy of 0.45. F_c is the fractional vegetation coverage (dimensionless, varies from 0 to 1) and is quantified from remotely sensed leaf area index (*LAI*) as:

$$F_c = 1 - \exp\left(\frac{-0.5\Omega LAI}{\cos\theta}\right) \tag{22}$$

where Ω is a factor indicating the degree to which vegetation is clumped and θ is the radiometer view zenith angle.

Then, a two-patch approach is adopted in TSEB to partition available energy into latent heat, sensible heat, and ground heat fluxes for each component. The surface energy budget for each component is given by:

$$R_{n,c} = (1 - F_c)(\lambda E_c + H_c) \tag{23}$$

$$R_{n,s} - G = (1 - F_c)(\lambda E_s + H_s) \tag{24}$$

where subscripts c and s stand from vegetation component and soil component, respectively.

As a result, latent heat flux can be estimated as a residual in the surface energy balance equation for each component:

$$F_c \lambda E_c = R_{n,c} - F_c H_c = R_{n,c} - F_c \rho_a C_p \frac{T_{s,c} - T_a}{r_a^c} \tag{25}$$

$$\begin{aligned}(1 - F_c)\lambda E_s &= R_{n,s} - G - (1 - F_c) H_s \\ &= R_{n,s} - G - (1 - F_c)\rho_a C_p \frac{T_{s,s} - T_a}{r_a^a + r_a^s}\end{aligned} \tag{26}$$

where r_a^a is the aerodynamic resistance to heat transfer between $Z_{om} + d$ (Z_{om} is the canopy roughness length for momentum transfer, and d is zero displacement height, usually calculated as 0.67 times vegetation height) and the reference height (s m^{-1}); r_a^c is the aerodynamic resistance to heat transfer between the canopy and the reference height (s m^{-1}); and r_a^s is the aerodynamic resistance to heat transfer in the boundary layer immediately above the soil surface (s m^{-1}). Equations for the three resistance terms above can be found in Sánchez et al. (2008). The ground heat flux in TSEB is assumed to be a constant ratio of $R_{n,s}$, i.e., $G = 0.35R_{n,s}$ (Norman et al. 1995).

To obtain estimates of H_s and H_c, surface temperatures for soil ($T_{s,s}$) and canopy ($T_{s,c}$) are needed. However, this information is unattainable directly from remote sensing data because T_s observed by remote sensing is a single temperature over heterogeneous surfaces. The temperature difference between vegetation and soil components can be as high as >20 °C (Norman et al. 1995; Yang et al. 2013b). In TSEB, the component surface temperatures are estimated by using an iterative approach. It firstly assumes a potential transpiration rate for the canopy component, and estimates λE_c using a modified Priestley-Taylor parameterization, given as:

$$\lambda E_c = \alpha_{PT} LE_{eq} = \alpha_{PT} \frac{\Delta}{\Delta + \gamma} R_{n,c} \tag{27}$$

where α_{PT} is the Priestley-Taylor coefficient and LE_{eq} is the equilibrium transpiration for the canopy component (Agam et al. 2010). For the first estimation of LE_c, the initial value of α_{PT} is taken to be 1.3 (or 1.26; Norman et al. 1995).

Canopy surface temperature can then be estimated from Eq. 25 based on λE_c estimates from Eq. 27, and a nonlinear relationship between T_s and component temperatures is used to get the first estimation of $T_{s,s}$:

$$T_{\mathrm{s}} = \left[F_{\mathrm{c}} T_{\mathrm{s,c}}^{4} + (1 - F_{\mathrm{c}}) T_{\mathrm{s,s}}^{4}\right]^{1/4} \tag{28}$$

Combining Eqs. 26 and 28 gives the first estimation of λE_{s}. However, λE_{s} from Eq. 26 may reflect some unrealistic conditions, such as condensation at the soil surface (i.e., $\lambda E_{\mathrm{s}} < 0$) under daytime convective conditions with a soil moisture deficit, which can be mainly ascribed to the initial estimate of λE_{c} given by the Priestley-Taylor parameterization for potential transpiration, which overestimates the actual transpiration. If this condition is encountered by the TSEB model, the Priestley-Taylor coefficient is iteratively reduced until λE_{s} approaches zero.

Vegetation Index-Land Surface Temperature Space Model

The vegetation index (VI)-land surface temperature (T_{s}) space model determines the partitioning of surface available energy into H and λE based on the interpretation of the image (pixel) distribution in VI-T_{s} space. As reviewed by Carlson (2007), if a sufficiently large number of pixels are present and when contaminated pixels and outliers (e.g., clouds, surface water, sloping terrain, and shading) are removed, the shape of the pixel envelope constitutes a meaningful triangle. Moran et al. (1994) claimed that the triangle space does not account for the effect of water stresses on canopy transpiration and therefore replaced the triangle by using a trapezoid. In such a space, a higher VI value generally corresponds to a lower T_{s} for a pixel where higher E would occur and vice versa. Similar to SEB models, the VI-T_{s} can also be categorized into single-source models that treat the land surface as a whole and multisource models that divide the land surface into different components. Here we only focus on two-source models; an example of a four-source VI-T_{s} model can be found in Merlin et al. (2014).

Single-Source Model: Triangle Model

A series of single-source remote sensing-based VI-T_{s} models for estimating E has been developed during the past three decades (e.g., Carlson et al. 1995; Hope et al. 1986; Jiang and Islam 2001; Nemani and Running 1989; Nishida et al. 2003; Wang et al. 2006). Despite their differences in model details, these models all share a similarity in interpreting the relationship between VI and T_{s}. In general, envelopes of the VI-T_{s} space constitute a triangle (Fig. 3), with the upper boundary of the VI-T_{s} space (also known as the hot edge; line AB in Fig. 3a) representing the highest T_{s} for any VI and the lower boundary of that space (known as the cold edge; line BC in Fig. 3a) indicating the lowest T_{s} for a specified VI value. According to Eq. 6, when all other variables are held constant, H increases with T_{s}. This follows that λE on the hot edge is minimum (e.g., zero) and λE on the cold edge is maximum (e.g., potential E). For other pixels within the VI-T_{s} space, their E rate is determined by their relative distance to the hot edge (i.e., b in Fig. 3a) and cold edge (i.e., a in Fig. 3a) for each VI value.

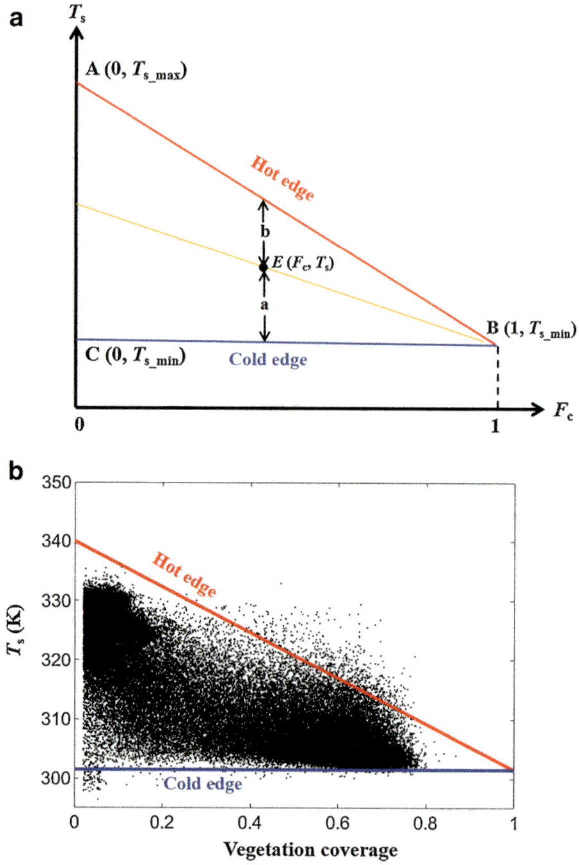

Fig. 3 The VI-T_s triangle space. (**a**) Conceptual plot of the VI-T_s triangle space and (**b**) scatterplot of F_c and T_s derived from ASTER imagery the MUSOEXE-12 campaign region on July 10, 2012

In this chapter, the triangle VI-T_s model developed in Jiang and Islam (1999) is illustrated as an example. Other triangle VI-T_s models follow a similar general procedure of Jiang and Islam (1999) yet differ in details. In Jiang and Islam (1999), λE is estimated based on the Priestley-Taylor parameterization:

$$\lambda E = \Phi\left(\frac{\Delta}{\Delta + \gamma}\right)(R_n - G) \tag{29}$$

where Φ is a parameter that integrates effects of soil moisture availability, vapor pressure gradient, and aerodynamic and surface resistances on λE, which is expressed as:

$$\Phi = \frac{T_{s\_max} - T_s}{T_{s\_max} - T_{s\_min}}(\Phi_{max} - \Phi_{min}) + \Phi_{min} \tag{30}$$

where T_{s_max} and T_{s_min} are the maximum and minimum T_s for each F_c value. Φ_{max} is the maximum value of parameter Φ, which represents conditions at which E is not water limited. In such conditions, Φ equals to the original Priestley-Taylor coefficient (i.e., $\Phi = 1.26$). By contrast, Φ_{min} is the minimum value of parameter Φ, which is determined for each F_c value by assuming that Φ_{min} is proportional to F_c:

$$\Phi_{min} = 1.26F_c/F_{c\_max} \tag{31}$$

where assumes that Φ_{min} is equal to zero when $F_c = 0$ and equal to 1.26 when F_c is at its minimum value across a scene (i.e., F_{c_max}).

One obvious issue of simplifying the VI-T_s scatter space as a triangle is that the triangle simplification implicitly assumes an identical surface temperature for fully vegetated surfaces under non-water-limited and fully water-limited conditions (i.e., point B in Fig. 3a). However, theoretically, when there is a water limitation, the surface available energy will be less partitioned into latent heat flux and more partitioned into sensible heat flux that increases surface temperature. Consequently, surface temperature for fully vegetated surface under water-limited conditions should be higher than that under non-water-limited conditions. Moran et al. (1994) pointed out this issue and replaced the triangle by using a trapezoid (Fig. 4). Similar to the triangle configuration, the trapezoid method also assumes a zero E rate along the hot edge and a maximum (i.e., potential) E rate on the cold edge. For other pixels within the trapezoidal space, their E rate is determined by their relative distance to the hot edge (i.e., b in Fig. 4a) and cold edge (i.e., a in Fig. 4a) for each VI value.

Determination of cold and hot edges. Accurate determination of the extreme boundaries (i.e., the cold and hot edges) is the key for the VI-T_s model, as the resultant latent/sensible fluxes are linearly sensitive to the position of the extreme boundaries. Traditional approaches for determining the extreme boundaries of the triangle (or trapezoid) space are based on either visual or automatic examination of the VI-T_s data points, which implicitly requires that both conditions of extreme dry and extreme wet points are present in a study scene to ensure the validity of the triangle (or trapezoid) method. It is worthwhile noting that the SEBAL model introduced in section "Single Source Model: SEBAL" subjects to the same requirement of the study scene. Unfortunately, this requirement is often not satisfied in reality. For wet regions or dry regions short after a rainfall event, extreme dry pixels are not necessarily present. On the contrary, for dry deserts, cold pixels usually do not exist. To overcome this difficulty in determining the extreme boundaries, researchers have developed methods for determining the extreme boundaries from a theoretical perspective.

One theoretical solution to the extreme boundaries is derived by Long et al. (2012) and Zhang et al. (2005), where a trapezoidal conceptualization of the VI-T_s space is adopted. Firstly, for bare soil surfaces, the surface radiation balance and energy budget can be respectively described as:

$$R_{n,s} = (1 - \alpha_s)R_{s,i} + \varepsilon_s\varepsilon_a\sigma T_a^4 - \varepsilon_s\sigma T_s^4 \tag{32}$$

$$R_{n,s} - G = H_s + \lambda E_s = \rho_a C_p\left(\frac{T_s - T_a}{r_a^a + r_a^s}\right) + \lambda E_s \tag{33}$$

Fig. 4 The VI-T_s trapezoid space. (**a**) Conceptual plot of the VI-T_s trapezoid space and (**b**) scatterplot of F_c and T_s derived from ASTER imagery the MUSOEXE-12 campaign region on July 10, 2012

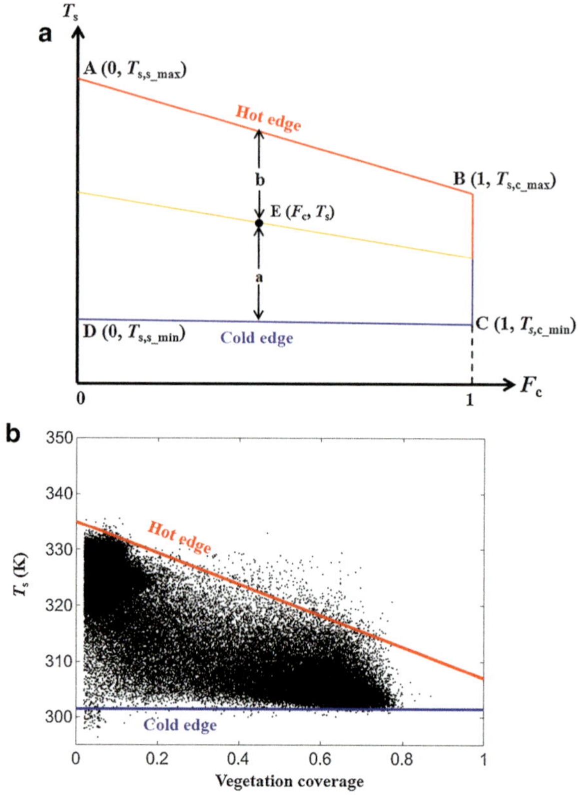

where α_s is the bare surface albedo; σ is the Stefan-Boltzmann constant (=5.67 × 10^{-8} W m^{-2} K^{-4}); ε_s is the bare surface emissivity (e.g., 0.95); and ε_a is the atmospheric emissivity, which is a function of air pressure (P_a) and vapor pressure (e_a).

The first two terms of Taylor's formula of the upward longwave radiation ($\varepsilon_s \sigma T_s^4$) at T_a are written as:

$$\varepsilon_s \sigma T_s^4 \approx \varepsilon_s \sigma T_a^4 + 4\varepsilon_s \sigma T_a^3 (T_s - T_a) \tag{34}$$

Substituting Eq. 34 into Eq. 32 and combining it with Eq. 33, we can get an explicit expression of T_s:

$$T_s = \frac{(1-\alpha_s)R_{s,i} + \varepsilon_s \varepsilon_a \sigma T_a^4 - \varepsilon_s \sigma T_a^4 - \lambda E_s/(1 - G/R_{n,s})}{4\varepsilon_s \sigma T_a^3 + \rho_a C_p / [(r_a^a + r_a^s)(1 - G/R_{n,s})]} + T_a \tag{35}$$

For extreme dry surfaces where there is the largest water stress, E is completely suppressed ($\lambda E_s = 0$). As a result, the surface temperature for the driest bare surface (point A in Fig. 1), $T_{s,s,max}$, is:

$$T_{s,s\_max} = \frac{(1-\alpha_s)R_{n,i} + \varepsilon_s\varepsilon_a\sigma T_a^4 - \varepsilon_s\sigma T_a^4}{4\varepsilon_s\sigma T_a^3 + \rho_a C_p/\left[\left(r_a^a + r_a^s\right)\left(1 - G/R_{n,s}\right)\right]} + T_a \tag{36}$$

where the ratio of $G/R_{n,s}$ is taken to be 0.35 for bare soil surface.

In a similar vein, the temperature for the driest fully vegetated surface (point B in Fig. 1), $T_{s,c,max}$, can be derived as:

$$T_{s,c\_max} = \frac{(1-\alpha_c)R_{n,i} + \varepsilon_c\varepsilon_a\sigma T_a^4 - \varepsilon_c\sigma T_a^4}{4\varepsilon_c\sigma T_a^3 + \rho_a C_p/r_a^c} + T_a \tag{37}$$

where α_c is the albedo for fully vegetated surfaces and ε_c is the vegetation emissivity (0.98).

The two albedo terms (α_s and α_c) can be estimated either by extending the upper envelope of the VI – albedo space intersecting with VI = 0 and VI = 1, respectively – or from the look-up table for each specific type of surface provided by Campbell and Norman (1998).

For the cold edge, the highest latent heat corresponds to the lowest sensible heat, and therefore the temperature gradient between the land surface and reference height would approach zero. As such, air temperature is taken as the surface temperature for the lower theoretical limiting edge (i.e., $T_{s,s\_min} = T_{s,c\_min} = T_a$). This simplification is an operational way to determine the lower theoretical limiting edge. However, this simplification may fail over surfaces with either extremely high vapor pressure deficit or strong advection effect (Yang et al. 2015).

Dual-Source Model: HTEM

The hybrid dual-source scheme and trapezoid framework-based ET model (HTEM) is a two-source (i.e., hybrid dual-source approach) VI-T_s model that estimates E_s and E_c separately by resolving the energy balance for both the soil and vegetation surfaces (Yang and Shang 2013). The energy allocation and resistance network of HTEM is shown in Fig. 1d. In the hybrid dual-source scheme, a layer approach is used to allocate the available energy for the soil and canopy component based on Beer's law:

$$R_{n,c} = R_n[1 - \exp(-k_c LAI)] \tag{38}$$

$$R_{n,s} = R_n[\exp(-k_c LAI)] \tag{39}$$

where k_c is the extinction coefficient of radiation attenuation within the canopy. For partially vegetated surfaces, the value of k_c is determined by linear interpolation of the value of k_c between that for full vegetation cover and bare soil in terms of vegetation coverage. The value of k_c for full vegetation cover depends on canopy structure and leaf shape and oriental angle and falls between 0.5 and 1.0 for most of plant species.

A patch approach is then adopted by HTEM to partition available energy into the latent heat and sensible heat fluxes (i.e., Eqs. 23 and 24), similar to that used in the TSEB model. In HTEM, surface temperatures over vegetation ($T_{s,c}$) and soil ($T_{s,s}$) components are determined by interpreting the VI-T_s trapezoid space, where the extreme boundaries are quantified theoretically, as described above.

Another important feature of the VI-T_s space is that there exist soil wetness isolines representing the same soil water content in the VI-T_s space (Fig. 5). Assuming a uniform soil texture for each pixel, the soil surface radiative temperature is solely affected by soil moisture. Thus, HTEM assumes that pixels along one soil wetness isoline are also isothermal. To determine $T_{s,s}$ for each pixel, the slope of each isoline is firstly derived by interpolating the slope of the warm edge and that of the cold edge in terms of the temperature difference between the pixel and the cold edge (a in Fig. 5) and that between the pixel and the warm edge (b in Fig. 5) (Long and Singh 2012; Zhang et al. 2005). As a result, $T_{s,s}$ for each pixel can be calculated from:

$$T_{\mathrm{s,s}} = F_{\mathrm{c}}\frac{a}{a+b}\left(T_{\mathrm{s,s\_max}} - T_{\mathrm{s,c\_max}}\right) + T_{\mathrm{s}} \tag{40}$$

in which

$$a = T_{\mathrm{s}} - T_{a} \tag{40a}$$

$$b = (1 - F_{\mathrm{c}})\left(T_{\mathrm{s,s\_max}} - T_{\mathrm{s,c\_max}}\right) + T_{\mathrm{s,c\_max}} - T_{\mathrm{s}} \tag{40b}$$

Then, a linear assumption is used in HTEM to explain soil moisture isolines; accordingly, canopy surface temperatures can be determined from:

$$T_{\mathrm{s,c}} = \frac{1}{F_{\mathrm{c}}}\left[T_{\mathrm{s}} - (1 - F_{\mathrm{c}})T_{\mathrm{s,s}}\right] \tag{41}$$

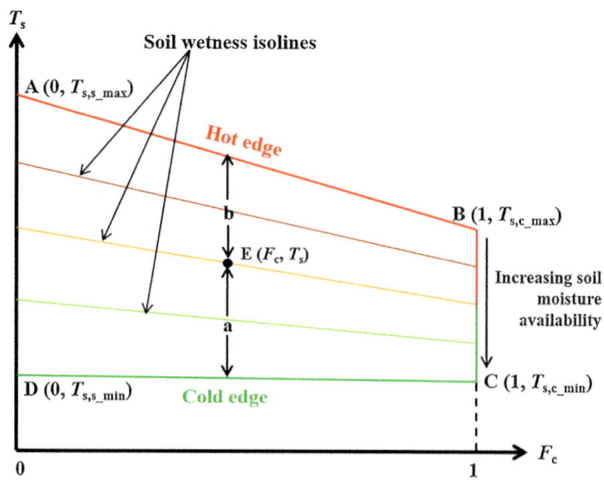

Fig. 5 A sketch of the trapezoidal $F_c - T_s$ space in HTEM. (Modified from Fig. 2 in Yang and Shang 2013)

Penman-Monteith-Based Model

The Penman-Monteith-based remote sensing E models directly estimate E (or λE) from Penman-Monteith type equations over the entire surface or for each surface component (e.g., soil and vegetation). The Penman-Monteith equation is expressed as (Monteith 1965):

$$\lambda E = \frac{\Delta(R_\mathrm{n} - G) + \rho_\mathrm{a} C_\mathrm{p} \dfrac{e_s - e_a}{r_a}}{\Delta + \gamma\left(1 + \dfrac{r_\mathrm{s}}{r_\mathrm{a}}\right)} \tag{42}$$

where r_a is the aerodynamic resistance and r_s is the (bulk) surface resistance.

Different from SEB models and VI-T_s space models, the Penman-Monteith-based remote sensing E models usually do not require remotely sensed land surface temperature as a model input. Instead, it utilizes remotely sensed VI in combination with concurrent meteorological observations to quantify resistance (or its reciprocal, conductance) terms required in the Penman-Monteith equation.

MODIS Terrestrial Evapotranspiration Model (MOD16)

The MODIS (MODerate resolution Imaging Spectroradiometer) terrestrial evapotranspiration algorithm (also known as the MOD16 algorithm) is a typical Penman-Monteith-based remote sensing E model (Mu et al. 2007, 2011). The MOD16 algorithm calculates actual evaporation from soil (E_s) and wet canopy (i.e., canopy interception; E_wet) and plant transpiration (E_c) individually based on the Penman-Monteith type equations, and the total E is considered as the sum of the three components:

$$E = E_\mathrm{wet} + E_\mathrm{s} + E_\mathrm{c} \tag{43}$$

in which

$$E_\mathrm{wet} = \frac{(\Delta A_\mathrm{c} + \rho_\mathrm{a} C_\mathrm{P} F_\mathrm{c} D / rhrc) \times F_\mathrm{wet}/\lambda}{\Delta + (P_\mathrm{a} \times C_\mathrm{P} \times rcv)/(\lambda \times \varepsilon \times rhrc)} \tag{43a}$$

$$E_\mathrm{s} = \frac{\left[\Delta A_\mathrm{s} + \rho_\mathrm{a} C_\mathrm{P}(1 - F_\mathrm{c}) D / r_a^s\right]/\lambda}{\Delta + \gamma \times r_\mathrm{tot}/r_a^s} \times \left[F_\mathrm{wet} + (1 - F_\mathrm{wet}) \times f_\mathrm{sm}\right] \tag{43b}$$

$$E_c = \frac{\left(\Delta A_\mathrm{c} + \rho_\mathrm{a} C_\mathrm{p} F_\mathrm{c} D / r_a\right) \times (1 - F_\mathrm{wet})/\lambda}{\Delta + \gamma\left(1 + r_s^c\right)/r_a^c} \tag{43c}$$

In the above equations (i.e., Eqs. 43a, 43b, and 43c), P_a is the atmospheric pressure (kP_a), D is the atmospheric vapor pressure deficit (kP_a), and ε is the ratio of the molecular weight of water to dry air (i.e., 0.622). F_wet is the relative surface wetness (dimensionless). A_s and A_c are available energy allocated to soil and vegetation component, respectively, and are determined using two-patch approach:

$$A_c = F_c \times R_n \tag{44}$$

$$A_s = (1 - F_c) \times R_n - G \tag{45}$$

For estimating E_{wet} (i.e., Eq. 43a), rcv and $rhrc$ represent wet canopy resistance (s m^{-1}) and aerodynamic resistance (s m^{-1}) to evaporated water on the wet canopy surface, respectively, and are quantified by the following equations:

$$rvc = \frac{1}{gl\_e \times LAI \times F_{wet}} \tag{46}$$

$$rhrc = \frac{rhc \times rrc}{rhc + rrc} \tag{47}$$

$$rhc = \frac{1}{gl\_sh \times LAI \times F_{wet}} \tag{48}$$

$$rrc = \frac{\rho_a \times C_P}{4 \times \sigma \times T_i^3} \tag{49}$$

where T_i presents daytime (or nighttime) average temperature (°C), LAI is the leaf area index, σ is the Stefan-Boltzmann constant (=5.67 × 10^{-8} W m^{-2} K^{-4}), rhc is the wet canopy resistance to sensible heat, and rrc is the resistance to radiative heat transfer through air (s m^{-1}). The two parameters, i.e., gl_e and gl_sh, present leaf conductance to evaporated water vapor and sensible heat per unit LAI, respectively (m s^{-1}).

For estimating E_s (i.e., Eq. 43b), r_{tot} is the total aerodynamic resistance (s m^{-1}), which represents the sum of surface resistance and the aerodynamic resistance for vapor transport, and r_a^s is the aerodynamic resistance at the soil surface (s m^{-1}):

$$rtot = rtotc \times rcorr \tag{50}$$

$$rtotc = \begin{cases} rbl_{min} & D \le D_{open} \\ rbl_{max} - \dfrac{(rbl_{max} - rbl_{max})(D_{close} - D)}{D_{close} - D_{open}} & D_{open} \le D \le D_{close} \\ rbl_{max} & D \ge D_{open} \end{cases} \tag{51}$$

where $rtotc$ is an initialized value for $rtot$, which can be calculated by a segment function of D with a maximum (rbl_{max}) and a minimum (rbl_{min}) limit (Table 1). The aerodynamic resistance to soil surface (r_a^s) is basically derived from rrs (the resistance to radiative heat transfer, same as rrc and rr in former sections) and rhs (the resistance to convective heat transfer, equal to $rtot$).

The term f_{sm} is the soil moisture constraint, which is estimated based on the complementary relationship as a function of relative humidity (RH, %) and D:

$$f_{sm} = \mathrm{RH}^{D/\beta} \tag{52}$$

where β is the sensitivity of f_{sm} to D and was set as a constant of 0.2 (kP_a) in Mu et al. (2011). However, previous studies have pointed out that the value of β should change with land cover types and climate conditions (Fisher et al. 2008; Yang et al. 2015). For examples, Zhu et al. (2016) have shown that the model performs better using a low value of β for water-limited ecosystems.

For estimating E_c (Eq. 43c), r_a^c is the aerodynamic resistance between mean canopy height (s m^{-1}) and the air above the canopy and r_s^c is the canopy surface resistance (s m^{-1}), which are estimated by:

$$r_a^c = \frac{rh \cdot rr}{rh + rr} \tag{53}$$

$$r_s^c = \frac{1}{Cc} \tag{54}$$

$$rh = \frac{1}{gl\_bl} \tag{55}$$

$$Cc = \frac{gl\_sh(Gs + Gcu)}{Gs + gl\_sh + Gcu} LAI(1 - F_{wet}) \tag{56}$$

$$Gs = Cl \times m(T_{min}) \times m(D) \times rcorr \tag{57}$$

$$Gcu = g\_cu \times rcorr \tag{58}$$

$$rcorr = \frac{P_a}{101.3}\left(\frac{T_i + 273.15}{293.15}\right)^{-1.75} \tag{59}$$

where rr is the radiative heat transfer resistance same as rrc (s m^{-1}), rh is the convective heat transfer resistance (s m^{-1}), gl_bl is the leaf boundary layer conductance (m s^{-1}) that equals gl_sh in the last section, Cc is the surface conductance (m s^{-1}), Gs is the stomatal conductance (m s^{-1}), g_cu is the cuticular conductance per unit LAI (m s^{-1}), Cl is the mean potential stomatal conductance per unit LAI (m s^{-1}), and $rcorr$ is a correction coefficient based on air pressure and temperature. $m(T_{min})$ and $m(D)$ are two scaling factors representing influences from air temperature (i.e., minimum air temperature) and D that reduce stomatal conductance under unfavorable conditions and are calculated as:

$$m(T_{min}) = \begin{cases} 0.1 & T_{min} \le T_{min\_close} \\ \dfrac{T_{min} - T_{min\_close}}{T_{min\_open} - T_{min\_close}} & T_{min\_close} < T_{min} < T_{min\_open} \\ 1.0 & T_{min} \ge T_{min\_open} \end{cases} \tag{60}$$

$$m(D) = \begin{cases} 1.0 & D \le D_{open} \\ \dfrac{D_{close} - D}{D_{close} - D_{open}} & D_{open} < D < D_{close} \\ 0.1 & D \ge D_{open} \end{cases} \tag{61}$$

where subscript close indicates nearly complete inhibition (full stomatal closure) and open indicates no inhibition to stomatal opening and thus transpiration. The biome-dependent model parameter values are given in Table 1.

The Penman-Monteith-Leuning (PML) Model

The Penman-Monteith-Leuning (PML) model combines the Penman-Monteith type equation for plant transpiration, the Leuning model for canopy conductance, and the Priestley-Taylor equation for soil evaporation (Gan et al. 2018; Zhang et al. 2016). In PML, λE is estimated as:

$$\lambda E = \lambda E_{\mathrm{c}} + \lambda E_{\mathrm{s}} \tag{62}$$

in which

$$\lambda E_{\mathrm{c}} = \frac{\Delta R_{\mathrm{n,c}} + \left(\rho_{\mathrm{a}} C_{\mathrm{p}}/\gamma\right) D G_{\mathrm{a}}}{\Delta + 1 + G_{\mathrm{a}}/G_{\mathrm{c}}} \tag{62a}$$

$$\lambda E_{s} = \frac{f\Delta\left(R_{\mathrm{n,s}} - G\right)}{\Delta + 1} \tag{62b}$$

where G_{a} is the aerodynamic conductance (reciprocal of aerodynamic resistance). G_{c} is the canopy conductance, which is estimated based on the stomatal conductance model of Leuning (1995):

$$\begin{aligned} G_{\mathrm{c}} &= \int_{0}^{LAI} g_{\mathrm{s}} dl \\ &= m \frac{P_1}{k_{\mathrm{c}}(P_2 + P_4)} \left[k_{\mathrm{c}} LAI + \ln \frac{P_2 + P_3 + P_4}{P_2 + P_3 \exp(k_{\mathrm{c}} LAI) + P_4}\right] \frac{1}{1 + D/D_0} \end{aligned} \tag{63}$$

where g_{s} is the leaf-level stomatal conductance (m s^{-1}), m is a stomatal conductance coefficient, and D_0 is a parameter that represents the sensitivity of g_{s} to D (at the leaf surface). The terms P_1, P_2, P_3, and P_4 are, respectively, expressed as:

$$P_1 = V_m q I_0 \eta; P_2 = V_m q I_0; P_3 = V_m \eta C_{\mathrm{a}}; P_4 = q I_0 \eta C_{\mathrm{a}}$$

where q is the initial slope of the photosynthesis light response cover (i.e., quantum efficiency; μmol CO_2 (μmol PAR)$^{-1}$, where PAR is the photosynthetic active radiation); η is the initial slope of the photosynthesis CO_2 response curve (i.e., carboxylation efficiency; μmol m^{-2} s^{-1} (μmol m^{-2} s^{-1})$^{-1}$); C_{a} is the atmospheric CO_2 concentration (μmol); and I_0 is the flux density of PAR at the top of the canopy (μmol m^{-2} s^{-1}), which approximates half of the incident solar radiation. V_{m} is a notional capacity to represent the light- and carbon-saturated photosynthesis rate (μmol m^{-2} s^{-1}) and is estimated as a function of T_{a}:

Table 1 The biome look-up table for model parameters. *ENF*, evergreen needleleaf forest; *EBF*, evergreen broadleaf forest; *DNF*, deciduous needleleaf forest; *DBF*, deciduous broadleaf forest; *MF*, mixed forest; *WL*, woody savannas; *SV*, savannas; *CSH*, closed shrubland; *OSH*, open shrubland; *Grass*, grassland, urban and built-up, barren or sparsely vegetated; *Crop*, cropland. (Modified from Table 1 in Mu et al. 2011)

Parameter	ENF	EBF	DNF	DBF	MF	CSH	OSH	WL	SV	Grass	Crop
T_{min_open}	8.31	9.09	10.44	9.94	9.50	8.61	8.80	11.39	11.39	12.02	12.02
T_{min_close}	−8.0	−8.0	−8.0	−6.0	−7.0	−8.0	− 8.0	− 8.0	− 8.0	− 8.0	− 8.0
D_{close}	3000	4000	3500	2900	2900	4300	4400	3500	3600	4200	4500
D_{open}	650	1000	650	650	650	650	650	650	650	650	650
gl_sh	0.04	0.01	0.04	0.01	0.04	0.04	0.04	0.08	0.08	0.02	0.02
gl_e	0.04	0.01	0.04	0.01	0.04	0.04	0.04	0.08	0.08	0.02	0.02
Cl	0.0032	0.0025	0.0032	0.0028	0.0025	0.0065	0.0065	0.0065	0.0065	0.0070	0.0070
rbl_{min}	65.0	70.0	65.0	65.0	65.0	20.0	20.0	25.0	25.0	20.0	20.0
rbl_{max}	95.0	100.0	95.0	100.0	95.0	55.0	55.0	45.0	45.0	50.0	50.0

$$V_{\mathrm{m}} = \frac{V_{\mathrm{m,25}}\exp[0.031(T_{\mathrm{a}} - 25)]}{1 + \exp[0.115(T_{\mathrm{a}} - 41)]} \tag{64}$$

where $V_{\mathrm{m,25}}$ is the value of V_{m} when $T_{\mathrm{a}} = 25$ °C. The biome type-dependent model parameters and details in the derivation of Eq. 64 can be found in Gan et al. (2018).

For estimating soil evaporation, the PML model adopts the Priestley-Taylor equation with a soil evaporation coefficient f to account for the water limitation on evaporation:

$$f = \min\left(\frac{\Sigma_{\mathrm{i-n}}^{\mathrm{i}} P_{\mathrm{i}}}{\Sigma_{\mathrm{i-n}}^{\mathrm{i}} E_{\mathrm{eq,s,i}}}, 1\right) \tag{65}$$

where n is the length of the "time lag" used to balance soil water content after precipitation, P_{i} is precipitation in the ith day (mm day^{-1}), and $E_{\mathrm{eq,s,i}}$ is the equilibrium E_{s} (mm day^{-1}). For each of the ith day in the time series, f is calculated as the accumulative proportion of precipitation to soil evaporative demand of the previous n days (Zhang et al. 2010a).

Model Comparison During MUSOEXE-12 Campaign

To provide an intuitive understanding of the three types of the abovementioned remote sensing E models, a comparison between three two-source models (i.e., TSEB, HTEM, and MOD16) is conducted in the following. Some adjustments were made to the three models to ensure a more reasonable comparison. These include using patch approach for radiation and energy partitioning and adopting the same sets of equations for estimating R_{n} (Allen et al. 1998) and G (i.e., $G = 0.35R_{\mathrm{n,s}}$). Additionally, the suggested parameterization scheme of each model was used in this comparison without performing local calibrations. Therefore, results from this comparison are generalizable, although it does not necessarily result in the best model performance.

Study Region and Remote Sensing Data

The MUSOEXE-12 Campaign: The Multi-Scale Observation Experiment on Evapotranspiration over heterogeneous land surfaces 2012 (MUSOEXE-12) field campaign took place in the middle reach of the Heihe River Basin from May to September 2012 (100°18'E~100°35'E, 38°42'N~39°00'N, Fig. 6).This region is characterized by a cold and dry landscape, with annual mean air temperature and precipitation of 7.2 °C and 126.7 mm (1960–2000), respectively. During the campaign, daily mean air temperature ranged from 12 °C to 28 °C and total precipitation was ~100 mm. Most of the region consists of an elevated plain at ~1550 m a.s.l. with the elevation slightly higher in the southwest and lower in the northeast.

The MUSOEXE-12 campaign is the first thematic experiment in the Heihe Watershed Allied Telemetry Experiment Research project (HiWATER; Li et al.

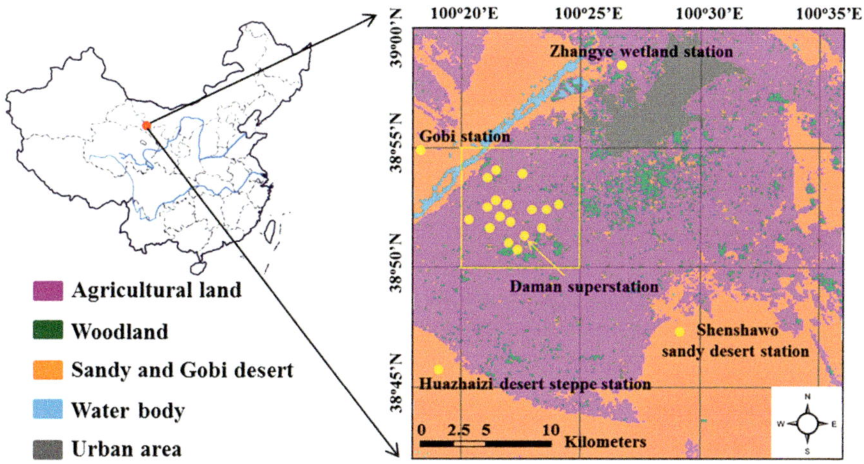

Fig. 6 Location and the land use map of the study area. Circles in the right figure represent eddy covariance towers. The highlighted area within the yellow rectangle denotes a focused area of the field campaign (i.e., the small experimental area) (Yang et al. 2015)

2013), comprising two nested observation matrices for surface energy budget (Fig. 6). One was a large experimental area of ~35 km^2, situating four eddy-covariance (EC) stations with underlying surfaces of desert, desert steppe, Gobi desert, and wetland (i.e., the Shenshawo sandy desert station, the Huazhaizi desert steppe station, the Gobi station, and the Zhangye wetland station, Fig. 6). Another was a small experimental area of ~11 km^2, with agriculture being the major land use type and 17 EC stations scattered in the area. For each EC station, an EC system was employed to measure H and λE at 30 min intervals, and an automatic weather station was installed to record meteorological variables (i.e., R_n, T_a, wind speed and direction, P_a, and RH) at 10-min intervals. Ground heat flux was measured by three soil heat flux plates buried at a depth of 5 cm around each flux tower. A residual method was used to perform the energy balance closure of the EC towers (Twine et al. 2000):

$$\lambda E = R_n - G - H \tag{66}$$

where R_n, G, and H are observations and λE is calculated as the residual in the surface energy budget.

The ratio of E_c over total E was also measured at the Daman superstation site (Fig. 6) using a stable isotope-based method (Wen et al. 2012), which enables the validation of E_c and E_s estimates from the three two-source models. In situ LAI, F_c, and crop height were also measured routinely at each EC station. LAI was observed with a vegetation canopy analyzer (LAI-2000, LI-COR), F_c was measured based on a digital photography method (Liu et al. 2012), and crop height was measured with a tape measure. Detailed description of the MUSOEXE-12 campaign can be found in Li et al. (2013).

Remote Sensing Data: Advance Spaceborne Thermal Emission and Reflection Radiometer (ASTER) images with a spatial resolution of 90 m for the thermal infrared band were used in this study. During the MUSOEXE-12 campaign, ASTER images (visible and near-infrared bands with a spatial resolution of 15 m and thermal infrared bands with a spatial resolution of 90 m) on six cloud-free days (i.e., July 10, Aug 02, Aug 11, Aug 18, Aug 27, and Sep 03) were acquired from NASA's Earth Observing System and Data and Information System (EOSDIS) (http://reverb.echo.nasa.gov/). The local overpass time for ASTER data was between 12:10 and 12:20 pm.

Variable Derivation: Spatial distributions of meteorological variables were obtained by interpolating observations at 21 sites with the inverse distance weighting method (Franke 1982). T_s was retrieved from the thermal infrared bands of the ASTER images using the "temperature and emissivity separation" method provided in Gillespie et al. (1998) and Schmugge et al. (1998). The broadband surface emissivity (ε) was calculated as a linear combination of ASTER channel emissivity using the coefficients provided by Ogawa and Schmugge (2004):

$$\varepsilon = 0.121\varepsilon_{11} + 0.194\varepsilon_{12} + 0.323\varepsilon_{13} + 0.113\varepsilon_{14} + 0.242 \tag{67}$$

where ε_{11}, ε_{12}, ε_{13}, and ε_{14} are channel emissivity provided in the ASTER data of thermal infrared bands 11, 12, 13, and 14.

Broadband surface albedo (α) was derived from the six narrowband reflectances following the method given by Liang (2001):

$$\begin{aligned}\alpha = {} & 0.484\alpha_1 + 0.335\alpha_3 - 0.324\alpha_5 + 0.551\alpha_6 + 0.305\alpha_8 - 0.367\alpha_9 \\ & - 0.0015\end{aligned} \tag{68}$$

where α_1, α_3, α_5, α_6, α_8, and α_9 are narrowband reflectance corresponding to bands 1, 3, 5, 6, 8, and 9 in the ASTER data, respectively.

The Normalized Difference Vegetation Index (NDVI) was calculated from the red (band 3) and near-infrared bands (band 2) following Huete et al. (2002):

$$\mathrm{NDVI} = \frac{\alpha_2 - \alpha_3}{\alpha_2 + \alpha_3} \tag{69}$$

Maps for vegetation parameters (i.e., *LAI*, crop height, and F_c) were derived based on the parametric relationships between NDVI and in situ observations of the corresponding variable (Fig. 7).

Model Performance and Comparison

For comparison with observations from the tower network, flux estimates were averaged over an upwind source area (1–2 pixels/~180 m) for each flux tower (Li et al. 2008). Moreover, all fluxes and meteorological measurements were linearly interpolated to the time of satellite overpass using the two bounding measurements.

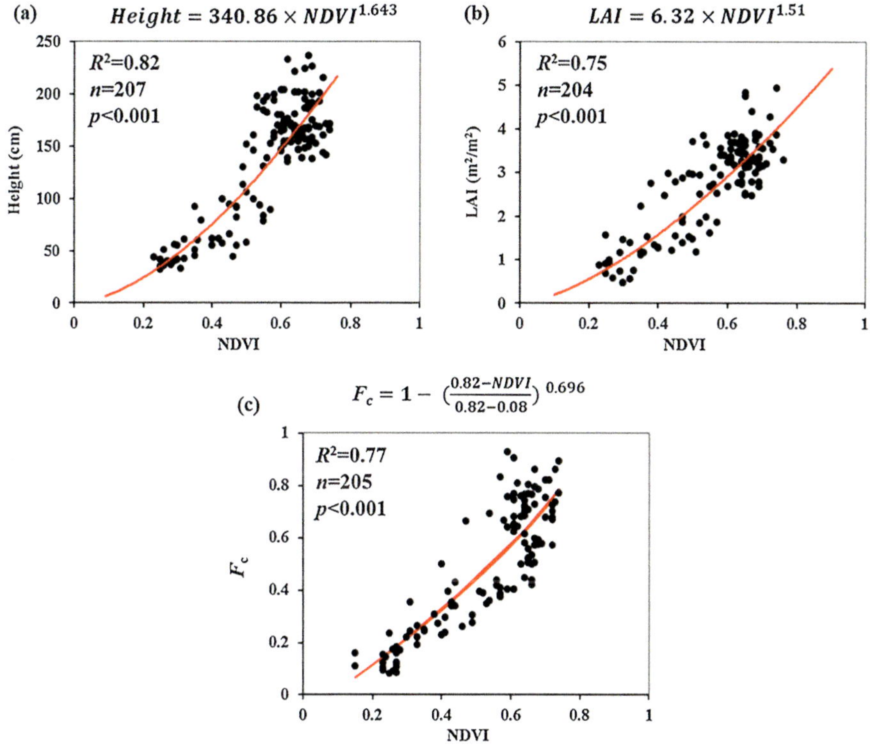

Fig. 7 Relationships between NDVI and (**a**) crop height, (**b**) LAI, (**c**) F_c. n is the sample size, and p-value represents the significance level of the overall regression based on F-test (Yang et al. 2015)

Comparison with EC Tower Measurements: Comparison of estimated sensible and latent heat fluxes from the three models and those from the EC towers is shown in Fig. 8. It shows that the three models performed very differently in estimating H and λE despite using the same net radiation forcing and ground heat flux. HTEM performed generally better than the other two models and the MOD16 algorithm performed worst. The RMSE of H and λE from the HTEM are 32.5 W m^{-2} and 42.3 W m^{-2}, respectively. The TESB performed similar with the HTEM, with an RMSE of 43.1 W m^{-2} for H and an RMSE of 49.8 W m^{-2} for λE.

Different from HTEM and TSEB, the MOD16 algorithm directly estimates λE, whereas H is considered to be a residual in the surface energy budget. However, as shown in Fig. 8f, λE simulated from the MOD16 are generally restricted within a smaller range. The slope of the best fit linear regression between MOD16 λE and observed λE is only 0.36, and the RMSE and mean bias of MOD16 λE are 95.3 W m^{-2} and −34.5 W m^{-2}, respectively. Consequently, H fluxes from the MOD16 algorithm also show large errors, and even no relationship can be found between the estimated H and the observed ones (Fig. 8e).

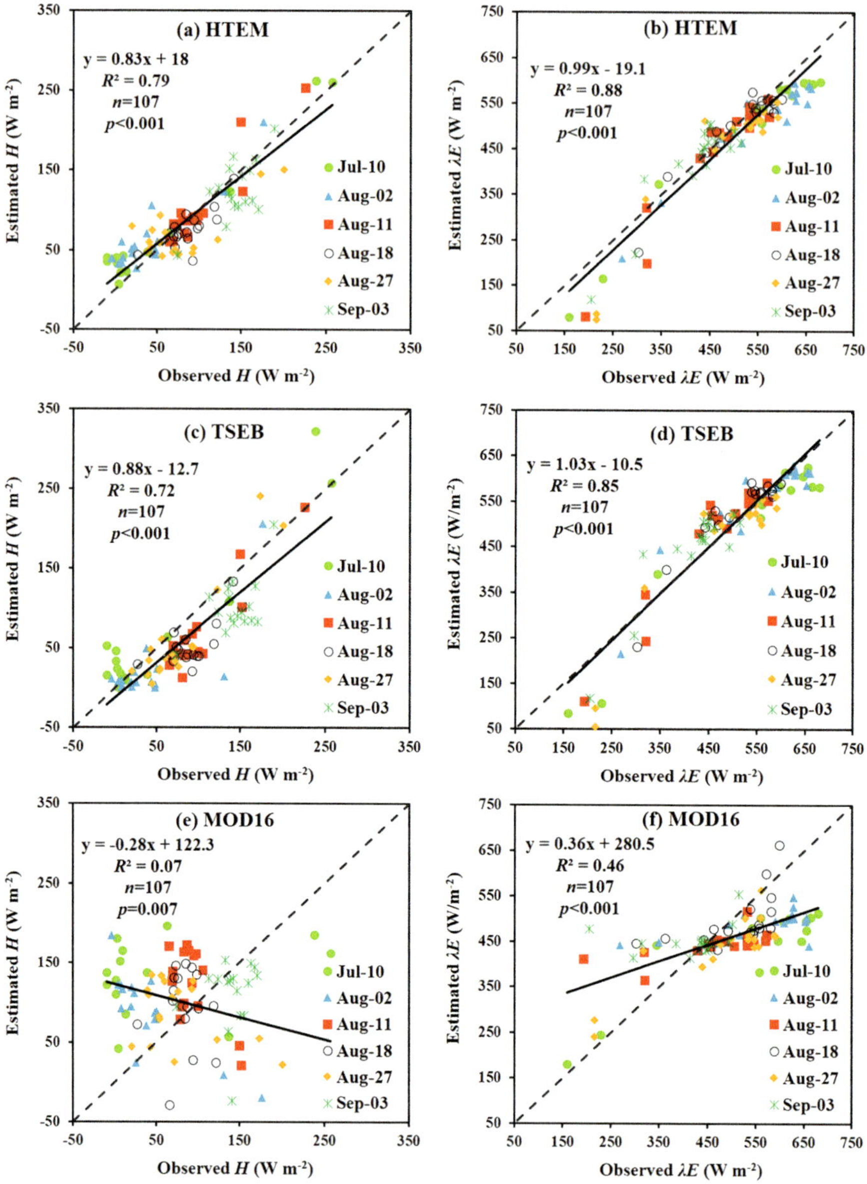

Fig. 8 Comparison of observed H and λE with those estimated from the HTEM (**a** and **b**), the TSEB (**c** and **d**), and the MOD16 ET algorithms (**e** and **f**). The solid lines represent the best fit linear relationship and the dashed lines are the 1:1 line. n is the sample size and p-value represents the significance level of the overall regression based on F-test (Yang et al. 2015)

Spatial Distribution of λE: Taking July 10 as an example, Fig. 9 shows the spatial pattern and frequency distributions of λE estimated from the three models. The λE maps from the HTEM and TSEB models are fairly similar, while comparisons with the MOD16 ET algorithm yield larger differences. λE distributions from HTEM and TSEB span a full range of 0 to >600 W m^{-2} and show a bimodal pattern, reflecting the large difference in land cover types in the area. In contrast, the distribution in λE from the MOD16 algorithm does not show an obvious bimodal pattern and is constrained within a narrower range, extending up to only ~450 W/m^2. Statistics also show a similar performance between the HTEM and TSEB models but, again, considerably different from the MOD16 algorithm. The regional average and standard deviation of λE for HTEM are 401.6 W m^{-2} and 126.9 W m^{-2} and are 440.8 W m^{-2} and 138.9 W m^{-2} for TSEB, respectively, which are remarkably higher than those from the MOD16 algorithm (i.e., 316.3 W m^{-2} and 70.2 W m^{-2}). Similar results are also obtained for other days.

For different land cover types, all three models produced relatively higher λE in better-vegetated surfaces (i.e., agricultural land and woodland) and lower λE in sandy and Gobi deserts as well as in Zhangye City which lies in the upper-middle part of the study area (Table 2). Consistent with the regional distributions of λE shown in Fig. 9, λE estimated from the HTEM is very similar to that from the TSEB model over different land use types, with λE from TSEB slightly higher than that from HTEM. However, λE from the MOD16 algorithm is greatly different from that generated by the other two models, especially for desert areas (Table 2). In addition, disparities in λE over different land use types are also similar for both the HTEM and TSEB models, but much smaller from the MOD16 algorithm, supporting the results in Fig. 9 that the spatial variability in estimated λE is lower in MOD16 than for the other two models.

Evapotranspiration Partitioning: One advantage of two-source models over one-source models is its ability to distinguish evaporation from soil and transpiration from vegetation canopy. The comparison of the ratio of transpiration/evapotranspiration (or $\lambda E_c/\lambda E$) between model estimates and isotope-based measurements at the Daman superstation site is shown in Fig. 10. As we can see, the ratio of $\lambda E_c/\lambda E$ is fairly high (i.e., 87.1% averaged over the 6 days), reflecting high LAI and vegetation cover at the site during the study period. For the three models, the HTEM model produced the closest $\lambda E_c/\lambda E$ ratio (i.e., 86%) with the measurements, whereas the TSEB slightly overestimated $\lambda E_c/\lambda E$ (i.e., 91.2%) and the MOD16 algorithm significantly underestimated it (66.6%).

Results of the comparison at the site correspond well with those from the spatial distribution analysis (using July 10 as an example), with TSEB producing the highest spatial averaged $\lambda E_c/\lambda E$ and MOD16 giving the lowest $\lambda E_c/\lambda E$ (Figs. 11 and 12 and Table 2), which are results of higher λE_c and lower λE_s estimated from the TSEB model and significantly lower λE_c estimated from the MOD16 algorithm (Figs. 11 and 12). Although the three models have similar standard deviations for $\lambda E_c/\lambda E$, the spatial patterns of the two λE components (i.e., λE_c and λE_s) are very different. λE_c and λE_s patterns produced by HTEM and TSEB are quite similar, while those from the MOD16 algorithm show significant smaller spatial variability. In addition, both HTEM and TSEB successfully captured high evaporation rates from

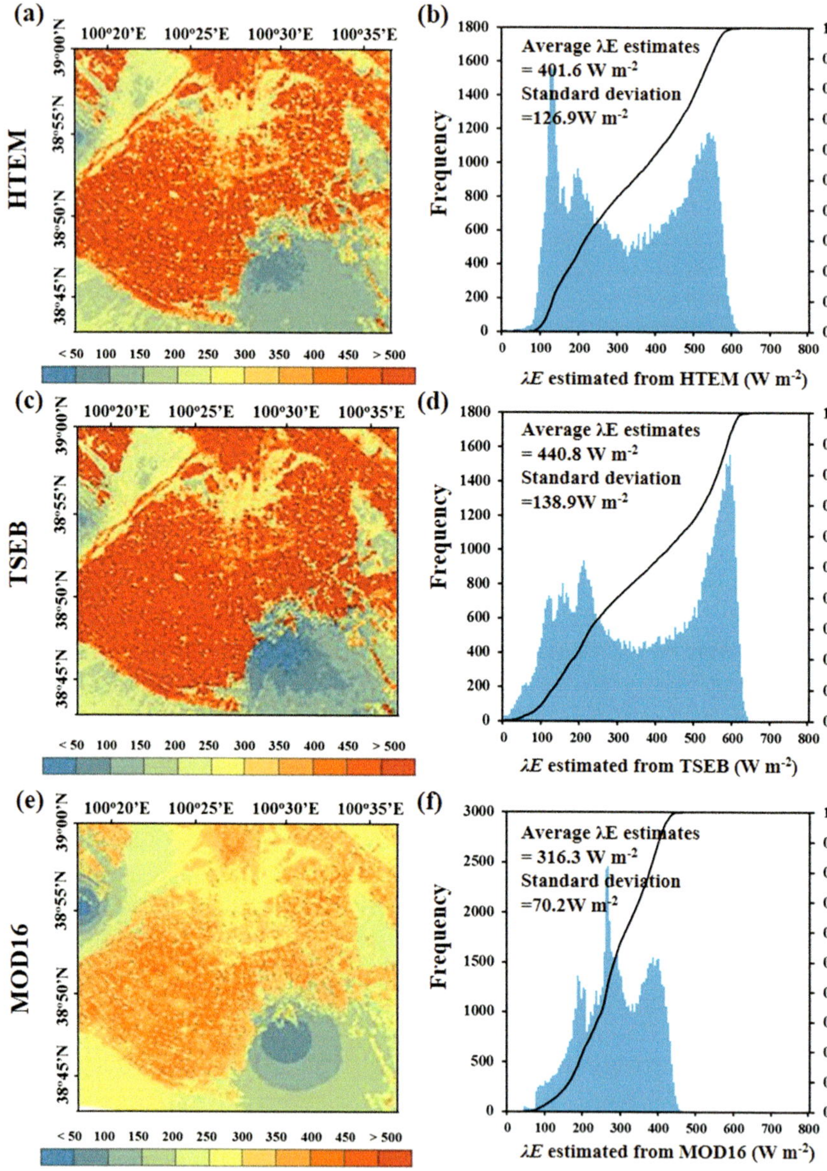

Fig. 9 Spatial patterns, frequency distributions (bar), and cumulative frequencies (curve) of λE estimated from the HTEM (**a** and **b**), the TSEB (**c** and **d**), and the MOD16 algorithms (**e** and **f**) on July 10, 2012. A bin size of 5 W m^{-2} was specified for the frequency distributions; the statistics show the spatially averaged λE and standard deviation (Yang et al. 2015)

Table 2 Mean estimated LE (W/m^2) and LE_c/LE over the six study days for each land use type

	HTEM		TSEB		MOD16	
Land use	λE	$\lambda E_c/\lambda E$	λE	$\lambda E_c/\lambda E$	λE	$\lambda E_c/\lambda E$
Agricultural	399.1	0.72	426.8	0.76	436.4	0.56
Woodland	392.7	0.65	420.9	0.68	426.4	0.51
Urban	263.6	0.49	295.1	0.58	367.1	0.39
Desert	157.0	0.05	179.4	0.13	329.2	0.03
Water	477.1	–	485.2	–	397.1	–

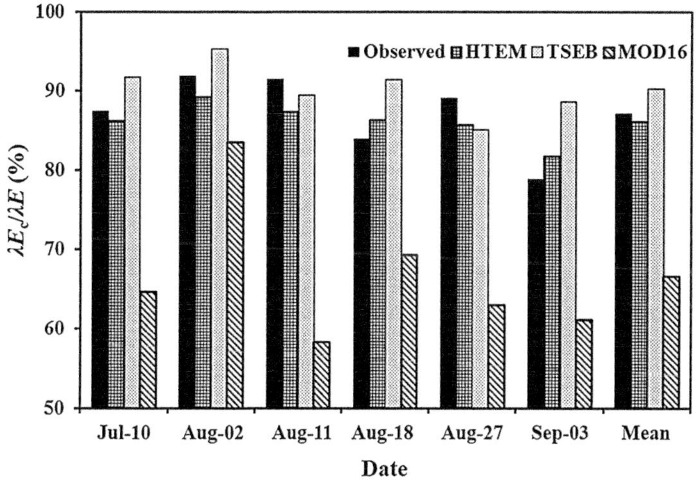

Fig. 10 Comparison of the ratio of transpiration to evapotranspiration (i.e., $\lambda E_c/\lambda E$) estimated from the three models and that measured by isotope analysis at the Daman superstation site. Mean represents averaged $\lambda E_c/\lambda E$ ratios over the 6 days (Yang et al. 2015)

water surfaces located in the upper-left part of the domain. Unfortunately, the MOD16 algorithm again seriously underestimated these expected high evaporative fluxes (Table 2 and Fig. 11).

Why MOD16 Is So Different? The above results clearly show that the MOD16 algorithm performed very differently to HTEM and TSEB. In particular, MOD16 produced a much smaller spatial variation in both λE and $\lambda E_c/\lambda E$ than the other two models. This is because the effects of soil moisture restriction on evaporation and transpiration processes in MOD16 are reflected solely by meteorological forcing (i.e., Eqs. 52 and 57), resulting in a slower response of variations in energy and heat fluxes than the thermal infrared remote sensing-based E models (e.g., HTEM and TSEB). Surface radiative temperature is physically related to soil moisture, with higher/lower T_s generally indicating lower/higher moisture conditions. However, the interactions between T_a or D and soil moisture are usually very weak, even at relatively long time scales. On the other hand, the spatial distributions of meteorological variables (e.g., T_a and D) are generally much more uniform than that of soil moisture, especially for areas that are relatively small but with heterogeneous land use types. Thus, using meteorological forcing solely cannot completely account for the effect of soil moisture restriction on E processes on a physical basis and therefore could result in a smaller variability in predicted λE and $\lambda E_c/\lambda E$. This limitation is expected to be most obvious in areas around transitions between two different land use types. Model intercomparison shows that the difference between λE from MOD16 and that from HTEM and TSEB is closely related to T_s (Fig. 13), highlighting the importance of using T_s to explicitly account for soil moisture restriction on E in remote sensing E models.

Combination of Energy Balance and Penman-Monteith Models: ETWatch

ETWatch is an integrated innovation of the energy balance approach and the Penman-Monteith-based approach. Firstly, single-source energy balance models

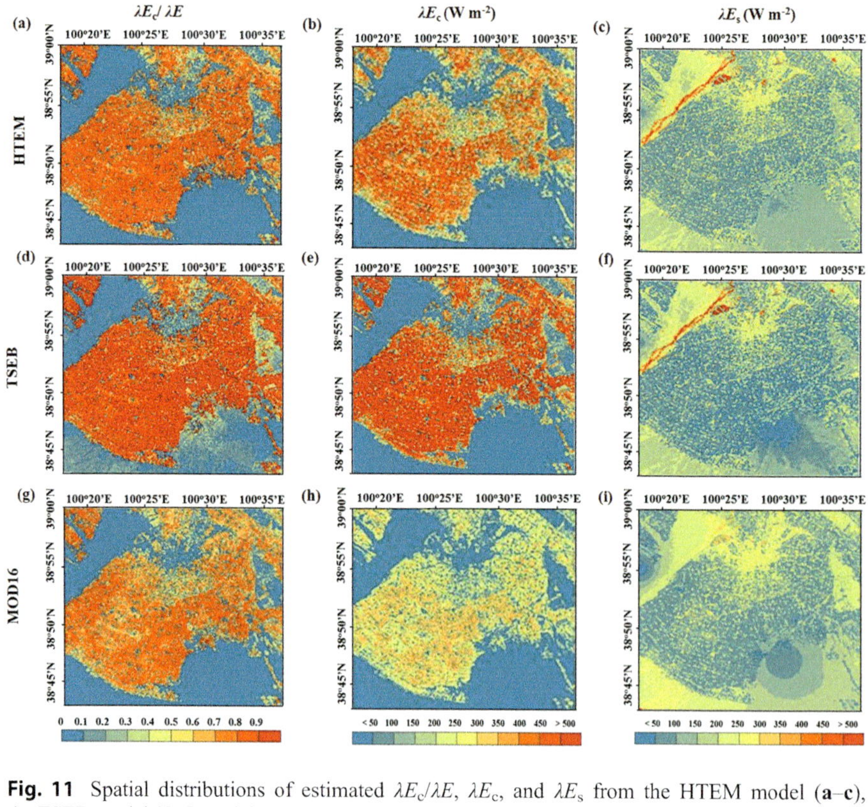

Fig. 11 Spatial distributions of estimated $\lambda E_c/\lambda E$, λE_c, and λE_s from the HTEM model (**a**–**c**), the TSEB model (**d**–**f**), and the MOD16 algorithm (**g**–**i**) for the study region on July 10, 2012 (Yang et al. 2015)

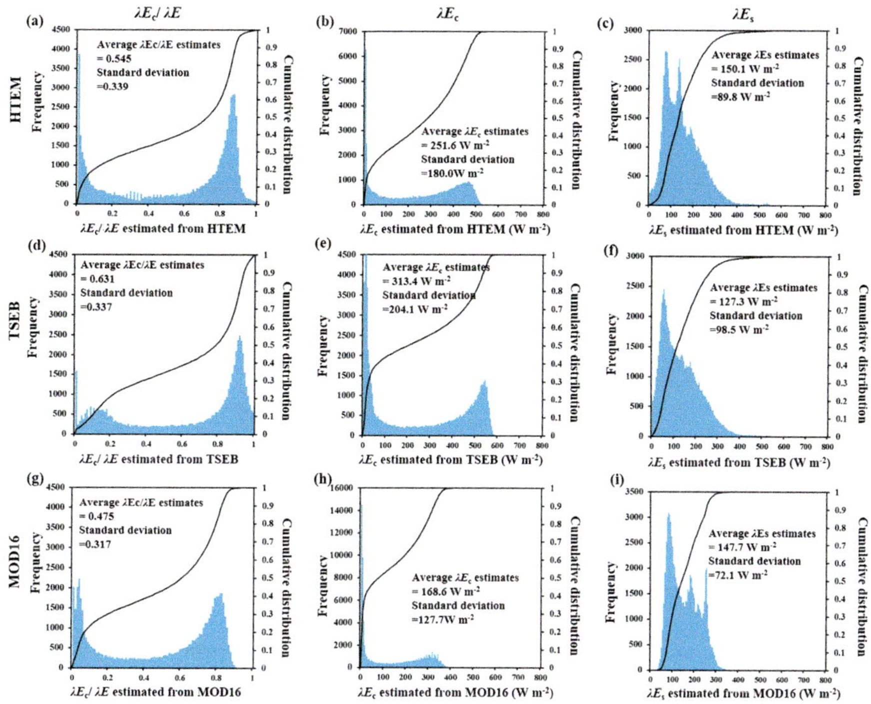

Fig. 12 Frequency distributions (bar) and cumulative frequencies (curve) of estimated $\lambda E_c/\lambda E$, λE_c, and λE_s from the HTEM model (**a**–**c**), the TSEB model (**d**–**f**), and the MOD16 algorithm (**g**–**i**) for the study region on July 10, 2012. A bin size of 5 W m^{-2} was specified for the frequency distributions; the statistics show the spatially averaged λE and standard deviation (Yang et al. 2015)

(i.e., SEBAL and/or SEBS) are used to calculate *E* from vegetated surfaces at satellite overpassing time. Then, surface resistance is determined by inverting the Penman-Monteith model with energy balance model estimated *E* and standard meteorological variables. Then, relationships between surface resistance and other remotely sensed land surface properties (i.e., LAI) at the instantaneous scale are established and used to extrapolate surface resistance to daily scale. The daily surface resistance is then applied in the Penman-Monteith model to calculate E at the daily scale. A data fusion algorithm is later applied to integrate *E* maps at different resolutions. A flowchart of the ETWatch model is given in Fig. 14, and detailed description of the ETWatch model can be found in Wu et al. (2008, 2011).

Other Models

Empirical Model

A common feature in empirical models lies in the use of statistical methods, ranging from simple single-parameter or multivariate methods to complex machine learning

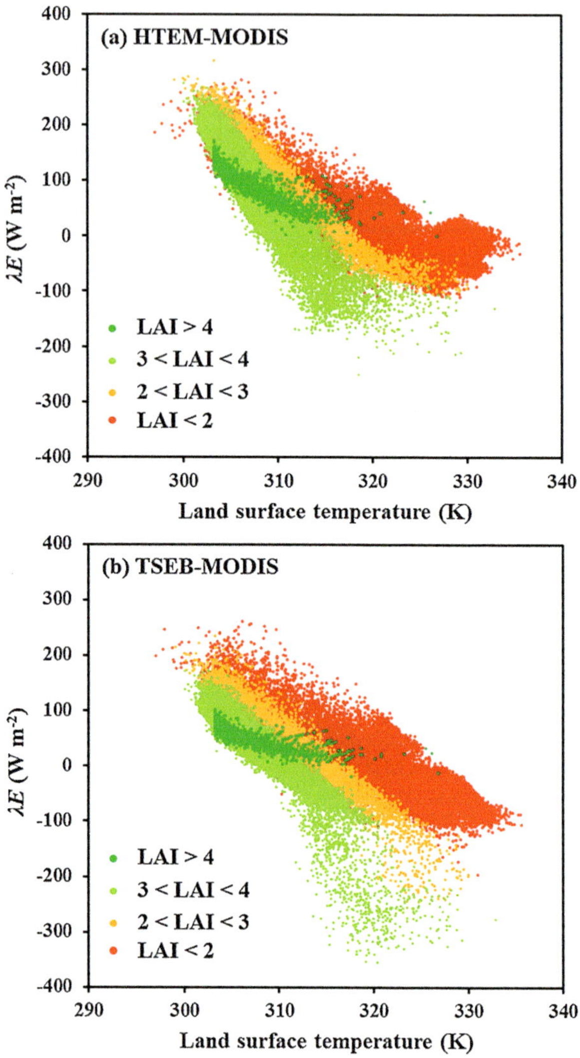

Fig. 13 Difference in estimated λE between (**a**) the HTEM and MOD16 model and between (**b**) the TSEB and MOD16 model, as a function of land surface temperature on July 10, 2012. Color scale represents differences in LAI (Yang et al. 2015)

approaches, to link E with explanatory variables that can be readily obtained from remote sensing techniques. The model of Jackson et al. (1977) is the first empirical model of estimating E from remote sensing, developed over an irrigated wheat farmland in Phoenix, Arizona (USA). The model of Jackson et al. (1977) has a form of:

$$\lambda E = R_{\mathrm{n}} - B(T_{\mathrm{s}} - T_a)^n \tag{70}$$

where λE and R_{n} are daily values (W m^{-2}) and the two temperature variables (i.e., T_{s} and T_{a}) are the temperature at about 13:30 to 14:00 local time. B and n are two model

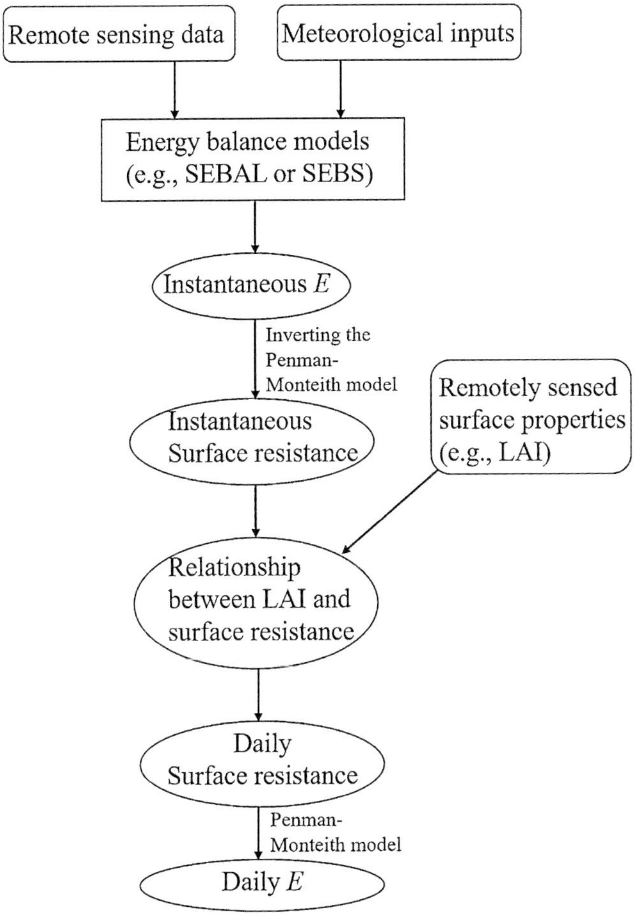

Fig. 14 Flowchart of the key processes in ETWatch

parameters that require local calibration. Carlson et al. (1995) provided explicit expressions of B and n, both as a linear function of F_c, given by:

$$B = 0.0175 + 0.05F_c \quad (\pm 0.002) \tag{71}$$

$$n = 1.004 - 0.335F_c \quad (\pm 0.053) \tag{72}$$

Another empirical model of estimating E at a daily time scale based on remote sensing data and a simple regression is developed by Wang et al. (2007):

$$\lambda E = R_n(a_0 + a_1 \mathrm{VI} + a_2 T) \tag{73}$$

where VI is the remotely sensed vegetation index, which can either be NDVI or EVI (Enhanced Vegetation Index; Huete et al. 2002). T is daytime average (or daily maximum) temperature, which can either be T_a or T_s. a_0, a_1, and a_2 are three model parameters and require local calibration.

Water-Carbon Relation Model

The water-carbon relation models estimate E by exploiting the linkage between the water and the carbon cycles. The scientific basis underpinning this method is that the terrestrial water and carbon cycles are intrinsically coupled through the control of plant stomata. More specifically, on one hand, plant assimilates carbon through photosynthesis. The overall photosynthetic fixation of carbon per unit space and time is called the gross primary production or GPP. On the other hand, plant also loss water to the atmosphere via transpiration at the same time of photosynthesizing. The ratio of GPP over E, known as the ecosystem water use efficiency (WUE), is a relative conservative variable over time for a certain ecosystem (Law et al. 2002).

In the water-carbon relation model developed by Yang et al. (2013a), WUE is estimated as:

$$\mathrm{WUE} = a_1\theta_F + a_2\left(1 - e^{k_c LAI_{\max}}\right) \tag{74}$$

where θ_F is volumetric soil water content at field capacity and $LAI_{\max}$ is the maximum leaf area index during the modeling period. In practical applications, θ_F can be estimated as the water content retained in the soil at −0.02 MPa of suction pressure, which is midway of most reported θ_F values (−0.01 to −0.033 MPa) (Haise et al. 1955). a_1 and a_2 are regression coefficients and are set to −5.55 and 4.89, respectively, based on calibration of Eq. 74 against observations at 20 Ameriflux sites.

GPP is estimated from the Temperature-Greenness (TG) model proposed by Sims et al. (2008). The TG model estimates GPP using a combination of MODIS T_s and EVI products at the 16-day interval as:

$$\mathrm{GPP} = m\_TG \times \mathrm{R\_EVI} \times \mathrm{R}\_T_s \tag{75}$$

in which

$$\mathrm{R\_EVI} = \mathrm{EVI} - 0.1 \tag{75a}$$

$$\mathrm{R}\_T_s = \min\left[\left(\frac{T_s}{30}\right), (2.5 - 0.05T_s)\right] \tag{75b}$$

where m_TG is the coefficient, which can be estimated as a function of annual mean nighttime T_s (Sims et al. 2008). Equation 75a indicates a zero GPP when EVI is smaller than 0.1, and Eq. 75b indicates the value of R_T_s changes linearly between one when T_s is 30 °C and zero when T_s is 0 °C or 50 °C.

Then, E is quantified by:

$$E = \mathrm{GPP}/\mathrm{WUE} \tag{76}$$

Water Balance Model

The water balance models calculated E as a residual term in the catchment water balance equation:

$$E = P - R - \Delta S \tag{77}$$

where P is the precipitation, which can be quantified by remote sensing techniques, rain gauge-based observations, or climate reanalysis. R is the runoff, which is often quantified by dividing observed streamflow at the catchment outlet with the catchment area. ΔS is the catchment storage change, which can be quantified by gravity satellite observations (i.e., the Gravity Recovery and Climate Experiment satellites, GRACE).

Temporal and Spatial Scaling

Satellite remote sensing offers measurements of land surface properties at a certain spatial and temporal resolutions. Often, increased spatial resolution comes at the expense of decreased temporal coverage. For most of the aforementioned models, in particular the SEB models and VI-T_s space models, they can only provide instantaneous E retrievals at the time of satellite overpassing. Therefore, it is necessary to develop methods that upscale instantaneous E into longer time scales. In addition, methods of combining data from different satellite sensors are warranted to provide E estimates at the highest available spatial and temporal resolutions, so that to fascinate various hydrological and water resources management applications.

Temporal Upscaling from Instantaneous *E* to Daily *E*

The Sine Function. The sine function method relates the ratio of instantaneous E to daily E to the diurnal trend of solar radiation with the following form (Jackson et al. 1983):

$$\frac{E_d}{E_i} = \frac{R_{sd}}{R_{si}} = \frac{2N}{\pi \sin(\pi t/N)} \tag{78}$$

where R_s is the incoming solar radiation. Subscripts d and i indicate the daily and instantaneous values, respectively. t is the time of day and N is the duration of daytime, which is quantified as:

$$N = 0.945\{a + b \sin^2[\pi(D + 10)/365]\} \tag{79}$$

in which

$$a = 12 - 5.69 \times 10^{-2} lat - 2.02 \times 10^{-4} lat^2 + 8.25 \times 10^{-6} lat^3 - 3.15 \times 10^{-7} lat^4 \tag{79a}$$

$$b = -0.123 lat - 3.1 \times 10^{-4} lat^2 + 8 \times 10^{-7} lat^3 + 4.99 \times 10^{-7} lat^4 \tag{79b}$$

where D is the day of year and lat is the geographical latitude between 60°S and 60°N. It should be noted that the above equations (i.e., Eqs. 78 and 79) work reasonably well for cloud-free days, and improvements accounting for the mount and temporal coverage of the cloud cover are needed in Eq. 79 for cloudy days.

Constant Evaporative Fraction. The constant evaporative fraction (EF, defined as the ratio of λE over $R_n - G$) method assumes that EF is constant during the daytime (Sugita and Brutsaert 1991). Then, daily λE can be written as:

$$\lambda E_d = (R_n - G)_d EF_i = (R_n - G)_d \frac{\lambda E_i}{(R_n - G)_i} \tag{80}$$

A number of studies have suggested that E during nighttime could account as many as ~10% of the daily total. To account for this effect, Anderson et al. (1997) revised Eq. 80 as:

$$\lambda E_d = (R_n - G)_d \frac{1.1 \lambda E_i}{(R_n - G)_i} \tag{81}$$

Constant Reference Crop Evaporative Fraction. Reference crop evaporative fraction (EF_0) is defined as the ratio of E_i over reference crop E (i.e., E_0; see Eq. 8). Compared with the constant EF method, the constant EF_0 method claims to be better able to capture impacts of advection and changing wind and humidity conditions during a day (Allen et al. 2007). The constant EF_0 method calculates daily E as:

$$E_d = E_{0d} \times EF_{0i} = E_{0d} \times \frac{E_i}{E_{0i}} \tag{82}$$

Modified Constant Evaporative Fraction. The modified constant evaporative fraction method is developed by Ryu et al. (2012) to upscale instantaneous E into 8-day means. Ryu et al. (2012) hypothesized that the mean of diurnal variation of E over an 8-day interval scales with the respective mean diurnal variation of potential solar radiation (R_{sPOT}) over the same interval. The 8-day mean daily E (E_{8d}) is calculated as:

$$E_{8d} = \frac{1}{8} \sum_{d=1}^{8} \frac{E_i}{SF_d} \tag{83}$$

in which

$$SF_d = \frac{E_i}{\int_d E_i dt} \approx \frac{R_{\mathrm{sPOT}i}}{\int_d R_{\mathrm{sPOT}i} dt} \tag{83a}$$

$$R_{\mathrm{sPOT}} = S_{\mathrm{sc}} \times \left[1 + 0.033 \cos\left(\frac{2\pi D}{365}\right)\right] \cos\beta \tag{83b}$$

where t is the time of day and D is the day of the year. S_{sc} is the solar constant (1360 W m^{-2}) and β is the solar zenith angle.

Space-Time Blending

ESTARFM. The enhanced spatial and temporal adaptive reflectance fusion model (ESTARFM) was developed to effectively blend remote sensing images having a high spatial but low temporal resolution (e.g., Landsat images) with those having a high temporal but low spatial resolution (e.g., MODIS images) so that to provide remote sensing data with both high spatial and temporal resolutions (Zhu et al. 2010). For a pure, homogeneous coarse-resolution pixel (e.g., covered by one land use type), the relationship of reflectance (e.g., band B) between the resampled coarse-resolution pixel (the coarse-resolution pixel is firstly resampled to the same spatial resolution, size, and extent of the fine-resolution pixel) and fine-resolution pixel is described by a linear model expressed as:

$$F_{\mathrm{in}} = h_{\mathrm{i}} \times \Delta t + F_{\mathrm{im}} \tag{84}$$

where F and C denote the fine-resolution reflectance and coarse-resolution reflectance, respectively. (x,y) is a given pixel location for both fine- and coarse-resolution images, t_{k} is the acquisition date, and a and b are locally calibrated coefficients of the linear regression model.

For a fine-resolution image and coarse-resolution image obtained at t_0 and another coarse-resolution image obtained at t_{p}, if the land use type and sensor calibration remained unchanged between t_0 and t_{p}, Eq. 84 can be written as Eq. 85 at t_0 and Eq. 86 at t_{p}:

$$F(x,y,t_0,B) = aC(x,y,t_0,B) + b \tag{85}$$

$$F(x,y,t_{\mathrm{p}},B) = aC(x,y,t_{\mathrm{p}},B) + b \tag{86}$$

Combining Eqs. 85 and 86, one can obtain a fine-resolution image at t_{p} by:

$$F(x,y,t_{\mathrm{p}},B) = F(x,y,t_0,B) + a\left[C(x,y,t_{\mathrm{p}},B) - C(x,y,t_0,B)\right] \tag{87}$$

For mixed (heterogeneous) coarse-resolution pixels, a simple linear regression model (i.e., Eq. 84) is not valid anymore. In such conditions, ESTARFM assumes that the reflectance of a mixed pixel can be modelled as a linear combination of the reflectance of the different land cover components present in that pixel weighted by their fractional area coverage, and the changes in reflectance of a mixed pixel between two dates represent the weighted sum of changes in reflectance for each land cover type within the pixel. Further assuming that the proportions of land cover types contained in the mixed coarse-resolution pixel are not changed from date t_m to t_n, the reflectance of a mixed coarse-resolution pixel can be described as the following according to linear mixture model:

$$C_m = \sum_{i=1}^{M} f_i \left(\frac{1}{a} F_{im} - \frac{b}{a} \right) + \varepsilon \tag{88}$$

$$C_n = \sum_{i=1}^{M} f_i \left(\frac{1}{a} F_{in} - \frac{b}{a} \right) + \varepsilon \tag{89}$$

where C_m and C_n are the reflectance of mixing coarse-resolution pixel at date t_m and t_n, respectively; f_i is the fraction of ith land use type (ith endmember); F_{im} and F_{in} are reflectances of ith endmember at date t_m and t_n obtained in fine-resolution image, respectively; M is the total number of endmembers; ε is the residual error; and a and b are the coefficients of relative calibration between coarse and fine-resolution reflectance as used in Eqs. 85 and 86. With Eqs. 88 and 89, changes in coarse-resolution reflectance from t_m to t_n can be written as:

$$C_n - C_m = \sum_{i=1}^{M} \frac{f_i}{a} (F_{in} - F_{im}) \tag{90}$$

ESTARFM also assumes that the change of reflectance of each endmember is linear from t_m to t_n, and then the reflectance of ith endmember at t_n can be described by the reflectance at t_m:

$$F_{in} = h_i \times \Delta t + F_{im} \tag{91}$$

where h_i is the change rate and can be thought as stable during a period. $\Delta t = t_n - t_m$, which can also be written as Eq. 92 given the reflectance of kth endmember at date t_m and t_n, is known:

$$\Delta t = \frac{F_{kn} - F_{km}}{h_k} \tag{92}$$

where h_k is the change rate of kth endmember reflectance.

Introducing Eqs. 92 and 91 into Eq. 90, one can obtain:

$$\frac{F_{kn} - F_{km}}{C_n - C_m} = \frac{h_k}{\sum_{i=1}^{M} \frac{f_i h_i}{a}} = v_k \tag{93}$$

The right part of Eq. 93 is a constant given the prior assumptions that the proportion of each endmember and the reflectance change rate of each endmember are stable. Thus, v_i indicates the ratio of the change of reflectance for kth endmember to the change of reflectance for a mixed coarse-resolution pixel. With this and assuming one pair of fine-resolution and coarse-resolution images at t_0 and another coarse-resolution image at t_p has been acquired, the unknown reflectance of fine-resolution pixel at t_p can be predicted from:

$$F(x,y,t_p,B) = F(x,y,t_0,B) + v(x,y) \times \left[C(x,y,t_p,B) - C(x,y,t_0,B)\right] \quad (94)$$

Details of the derivation of ESTARFM and examples of applying ESTARFM for blending Landsat and MODIS images can be found in Zhu et al. (2010).

Global Remote Sensing Evapotranspiration Products

Over the past decade, a number of global terrestrial E products have been produced. Although these global E products employ different underlying models/algorithms and satellite remote sensing data and vary in spatial and temporal resolutions and temporal coverage, they have been extensively used in various kinds of hydrological, agricultural, meteorological, and ecological scientific studies and practical applications. The major global E products and their key characteristic are summarized in Table 3. It should be noted that only E products that have a global coverage are introduced here; there are also many regional E products that could potentially have a higher regional accuracy and/or spatial and temporal resolutions (e.g., the LAS-SAF E product for Europe and Africa, the ALEXI/DisALEXI E product for the USA, and the FAO-WaPOR E product for Africa).

MOD16. The MODIS terrestrial E product (termed MOD16) provides 8-day (the highest temporal resolution, as it also provides monthly and annual data), 1 km spatial resolution E estimates from 2000 to the near present. It uses a modified Penman-Monteith model (see section "MODIS Terrestrial Evapotranspiration Model (MOD16)"), forced with MODIS remote sensing data (i.e., global 1 km MODIS land cover type, 1 km MODIS vegetation index, and 0.05° CMG MODIS albedo), and a daily meteorological reanalysis data from NASA's Global Modeling and Assimilation Office (GMAO) at $1° \times 1.25°$ spatial resolution. The MOD16 product can be accessed via http://www.ntsg.umt.edu/project/modis/default.php.

AVHRR-E. The AVHRR-E product produces daily 8 km spatial resolution E estimates from 1983 to 2013. It quantifies canopy transpiration and soil evaporation using a modified Penman-Monteith approach with biome-specific canopy conductance determined from the NDVI, and quantifies open water evaporation using a Priestley-Taylor approach, forced with AVHRR GIMMS NDVI, NCEP/NCAR Reanalysis (NNR) daily surface meteorology, and NASA/GEWEX Surface Radiation Budget (SRB) solar radiation inputs. The AVHRR-E product can be accessed via http://www.ntsg.umt.edu/project/global-et.php.

Table 3 Summary of major global E products

Product name	Model scheme	Temporal coverage	Temporal resolution	Spatial resolution	E components	Reference
MOD16	Modified Penman-Monteith	2000–now	8 days	1 km	No	Mu et al. (2011)
AVHRR-E	Modified Penman-Monteith	1983–2013	Daily	8 km	No	Zhang et al. (2010b)
PML-E	Modified Penman-Monteith	1982–2012	Monthly	0.5°	Yes	Zhang et al. (2016)
LST-E	Modified Penman-Monteith	2000–2013	Daily	0.05°	No	Raoufi and Beighley (2017)
PT-JPL	Modified Priestley-Taylor	1986–1993	Monthly	1°	No	Fisher et al. (2008)
GLEAM	Modified Priestley-Taylor	1980–2016	Daily	0.25°	Yes	Miralles et al. (2011)
WB-E	Water balance	1982–2009	Annual	0.5°	No	Zeng et al. (2012)
MTE-E	Model tree ensemble	1982–2011	Monthly	0.5°	No	Jung et al. (2011)
GLASS-E	Bayesian ensemble	2000–now	8 days	1 km	No	Yao et al. (2014)
		1982–now	8 days	5 km	No	Yao et al. (2014)

PML-E. The PML-E product provides global estimates of plant transpiration, soil evaporation, and canopy interception at 0.5° spatial resolution and monthly time step from 1981 to 2012. It employs a modified Penman-Monteith model forced with AVHRR GIMMS NDVI and meteorological data from two sources, i.e., the Princeton Global Forcing data and the WATCH ERA-Interim Forcing data. The PML-E is further constrained by the Budyko framework using global streamflow observation at the mean annual scale to ensure the internal water balance. The PML-E data can be accessed via https://data.csiro.au/dap/landingpage?pid=csiro%3A17375.

LST-E. The LST-E product provides daily 0.05° spatial resolution E estimates for the period of 2000 to 2013. It estimates E based on the Penman-Monteith equation with radiation and vapor pressures derived using remotely sensed land surface temperature from MODIS Aqua and Terra satellites.

PT-JPL. The PT-JPL product provides monthly 1° spatial resolution E estimates from 1986 to 1993. It adopts the Priestley-Taylor model to firstly estimate potential

E and then a biometeorological approach for translating potential *E* into actual *E*. It uses AVHRR GIMMS NDVI and meteorological data from the International Satellite Land-Surface Climatology Project, Initiative II. The PT-JPL global *E* product can be accessed via https://ecostress.jpl.nasa.gov/products.

GLEAM. The GLEAM provides daily estimates of plant transpiration, soil evaporation, and canopy interception at 0.25° spatial resolution from 1980 to 2016. Actual *E* in GLEAM is calculated as the Priestley-Taylor potential *E* (except for canopy interception) multiplied by a water stress factor. Different from Penman-Monteith-based approaches (e.g., MOD16, AVHRR-E, PML-E) where water stress on *E* is estimated based on meteorological variables (e.g., vapor pressure deficit), GLEAM explicitly considers soil water constraint on *E* through incorporating microwave satellite-based measurement of surface soil moisture. The GLEAM global *E* product can be accessed via https://www.gleam.eu/#home.

WB-E. The WB-E product provides global annual estimates of *E* at 0.5° spatial resolution from 1982 to 2009. It firstly calculates catchment *E* as a residual in catchment water balance equation based on the Climate Research Unit precipitation data, streamflow observations from 59 global large basins, and GRACE-derived total storage change. Then, a multi-regression model is used to extrapolate catchment *E* into the entire global land surface.

MTE-E. The MTE-E product provides global monthly estimates of *E* at 0.5° spatial resolution from 1982 to 2011 by upscaling FLUXNET observation of *E* to the global scale using a data-adaptive machine learning approach (i.e., the model tree ensemble (MTE) approach). The key satellite input in the MTE-E is the AVHRR GIMMS NDVI-derived f_{PAR} (fractional of photosynthetic active radiation). The MTE-E product can be accessed via https://www.bgc-jena.mpg.de/geodb/projects/Data.php.

GLASS-E. The GLASS-E product provides global 8-day estimates of *E* at 1 km (2000 onward based on MODIS data) or 5 km (1982 onward based on AVHRR data) by merging *E* estimates from five process-based and/or semiempirical models using a Bayesian model averaging method. The five models used include the MOD16 model, the revised Penman-Monteith model (Yuan et al. 2010), the Priestley-Taylor-based model (Fisher et al. 2008), the modified Priestley-Taylor model (Yao et al. 2013), and the semiempirical Penman model of University of Maryland (Wang et al. 2010).

Conclusion and Outlook

Since the late 1970s, satellite remote sensing has been widely used in *E* modelling by providing critical variables depicting characteristics and/or the state of the land surface. The basis of satellite-based *E* modelling approaches is reliant on physical and mathematical description and/or simplification of the interactions of water and heat fluxes between vegetation and soil components. Although different modelling approaches entail different assumptions, configurations, and coupling between vegetation and soil, they all primarily rely on two remotely sensed information, that is, land surface temperature and vegetation index.

There is always a trade-off between data requirement, model complexities, and uncertainties, as well as purposes of studies and applications. Many satellite-based E models were developed with the intention to reduce parameters and forcing data so as to be applicable over large heterogeneous areas. It is worthwhile highlighting that for large-scale studies, use of E models/products with the law of parsimony and simplicity is necessary. Therefore, incorporating remotely sensed vegetation indices as was done by the MOD16 E product is feasible and should be very useful on global and continental scales. For such global- and continental-scale remote sensing E models/products, evaluation of their validity and accuracy should focus on the spatial and temporal patterns as well as the magnitude over large regions, rather than pursuing local- or site-level accuracies. For regional, watershed, and field scales, use of SEB and/or VI-T_s approaches by incorporating more or less a priori knowledge and information on soil and vegetation is necessary and should be able to provide more realistic E and partitioning between E_c and E_s.

In addition, it is also widely noticed in the literature that satellite-based E models often perform relatively poor at large scales (e.g., Chen et al. 2014; Long et al. 2014). This is partly due to the fact that the majority of satellite-based E models (except for water balance-based approaches) are based on energy balance, whereas water balance is often under-attended in these models (Long et al. 2014). This remains a challenge for future studies. One possible solution would be to incorporate the E equations into hydrological model and perform parameter estimation so that the water balance within a given domain (e.g., river basin) can be satisfied. Another possible solution would be to constrain remote sensing-estimated E with catchment water balance observations at long-term scales.

Finally, traditional satellites (e.g., Landsat, MODIS, AVHRR) only have moderate-to-low temporal coverages, which require an additional temporal scaling method to upscale instantaneous E retrievals at satellite-overpassing time to longer time scales. This upscaling procedure often introduces additional uncertainties in the final E estimates. With recent development in geostationary satellites, which provide temporally continuous observations of multiple land surface variables (e.g., the MSG satellite with 15-min temporal resolution; https://www.esa.int/Our_Activities/Observing_the_Earth/Meteosat/MSG_overview2), this represents an effective and promising way to overcome uncertainties raised by traditional upscaling methods and should be considered as a key focus in future studies as we continue on mapping E from satellite remote sensing.

References

N. Agam, W.P. Kustas, M.C. Anderson, J.M. Norman, P.D. Colaizzi, T.A. Howell, J.H. Prueger, T.P. Meyers, T.B. Wilson, Application of the Priestley–Taylor Approach in a Two-Source Surface Energy Balance Model, J. Hydrometeorol. **11**, 185 (2010)

R.G. Allen, L.S. Pereira, D. Raes, M. Smith, *Crop Evapotranspiration*. FAO Irrigation and Drainage Paper 56, Rome, (1998)

R.G. Allen, M. Tasumi, A. Morse, R. Trezza, J.L. Wright, W. Bastiaanssen, W. Kramber, I. Lorite, C.W. Robison, Satellite-Based Energy Balance for Mapping Evapotranspiration with Internalized Calibration (METRIC) - Applications, J. Irrig. Drain. Eng. **133**, 395 (2007)

M.C. Anderson, J.M. Norman, G.R. Diak, W.P. Kustas, J.R. Mecikalski, A two-source time-integrated model for estimating surface fluxes using thermal infrared remote sensing, Remote Sens. Environ. **60**, 195 (1997)

M.C. Anderson, J.M. Norman, J.R. Mecikalski, J.A. Otkin, W.P. Kustas, A climatological study of evapotranspiration and moisture stress across the continental United States based on thermal remote sensing: 1. Model formulation, J. Geophys. Res. Atmos. **112** (2007)

W.G.M. Bastiaanssen, M. Menenti, R.A. Feddes, A.A.M. Holtslag, A remote sensing surface energy balance algorithm for land (SEBAL). 1. Formulation, J. Hydrol. **212–213**, 198 (1998)

G.B. Bonan, *Ecological Climatology: Concepts and Applications* (Cambridge University Press, New York, 2002)

G.S. Campbell, J.M. Norman, *Introduction to Environmental Biophysics* (Springer, New York, 1998)

T. Carlson, An overview of the "Triangle Method" for estimating surface evapotranspiration and soil moisture from satellite imagery, Sensors **7**, 1612 (2007)

T.N. Carlson, W.J. Capehart, R.R. Gillies, A new look at the simplified method for remote sensing of daily evapotranspiration, Remote Sens. Environ. **54**, 161 (1995)

Y. Chen et al., Comparison of satellite-based evapotranspiration models over terrestrial ecosystems in China, Remote Sens. Environ. **140**, 279 (2014)

X.P. Deng et al., Improving agricultural water use efficiency in arid and semiarid areas of China, Agric. Water Manag. **80**, 23 (2006)

J.B. Fisher, K.P. Tu, D.D. Baldocchi, Global estimates of the land–atmosphere water flux based on monthly AVHRR and ISLSCP-II data, validated at 16 FLUXNET sites, Remote Sens. Environ. **112**, 901 (2008)

R. Franke, Scattered Data Interpolation: Tests of Some Method, Math. Comput. **38**, 181 (1982)

R. Gan, Y. Zhang, H. Shi, Y. Yang, D. Eamus, L. Cheng, F.H.S. Chiew, Q. Yu, Use of satellite leaf area index estimating evapotranspiration and gross assimilation for Australian ecosystems, Ecohydrology **0**, e1974 (2018)

A. Gillespie, S. Rokugawa, T. Matsunaga, J.S. Cothern, S. Hook, A.B. Kahle, A temperature and emissivity separation algorithm for Advanced Spaceborne Thermal Emission and Reflection Radiometer (ASTER) images, IEEE Trans. Geosci. Remote Sens. **36**, 1113 (1998)

H. Guan, J.L. Wilson, A hybrid dual-source model for potential evaporation and transpiration partitioning, J. Hydrol. **377**, 405 (2009)

H.R. Haise, H.J. Haas, L.R. Jensen, Soil moisture studies of some Great Plains soil: II Field capacity as related to 1/3-atmosphere percentage, and "minimum point" related to 15- and 26- atmosphere percentages, Soil Sci. Soc. Am. J. **19**, 20 (1955)

A.S. Hope, D.E. Petzold, S.N. Goward, R.M. Ragan, Simulated relationships between spectral reflectance, thermal emissions, and evapotranspiration of a soybean canopy, J. Am. Water Resour. Assoc. **22**, 1011 (1986)

A. Huete, K. Didan, T. Miura, E.P. Rodriguez, X. Gao, L.G. Ferreira, Overview of the radiometric and biophysical performance of the MODIS vegetation indices, Remote Sens. Environ. **83**, 195 (2002)

R.D. Jackson, R.J. Reginato, S.B. Idso, Wheat canopy temperature: A practical tool for evaluating water requirements, Water Resour. Res. **13**, 651 (1977)

R.D. Jackson, J.L. Hatfield, R.J. Reginato, S.B. Idso, P.J. Pinter, Estimation of daily evapotranspiration from one time-of-day measurements, Agric. Water Manag. **7**, 351 (1983)

L. Jiang, S. Islam, A methodology for estimation of surface evapotranspiration over large areas using remote sensing observations, Geophys. Res. Lett. **26**(17), 2773–2776 (1999)

L. Jiang, S. Islam, Estimation of surface evaporation map over Southern Great Plains using remote sensing data, Water Resour. Res. **37**, 329 (2001)

M. Jung et al., Global patterns of land-atmosphere fluxes of carbon dioxide, latent heat, and sensible heat derived from eddy covariance, satellite, and meteorological observations, J. Geophys. Res. Biogeosci. **116** (2011)

B.E. Law et al., Environmental controls over carbon dioxide and water vapor exchange of terrestrial vegetation, Agric. For. Meteorol. **113**, 97 (2002)

R. Leuning, A critical appraisal of a combined stomatal-photosynthesis model for C3 plants, Plant Cell Environ. **18**, 339 (1995)

F. Li, W.P. Kustas, M.C. Anderson, J.H. Prueger, R.L. Scott, Effect of remote sensing spatial resolution on interpreting tower-based flux observations, Remote Sens. Environ. **112**, 337 (2008)

X. Li et al., Heihe Watershed Allied Telemetry Experimental Research (HiWATER): Scientific Objectives and Experimental Design, Bull. Am. Meteorol. Soc. **94**, 1145 (2013)

S. Liang, Narrowband to broadband conversions of land surface albedo I: Algorithms, Remote Sens. Environ. **76**, 213 (2001)

C.M. Liu, X. Zhang, Y. Zhang, Determination of daily evaporation and evapotranspiration of winter wheat and maize by large-scale weighing lysimeter and micro-lysimeter, Agric. For. Meteorol. **111**, 109 (2002)

Y. Liu, X. Mu, H. Wang, G. Yan, G. Henebry, A novel method for extracting green fractional vegetation cover from digital images, J. Veg. Sci. **23**, 406 (2012)

D. Long, V. P. Singh, A two-source trapezoid model for evapotranspiration (TTME) from satellite imagery, Remote Sens. Environ. **121**, 370–388 (2012).

D. Long, V.P. Singh, B.R. Scanlon, Deriving theoretical boundaries to address scale dependencies of triangle models for evapotranspiration estimation, J. Geophys. Res. Atmos. **117**, (2012)

D. Long, L. Longuevergne, B.R. Scanlon, Uncertainty in evapotranspiration from land surface modeling, remote sensing, and GRACE satellites, Water Resour. Res. **50**, 1131 (2014)

O. Merlin, J. Chirouze, A. Olioso, L. Jarlan, G. Chehbouni, G. Boulet, An image-based four-source surface energy balance model to estimate crop evapotranspiration from solar reflectance/thermal emission data (SEB-4S), Agric. For. Meteorol. **184**, 188 (2014)

D.G. Miralles, R.A.M. de Jeu, J.H. Gash, T.R.H. Holmes, A.J. Dolman, Magnitude and variability of land evaporation and its components at the global scale, Hydrol. Earth Syst. Sci. **15**, 967 (2011)

J. Monteith, Evaporation and environment, Symp. Soc. Exp. Biol. **19**, 205 (1965)

M.S. Moran, T.R. Clarke, Y. Inoue, A. Vidal, Estimating crop water deficit using the relation between surface-air temperature and spectral vegetation index, Remote Sens. Environ. **49**, 246 (1994)

Q. Mu, F.A. Heinsch, M. Zhao, S.W. Running, Development of a global evapotranspiration algorithm based on MODIS and global meteorology data, Remote Sens. Environ. **111**, 519 (2007)

Q. Mu, M. Zhao, S.W. Running, Improvements to a MODIS global terrestrial evapotranspiration algorithm, Remote Sens. Environ. **115**, 1781 (2011)

R.R. Nemani, S.W. Running, J. Appl. Meteorol. **28**, 276 (1989)

K. Nishida, R.R. Nemani, J.M. Glassy, S.W. Running, Development of an evapotranspiration index from Aqua/MODIS for monitoring surface moisture status, IEEE Trans. Geosci. Remote Sens. **41**, 493 (2003)

J.M. Norman, W.P. Kustas, K.S. Humes, Source approach for estimating soil and vegetation energy fluxes in observations of directional radiometric surface temperature, Agric. For. Meteorol. **77**, 263 (1995)

J.M. Norman, M.C. Anderson, W.P. Kustas, A.N. French, J. Mecikalski, R. Torn, G.R. Diak, T.J. Schmugge, B.C.W. Tanner, Remote sensing of surface energy fluxes at 101-m pixel resolutions, Water Resour. Res. **39**, (2003)

K. Ogawa, T. Schmugge, Mapping Surface Broadband Emissivity of the Sahara Desert Using ASTER and MODIS Data, Earth Interact. **8**, 1 (2004)

T. Oki, S. Kanae, Global hydrological cycle and world water resources, Science **25**, 1068 (2006)

C.A. Paulson, The Mathematical Representation of Wind Speed and Temperature Profiles in the Unstable Atmospheric Surface Layer, J. Appl. Meteorol. **9**, 857 (1970)

R. Raoufi, E. Beighley, Estimating Daily Global Evapotranspiration Using Penman–Monteith Equation and Remotely Sensed Land Surface Temperature, Remote Sens. **9**, 1138 (2017)

G.J. Roerink, Z. Su, M. Menenti, S-SEBI: A simple remote sensing algorithm to estimate the surface energy balance, Phys. Chem. Earth Part B: Hydrol. Oceans Atmos. **25**, 147 (2000)

R.D. Rosen, *The Global Energy Cycle* (Cambridge University Press, Cambridge, 1999)

Y. Ryu et al., On the temporal upscaling of evapotranspiration from instantaneous remote sensing measurements to 8-day mean daily-sums, Agric. For. Meteorol. **152**, 212 (2012)

J.M. Sánchez, W.P. Kustas, V. Caselles, M.C. Anderson, Modelling surface energy fluxes over maize using a two-source patch model and radiometric soil and canopy temperature observations, Remote Sens. Environ. **112**, 1130 (2008)

T. Schmugge, S.J. Hook, C. Coll, Recovering Surface Temperature and Emissivity from Thermal Infrared Multispectral Data, Remote Sens. Environ. **65**, 121 (1998)

D.A. Sims et al., A new model of gross primary productivity for North American ecosystems based solely on the enhanced vegetation index and land surface temperature from MODIS, Remote Sens. Environ. **112**, 1633 (2008)

Z. Su, The Surface Energy Balance System (SEBS) for estimation of turbulent heat fluxes, Hydrol. Earth Syst. Sci. **6**, 85 (2002)

M. Sugita, W. Brutsaert, Daily evaporation over a region from lower boundary layer profiles measured with radiosondes, Water Resour. Res. **27**, 747 (1991)

Z. Sun, Q. Wang, B. Matsushita, T. Fukushima, Z. Ouyang, M. Watanabe, Development of a Simple Remote Sensing EvapoTranspiration model (Sim-ReSET): Algorithm and model test, J. Hydrol. **376**, 476 (2009)

T.E. Twine, W.P. Kustas, J.M. Norman, D.R. Cook, P.R. Houser, T.P. Meyers, J.H. Prueger, P.J. Starks, M.L. Wesely, Correcting eddy-covariance flux underestimates over a grassland, Agric. For. Meteorol. **103**, 279 (2000)

K. Wang, Z. Li, M. Cribb, Estimation of evaporative fraction from a combination of day and night land surface temperatures and NDVI: A new method to determine the Priestley–Taylor parameter, Remote Sens. Environ. **102**, 293 (2006)

K. Wang, P. Wang, Z. Li, M. Cribb, M. Sparrow, A simple method to estimate actual evapotranspiration from a combination of net radiation, vegetation index, and temperature, J. Geophys. Res. Atmos. **112** (2007)

K. Wang, R. Dickson, M. Wild, S. Liang, Evidence for decadal variation in global terrestrial evapotranspiration between 1982 and 2002: 1. Model development, J. Geophys. Res. **115**, D20112 (2010)

R. Waters, R. Allen, M. Tasumi, R. Trezza, W. Bastiaanssen, SEBAL Advanced Training and Users Manual version 1.0, Idaho (2002)

E.K. Webb, Profile relationships: The log-linear range, and extension to strong stability, Q. J. R. Meteorol. Soc. **96**, 67 (1970)

X.F. Wen, X. Lee, X.M. Sun, J.L. Wang, Y.K. Tang, S.G. Li, G.R. Yu, Intercomparison of Four Commercial Analyzers for Water Vapor Isotope Measurement, J. Atmos. Ocean. Technol. **29**, 235 (2012)

B. Wu, J. Xiong, N. Yan, ETWatch for monitoring regional evapotranspiration with remote sensing, Adv. Water Sci. **19**, 671–678 (2008)

B. Wu, J. Xiong, N. Yan, ETWatch: A method of multi-resolution ET data fusion, J. Remote Sens. **15**, 224–239 (2011)

Y. Yang, *Evapotranspiration Over Heterogeneous Vegetated Surfaces: Models and Applications* (Springer, Berlin, 2015)

Y. Yang, S. Shang, A hybrid dual-source scheme and trapezoid framework–based evapotranspiration model (HTEM) using satellite images: Algorithm and model test, J. Geophys. Res. Atmos. **118**, 2284 (2013)

Y. Yang, D. Long, S. Shang, Remote estimation of terrestrial evapotranspiration without using meteorological data, Geophys. Res. Lett. **40**, 3026 (2013a)

Y.Yang, R.L. Scott, S. Shang, Modeling evapotranspiration and its partitioning over a semiarid shrub ecosystem from satellite imagery: a multiple validation, (SPIE, 2013b), p. 16

Y. Yang, D. Long, H. Guan, W. Liang, C. Simmons, O. Batelaan, Comparison of three dual-source remote sensing evapotranspiration models during the MUSOEXE-12 campaign: Revisit of model physics, Water Resour. Res. **51**, 3145 (2015)

Y. Yao et al., MODIS-driven estimation of terrestrial latent heat flux in China based on amodified Priestley–Taylor algorithm, Agric. For. Meteorol. **171–172**, 187–202 (2013)

Y. Yao et al., Bayesian multimodel estimation of global terrestrial latent heat flux from eddy covariance, meteorological, and satellite observations, J. Geophys. Res.-Atmos. **119**, 4524–4545 (2014)

W. Yuan et al., Global estimates of evapotranspiration and gross primary production based on MODIS and global meteorology data, Remote Sens. Environ. **114**, 1416–1431 (2010)

Z.Z. Zeng, S.L. Piao, X. Lin, G.D. Yin, S.S. Peng, P. Ciais, R.B. Myneni, Global evapotranspiration over the past three decades: estimation based on the water balance equation combined with empirical models, Environ. Res. Lett. **7**, 014026 (2012)

R.H. Zhang, X.M. Sun, W.M. Wang, J.P. Xu, Z.L. Zhu, J. Tian, An operational two-layer remote sensing model to estimate surface flux in regional scale: Physical background, Sci. China Earth Sci. **48**, 225 (2005)

Y. Zhang, R. Leuning, L.B. Hutley, J. Beringer, I. McHugh, J.P. Walker, Using long-term water balances to parameterize surface conductances and calculate evaporation at 0.05° spatial resolution, Water Resour. Res. **46**, W05512 (2010a)

K. Zhang, J.S. Kimball, R.R. Nemani, S.W. Running, A continuous satellite-derived global record of land surface evapotranspiration from 1983 to 2006, Water Resour. Res. **46** (2010b)

Y. Zhang et al., Multi-decadal trends in global terrestrial evapotranspiration and its components, Sci. Rep. **6**, 19124 (2016)

X. Zhu, J. Chen, F. Gao, X. Chen, J.G. Masek, An enhanced spatial and temporal adaptive reflectance fusion model for complex heterogeneous regions, Remote Sens. Environ. **114**, 2610 (2010)

G.F. Zhu, K. Zhang, X. Li, S.M. Liu, Z.Y. Ding, J.Z. Ma, C.L. Huang, T. Han, J.H. He, Evaluating the complementary relationship for estimating evapotranspiration using the multi-site data across north China, Agric. For. Meteorol. **230–231**, 33 (2016)

Micrometeorological Methods to Determine Evapotranspiration

Shaomin Liu and Ziwei Xu

Contents

Abstract

Evapotranspiration (ET) is a combination of two distinct processes, soil or water evaporation (E) and plant transpiration (T), that occur between plants and the atmosphere, soil and the atmosphere, and water and the atmosphere. ET is also an important link between the terrestrial ecosystem and hydrological processes. In this chapter, we focus primarily on ET measurements using micrometeorological methods. Three typical ET measurement techniques, namely, the Bowen ratio-energy balance, eddy covariance, and scintillometer methods, which have a long history and are used widely throughout the world, are introduced. A brief review of their theoretical background, installation and maintenance, data processing and quality control, and footprint is presented, in addition to a brief summary of the advantages and disadvantages of each method. Additionally, ET measurements at

S. Liu (✉) · Z. Xu
State Key Laboratory of Earth Surface Processes and Resource Ecology, School of Natural Resources, Faculty of Geographical Science, Beijing Normal University, Beijing, China
e-mail: smliu@bnu.edu.cn

X. Li, H. Vereecken (eds.), *Observation and Measurement of Ecohydrological Processes*, Ecohydrology, https://doi.org/10.1007/978-3-662-48297-1_7

observational networks and intensive experiments are presented. The ET measurement methods differ in observational theory, temporal–spatial scales, and precision. Researchers can select a suitable method according to their research objectives.

Keywords
Evapotranspiration · Bowen ratio-energy balance method · Eddy covariance method · Scintillometer method

Introduction

Several issues, such as a shortage of fresh water resources and deterioration of water quality, have led to a new interdiscipline, eco-hydrology. One of its core research goals is to reveal the factors and regulation mechanisms influencing the change in the ecological environment related to the water cycle. Evapotranspiration (ET) is a combination of two separate processes, soil or water evaporation (E) and plant transpiration (T), that occur between plants and the atmosphere, soil and the atmosphere, and water and the atmosphere. ET is an essential component of energy and water budgets and is an important process in the soil-plant-atmosphere continuum (SPAC), the water and energy source for crop growth, and an important link between terrestrial ecosystems and hydrological processes. Some important ecosystem parameters and processes, such as soil moisture, vegetation productivity, ecosystem energy, water, and nutrient budgets, are influenced by ET (Wever et al. 2002).

ET measurements have been performed for more than 300 years throughout the world. The first lysimeter to study water use can trace its history back to the seventeenth century (Howell et al. 1991), and the first self-recording lysimeter was developed in Ohio, USA, in 1937 (Garstka 1937). The Bowen ratio-energy balance method (Bowen 1926) and aerodynamics method (Thornthwaite and Holzman 1939) were subsequently developed. Swinbank (1951) proposed the eddy covariance method to directly measure ET, which is currently widely used around the world. The zero-flux plane (ZFP) method was originally used to calculate the infiltration of soil moisture and has been used to measure ET since the 1970s (Cooper 1979). The scintillometer technique to measure surface fluxes was proposed in the 1970s (Wang et al. 1978), and it has been widely used in land surface process experiments since the 1990s. In addition, quick-weighting (checking the weight quickly by two times under no wind condition and recording the water loss of the whole or part of plant and then calculating the transpiration rate), ventilated chambers, and several thermal-based methods (e.g., packaged and probe sap flow) have been used to obtain quantitative estimates of whole-plant water use since the mid-1930s (Wullschleger et al. 1998; Allen et al. 2011).

Evapotranspiration is not easy to measure due to the numerous influencing factors. The methods used to measure ET are based on different measuring principles, for example, hydrological approaches (such as lysimeter and ZFP), plant physiological approaches (e.g., quick-weighting, chambers system, and sap flow method), and micrometeorological methods (e.g., Bowen ratio-energy balance, aerodynamics, eddy covariance, and scintillometer). In the hydrological approaches (lysimeter and ZFP), the water balance equation is the basic equation used to estimate ET by determining all other components. The lysimeter method can derive reliable ET estimates with well-defined surfaces and controlling the lower boundary conditions. The main disadvantages of the lysimeter are construction and maintenance and its limited areal extent especially when the surrounding crops are not uniform. The ZFP method relies on locating a plane of zero hydraulic gradient in the soil profile. However, the disadvantage of the ZFP method is that it cannot be used in areas with high precipitation, i.e., where the zero-flux plane is not stable (Khalil et al. 2003). The plant physiology approaches, such as quick-weighting, ventilated chambers, and sap flow methods, can measure one component of ET (transpiration), and they can be used in complex terrain and spatially heterogeneous environments. However, the quick-weighting method cannot obtain continuous data, and it is a destructive measurement; the chambers method creates an artificial environment and is characterized by technical difficulties, especially when measuring high flow flux rates (Dragoni et al. 2005). Radial gradients in sapwood can result in errors in the sap flow method and require scaling procedures to extrapolate from individual trees (Wilson et al. 2001). Micrometeorological methods are widely used to determine ET and energy partitioning in a number of ecosystems. In this chapter, we focus on these types of approach, including the Bowen ratio-energy balance, eddy covariance, and scintillometer methods.

Micrometeorological Methods to Measure Evapotranspiration

Bowen Ratio-Energy Balance Method

The Bowen ratio-energy balance (BREB) is a micrometeorological method based on the energy balance equation that measures two-layer gradients in the collocated air temperature and vapor pressure in the near-surface layer above the evaporating surface (Bowen 1926). This method is often used to estimate the sensible and latent heat flux because of its simplicity, robustness, and low cost. This method also compares favorably with other methods, such as the weighing lysimeters (Prueger et al. 1997), eddy covariance (Cellier and Olioso 1993), and water balance methods (Malek and Bingham 1993).

Theory

The BREB method estimates sensible and latent heat flux from a surface based on measurements of air temperature and humidity gradients, net radiation, and soil heat flux (Fritschen and Simpson 1989).

The sensible heat (H) and latent heat (LE) flux in one-dimension can be described in terms of flux–gradient relationships:

$$H = -\rho c_p k_h \frac{\partial T}{\partial z} \tag{1}$$

$$LE = -\frac{\rho c_p}{\gamma} k_v \frac{\partial e}{\partial z} \tag{2}$$

where ρ is the air density (kg m^{-3}); c_p is the heat capacity of air at a constant pressure (J kg^{-1} °C^{-1}); k_h and k_v are the turbulent transfer coefficients for heat and water vapor (m^2 s^{-1}), respectively; T is the air temperature (°C); e is the actual vapor pressure (kPa); and γ is the psychrometric constant (kPa °C^{-1}).

By assuming $k_h = k_v$ (Verma et al. 1978) and measuring the temperature and vapor pressure gradients between two levels within the constant-flux layer, β is obtained as:

$$\beta = \frac{H}{LE} = \gamma \frac{\partial T/\partial z}{\partial e/\partial z} = \gamma \frac{\Delta T}{\Delta e} \tag{3}$$

where ΔT and Δe are the temperature (°C) and vapor pressure (kPa) differences between the two measurement levels, respectively.

The energy balance equation is usually expressed as:

$$R_n = H + LE + G_0 \tag{4}$$

where R_n is the net radiation (W m^{-2}), and G_0 is the surface soil heat flux (W m^{-2}). The sensible heat and latent heat flux can be derived by combining Eqs. 3 and 4.

$$H = \frac{\beta}{1+\beta}(R_n - G_0) \tag{5}$$

$$LE = \frac{R_n - G_0}{1+\beta} \tag{6}$$

The BREB method relies on several assumptions (Fritschen and Simpson 1989). Transport is assumed to be one-dimensional, with no horizontal gradients. The sensors that measure gradients are assumed to be located within the constant-flux layer, where fluxes are assumed to be constant with height. The surface is assumed to be homogeneous with respect to heat sources and sinks, water vapor, and momentum. The ratio of the turbulent transfer coefficients for heat (k_h) and water vapor (k_v) is assumed to be 1.

Installation and Maintenance

BREB system is composed of air temperature and humidity sensors to measure gradients, a radiometer (net radiometer or four-component radiometer), soil heat flux plates, and soil temperature and moisture sensors in the near-surface layer. To reduce the observed errors, the sensors must maintain high sensitivity and precision, especially the air temperature and humidity sensors. Generally, BREB systems can

Fig. 1 BREB systems. (**a**) Normal, (**b**) interchange, and (**c**) pumping

be categorized as normal (two layers of air temperature and humidity observations in a fixed position), interchange (two layers of air temperature and humidity sensors interchanged in a period), and pumping (water vapor was pumped in the two layers of the channels to a gas analyzer (i.e., Li 840) and then analyzed) (Payero et al. 2003; Fritschen and Fritschen 2005; Allen et al. 2011) (Fig. 1).

When the BREB system is used in an experiment, it must be installed in the downwind of the surfaces of interest. The two layers of air temperature and humidity sensors must maintain a certain height distance to ensure that the differences in temperature and humidity are greater than their accuracy. The lower sensor should be at least 0.3 m above the crop canopy for relatively homogeneous surfaces, and the upper sensor should be 1–2 m above the lower sensor (Allen et al. 2011). The sampling interval is usually 10 s or less, and data are typical stored every 15 min (for the interchange type) or 30 min (for the normal and pumping type). Half-hourly values are calculated as the average of two corresponding 15-min periods (for the interchange type). The net radiometer, soil heat flux plate, and soil temperature and moisture measurements are recorded every 10 s, and the average values are calculated and registered every 30 min.

Sensor drift, calibration errors, and environmental effects may cause errors in the air temperature and vapor pressure gradients when using the BREB method. When the gradients are small, the errors become significant and may even change the sign of the observed fluxes. To obtain valid air temperature and vapor pressure gradients, it is essential to periodically switch the locations of the upper and lower sensors using an automatic exchange mechanism. When an exchange mechanism is used, the sensors can be exchanged periodically, for example, every 15 min. After each exchange, sampling must stop for a period to allow the sensors to equilibrate to the conditions of the new location (Fritschen and Fritschen 2005).

The BREB system is closely supervised, and general maintenance is performed at least once a month. Maintenance includes cleaning and adjusting the heights of the lower and upper Bowen ratio arms in response to increased crop height. For the pumping type, the air intake filters must be regularly cleaned or changed. In addition, the net radiometer maintenance includes cleaning the domes, checking the desiccant, and making sure it is properly leveled.

Data Processing and Quality Control

The BREB method is an indirect method compared to methods such as eddy covariance system or lysimeter (Todd et al. 2000), and it also relies on several assumptions. Therefore, the data processing and quality control must be carefully implemented.

First, the quality of R_n should be assured. Current radiometers have high precision; it is important that the radiometer observations represent the average surface conditions similar to that of the source area (the primary contribution of the upwind area to the observed values) of the flux measurement as closely as possible. Second, the soil heat flux must be corrected to the surface. The soil heat plate is usually buried a few centimeters under the soil, and the measurements must be corrected to the surface. There are several correction methods, including Plate Cal (combination of heat flux plate measurements and calorimetry), TCAV (averaging soil thermocouple probe), TDEC (thermal diffusion equation and correction), and HM (harmonic analysis method) (Heusinkverld et al. 2004; Liebethal et al. 2005; Yang and Wang 2008). A suitable method can be selected according to the observation conditions.

The accuracy of the observed latent and sensible heat fluxes depends on the accuracy of the Bowen ratio. In many works, data within the instrumental error of the Bowen ratio system are excluded to avoid serious errors in the flux measurements. For example, the temperature and humidity gradients should be greater than the resolution of the sensors (Perez et al. 1999). For cases in which the β values are close to -1, some authors eliminate values less than -0.75 or values in the range $-1.3 < \beta < -0.7$ (Unland et al. 1996). After careful analysis and deduction, Perez et al. (1999) proposed the following to be satisfied by the BREB method under non-advective conditions (Table 1).

Footprint

The footprint of a measurement is the transfer function between the measured value and the set of forcings on the surface–atmosphere interface, while the source area (the upwind area with the primary contribution to the measurement) can be interpreted as the integral of the footprint function over a specified domain (Schmid 1994, 2002) (Fig. 2). The size and extent of source area depend on many factors, such as the measurement height, atmospheric stability, wind speed and direction, and surface roughness length.

The measurement scales of R_n, G_0, H, and LE are different. The source area of H and LE is on the scale of a hundred meters, and the accuracy relies on the representativeness of the R_n and G_0 measurements.

The observation area of G_0 is limited to a few square meters. Therefore, several sampling sites may be required to represent the area sensed by the other sensors, especially if the surface is heterogeneous. Net radiometers have an observation area

Table 1 Requirements to be satisfied by the BREB method under non-advective conditions

Available energy	Vapor pressure difference	Bowen ratio	Heat fluxes
$R_n - G_0 > 0$	$\Delta e > 0$	$\beta > -1$	$LE > 0$ and $H \leq 0$ for $-1 < \beta \leq 0$ or $H > 0$ for $\beta > 0$
	$\Delta e < 0$	$\beta < -1$	$LE < 0$ and $H > 0$
$R_n - G_0 < 0$	$\Delta e > 0$	$\beta < -1$	$LE > 0$ and $H < 0$
	$\Delta e < 0$	$\beta > -1$	$LE < 0$ and $H \geq 0$ for $-1 < \beta \leq 0$ or $H < 0$ for $\beta > 0$

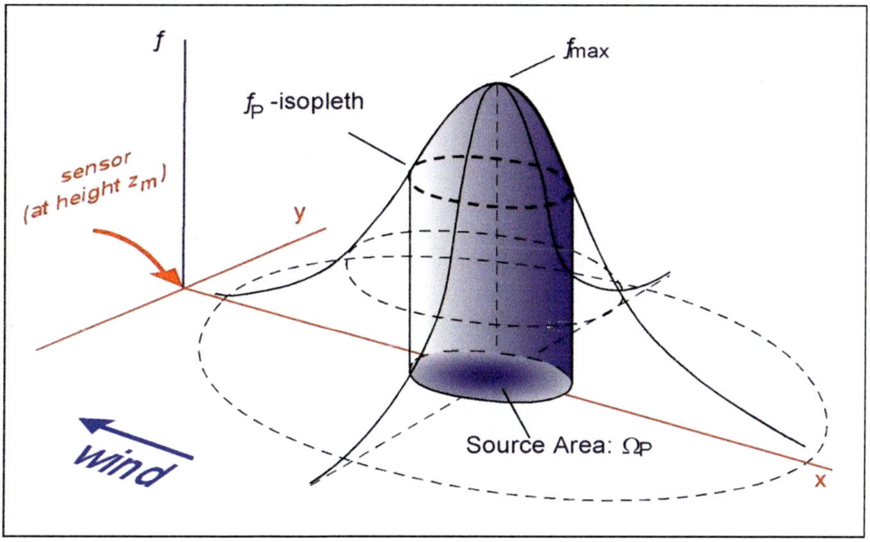

Fig. 2 The source area and its relation to the footprint function (From Schmid 1994)

on the scale of dozens of meters, representing the underlying surface conditions, and they also require more sampling observations for heterogeneous surfaces.

Under homogeneous conditions, the ratio of the source area to the upper measurement height is on the order of 1:100 for the BREB method (Horst and Weil 1992). Fritschen et al. (1983) found that a ratio (fetch/observation height) of 85 was adequate over irrigated cotton with an upwind desert surface. Heilman et al. (1989) reported BREB measurements with a fetch to height ratio of 15 over 0.03 m Bermuda grass with upwind cotton. Stannard (1997) proposed a theoretical model to determine the Bowen ratio fetch and tested it on a Bermuda grass surface.

Summary

Valid Bowen ratio data require careful instrument location, installation, and on-site supervision (Payero et al. 2003). This method assumes that the turbulent transfer coefficients for heat and water vapor are identical, neglecting the heat storage and advection, and that the surface energy is balanced, which is true in neutral conditions but may not be valid in strongly stable or unstable conditions and when advection appears under heterogeneous surfaces (Cellier and Brunet 1992; Perez et al. 1999).

The BREB method has several advantages and disadvantages (Todd et al. 2000). Its advantages include that it is based on straightforward, simple measurements, it requires no information about the aerodynamic characteristics of the interested surface, it can integrate latent heat fluxes over large areas (scale of hundreds of meters), it can observe fluxes on fine time scales (e.g., 10 min and 30 min), and it can provide continuous, unattended measurements. The disadvantages include sensitivity to the biases of the instruments that measure the gradients and energy balance terms, the possibility of discontinuous data when the Bowen ratio approaches −1, the heavy reliance of the ET observation on the accuracy and representativeness of the R_n and G_0 measurements, and the requirement for an adequate upwind source area to establish an equilibrium boundary layer where the temperature and vapor gradients are constant in horizontal space.

Eddy Covariance Method

The eddy covariance method can observe surface exchanges of momentum, energy, and mass. It is based on the Reynolds decomposition concept, and such exchanges can be derived from the high-frequency fluctuation of three-dimensional velocity, temperature, water vapor, carbon dioxide, etc. Due to the relative robustness of both its theoretical underpinnings and modern environmental sensors, the eddy covariance is widely used to measure vertical turbulent fluxes of energy, water, and CO_2 over various terrestrial ecosystems. Eddy covariance provides a direct measure of flux and enables near continuous coverage over time. Furthermore, it is the predominant method in FLUXNET, the cornerstone for nearly all networks.

Theory

The theoretical basis of the eddy covariance (EC) method, i.e., Reynolds decomposition to separate a quantity into the average and fluctuating parts, was established as early as the year 1895. In 1938, Taylor's frozen turbulence hypothesis (Taylor 1938) was proposed, assuming that the measured changes in time at a point were the result of a frozen turbulence field passing by, which enables derivation of the spatial pattern of turbulence according to its temporal description. Turbulent flux determination based on the EC method was first proposed in the 1950s (Swinbank 1951; Foken 2008). However, the deficiency of observation instruments restricted the development and application of the EC method. The possibility of continuous eddy flux measurements arose in the 1990s with the development of a new generation of sonic anemometers and infrared gas analyzers for water vapor and carbon dioxide (Foken and Oncley 1995). The EC method is based on the transfer equations for momentum, heat, humidity, and trace gases.

Based on the mass conservation equation and Reynolds decomposition, neglecting the horizontal turbulent flux divergence terms and the horizontal variation of the vertical flux, applying continuity, and assuming a horizontally homogeneous concentration gradient (Feigenwinter et al. 2004), the equation is:

$$\int_0^{z_r} S(t,z)dz = \underbrace{\int_0^{z_r} \frac{\partial \overline{c}(z)}{\partial t} dz}_{I} + \underbrace{\overline{w'c'}(z_r)}_{II} + \underbrace{\int_0^{z_r} \overline{w}(z) \frac{\partial \overline{c}(z)}{\partial z} dz}_{III} + \underbrace{\int_0^{z_r} \left(\overline{u}(z) \frac{\partial \overline{c}(z)}{\partial x} + \overline{v}(z) \frac{\partial \overline{c}(z)}{\partial y} \right) dz}_{IV} \tag{7}$$

where S is the source/sink strength term; u, v, and w are the wind velocity components (m s^{-1}); c is a scalar; x, y, and z are the coordinate system; and z_r is a control volume of height (m). Term I is the storage change, II is the turbulent vertical flux, III is the vertical advection, and IV is the horizontal advection. The overbar represents the Reynolds averaging operator.

Under the assumptions of (1) fully developed turbulence, (2) stationary state, (3) horizontal homogeneity (no advection), (4) constant-flux surface layer, (5) all scales of turbulence being included, and (6) representative to the specific surface, the equation can be simplified as:

$$\int_0^{z_r} S(t,z)dz = \overline{w'c'}(z_r) \tag{8}$$

The sensible heat (H), latent heat (LE), and carbon dioxide (F_c) fluxes are then:

$$H = \rho C_p \overline{w'T'} \tag{9}$$

$$LE = \lambda \overline{w'q'} \tag{10}$$

$$F_c = \overline{w'CO_2{}'} \tag{11}$$

where ρ is the air density (kg m^{-3}), c_p is the heat capacity of air at a constant pressure (J kg^{-1} °C^{-1}), λ is the latent heat of vaporization for water (J kg^{-1}), T is the air temperature (°C), q is the water vapor density (g m^{-3}), and CO_2 is the carbon dioxide concentration (mg m^{-3}).

Installation and Maintenance

Eddy covariance systems (EC) usually consist of a three-dimensional sonic anemometer that measures wind speed fluctuations in three directions and virtual temperature and an infrared gas analyzer (IRGA) that measures CO_2 and water vapor density fluctuations. The most common sonic anemometers are CSAT3 (Campbell, USA) and WindMaster/R3/HS (Gill, UK); the most common types of IRGA are Li7500/Li7500A/Li7500RS/Li7700 (CH4 analyzer) (LI-COR, USA)/ EC150 (Campbell, USA) for open path, Li7200RS (LI-COR, USA) and EC155 (Campbell, USA) for enclosed path, and the integrated CO_2 and H_2O open-path gas analyzer and sonic anemometer IRGASON (Campbell, USA) (Fig. 3).

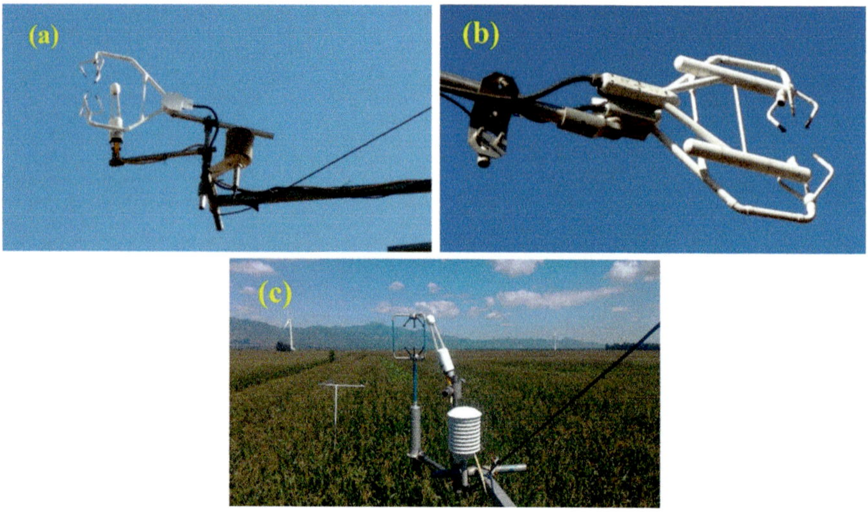

Fig. 3 Eddy covariance system. (**a**) CSAT3 and Li7500A, (**b**) CSAT3 and EC150, and (**c**) Gill and Li7500A

The choice of observation site is very important because it affects the quality of the observed data. Site selection mainly depends on the specific scientific issue of interest. Generally, the selected site should maintain the same prevailing wind direction from the surface of interest during the observation period. Therefore, the prevailing wind direction should be determined before installing the instruments. When the prevailing wind direction is not clear, a common approach is to install the instrument in the center of study area, especially when the surface is homogeneous. Once the observational object is ascertained, the flux tower should be placed in a location that is disturbed by the nearby ecosystem as minimally as possible. The installed height of the EC system is determined according to the research objective and is usually installed at least above 1.5 m or more above the surface or canopy. The sonic anemometer should be aligned to the prevailing wind direction. If the EC system is installed on the tower, the arm length should be two times larger than the tower structure (i.e., the diameter of the tower) to reduce the influence of the tower on the flow distribution. The IRGA sensor should be installed near the sonic anemometer (usually within 20 cm) and slanted a certain angle (15–30°) horizontally.

The raw data for the EC system are high-frequency data (usually 10–20 Hz). The datalogger can store the raw 10-Hz or 20-Hz data (for post-processing) and can process the data into a flux file (usually 30-min period). The raw data (10-Hz or 20-Hz) and flux file can be stored in the datalogger using a CF card. The flux file can be transferred to the researcher via a wireless network (GPRS) in real time so that the data can be regularly checked. The sonic anemometer should remain level (the horizontal bubble in the center), and the IRGA must be kept clean and be calibrated at least twice a year.

Data Processing and Quality Control

The theory of the EC method is based on a series assumption, and it can usually produce reliable results for uniform, flat surfaces. However, the conditions of stationarity in time and homogeneity in space under which the original theories were established are never fully satisfied in practice (Moncrieff et al. 2004). Therefore, careful data processing and quality assurance are important, challenging issues.

There is no uniform scheme for quality control in eddy covariance measurements (Foken et al. 2004). Common quality control measures for EC data include spike detection, lag correction of H_2O/CO_2 relative to the vertical wind component, sonic virtual temperature correction (Schotanus et al. 1983), rotation coordination using the planar fit method (Wilczak et al. 2001), angle of attack error correction (for the Gill-WindMaster sonic anemometer, Nakai and Shimoyama 2012), corrections for density fluctuation (WPL correction, Webb et al. 1980), and frequency response correction.

In addition to these processing steps, half-hourly flux data are screened according to the following criteria: (i) data are rejected when the sensor is malfunctioning (e.g., when there is a fault diagnostic signal), (ii) data are rejected when precipitation occurs within 1 h before or after collection, (iii) incomplete 30-min data are rejected when the percentage of missing data is greater than 3% in the 30-min raw record, and (iv) data are rejected at night during weak turbulence (using the threshold of friction velocity as the criterion) (Liu et al. 2011, 2013).

To be useful, the final data archive must include the data that passes the quality control verification. Measurements are normally flagged according to an index evaluation system. Foken and Wichura (1996) classified tests as stationary tests and integral turbulence characteristics test and proposed a flag for each 30-min flux data run. Three classes were used in the current research (0, best; 1, medium; and 2, poor quality and discarded).

The EC method suffers from an energy imbalance issue. The sum of turbulent latent and sensible heat flux measured by EC method are usually less than the available energy (net radiation minus soil heat flux). An imbalance of approximately 10–30% is traditionally reported. The underlying cause of the imbalance is attributed to measurement errors, an incompletely considered storage term, mismatch between the scales of the energy balance components, and large eddy transport or secondary circulations not captured by EC. Foken (2008) pointed out that the energy imbalance problem is a scale problem, and Foken et al. (2010) concluded that mesoscale circulations resulting from landscape heterogeneity are likely responsible for the energy imbalance at the tower measurement level. Landscape-level heterogeneity in vegetation and topography can also result in an energy imbalance (Stoy et al. 2013), and the EBR would decrease with increasing land surface heterogeneity. There are general two ways to correct the energy imbalance in EC measurements (Twine et al. 2000): (1) the residual method, which assumes that the sensible heat flux is accurate and attributes the residual energy to the latent heat flux using the energy balance equation, and (2) the Bowen ratio correction method, which recomputed the fluxes according to the Bowen ratio values. Charuchittipan et al. (2014) proposed a new method, named the buoyancy flux ratio, to correct the energy imbalance, which uses the relative contribution of the sensible heat flux to the buoyancy flux to correct the sensible and latent heat fluxes. Flux measurements via the EC method are also subject to systematic and random errors (Wang et al. 2015). A landmark paper by Lenschow et al. (1994) defined systematic and random errors associated with limited samplings of EC measurements. Subsequently, with the continued extensive use of EC systems, several methods have emerged for flux uncertainty estimation (Mann and Lenschow 1994; Finkelstein and Sims 2001; Hollinger and Richardson 2005). The normalized uncertainties range from 10% for sensible heat to 25–30% for latent heat and carbon dioxide fluxes (Finkelstein and Sims 2001).

In long-term observations, missing data will occur due to instrument malfunction, poor maintenance, and bad weather conditions, as well as rejected bad data. In order to obtain continuous ET or carbon measurements, a suitable gap-filling method should be used. Several gap-filling methods are used to complete flux measurement data, such as look-up table (LUT), mean diurnal variations (MDV), nonlinear regression (Regr.), artificial neural networks (ANNs), the Kalman filter (KF), and multiple imputation (MI) (Falge et al. 2001; Hui et al. 2004; Alavi et al. 2006). The general EC data processing flowchart is shown in Fig. 4.

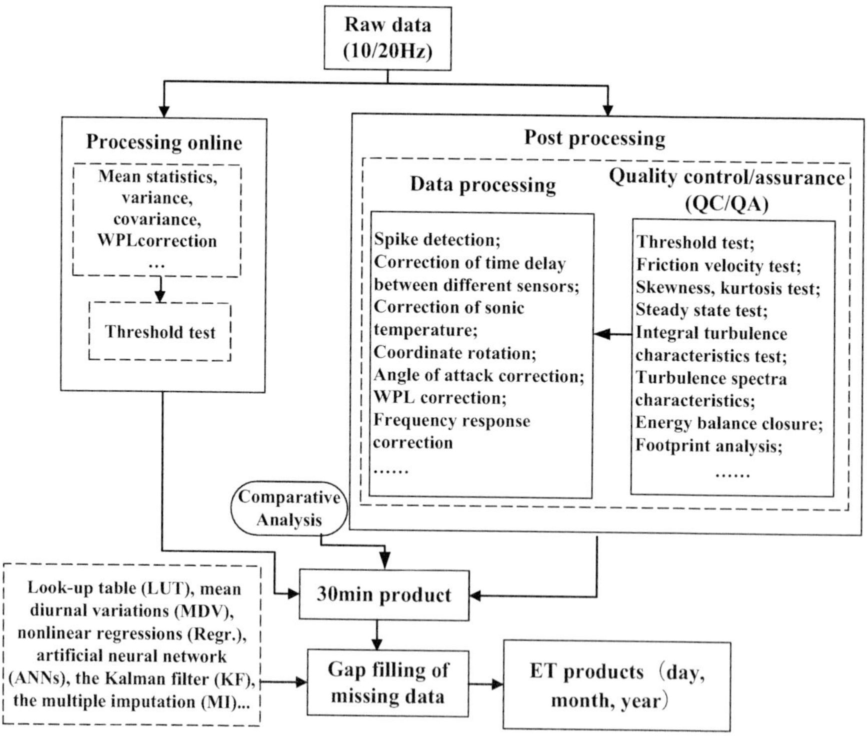

Fig. 4 EC data processing flowchart

There are a variety of software programs for flux calculations, including ECPack from Wageningen University, TK2/3 from the University of Bayreuth, EddySoft developed at the Max Planck Institute for Biogeochemistry in Jena, EdiRE software from the University of Edinburgh, and EddyPro software developed by LI-COR Company. The software programs include the necessary procedures, e.g., spike detection, sonic temperature correction, coordinate rotation, frequency response correction, WPL correction, quality control, and uncertainty assessment (in ECPack, TK3, EddyPro). The software programs can process EC raw data well and show common agreement (Mauder et al. 2008). Researchers can choose any suitable software to process EC data.

Footprint

Footprint analysis is now a recognized part of the establishment and placement of flux towers and the analysis of their output (Finnigan 2004). The turbulent fluxes obtained from EC measurements reflect the influence of the underlying surface on the turbulent exchange (Schmid 2002). It is necessary to determine the source area of the EC measurements using the footprint model before analyzing the characteristics of the surface fluxes. Generally, footprint models can be divided into two categories: numerical models, which emphasize simulation research and are complicated (Flesch et al. 1995), and analytical or semiempirical parameterized models, which emphasize applicability and are more concise (Horst and Weil 1992; Hsieh et al. 2000; Kormann and Meixner 2001).

The Eulerian analytic flux footprint model (Kormann and Meixner 2001) is commonly used to obtain the flux footprint of EC flux measurements, $f(x, y, z_m)$, as follows:

$$f(x, y, z_m) = D_y(x, y)f^y(x, z_m) \tag{12}$$

where x is the downwind distance pointing against the average horizontal wind direction, y is the crosswind wind distance, z_m is the measurement height, $f^y(x, z_m)$ is the crosswind integrated footprint, and $D_y(x, y)$ is the Gaussian crosswind distribution function of the lateral dispersion. The parameters of stability (Obukhov length), friction velocity, standard deviation of lateral wind speed, wind speed and direction, and measurement height are used as the input to the model.

In application, the most important issue for researchers is the source area. Generally, the source area is set to the P%, which is the fraction of the calculated footprint weight function with the most contribution to the total cumulative value within the horizontal source area. The equation can be expressed as:

$$P = \frac{\iint\limits_{\Omega_P} f(x, y, z_m)dxdy}{\int_{-\infty}^{\infty}\int_{-\infty}^{\infty} f(x, y, z_m)dxdy} \tag{13}$$

EC systems are usually installed at a height of several meters with a source area generally covering several hundred meters. This concept is a robust tool to help researchers to perform specific investigations, such as experimental design for flux measurements (Schmid 1997), analyzing the long-term ET variations (investigating specific surfaces, Liu et al. 2011, 2013), and validating remotely sensed ET at medium–high spatial resolution (choosing validated pixels, Jia et al. 2012; Bai et al. 2015).

Summary

The EC method has become increasingly common since the 1990s for measuring the water, heat, and carbon dioxide exchange between an ecosystem and the atmosphere. Some milestones in the development of the EC method can be found in Table 1.1 of Foken et al. (2012).

The advantages of the EC method include (1) direct measurement of the sensible heat, latent heat, CO_2, and momentum fluxes and other scalars, (2) continuous collection of fine temporal (e.g., 30 min, 1 h) and spatial (hundreds of meters) resolution data, and (3) automatic operation in the field. The disadvantages are (1) the requirement for careful data processing and corrections, (2) the requirement of a substantial source area, (3) the "unclosed" energy balance in the surface layer due to numerous reasons, and (4) the relatively large power consumption and requirement for regular calibration.

Scintillometer Method

A scintillometer is a device consisting of a transmitter and a receiver separated by a distance of several hundred meters to several kilometers that can measure the area-averaged sensible and latent heat fluxes based on Monin–Obukhov similarity theory (MOST). Scintillometers have now become common in hydrological and meteorological research (De Bruin 2009). As the measurement scale matches well with satellite pixels and model grids, this method has broad application prospects.

The transmitter of a scintillometer emits electromagnetic radiation that is scattered by the turbulent atmosphere, and the receiver detects and evaluates the intensity fluctuations of the transmitted signal. The magnitude of the fluctuations in the refractive index is usually measured in terms of the structure parameter C_n^2, which is the spectral amplitude of the refractive index fluctuations in the inertial subrange of turbulence (Wang et al. 1978). The structure parameter depends on the wavelength of radiation as a representation of the "turbulent strength" of the atmosphere. The National Oceanic and Atmospheric Administration (NOAA) scintillometer group first developed an instrument based on this method. Optical/near-infrared and microwave scintillometers have emerged since the 1980s and have been applied widely in field experiments since the mid-1990s (a commercial microwave scintillometer was developed in 2014 (RPG Radiometer Physics GmbH 2014) and applied worldwide since then). More detailed information about the history of scintillometry can be found in De Bruin and Wang (2017) and Ward (2017).

Theory

Optical/near-infrared scintillometers measure the sensible heat flux by relating the structure parameter to a temperature structure parameter and the Monin–Obukhov stability parameters. The microwave scintillometer (MWS) is sensitive to both the humidity and temperature structure parameters; however, the humidity structure parameter is the dominating influencing factor, which can be used to infer evapotranspiration. The combination of optical/near-infrared and microwave scintillometer (OMS), together with the related meteorological parameter and MOST, can be used to measure the area-averaged sensible and latent heat fluxes simultaneously.

The structure parameter of the refractive index of air, C_n^2 ($m^{-2/3}$), is calculated from the variance of the natural logarithm of the intensity fluctuations ($\sigma_{\ln I}^2$) using the following equations (Wang et al. 1978):

$$C_{n,LAS}^2 = 1.12\sigma_{\ln(I_{LAS})}^2 D^{7/3} L^{-3} \tag{14}$$

$$C_{n,MWS}^2 = 2.015\sigma_{\ln(I_{MWS})}^2 k^{-7/6} L^{-11/6} \tag{15}$$

where D is the aperture diameter (m), L is the path length (m), and k is the wave number. The refractive index of air is primarily affected by air temperature (T) and specific humidity (q) fluctuations, which are driven by sensible heat and latent heat fluxes. The index C_n^2 ($C_{n,\ LAS}^2$, $C_{n,\ MWS}^2$) is related to the temperature structure parameter, C_T^2 ($K^2\ m^{-2/3}$); the humidity structure parameter, C_q^2 ($kg^2\ m^{-6}\ m^{-2/3}$); and a covariant term, C_{Tq} ($K\ kg\ m^{-3}\ m^{-2/3}$), and can be expressed as:

$$C_n^2, LAS = \frac{{A_{T,LAS}}^2}{\overline{T}^2} C_T^2 + 2\frac{A_{T,LAS}A_{q,LAS}}{\overline{Tq}} C_{Tq} + \frac{{A_{q,LAS}}^2}{\overline{q}^2} C_q^2 \tag{16}$$

$$C_n^2, MWS = \frac{{A_{T,MWS}}^2}{\overline{T}^2} C_T^2 + 2\frac{A_{T,MWS}A_{q,MWS}}{\overline{Tq}} C_{Tq} + \frac{{A_{q,MWS}}^2}{\overline{q}^2} C_q^2 \tag{17}$$

where the A-coefficients include the partial derivatives of the refractive index depending mainly on the temperature and humidity, which are obtained in related experiments. There are three variables (C_T^2, C_q^2, C_{Tq}) in the two equations. Two methods can be used to solve for the variables:

1. Setting the parameter of R_{Tq}. Assume a correlation coefficient of temperature and humidity R_{Tq}, defining $C_{Tq} = R_{Tq}\sqrt{C_T^2 C_q^2}$ (Hill et al. 1988; Andreas 1989).
2. Bichromatic method based on near-infrared and MWS measurements (Lüdi et al. 2005).

After C_T^2, C_q^2, and C_{Tq} are obtained, the sensible heat and latent heat fluxes, H_{OMS} and LE_{OMS}, can be calculated according to MOST and the related meteorological parameters using the following equations:

$$\frac{C_T^2(z_{\mathrm{OMS}} - d)^{2/3}}{T_*^{\;2}} = f_T\left(\frac{z_{\mathrm{OMS}} - d}{L_{\mathrm{Ob}}}\right) \tag{18}$$

$$\frac{C_q^2(z_{\mathrm{OMS}} - d)^{2/3}}{q_*^{\;2}} = f_q\left(\frac{z_{\mathrm{OMS}} - d}{L_{\mathrm{Ob}}}\right) \tag{19}$$

$$H_{\mathrm{OMS}} = -\rho C_p u_* T_* \tag{20}$$

$$LE_{\mathrm{OMS}} = -\rho L_v u_* q_* \tag{21}$$

$$u_* = \frac{k_v u}{\ln\left(\frac{z-d}{z_{0m}}\right) - \Psi_m\left(\frac{z}{L_{\mathrm{Ob}}}\right) + \Psi_m\left(\frac{z_{0m}}{L_{\mathrm{Ob}}}\right)} \tag{22}$$

where z_{OMS} is the effective height of the scintillometer (m), d is the zero-plane displacement height (m), L_{Ob} is the Obukhov length (m) calculated via an iterative process, and $f_T$and f_q are the stability functions of temperature and specific humidity, respectively. There are several empirical functions under unstable (i.e., $L_{\mathrm{Ob}} < 0$) and stable (i.e., $L_{\mathrm{Ob}} > 0$) conditions (Andreas 1989; Thiermann and Grassl 1992; Hartogensis and De Bruin 2005). c_p is the specific heat capacity of air at a constant pressure (J kg^{-1} $°\mathrm{C}^{-1}$), ρ is the density of air (kg m^{-3}), u_* is the friction velocity (m s^{-1}), T_* is the temperature scale (°C), q_* is the specific humidity scale (kg kg^{-1}), k_v is the von Kármán constant (0.40), u is the wind speed (m s^{-1}), z is the wind speed measurement height (m), z_{0m} is the aerodynamic roughness length (m), and Ψ_m is the stability correction function for the momentum transfer.

Installation and Maintenance

Scintillometers can be categorized as near-infrared (common wavelengths of 850 nm, 880 nm, and 940 nm) and microwave (common wavelengths of 1.86 mm, 3.19 mm, and 11.11 mm) according to the emission spectrum of the transmitter. The near-infrared range is widely used to measure the sensible heat flux, and the devices utilizing this range can be categorized as small-aperture scintillometer (SAS), large-aperture scintillometer (LAS), and extra-large-aperture scintillometer (XLAS) according to the aperture size. The main manufacturers are the Kipp & Zonen Company, Netherlands, and the Scintec Company, Germany. The MWS can measure latent heat flux directly, and the commercial manufacturer is Radiometer Physics GmbH (RPG) Company, Germany (Fig. 5).

The following section uses the most common near-infrared scintillometer (LAS) as an example to illustrate the installation and maintenance and also with some additional introductions to the MWS. The choice of observation site depends on the specific scientific issue of interest. After the observation site is selected, the LAS/MWS should be installed in the location that is least disturbed by the nearby ecosystem. The transmitter and receiver are usually placed in a north and south orientation to avoid direct sunlight at low solar elevation angles. LAS/MWS measurements are primarily effected by the center area of the path length. Therefore, the

Fig. 5 Scintillometer (LAS (BLS450, Scintec) and MWS; the picture on the *right* is from the RPG manual)

installation of the transmitter and receiver is fixed as firm as possible, and the influence of the platform is not considered. The platform can be a tripod, tower, building, mountain, etc., depending on the actual conditions. The path length of LAS/MWS should be perpendicular to the prevailing wind direction to ensure the largest source area of the observed fluxes. The automatic weather station can be installed in the center of the scintillometer's path length so that the meteorological elements (e.g., radiation, soil heat flux, air temperature and humidity, wind speed) represent the average conditions of the scintillometer's source area. Meanwhile, the height of the wind speed and air temperature sensors (used in the scintillometer calculation) should be the same as the effective height of the scintillometer. The effective height is an important parameter that is used to derive the sensible and latent heat fluxes. Both near-infrared and microwave scintillometer have their own weighting functions, and the OMS system also has a combined weighting function. The central area makes the greatest contribution to the near-infrared, microwave scintillometer and the OMS system. When the scintillometer system is operated at almost constant beam height over flat terrain, the installation height can be considered as the effective height. However, for slanted paths and over varying topographies, the effective height should be carefully calculated according to the atmosphere stability, topography, and weight function. The readers can refer to Hartogensis et al. (2003) for LAS and to Evans and De Bruin (2011) for the OMS system to obtain more information about the effective height. In field observations, the "saturation" phenomenon occurs in LAS measurements, especially when subjected to strong turbulence and scattering. To avoid this phenomenon, the relationship among the path length, effective height, and local sensible heat flux should be comprehensively considered. Usually, a longer path length is accompanied by a higher effective height when the sensible heat flux is constant; otherwise, a shorter path length corresponds to a lower effective height. Details can be found in the scintillometer manual.

The emission frequencies for the Boundary-Layer Scintillometer (BLS) series are 1 Hz, 5 Hz, 25 Hz, and 125 Hz, and those for the Kipp and Zonen LAS are 6.5–7.5 kHz. The emission frequency for MWS (RPG Company) is 160.8 GHz. The raw data for LAS is usually obtained during a 1-min collection (10-min or 30-

min post-processing period), and the data can be transferred to the researcher via wireless network (GPRS) in real time so that the data can be checked in a timely manner. The raw data for MWS are composed of several files, and the post-processed raw data for a 1-min or 10-min period can be transferred by GPRS. Routine scintillometer maintenance must also be performed regularly, i.e., at least monthly. Regular maintenance should include checking the signal, keeping the transmitter aligned to the receiver, and cleaning the mirror.

Data Processing and Quality Control

The determination of the sign of the heat flux is important. Previous near-infrared LAS measurements were unable to determine the sign and often used empirical methods (e.g., based on the sunrise and sunset, net radiation, or Obukhov length) (Lu et al. 2009). These methods are not required when using an OMS system. The co-spectrum provides the correlated structure parameter C_{Tq}, which can be used to calculate the correlation coefficient R_{Tq}:

$$R_{Tq} = \frac{C_{Tq}}{\sqrt{C_T^2 C_q^2}} \quad (23)$$

where R_{Tq} ranges between -1 and $+1$ and has the same sign as the $H \times LE$. The latent heat flux LE is almost always larger than 0, and even when it is less than 0, the absolute value is usually small. The OMS procedure assumes LE is always positive sign. In this case, R_{Tq} unambiguously provides the sign of the H.

Four steps are taken to ensure the quality of LAS data (Liu et al. 2013): (i) reject data for C_n^2 beyond the saturation criterion; (ii) reject data obtained during periods of precipitation; (iii) reject data when the demodulated signal is small; and (iv) reject data when the sensor is malfunctioning. The above procedures can also be used for MWS data.

The uncertainty of scintillometer measurements can be obtained by comparison with EC measurements on a relatively homogeneous surface. In general, good consistency is observed between scintillometer and EC measurements, with the sensible heat flux showing greater than 90% agreement (McAneney et al. 1995; Hoedjes et al. 2002; Kleissl et al. 2009; Xu et al. 2013). Scintillometer theory is based on MOST. To ascertain the scintillometer data quality (according to MOST), the observed values of $C_T^2(z_{\mathrm{OMS}} - d)^{2/3}/T_*^2$ and $C_q^2(z_{\mathrm{OMS}} - d)^{2/3}/q_*^2$ (C_T^2 and C_q^2 from the scintillometer) can be plotted against the observed values of $(z_{\mathrm{OMS}} - d)/L$, together with the theoretical line from previous studies, such as Andreas (1989), Thiermann and Grassl (1992), and Hartogensis and De Bruin (2005). The values of T_*, q_*, and L can be taken from the EC measurements. The better consistency of observed values and theoretical line indicates the better of the scintillometer data quality.

During long-term observation, missing data will occur due to instrument malfunction, poor maintenance, bad weather conditions, and rejected bad data. Gap-filling methods, such as free convection, nonlinear regressions (regr.), dynamic linear regression (DLR), and harmonic analysis of times series (HANTS) (Bai et al. 2010), can be

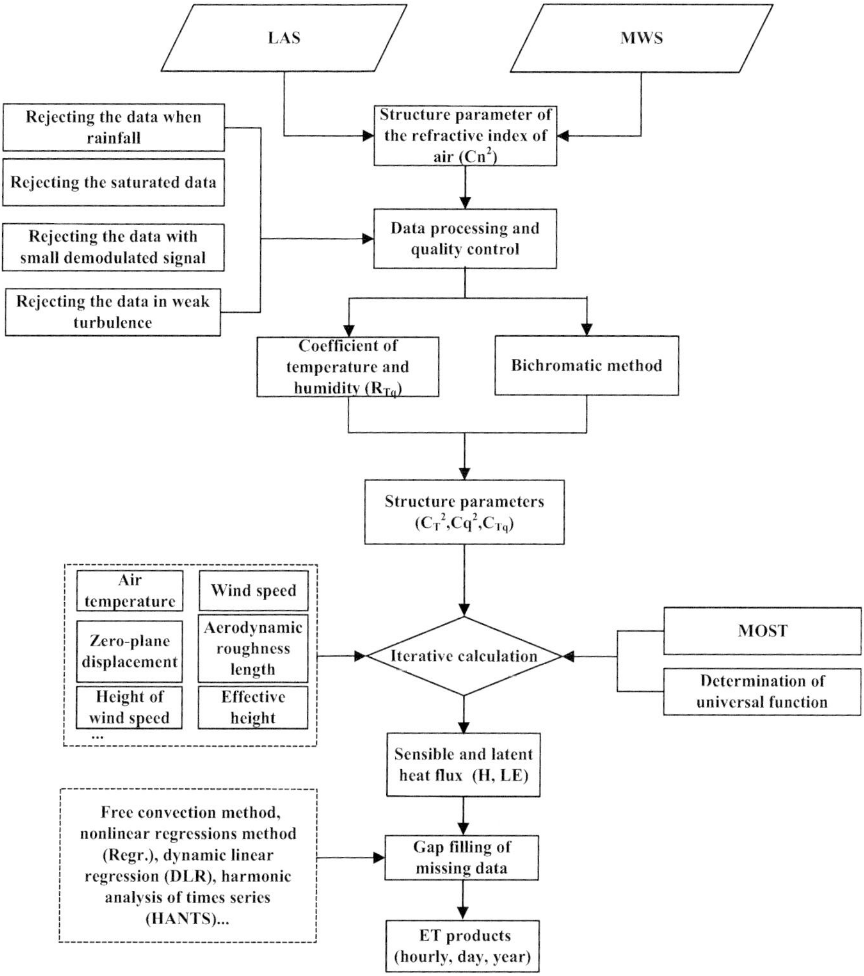

Fig. 6 Scintillometer data processing flowchart

applied to fill the gaps. The general scintillometer data processing flowchart is shown in Fig. 6.

Footprint

The equation for LAS flux observations can be obtained by combining the path-weighting function of the LAS (Meijninger et al. 2002) with the point flux footprint model (EC system):

$$f_{LAS}(x', y', z_m) = \int_{x_2}^{x_1} W(x) f(x - x', y - y', z_m) dx \tag{24}$$

where $W(x)$ is the path-weighting function of the LAS, x_1 and x_2 are the locations of the LAS transmitter and receiver, x and y denote points along the optical length of the LAS, and x' and y' are the coordinates upwind of each of the points (x and y).

The source area of LAS is usually on the scale of several kilometers, e.g., the LITFASS experiment in Germany (Meijninger et al. 2002) and the Heihe and Hai River Basin in China (Liu et al. 2011, 2013). These data can be applied to validate the ET at medium–low spatial resolution (Jia et al. 2012), for numerical model simulations (Xu et al. 2015), and for flux upscaling investigations (Liu et al. 2016). The source area of the OMS system uses similar equations that differ in the path-weighting function; however, it has not yet been reported in the literature.

Summary

Scintillometers can obtain the path-averaged surface fluxes of both sensible heat (near-infrared scintillometer) and latent heat (OMS system) and have the potential to bridge the gap between point observations and satellite pixel or model grid scales. Moreover, scintillometers are cheaper than the other instruments that are used to obtain area average fluxes, such as airborne EC and flux observation matrix. Additional advantages of scintillometers are that they can be used to acquire high temporal resolution averaged fluxes according to a specific objective and that they are relatively simple to operate and maintain in the field. Besides fluxes and structure parameters, scintillometers can also yield crosswind, rainfall, visibility, etc. (Ward 2017). Therefore, scintillometers have broad application prospects.

Nevertheless, scintillometers have several demerits. The primary disadvantage is that they depend on Monin–Obukhov similarity theory to calculate fluxes. Therefore, it must first be determined whether Monin–Obukhov similarity theory is applicable during data processing, especially in the stable boundary layer and roughness sublayer, as well as complex surfaces. The energy balance residual method, which requires the area-averaged net radiation and surface soil heat flux measurements, should be used to derive ET when using a near-infrared scintillometer. In the field observations, unsuitable meteorological conditions, such as precipitation, poor visibility, and weak turbulence, as well as tower vibrations and the power supply, restrict the application of scintillometers for long-term operation (Moene et al. 2009). In addition, the power consumption of MWS scintillometers is very large; it is better to use alternating current. The commercial MWS scintillometer has only recently been applied in field experiments; more research results should be summarized in the future.

Typical Evapotranspiration Observation Systems

A number of observation networks, observatories, or experiments have been carried out on various terrestrial surfaces throughout the world to determine ET.

ET Observation Network

The global network of FLUXNET was established in the mid-1990s; the first efforts of FLUXNET were to measure carbon, water, and energy fluxes on a quasi-continuous basis for a year or more. ET is one of the primary observation elements in various ecosystems with long-term observations, providing valuable datasets to study ET variations around the world (Wilson and Baldocchi 2000; Wever et al. 2002). The National Ecological Observatory Network (NEON), which is a continental-scale ecological observation facility founded in the USA, was completed in 2012 (Kuhlman et al. 2016). NEON has 81 field sites and will collect data for 30 years. These data can be applied to analyze long-term ET characteristics. At the catchment scale, ET is a major component in the water cycle, and it is the most challenging variable to measure at this scale due to the heterogeneity of the landscape. In the USA, the critical zone observatory (CZO) network was founded in 2007 (Anderson et al. 2008), including nine sub-CZOs. The function of CZO is to focus on the interconnected chemical, physical, and biological processes shaping Earth's surface at the catchment/watershed scale, including the overall watershed energy, carbon, and mass balance interactions. In Germany, the Terrestrial Environmental Observatories (TERENO) network, in which four typical terrestrial observatories were selected to monitor the land surface–atmosphere interactions, was established in 2008 (Bogena et al. 2006). The hydrological observatory (HOBE) was established in 2007 to investigate water resources using multi-instruments in the Skjern catchment in Denmark (Jensen and Illangasekare 2011). In China, a hydrometeorological observation network in the Heihe River Basin was established during the Watershed Allied Telemetry Experimental Research (WATER, 2008–2011) and completed in 2013 during the Heihe Watershed Allied Telemetry Experimental Research (HiWATER) projects (Li et al. 2009, 2013), which included three superstations and 18 ordinary stations to monitor the hydrological process in the Heihe watershed. In these typical watershed experiments, the ET measurements were essential or basic observations, and the primary observation techniques were micrometeorological methods, such as EC and LAS.

Global FLUXNET

In 1995, a network arose after discussions at the La Thuile workshop in Italy (Baldocchi et al. 1996). After the workshop, more flux sites and regional networks were established around the world. The flux measurement sites are linked across a confederation of regional networks in the USA, Europe, Canada, Oceania, Africa, Japan, Korea, and China, as well as other professional research organizations, in a global network, called FLUXNET. The FLUXNET project started circa 1997 with support from NASA and other organizations; it coordinates regional and global analyses of observations from micrometeorological tower sites. The flux tower sites mainly use the eddy covariance method to measure the exchanges of *ET*, *H*, and CO_2 between terrestrial ecosystems and the atmosphere.

The distribution of FLUXNET sites ranges from 70 °N to 30 °S, with underlying surfaces of boreal forest, tropical rainforest, evergreen broadleaf forest, deciduous broadleaf forest, evergreen needleleaf forest, deciduous needleleaf forest, mixed forest, woody savanna, savanna, temperate grassland, wetland, tundra, shrubland, cropland, barren, urban, etc. Till February 2017, the number of total registered sites in FLUXNET was 914, accounting for 7,479 site years (Fig. 7).

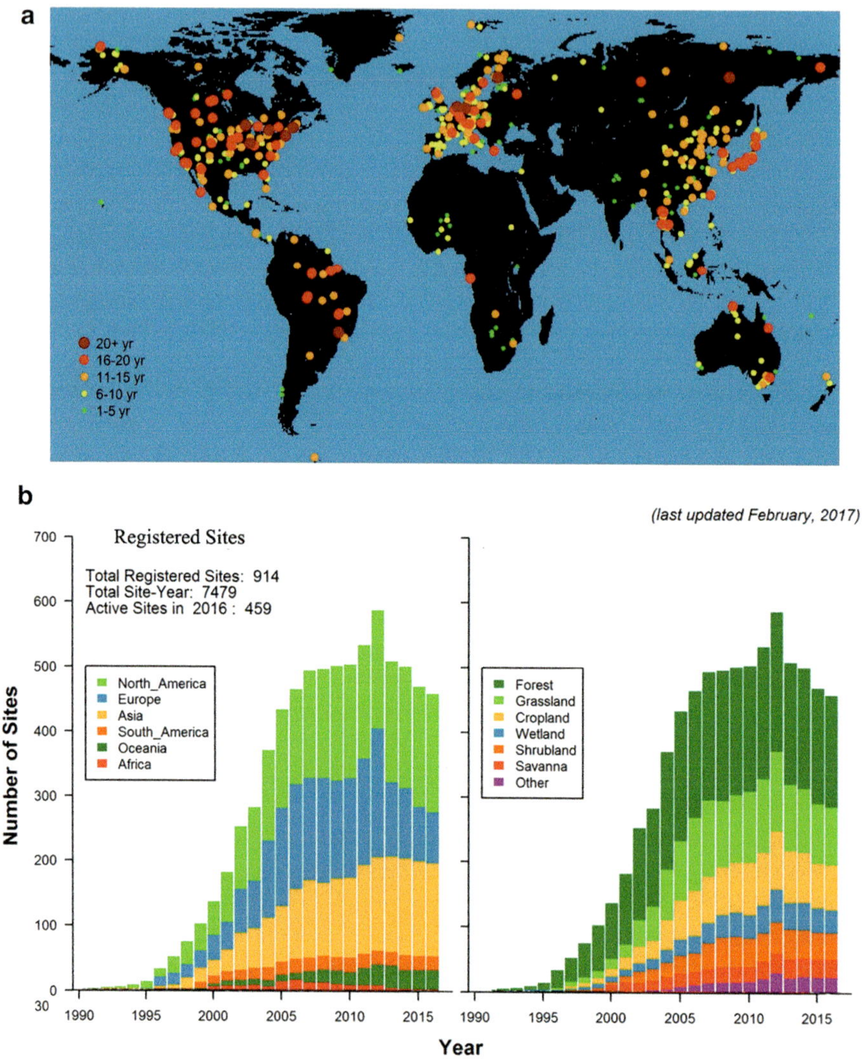

Fig. 7 (**a**) The spatial representativeness of the FLUXNET network and (**b**) summary of the tower sites that are registered in FLUXNET (data accessed in February 2017) (From http://fluxnet.fluxdata.org)

One of the main objectives of FLUXNET was to provide infrastructure to compile, archive, and distribute water, energy, and carbon flux measurements and meteorological data to the science community (Baldocchi et al. 2001). Several data synthesis activities have been conducted by the FLUXNET research community, and the most recently produced dataset is the FLUXNET2015 dataset (http://fluxnet.fluxdata.org/data/fluxnet2015-dataset/). Scientists around the world can use this dataset to conduct research. The dataset can be used to investigate the characteristics and controlling factors of ET (Wilson and Baldocchi 2000; Wever et al. 2002) and to validate and promote the development of evaporation at the regional and global scale by remote sensing estimations to better understand the water and energy cycles (Ershadi et al. 2014; Michel et al. 2016).

Watershed Observatories

The catchment scale has been adopted as the most appropriate scale for water legislation and management (Jensen and Illangasekare 2011). In this chapter, three watershed observatory systems in different countries, whose main objective was ET observations, were selected. In Germany, the TERENO project has established four observatories across Germany since 2008 (Bogena et al. 2006). In Denmark, a hydrological observatory, HOBE, was established in 2007 (Jensen and Illangasekare 2011). In China, the hydrometeorological observation network was established in 2008 during WATER and was completed in 2013 during the HiWATER projects (Li et al. 2009, 2013).

TERENO

In 2008, the Terrestrial Environmental Observatories (TERENO) network was established in Germany as an interdisciplinary research program to observe and explore the long-term ecological, social, and economic impacts of global change at the regional level (Zacharias et al. 2011). Within TERENO, four terrestrial observatories were selected as representing the highest vulnerability with respect to climate change effects in Germany (Fig. 8).

The observation instruments included EC, automatic weather station, isotope, sap flow (thermal dissipation probe, TDP), lysimeter, etc. The TERENO observations used 126 lysimeters at 12 different test sites and one or two EC sets in each catchment to monitor ET.

In addition, there were installed 150 wireless soil temperature/moisture networks in the Wüstebach catchment, as well as runoff, groundwater, and water quality measurements. Additionally, there were cosmic-ray sets in the Wüstebach (one set), Schäfertal (one set), and Rur (ten sets) catchments and 245 rain gauges in the Bode catchment. Furthermore, airborne experiments were also conducted in TERENO, and the airborne sensors included hyperspectral, infrared, microwave radiometer, and lidar. TERENO is a large-scale project, with an Earth observation network across Germany that extends from the North German lowlands to the Bavarian Alps. The observed data and additional information can be found on the website (http://teodoor.icg.kfa-juelich.de/overview-en).

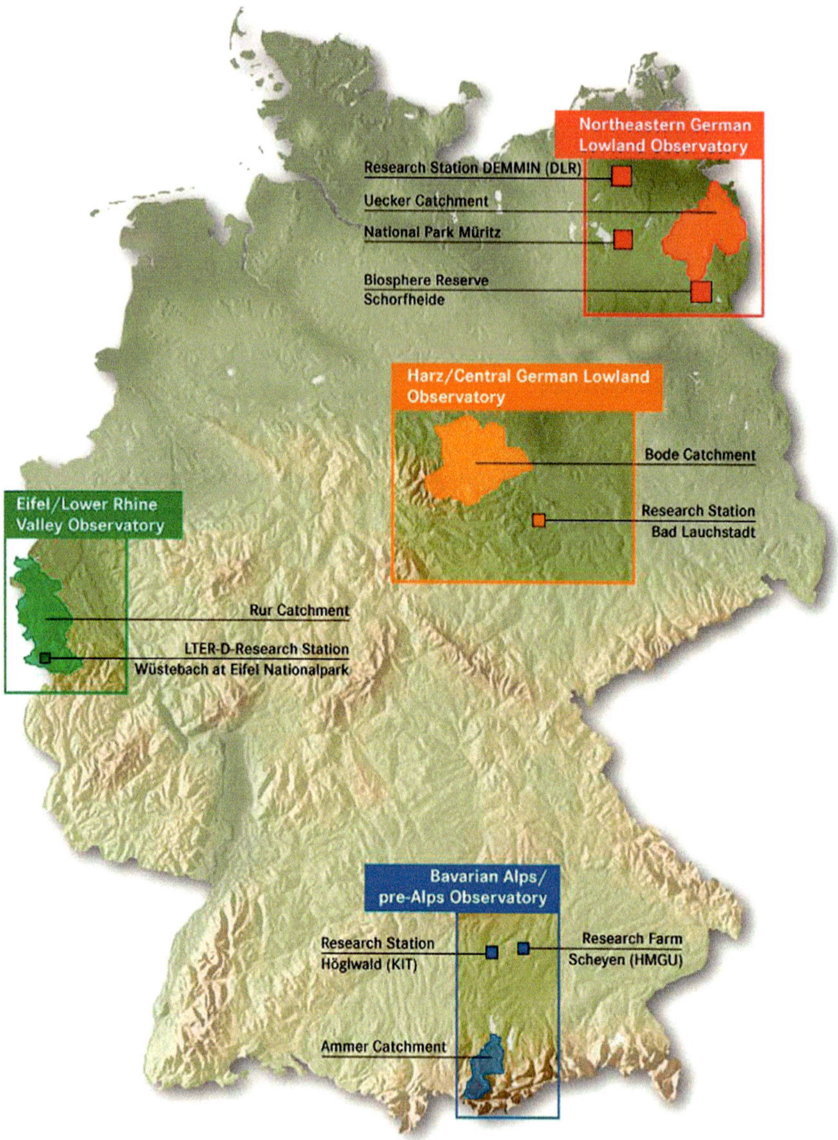

Fig. 8 TERENO observatories in Germany (From Zacharias et al. 2011)

HOBE

HOBE, a hydrological observatory, was established in 2007 in the Skjern catchment located in the western part of Denmark. The entire catchment covers 2,500 km^2 of mostly rural with land cover including grain, corn, grass, forest, heath, and urban (Fig. 9). The overall objectives of the research are to establish a hydrological

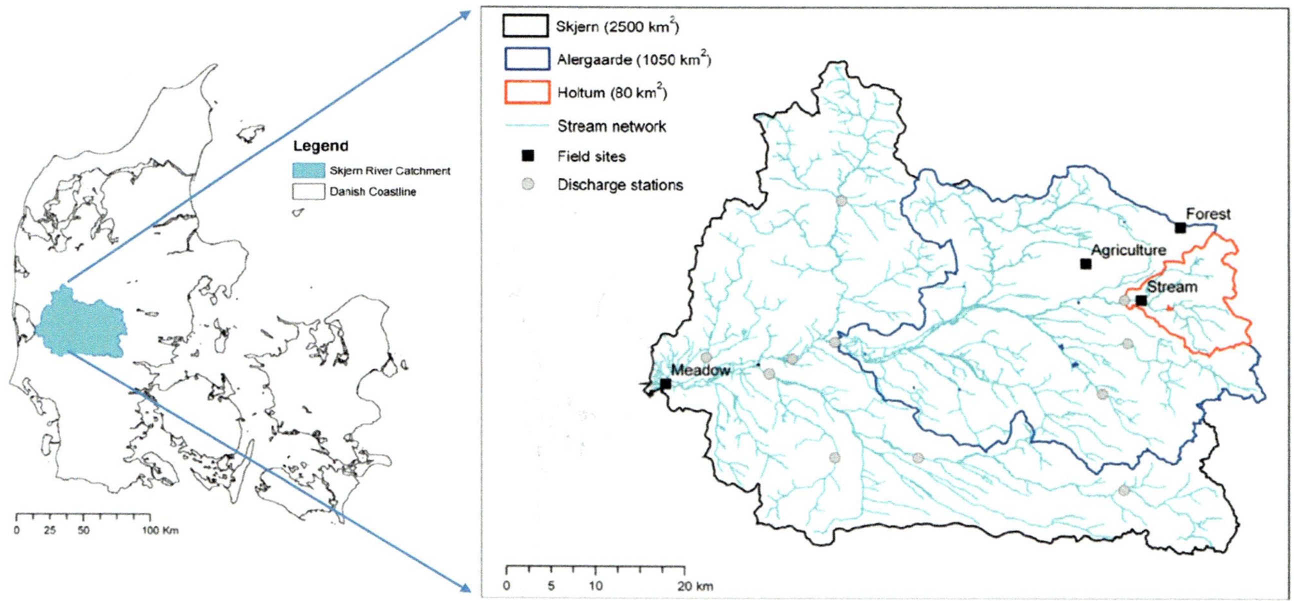

Fig. 9 Location of the Skjern River catchment in Western Denmark (From Jensen and Illangasekare 2011). The forest site, agricultural site, and meadow site were installed the eddy covariance systems

observation platform with an interdisciplinary focus and to better understand the water resources and reduce the uncertainty in the water balance closure at the catchment scale.

The observation instruments included three eddy covariance systems and automatic weather stations, 30 soil moisture and temperature wireless networks, three cosmic-ray measurement sites and isotope, etc. In addition, airborne campaigns (passive L-band microwave radiometer) were conducted to calibrate/validate the SMOS products.

Three eddy covariance flux stations were established to obtain the evapotranspiration and energy balance (agricultural land, spruce plantation, and wet grassland), representing the main land-use types of the catchment (Fig. 9). Measurements have been ongoing since 2008, and these measurements provide a continuous flux dataset for ET analysis and for modeling and scaling of ET to the entire catchment (Ringgaard et al. 2011). More information, including the data, can be found on the website (http://www.hobecenter.dk/).

Heihe Hydrometeorological Observatory

The Heihe River Basin (HRB) is a typical inland river basin, the second largest in China. The HRB is an area of approximately 143,000 km^2, covering Qilian County in Qinghai Province (upstream), Zhangye City in Gansu Province (midstream), and Ejin Banner in Inner Mongolia (downstream). The land cover is characterized by distinct cold and arid landscapes, glaciers, frozen soil, alpine meadow, forest, irrigated crops, riparian ecosystem, and desert, which are distributed from upstream to downstream (Li et al. 2013).

During the WATER experiment from 2008 to 2011 (Li et al. 2009), a prototype of a hydrometeorology observatory was established, in which three long-term EC systems (A'rou site, grassland; Yingke site, cropland; Guantan site, forest) were established in the upstream and midstream areas, in addition to one group of large-aperture scintillometer and three automatic weather stations (AWSs). The HiWATER project was launched in 2012 as a comprehensive eco-hydrological experiment in the framework of the Heihe Plan, based on the diverse needs of the interdisciplinary studies of the research plan and the existing observation infrastructure in the basin (Li et al. 2013). The overall objectives of HiWATER are to improve the observability of hydrological and ecological processes, to build a world-class watershed observation system, and to enhance the applicability of remote sensing in integrated eco-hydrological studies and water recourse management at the basin scale.

Through the HiWATER project, a comprehensive hydrometeorological observatory was established in 2013. The network covers the upstream, midstream, and downstream zones, including three superstations and 18 ordinary stations, covering the main underlying surfaces in the basin, including alpine meadow, cropland, desert, wetland, frozen soil, *Tamarix chinensis*, and *Populus euphratica*. The superstation is a multi-scale observation system for ET measurements, consisting of an LAS, EC, lysimeter, TDP, and meteorological element gradient observation system (seven layers of wind speed/direction and air temperature/humidity, precipitation, air pressure, infrared temperature, four-component radiation, photosynthetically active

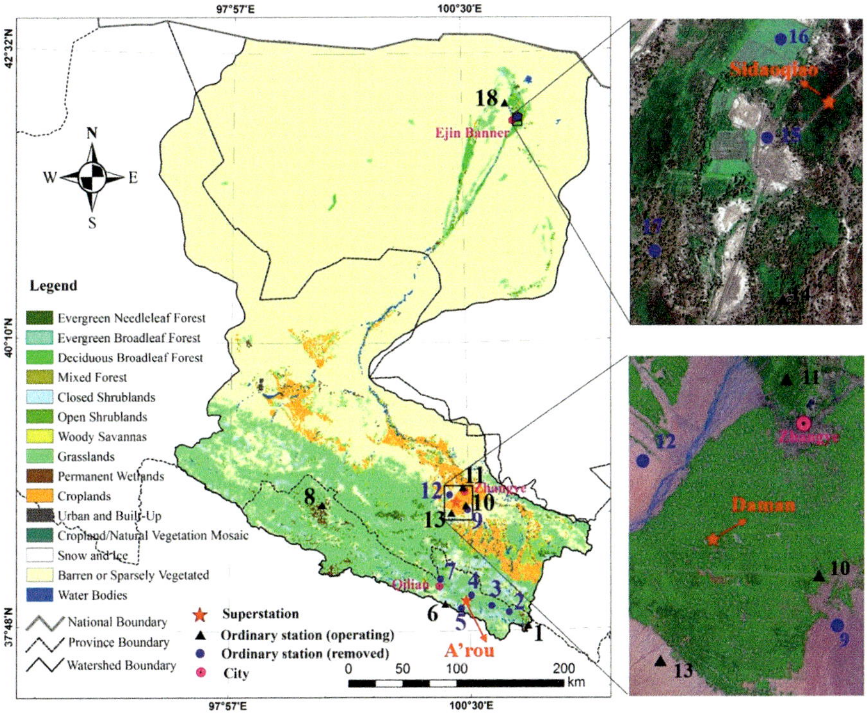

Fig. 10 The hydrometeorological observatory over the Heihe River Basin (*red star*, superstations; *black triangle*, ordinary stations in operation; *blue circles*, ordinary stations have been removed. Numbers *1–8* are stations of Jingyangling, E'Bao, Huangcaogou, A'rou sunny slope, A'rou shady slope, Yakou, Huangzangsi, and Dashalong, respectively, in upstream; *9–13* are stations of Shenshawo sandy desert, Heihe remote sensing, Zhangye wetland, Bajitan Gobi, and Huazhaizi desert steppe, respectively, in midstream; *14–18* are stations of mixed forest, barren land, cropland, *Populus* forest, and desert station, respectively, in downstream)

radiation, soil heat flux, soil temperature/moisture profile, etc.), as well as a cosmic ray, and a wireless soil temperature and moisture network. The ordinary stations comprise an EC and an AWS. In 2016, there were three superstations and eight ordinary stations (the other ten ordinary stations were removed, *blue circles* in Fig. 10) in operation after comprehensive consideration (Fig. 10).

Additionally, airborne remote sensing experiments were conducted during the HiWATER project. The airborne experiments in the upstream and midstream areas were conducted during June 29–August 29, 2012, which included an imaging spectrometer, multi-angle thermal infrared camera (CASI, TASI, WiDAS), light detection and ranging (lidar), microwave radiometer (PLMR), etc. The downstream airborne experiment was conducted from July 29 to August 4, 2014, which included lidar, hyperspectral imager, thermal imager, etc. Airborne lidar and a charge-coupled device (CCD) were also used in the upstream airborne experiments on October 1 and 2, 2014 (Li et al. 2013). The corresponding satellite remote sensing data products were produced, including the ET, fractional snow cover area, surface temperature/emissivity, soil moisture, leaf area index (LAI), fractional vegetation cover (FVC), compositing vegetation index (NDVI/EVI), fraction of absorbed photosynthetically active radiation (FPAR), crop phenology, net primary productivity (NPP), digital elevation model (DEM), land cover/use, etc. All the data were carefully processed, subjected to quality control, and released online at http://www.heihedata.org/hiwater (Chinese version) and http://card.westgis.ac.cn/hiwater (English version).

ET Observation Experiments

Several large land surface process experiments have been implemented worldwide, e.g., the First ISLSCP Field Experiment (FIFE, America; Sellers et al. 1988), the Hydrologic Atmospheric Pilot Experiment–Modélisation duBilan Hydrique (HAPEX-MOBILHY, France; André et al. 1986), the Hydrologic Atmospheric Pilot Experiment in the Sahel (HAPEX-Sahel, Niger; Goutorbe et al. 1994), the Northern Hemisphere Climate Processes Land Surface Experiment (NOPEX, Sweden; Halldin et al. 1999), the Boreal Ecosystem-Atmosphere Study (BOREAS, Canada; Sellers et al. 1995), Lindenberg Inhomogeneous Terrain-Fluxes Between Atmosphere and Surface: A Long-Term Study (LITFASS-98, LITFASS-2003, LITFASS-2009, Germany; Beyrich et al. 2002, 2012; Beyrich and Mengelkamp 2006), and the Multi-Scale Observation Experiment on Evapotranspiration over Heterogeneous Land Surfaces 2012 in the Heihe Watershed Allied Telemetry Experimental Research (HiWATER-MUSOEXE, China, Liu et al. 2016). In these experiments, the primary objective was to measure the interactions of the energy and water fluxes (especially ET) between the surface and the atmosphere. In the following section, several of these experiments are presented in greater detail.

LITFASS Experiment

LITFASS, an acronym for "Lindenberg Inhomogeneous Terrain-Fluxes Between Atmosphere and Surface: A Long-Term Study," was launched in 1995 near the Meteorological Observatory Lindenberg (MOL) in Northeast Germany. The objective of the project was to contribute to solving the problem of the area-averaging of fluxes over a heterogeneous landscape based on coupling between measurements and modeling activities and to develop a strategy for the operational determination of the area-averaged turbulence fluxes of heat, momentum, and water vapor over a heterogeneous landscape (Beyrich et al. 2002). LITFASS-98 and LITFASS-2003 are taken as examples to introduce the LITFASS experiment.

The initial field campaign, LITFASS-98, was conducted in May and June 1998 and was designed to focus on the experimental investigation of the atmospheric boundary-layer structure and the exchange of momentum, heat, and water vapor over a heterogeneous land surface (Beyrich et al. 2002). The study region covers a $20 \times 20\ km^2$ area with land-use types of forest and agricultural fields, lakes, villages, and traffic roads (Fig. 11).

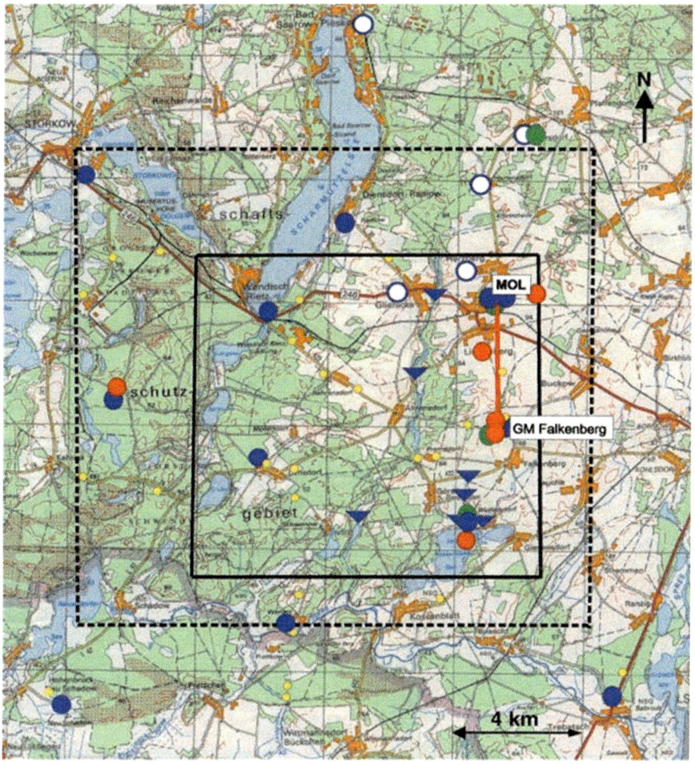

Fig. 11 The LITFASS area and the experimental setup during the LITFASS-98 experiment (From Beyrich et al. 2002) (*Bright areas mark* agricultural farmland, *green areas mark* forest, *blue areas* are lakes. The symbols are as follows: *red circles*, micrometeorological stations; *filled blue circles*, recording rain gauges; *open blue circles*, Hellmann-type rain gauges; *blue triangles*, water table gauges; *green circles*, infrared radiometers; *yellow points*, monitoring sites; *red line*, scintillometer path; *solid/dashed black boxes*, Helipod = Do128 flight pattern)

Micrometeorological and turbulence measurements (using an EC system) were conducted at five sites to characterize the interactions between the atmosphere and different types of underlying surfaces (grass, barley, triticale, forest, water). An LAS was operated at a distance of 4.7 km and a height of 45 m, covering a surface primarily composed of forest, grass and water. A special flux instrumentation intercomparison was performed after the main field phase of the LITFASS-98 experiment. Simultaneous flight measurements (fast temperature and humidity, airborne EC) were performed on June 18, 1998, using two research aircraft systems. Additionally, airborne lidar observations were conducted on June 3, 8, 13, and 18, 1998.

LITFASS-2003 was performed the same region in May and June 2003 (Beyrich and Mengelkamp 2006). The LITFASS measurement facilities were located at the MOL, including a boundary-layer field site (99 m meteorological tower, a sodar/

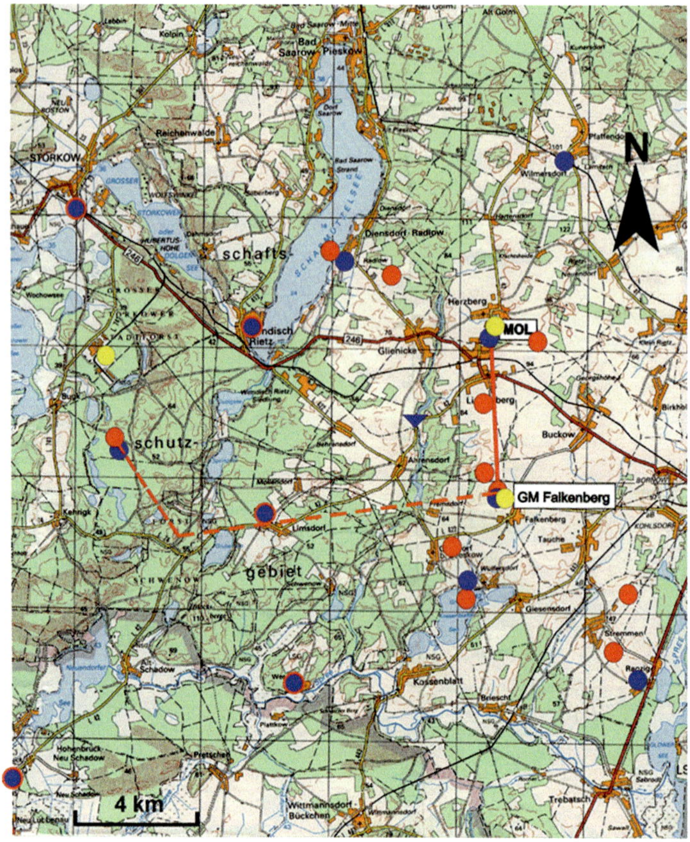

Fig. 12 The experimental setup and measuring strategy of the LITFASS-2003 experiment (From Beyrich and Mengelkamp 2006). *Red circle*, micrometeorological station; *yellow circle*, remote sensing site; *blue circle*, rain gauge; *blue circle with red ring*, rain gauge with global radiation sensor; *blue triangle*, water table measurement; *red solid/dashed line*, long-distance scintillometer path

RASS), a network of micrometeorological (flux) stations over different land-use classes, networks of automatically recording rain gauges and global radiation sensors, and a large-aperture scintillometer. The facility has been in continuous operation since summer 2001. More systems and instruments were added for the LITFASS-2003 experiment, including 14 micrometeorological and flux stations, laser scintillometers at five sites, three LAS systems, and an MWS. The 14 EC sets were composed of a flux observation matrix, including one set each in barley, triticale, and forest; two sets each in grass, maize, rye, and water; and three sets each in rape. The three LAS systems were operated over distances of 3–10 km to determine the area-averaged sensible heat flux, and an MWS was placed along one of the LAS paths over a distance of 4.7 km to determine the latent heat flux. Moreover, a Helipod hanging a turbulence probe recorded more than 60 flight hours (airborne EC) during May 23–June 17, 2013 (Fig. 12).

The LITFASS experiment provided a comprehensive and unique dataset on land surface-atmosphere interactions over a heterogeneous land surface. The turbulence intercomparison was performed in the experiment, and a complete quality-controlled time series of surface fluxes was created (Beyrich et al. 2002; Beyrich and Mengelkamp 2006).

HiWATER-MUSOEXE Experiment

In the HiWATER experiment, HiWATER-MUSOEXE in Zhangye City of Gansu Province, located in the middle reaches of the HRB, was established between May 3 and September 21, 2012, as the first thematic experiment: the Multi-Scale Observation Experiment on Evapotranspiration over Heterogeneous Land Surfaces (HiWATER-MUSOEXE). The objectives were to capture the 3-D dynamic characteristics of water and heat interactions between the surface and the atmosphere to better understand the ET process under the heterogeneous surface, to reveal the spatial heterogeneity of ET and its influence mechanism, to develop the upscaling methods for ET, and to provide ground truth values at the satellite pixel scale to develop and validate the ET remote sensing estimation model.

HiWATER-MUSOEXE was composed of two nested matrices: one large experimental area (30 km × 30 km) and one kernel experimental area (5.5 km × 5.5 km) (Liu et al. 2016). The large experimental area contained one superstation (within the oasis, cropland) and four ordinary stations (around the oasis), with underlying desert, desert steppe, Gobi, and wetland surfaces. The primary objectives were focused on studying the temporal–spatial variation of evapotranspiration (ET) and the effects of advection in the oasis-desert ecosystem. The kernel experimental area was located in the Yingke and Daman irrigation district, and the main surfaces were maize, residential area, vegetable, and orchard. These areas were separated into rows and columns by shelterbelts. Together, they represent the land cover and planting structure in the oasis of the middle reaches of the HRB. The heterogeneity of ET and the ET acquisition at the pixel scale were investigated in the kernel experimental area. Moreover, 17 elementary sampling plots were divided according to the distribution of crops, shelterbelts, residential areas, roads, and canals, as well as according to soil moisture and irrigation status. These divisions resulted in one residential area site, one orchard site, one vegetable site, and 14 maize sites. In each plot, one EC and one AWS were installed to observe the sensible heat flux, latent heat flux, and meteorological elements. Two EC sets and seven layers of meteorological gradient observation systems were installed at the Daman superstation. Moreover, the transpiration of shelterbelts with different heights and diameters at breast height (DBH) was measured using a TDP at three sites, representing the mean conditions of the shelter-forest in the study area. Three TDP probes were installed at a height of 1.3 m on three poplar trees at each site. Additionally, four groups of near-infrared LASs (eight sets with two sets in each group) were installed in the 3 × 3 and 2 × 1 MODIS pixels within the kernel experimental area (three groups in three 3 × 1 MODIS pixels, named LAS1 to LAS3 from west to east and one group in one 2 × 1 MODIS pixel, LAS4). In each LAS group, we primarily used the BLS series scintillometer data and only used the data measured by another scintillometer (zzLAS developed

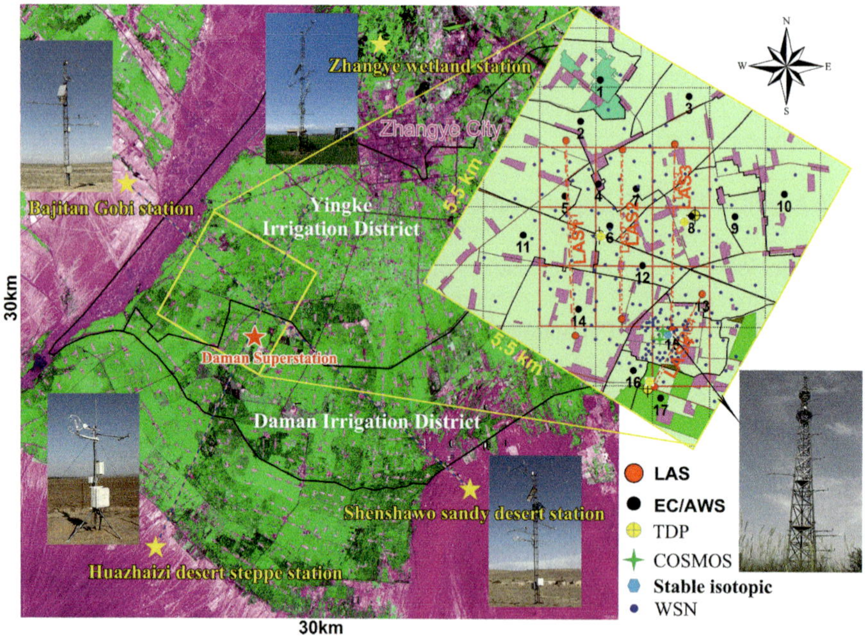

Fig. 13 Multi-Scale Observation Experiment on Evapotranspiration over Heterogeneous Land Surfaces (HiWATER-MUSOEXE)

by this chapter author's group or Kipp and Zonen LAS) if the BLS scintillometer measurements were missing. In addition, two cosmic-ray observations and a stable isotopic observation were conducted around the Daman superstation. There were also 180 wireless soil temperature and moisture network (WSN, 4 cm, 10 cm, 20 cm, and 40 cm) (Jin et al. 2014) and 50 leaf area index networks (LAInets) in the kernel experiment area (Qu et al. 2014) (Fig. 13).

Overall, 22 EC system sets, eight LAS sets, 21 AWS sets, two cosmic-ray sets, three group TDP sets, a stable isotopic observation, 180 WSNs, and 50 LAInets were included in the flux observation matrix. Additionally, the auxiliary parameters of ground-based observations were also measured, including GPS sounding and aerosol, spectral reflectance, emissivity, FVC, LAI/FPAR, photosynthesis, vegetation chlorophyll content, biomass, plant height, soil respiration, soil parameters, irrigation and field management information, etc.

Airborne remote sensing observations were also performed using imaging spectrometers (i.e., CASI and TASI system), a multi-angle thermal infrared camera (WiDAS), lidar, a CCD camera, and a microwave radiometer (L bands) during flight from June 29 to the end of August 2012. Satellite remote sensing data, including ASTER, Landsat ETM+, COSMO-SkyMed, SPOT, TerraSAR-X, WorldView, PROBA CHRIS, QuickBird, ZIYUAN-3 (ZY-3), and Radarsat-2, were also collected during HiWATER-MUSOEXE. The corresponding satellite remote sensing data products were produced, including ET, surface temperature/emissivity, LAI, FVC, NDVI/EVI, FPAR, crop phenology, and land cover/use. All the datasets were carefully screened and processed, and the dataset were released online at http://www.heihedata.org/hiwater (Chinese version) and http://card.westgis.ac.cn/hiwater (English version).

To evaluate the differences among these flux instruments, a comparison of the surface energy flux measurement systems was conducted during May 14–24, 2012, prior to HiWATER-MUSOEXE. The comparison field was located in the Bajitan Gobi desert (a nearly flat and open surface), west of Zhangye City. In total, 20 EC sets, seven LAS sets, and 18 radiometer sets from HiWATER-MUSOEXE were included in the comparison (Xu et al. 2013). The comparison results were used to assess the consistency and reliability of instruments used in HiWATER-MUSOEXE, to guide the layout and installation of instruments, and to aid the subsequent data processing and analysis.

Conclusions and Outlook

ET is affected by multiple factors, e.g., meteorological factors, vegetation factors, and soil water content, and it is difficult to measure accurately. Among the ET observation methods, micrometeorological methods are the most commonly used due to the high temporal resolution, large spatial observation scale, and relatively high accuracy. In this chapter, the typical meteorological methods measuring ET were presented, including the Bowen ratio-energy balance method, eddy covariance method, and scintillometer method. The development of each method was briefly introduced. Afterward, the theory, installation and maintenance, data processing and quality control, and footprint were expounded, and a short summary of the advantages and disadvantages of each method was presented. Additionally, ET measurements at observational networks and intensive experiments were introduced.

The observational theory, temporal–spatial scale, and precision of each ET measurement method are different. In field experiments, researchers can select a suitable method according to their study objectives. Observational site selection, installation and maintenance of the instrument, data processing, and quality control are the key factors and core contents of ET observations.

Ground ET measurements are usually considered as the ground truth to calibrate and validate remotely sensed ET products or results simulated via land surface models, hydrologic models, or ecological models. However, the observation source area of micrometeorological methods typically has a length of hundreds to thousands of meters, and the source area is affected by the measurement height, wind speed and direction, atmospheric stability, and surface roughness length. Therefore, the source area of flux measurements does not always fully cover one or several satellite pixels, resulting in the problem of scale mismatch between the ground measurements and the remotely sensed estimations. In the future, large-scale ET observational methods should be developed and improved, such as OMSs and airborne EC, which can directly obtain large-scale ground ET measurements. Additionally, flux observation matrices should be constructed to upscale multi-site ET to the satellite pixel scale or model grid scale using various upscaling methods.

References

N. Alavi, J.S. Warland, A.A. Berg, Filling gaps in evapotranspiration measurements for water budget studies: evaluation of a Kalman filtering approach. Agric. For. Meteorol. **141**, 57–66 (2006)

R.G. Allen, L.S. Pereira, T.A. Howell, M.E. Jensen, Evapotranspiration information reporting: I. Factors governing measurement accuracy. Agric. Water Manag. **98**, 899–920 (2011)

S.P. Anderson, R.C. Bales, C.J. Duffy, Critical zone observatories: building a network to advance interdisciplinary study of earth surface processes. Mineral. Mag. **72**(1), 7–10 (2008)

J.C. André, J.P. Goutorbe, A. Perrier, HAPEX–MOBLIHY: a hydrologic atmospheric experiment for the study of water budget and evaporation flux at the climatic scale. Bull. Am. Meteorol. Soc. **67**(2), 138–144 (1986)

E.L. Andreas, Two-wavelength method of measuring path-averaged turbulent surface heat fluxes. J. Atmos. Ocean. Technol. **62**, 280–292 (1989)

J. Bai, S.M. Liu, X.P. Ding, L. Lu, A study of the processing method of large aperture scintillometer observation data. Adv. Earth Science **25**(11), 1148–1165 (2010.) (In Chinese with English abstract)

J. Bai, L. Jia, S.M. Liu, Z.W. Xu, G.C. Hu, M.J. Zhu, Characterizing the footprint of eddy covariance system and large aperture scintillometer measurements to validate satellite-based surface fluxes. IEEE Geosci. Remote Sens. Lett. **12**(5), 943–947 (2015)

D. Baldocchi, R. Valentini, S. Running, W. Oechel, R. Dahlman, Strategies f or measuring and modelling carbon dioxide and water vapour fluxes over terrestrial ecosystems. Glob. Chang. Biol. **2**, 159–168 (1996)

D. Baldocchi, E. Falge, L.H. Gu, R. Olson, D. Hollinger, S. Running, P. Anthoni, C. Bernhofer, K. Davis, R. Evans, J. Fuentes, A. Goldstein, G. Katul, B. Law, X.H. Lee, Y. Malhi, T. Meyers, W. Munger, W. Oechel, K.T. Paw U, K. Pilegaard, H.P. Schmid, R. Valentini, S. Verma, T. Vesala, K. Wilson, S. Wofsy, FLUXNET: a net tool to study the temporal and spatial variability of ecosystem-scale carbon dioxide, water vapor, and energy flux densities. Bull. Am. Meteorol. Soc. **82**(11), 2415–2434 (2001)

F. Beyrich, H. Mengelkamp, Evaporation over a heterogeneous land surface: EVA_GRIPS and the LITFASS-2003 experiment – an overview. Bound.-Layer Meteorol. **121**, 5–32 (2006)

F. Beyrich, H.J. Herzog, J. Neisser, The LIFASS project of DWD and the LITFASS-98 experiment: the project strategy and the experimental setup. Theor. Appl. Climatol. **73**, 3–18 (2002)

F. Beyrich, J. Bange, O.K. Hartogensis, S. Raasch, M. Braam, D. van Dinther, D. Gräf, B. van Kesteren, A.C. van den Kroonenberg, B. Maronga, S. Martin, A.F. Moene, Towards a validation of scintillometer measurements: the LITFASS-2009 experiment. Bound.-Layer Meteorol. **144**, 83–112 (2012)

H. Bogena, K. Schulz, H. Vereeken, Towards a network of observatories in terrestrial environmental research. Adv. Geosci. **9**, 109–114 (2006)

I.S. Bowen, The ratio of heat losses by conduction and by evaporation from any water surface. Phys. Rev. **27**(6), 779–787 (1926)

P. Cellier, Y. Brunet, Flux–gradient relationships above tall plant canopies. Agric. For. Meteorol. **58**, 93–117 (1992)

P. Cellier, A. Olioso, A simple system for automated long term Bowen ratio measurement. Agric. For. Meteorol. **66**, 81–92 (1993)

D. Charuchittipan, W. Babel, M. Mauder, J.P. Leps, T. Foken, Extension of the averaging time in eddy-covariance measurements and its effect on the energy balance closure. Bound.-Layer Meteorol. **152**(3), 303–327 (2014)

J.D. Cooper, Water use of a tea estate from soil moisture measurements. East Afr. Agric. For. J. **43**, 102–121 (1979)

H.A.R. De Bruin, Time to think, reflections of a pre-pensioned scintillometer researcher. Bull. Am. Meteorol. Soc. **90**, ES17–ES26 (2009)

H.A.R. De Bruin, J.M. Wang, Scintillometry: a review. ResearchGate (2017). https://www.researchgate.net/pulications/316285424

D. Dragoni, A.N. Lakso, R.M. Piccioni, Transpiration of apple trees in a humid climate using heat pulse sap flow gauges calibrated with whole-canopy gas exchange chambers. Agric. For. Meteorol. **130**, 85–94 (2005)

A. Ershadi, M.F. McCabe, J.P. Evans, N.W. Chaney, E.F. Wood, Multi-site evaluation of terrestrial evaporation models using FLUXNET data. Agric. For. Meteorol. **187**, 46–61 (2014)

J. Evans, H.A.R. De Bruin, The effective height of a two-wavelength scintillometer system. Bound.-Layer Meteorol. **141**(1), 165–177 (2011)

E. Falge, D.D. Baldocchi, R. Olson, P. Anthoni, M. Aubinet, C. Bernhofer, G. Burba, R. Ceulemans, R. Clement, H. Dolman, A. Granier, P. Gross, T. Grünwald, D. Hollinger, N.O. Jensen, G. Katul, P. Keronen, A. Kowalski, C.T. Lai, B.E. Law, T. Meyers, J. Moncrieff, E. Moors, J.W. Munger, K. Pilegaard, ü. Rannik, A. Rebmann, C. Suyker, J. Tenhunen, K. Tu, S. Verma, T. Vesala, K. Wilson, S. Wofsy, Gap filling strategies for defensible annual sums of net ecosystem exchange. Agric. For. Meteorol. **107**, 43–69 (2001)

C. Feigenwinter, C. Bernhofer, R. Vogt, The influence of advection on the short term CO_2-budget in an above a forest canopy. Bound.-Layer Meteorol. **113**, 201–224 (2004)

P.L. Finkelstein, P.F. Sims, Sampling error in eddy correlation flux measurements. J. Geophys. Res. **106**(D4), 3503–3509 (2001)

J. Finnigan, The footprint concept in complex terrain. Agric. For. Meteorol. **127**, 117–129 (2004)

T.K. Flesch, J.D. Wilson, E. Yee, Backward-time Lagrangian stochastic dispersion models and their application to estimate gaseous emissions. J. Appl. Meteorol. **34**(6), 1320–1332 (1995)

T. Foken, The energy balance closure problem: an overview. Ecol. Appl. **18**(6), 1351–1367 (2008)

T. Foken, S.P. Oncley, Results of the workshop Instrumental and methodical problems of land surface flux measurements. Bull. Am. Meteorol. Soc. **76**, 1191–1193 (1995)

T. Foken, B. Wichura, Tools for quality assessment of surface-based flux measurements. Agric. For. Meteorol. **78**, 83–105 (1996)

T. Foken, M. Göckede, M. Mauder, L. Mahrt, B. Amiro, W. Munger, Post-field data quality control, in *Handbook of Micrometeorology*, ed. by X. Lee et al. (Kluwer, Dordrecht, 2004), pp. 181–208

T. Foken, M. Mauder, C. Liebethal, F. Wimmer, F. Beyrich, J.P. Leps, S. Raasch, H.A.R. De Bruin, W.M.L. Meijninger, J. Bange, Energy balance closure for the LITFASS-2003 experiment. Theor. Appl. Climatol. **101**, 149–160 (2010)

T. Foken, M. Aubinet, R. Leuning, The eddy covariance method, in Eddy Covariance: A Practical Guide to Measurement and Data Analysis, Springer Atmospheric Sciences, ed. by M. Aubinet et al. (Springer, Dordrecht, 2012). https://doi.org/10.1007/978-94-007-2351-1_1

L.J. Fritschen, C.L. Fritschen, Bowen ratio energy balance method, in *Micrometeorology in Agricultural Systems*, ed. by J.L. Hatfield, J.M. Baker. Agronomy Series, vol. 47 (American Society of Agronomy, Madison, 2005), pp. 397–405

L.J. Fritschen, J.R. Simpson, Surface energy balance and radiation systems: general description and improvements. J. Appl. Meteorol. **28**, 680–689 (1989)

L.J. Fritschen, L.W. Gay, J.R. Simpson, The effect of a moisture step change and advective conditions on the energy balance components of irrigated alfalfa, in *Proceedings of the 16th Conference on Agricultural and Forest Meteorology*, American Meteorological Society, Boston, 21–24 May 1983, pp. 83–86

W.U. Garstka, Design of the automatic recording in-place lysimeters near Coshocton, Ohio. Soil Sci. Soc. Am. Proc. **2**, 555–559 (1937)

J.P. Goutorbe, T. Lebel, A. Tinga, P. Bessemoulin, J. Brouwer, A.J. Dolman, E.T. Engman, J.H.C. Gash, M. Hoepffner, P. Kabat, Y.H. Kerr, B. Monteny, S. Prince, F. Said, P. Sellers, J.S. Wallace, HAPEX-Sahel: a large-scale study of land–atmosphere interactions in the semi-arid tropics. Ann. Geophys. **12**(1), 53–64 (1994)

S. Halldin, S.E. Gryning, L. Gottschalk, A. Jochum, L.C. Lundin, A.A. Van de Griend, Energy, water and carbon exchange in a boreal forest landscape – NOPEX experiences. Agric. For. Meteorol. **98–99**, 5–29 (1999)

O.K. Hartogensis, H.A.R. De Bruin, Monin–Obukhov similarity functions of the structure parameter of temperature and turbulent kinetic energy dissipation rate in the stable boundary layer. Bound.-Layer Meteorol. **116**(2), 253–276 (2005)

O.K. Hartogensis, C.J. Watts, J.C. Rodriguez, H.A.R. De Bruin, Derivation of an effective height for scintillometers: La Poza experiment in northwest Mexico. J. Hydrometeorol. **4**, 915–928 (2003)

J.L. Heilman, C.L. Brittin, C.M.U. Neale, Fetch requirements for Bowen ratio measurements of latent and sensible heat fluxes. Agric. For. Meteorol. **44**, 261–273 (1989)

B. Heusinkverld, A. Jacobs, A. Holtslag, et al., Surface energy balance closure in an arid region: role of soil heat flux. Agric. For. Meteorol. **122**, 21–37 (2004)

R.J. Hill, R.A. Bohlander, S.F. Clifford, R.W. McMillan, J.T. Priestly, W.P. Schoenfeld, Turbulence-induced millimeter-wave scintillation compared with micrometeorological measurements. IEEE Trans. Geosci. Remote Sens. **26**, 330–342 (1988)

J.C.B. Hoedjes, R.M. Zuurbier, C.J. Watts, Large aperture scintillometer used over a homogeneous irrigated area, partly affected by regional advection. Bound.-Layer Meteorol. **105**, 99–117 (2002)

D.Y. Hollinger, A.D. Richardson, Uncertainty in eddy covariance measurements and its application to physiological models. Tree Physiol. **25**, 873–885 (2005)

T.W. Horst, J.C. Weil, Footprint estimation for scalar flux measurements in the atmospheric surface layer. Bound.-Layer Meteorol. **59**, 279–296 (1992)

T.A. Howell, A.D. Schneider, M.E. Jensen, History of lysimeter design and use for evapotranspiration measurements, in *Lysimeters for Evapotranspiration and Environmental Measurements*, IR Diy/ASCE/Honolulu, 23–25 July 1991, pp. 1–9

C.I. Hsieh, G. Katul, T.W. Chi, An approximate analytical model for footprint estimation of scalar fluxes in thermally stratified atmospheric flows. Adv. Water Resour. **23**(7), 765–772 (2000)

D.F. Hui, S. Wan, B. Su, G. Katul, R. Monson, Gap-filling missing data in eddy covariance measurements using multiple imputation (MI) for annual estimation. Agric. For. Meteorol. **121**(1–2), 93–111 (2004)

K.H. Jensen, T.H. Illangasekare, HOBE: a hydrological observatory. Vadose Zone J. **10**, 1–7 (2011)

Z.Z. Jia, S.M. Liu, Z.W. Xu, Y.J. Chen, M.J. Zhu, Validation of remotely sensed evapotranspiration over the Hai River Basin, China. J. Geophys. Res. **117**, D13113 (2012). https://doi.org/10.1029/2011JD017037

R. Jin, X. Li, B.P. Yan, X.H. Li, W.M. Luo, M.G. Ma, J.W. Guo, J. Kang, Z.L. Zhu, A nested eco-hydrological wireless sensor network for capturing surface heterogeneity in the middle-reach of Heihe River Basin, China. IEEE Geosci. Remote Sens. Lett. **11**(11), 2015–2019 (2014)

M. Khalil, M. Sakai, M. Mizoguchi, T. Miyazaki, Current and prospective applications of Zero Flux Plane (ZFP) method. J. Jpn. Soc. Soil Phys. **95**, 75–90 (2003)

J. Kleissl, J. Gomez, S.H. Hong, Large aperture scintillometer intercomparison study. Bound.-Layer Meteorol. **128**, 133–150 (2009)

R. Kormann, F.X. Meixner, An analytic footprint model for neutral stratification. Bound.-Layer Meteorol. **99**, 207–224 (2001)

M.R. Kuhlman, H. Loescher, R. Leonard, R. Farnsworth, T.E. Dawson, E.F. Kelly, A new engagement model to complete and operate the national ecological observatory network. Bull. Ecol. Soc. Am. **97**, 283–287 (2016)

D. Lenschow, J. Mann, L. Kristensen, How long is long enough when measuring fluxes and other turbulence statistics? J. Atmos. Ocean. Technol. **11**(3), 661–673 (1994)

X. Li, X.W. Li, Z.Y. Li, M.G. Ma, J. Wang, Q. Xiao, Q. Liu, T. Che, E.X. Chen, G.J. Yan, Z.Y. Hu, L.X. Zhang, R.Z. Chu, P.X. Su, Q.H. Liu, S.M. Liu, J.D. Wang, Z. Niu, Y. Chen, R. Jin, W.Z. Wang, Y.H. Ran, X.Z. Xin, H.Z. Ren, Watershed allied telemetry experimental research. J. Geophys. Res. **114**, D22103 (2009). https://doi.org/10.1029/2008JD011590

X. Li, G.D. Cheng, S.M. Liu, Q. Xiao, M.G. Ma, R. Jin, T. Che, Q.H. Liu, W.Z. Wang, Y. Qi, J.G. Wen, H.Y. Li, G.F. Zhu, J.W. Guo, Y.H. Ran, S.G. Wang, Z.L. Zhu, J. Zhou, X.L. Hu, Z.W. Xu, Heihe watershed allied telemetry experimental research (HiWATER): scientific objectives and experimental design. Bull. Am. Meteorol. Soc. **94**(8), 1145–1160 (2013)

C. Liebethal, B. Huwe, T. Foken, Sensitivity analysis for two ground heat flux calculation approaches. Agric. For. Meteorol. **132**, 253–262 (2005)

S.M. Liu, Z.W. Xu, W.Z. Wang, J. Bai, Z.Z. Jia, M.J. Zhu, J.M. Wang, A comparison of eddy-covariance and large aperture scintillometer measurements with respect to the energy balance closure problem. Hydrol. Earth Syst. Sci. **15**(4), 1291–1306 (2011)

S.M. Liu, Z.W. Xu, Z.L. Zhu, Z.Z. Jia, M.J. Zhu, Measurements of evapotranspiration from eddy-covariance systems and large aperture scintillometers in the Hai River Basin, China. J. Hydrol. **487**, 24–38 (2013)

S.M. Liu, Z.W. Xu, L.S. Song, Q.Y. Zhao, Y. Ge, T.R. Xu, Y.F. Ma, Z.L. Zhu, Z.Z. Jia, F. Zhang, Upscaling evapotranspiration measurements from multi-site to the satellite pixel scale over heterogeneous land surfaces. Agric. For. Meteorol. **230–231**, 97–113 (2016)

L. Lu, S.M. Liu, Z.W. Xu, J.M. Wang, X.W. Li, Results from measurements of large aperture scintillometer over different surfaces. J. Appl. Meteorol. Sci. **20**(2), 171–178 (2009) (In Chinese with English abstract)

A. Lüdi, F. Beyrich, C. Matzler, Determination of the turbulent temperature–humidity correlation from scintillometric measurements. Bound.-Layer Meteorol. **117**, 525–550 (2005)

E. Malek, G.E. Bingham, Comparison of the Bowen ratio-energy balance and the water balance methods for the measurement of evapotranspiration. J. Hydrol. **146**, 209–220 (1993)

J. Mann, D.H. Lenschow, Errors in airborne flux measurements. J. Geophys. Res. **99**(7), 14519–14526 (1994)

M. Mauder, T. Foken, R. Clement, J.A. Elbers, W. Eugster, T. GrÜnwald, B. Heusinkveld, O. Kolle, Quality control of CarboEurope flux data – part 2: inter-comparison of eddy-covariance software. Biogeosciences **5**, 451–462 (2008)

K.J. McAneney, A.E. Green, M.S. Astill, Large-aperture scintillometry – the homogeneous case. Agric. For. Meteorol. **76**, 149–162 (1995)

W.M.L. Meijninger, A.E. Green, O.K. Hartogensis, W. Kohsiek, C.B. Hoedjes, R.M. Zuurbier, H.A.R. De Bruin, Determination of area-averaged water vapour fluxes with large aperture and radio wave scintillometers over a heterogeneous surface – Flevoland field experiment. Bound.-Layer Meteorol. **105**, 63–83 (2002)

D. Michel, C. Jiménez, D.G. Miralles, M. Jung, M. Hirschi, A. Ershadi, B. Bartens, M.F. McCabe, J.B. Fisher, Q. Mu, S.I. Seneviratne, E.F. Wood, D. Fernández-Prieto, The WACMOS-ET project – part 1: tower-scale evaluation of four remote-sensing-based evapotranspiration algorithms. Hydrol. Earth Syst. Sci. **20**, 803–822 (2016)

A.F. Moene, F. Beyrich, O.K. Hartogensis, Developments in scintillometer. Bull. Am. Meteorol. Soc. **90**(5), 694–698 (2009)

J. Moncrieff, R. Clement, J. Finnigan, T. Meyers, Averaging, detrending, and filtering of eddy covariance time series, in *Handbook of Micrometeorology*, ed. by X. Lee et al. (Kluwer, Dordrecht, 2004), pp. 7–31

T. Nakai, K. Shimoyama, Ultrasonic anemometer angle of attack errors under turbulent conditions. Agric. For. Meteorol. **162–163**, 14–26 (2012)

J.O. Payero, C.M.U. Neale, J.L. Wright, R.G. Allen, Guidelines for validating Bowen ratio data. Trans. ASAE **46**(4), 1051–1060 (2003)

P.J. Perez, F. Castellvi, M. Ibañez, J.I. Rosell, Assessment of reliability of Bowen ratio method for partitioning fluxes. Agric. For. Meteorol. **97**, 141–150 (1999)

J.H. Prueger, J.L. Hatfield, J.K. Aase, J.L. Pikul Jr., Bowen-ratio comparisons with lysimetric evapotranspiration. Agron. J. **89**, 730–736 (1997)

Y.H. Qu, Y.Q. Zhu, W.C. Han, J.D. Wang, M.G. Ma, Crop leaf area index observations with a wireless sensor network and its potential for validating remote sensing products. IEEE J. Sel. Top. Appl. Earth Obs. Remote Sens. **7**(2), 431–444 (2014)

R. Ringgaard, M. Herbst, T. Friborg, K. Schelde, A.G. Thomsen, H. Soegaard, Energy fluxes above three disparate surfaces in a temperate mesoscale coastal catchment. Vadose Zone J. **11**, 54–66 (2011)

RPG Radiometer Physics GmbH, RPG-MWSC-160 microwave scintillometer manual, version 1.02. Meckenheim (2014)

H.P. Schmid, Source areas for scalars and scalar fluxes. Bound.-Layer Meteorol. **67**, 293–318 (1994)

H.P. Schmid, Experimental design for flux measurements: matching scales of observations and fluxes. Agric. For. Meteorol. **87**, 179–200 (1997)

H.P. Schmid, Footprint modeling for vegetation atmosphere exchange studies: a review and perspective. Agric. For. Meteorol. **113**, 159–183 (2002)

P. Schotanus, F.T.M. Nieuwstadt, H.A.R. De Bruin, Temperature measurement with a sonic anemometer and its application to heat and moisture fluxes. Bound.-Layer Meteorol. **26**, 81–93 (1983)

P.J. Sellers, F.G. Hall, G. Asrar, D.E. Strebel, R.E. Murphy, The first ISLSCP field experiment (FIFE). Bull. Am. Meteorol. Soc. **69**(1), 22–27 (1988)

P. Sellers, F. Hall, H. Margolis, B. Kelly, D. Baldocchi, G. Hartog, J. Cihlar, M.G. Ryan, B. Goodison, P. Crill, K.J. Ranson, D. Lettenmaier, D.E. Wickland, The boreal ecosystem–atmosphere study (BOREAS): an overview and early results from the 1994 field year. Bull. Am. Meteorol. Soc. **76**, 1549–1577 (1995)

D.I. Stannard, A theoretically based determination of Bowen-ratio fetch requirements. Bound.-Layer Meteorol. **83**, 375–406 (1997)

P.C. Stoy, M. Mauder, T. Foken, M. Barbara, E. Boegh, A. Ibrom, M.A. Arain, A. Arneth, M. Aurela, C. Bernhofer, A. Cescatti, E. Dellwik, P. Duce, D. Gianelle, E. Gorsel, G. Kiely, A. Knohl, H. Margolis, H. McCaughey, L. Merbold, L. Montagnani, D. Papale, M. Reichstein, M. Saunders, P. Serrano-Ortiz, M. Sottocornola, D. Spano, F. Vaccari, A. Varlagin, A data-driven analysis of energy balance closure across FLUXNET research sites: the role of landscape scale heterogeneity. Agric. For. Meteorol. **171–172**, 137–152 (2013)

W.C. Swinbank, The measurement of vertical transfer of heat and water vapor by eddies in the lower atmosphere. J. Meteorol. **8**, 135–145 (1951)

G.I. Taylor, The spectrum of turbulence. Proc. R. Soc. **164**(919), 476–490 (1938)

V. Thiermann, H. Grassl, The measurement of turbulent surface-layer fluxes by use of bichromatic scintillation. Bound.-Layer Meteorol. **58**, 367–389 (1992)

C.W. Thornthwaite, B. Holzman, The determination of evaporation from land and water surfaces. Mon. Weather Rev. **67**(1), 4–11 (1939)

R.W. Todd, S.R. Evett, T.A. Howell, The Bowen ratio-energy balance method for estimating latent heat flux of irrigated alfalfa evaluated in a semi-arid, advective environment. Agric. For. Meteorol. **103**, 335–348 (2000)

T.E. Twine, W.P. Kustas, J.M. Norman, D.R. Cook, P.R. Houser, T.P. Meyers, J.H. Prueger, P.J. Starks, M.L. Wesely, Correcting eddy-covariance flux underestimates over a grassland. Agric. For. Meteorol. **103**, 279–300 (2000)

H.E. Unland, P.R. Houser, W.J. Shuttleworth, Z.L. Yang, Surface flux measurement and modelling at a semi-arid Sonoran Desert site. Agric. For. Meteorol. **82**, 119–153 (1996)

S.B. Verma, N.J. Rosenberg, B.L. Blad, Turbulent exchange coefficients for sensible heat and water vapor under advective conditions. J. Appl. Meteorol. **17**, 330–338 (1978)

T.I. Wang, G.R. Ochs, S.F. Clifford, A saturation–resistant optical scintillometer to measure C_n^2. J. Opt. Soc. Am. **69**, 334–338 (1978)

J.M. Wang, J.X. Zhuang, W.Z. Wang, S.M. Liu, Z.W. Xu, Assessment of uncertainties in eddy covariance flux measurement based on intensive flux matrix of HiWATER-MUSOEXE. IEEE Geosci. Remote Sens. Lett. **12**(2), 259–263 (2015)

H.C. Ward, Scintillometry in urban and complex environments: a review. Meas. Sci. Technol. **28**, 064005 (2017). https://doi.org/10.1088/1361-6501/aa5e85

E.K. Webb, G.I. Pearman, R. Leuning, Correction of flux measurements for density effects due to heat and water vapour transfer. Q. J. R. Meteorol. Soc. **106**, 85–100 (1980)

L.A. Wever, L.B. Flanagan, P.J. Carlson, Seasonal and interannual variation in evapotranspiration, energy balance and surface conductance in a northern temperate grassland. Agric. For. Meteorol. **112**, 31–49 (2002)

J.M. Wilczak, S.P. Oncley, S.A. Stage, Sonic anemometer tilt correction algorithms. Bound.-Layer Meteorol. **99**, 127–150 (2001)

K.B. Wilson, D.D. Baldocchi, Seasonal and interannual variability of energy fluxes over a broadleaved temperate deciduous forest in North America. Agric. For. Meteorol. **100**, 1–18 (2000)

K.B. Wilson, P.J. Hanson, P.J. Mulholland, D.D. Baldocchi, S.D. Wullschleger, A comparison of methods for determining forest evapotranspiration and its components: sap-flow, soil water budget, eddy covariance and catchment water balance. Agric. For. Meteorol. **106**, 153–168 (2001)

S.D. Wullschleger, F.C. Meinzer, R.A. Vertessy, A review of whole-plant water use studies in trees. Tree Physiol. **18**, 499–512 (1998)

Z.W. Xu, S.M. Liu, X. Li, S.J. Shi, J.M. Wang, Z.L. Zhu, T.R. Xu, W.Z. Wang, M.G. Ma, Intercomparison of surface energy flux measurement systems used during the HiWATER-MUSOEXE. J. Geophys. Res. **118**, 13140–13157 (2013)

Z.W. Xu, Y.F. Ma, S.M. Liu, W.J. Shi, J.M. Wang, Assessment of the energy balance closure under advective conditions and its impact using remote sensing data. J. Appl. Meteorol. Climatol. **56**(1), 127–140 (2017)

T.R. Xu, S.M. Liu, Z.W. Xu, S.L. Liang, L. Xu, A dual-pass data assimilation scheme for estimating surface energy fluxes with FY3-AVIRR land surface temperature. Sci. China Earth Sci. **58**(2), 211–230 (2015)

K. Yang, J. Wang, A temperature prediction–correction method for estimating surface soil heat flux from soil temperature and moisture data. Science **51**(5), 721–729 (2008)

S. Zacharias, H. Bogena, L. Samaniego, M. Mauder, R. Fuβ, T. Pütz, M. Frenzel, M. Schwank, C. Baessler, K. Butterbach-Bahl, O. Bens, E. Borg, A. Brauer, P. Dietrich, I. Hajnsek, G. Helle, R. Kiese, H. Kunstmann, S. Klotz, J.C. Munch, H. Paper, E. Priesack, H.P. Schmid, R. Steinbrecher, U. Rosenbaum, G. Teutsch, H. Vereeken, A network of terrestrial environmental observatories in Germany. Vadose Zone J. **10**, 955–973 (2011)

Surface Runoff

Yanhong Guo, Yinsheng Zhang, Teng Zhang, Kunxin Wang, Jie Ding, and Haifeng Gao

Contents

Y. Guo · T. Zhang · K. Wang · J. Ding · H. Gao
Key Laboratory of Tibetan Environment Changes and Land Surface Processes, Institute of Tibetan Plateau Research, Chinese Academy of Sciences, Beijing, China

Y. Zhang (✉)
Key Laboratory of Tibetan Environment Changes and Land Surface Processes, Institute of Tibetan Plateau Research, Chinese Academy of Sciences, Beijing, China

CAS Center for Excellence in Tibetan Plateau Earth Sciences, Beijing, China
e-mail: yszhang@itpcas.ac.cn

X. Li, H. Vereecken (eds.), *Observation and Measurement of Ecohydrological Processes*, Ecohydrology, https://doi.org/10.1007/978-3-662-48297-1_8

Abstract

Surface runoff, or overland flow, is a fundamental process of interest in hydrology. Surface runoff generation can occur at multiple scales, ranging from small pools of excess water that propagate downhill to stream networks that drain large catchments. Accurate quantification of runoff is vital to clarify the mechanisms and effects of overland flow and also indispensable to understand fundamental hydrological processes. In this chapter, four kinds of measurement techniques, including runoff plot method, curve number method, isotopic tracer method, and salt solution method, are introduced. Runoff plot experiments are often conducted to evaluate the rainfall–runoff processes and widely used to study runoff and/or sediment losses. The curve number method is used to estimate watershed direct-runoff volume by a curve number value which is developed based on measured watershed runoff and rainfall data. The isotopic tracer method is used to measure the surface runoff by separating its contribution from multicomponent based on the mass balance of stable isotopes. The salt solution method is usually used to measure the shallow water flow by detecting the movement of salt. Besides, models of surface runoff are also summarized in this chapter. The models can be classified into conceptual models and process-based models. The conceptual models are simple transfer functions describing a linear relationship between rainfall and surface runoff. While the process-based models take into account of the spatial variability of climate, soil, vegetation, and terrain, which are able to make a series of hydrological processes interconnected. Despite their complexities, the process-based models are very helpful to study the changes in hydrological processes caused by human activities. Furthermore, the vegetation has important impact on surface runoff. For instance, with the increase of vegetation coverage, surface runoff can be reduced effectively. And root induces macropores, which are of importance for runoff mitigation due to their large diameters and high connectivity, enhancing rapid rainfall infiltration and percolation to deeper soil layers. Finally, we put forward some challenges about the measurement and simulation of the surface runoff including the establishment of surface runoff observation network in different ecological system, the combination of land surface model and distributed hydrological model, and the coupling between ecological processes and the runoff process on different scales.

Keywords
Surface runoff · Runoff observation techniques · Runoff modeling · Vegetation · Root impact

Introduction

Surface Runoff Definition

The flow of a water layer over the surface and through the pores of soils and sediments that is coming out of the watershed is termed as a runoff. There are three components of the runoff from watersheds (Dingman 2002; Rumynin 2015): (i) surface runoff or overland flow (sometimes termed as direct runoff), (ii) subsurface runoff or interflow (throughflow), and (iii) groundwater runoff or baseflow (Fig. 1). The surface runoff is a two-dimensional flow occurring on slopes or in ephemeral drainage patterns. Different mechanisms involved in runoff generation are discussed below. Surface runoff rapidly reaches the nearest discharge zones, thus showing a quick response to a rain event and/or snow melting.

Surface Runoff Formation Mechanisms

The development of the basis for quantifying the transformation of rainfall to runoff has been a prime target of hydrologists for several generations. In general, two forms of runoff generation are widely accepted (Rumynin 2015): (i) the infiltration excess

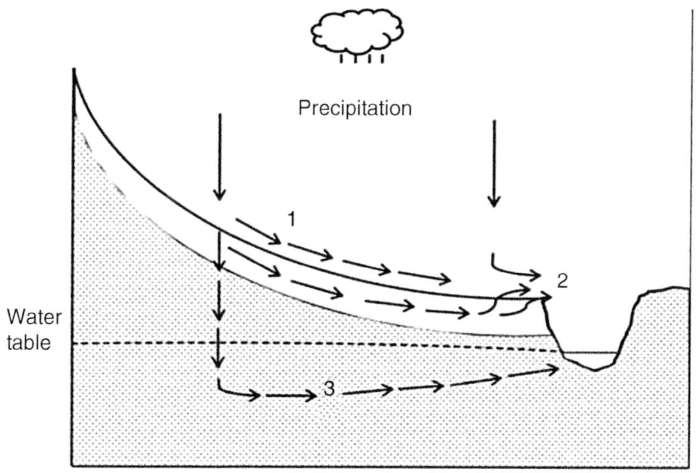

Fig. 1 Possible paths of water moving downhill: path 1 is Horton overland flow, path 2 is saturation overland flow, path 3 is groundwater flow. The unshaded zone indicates highly permeable topsoil, and the shaded zone represents less permeable subsoil

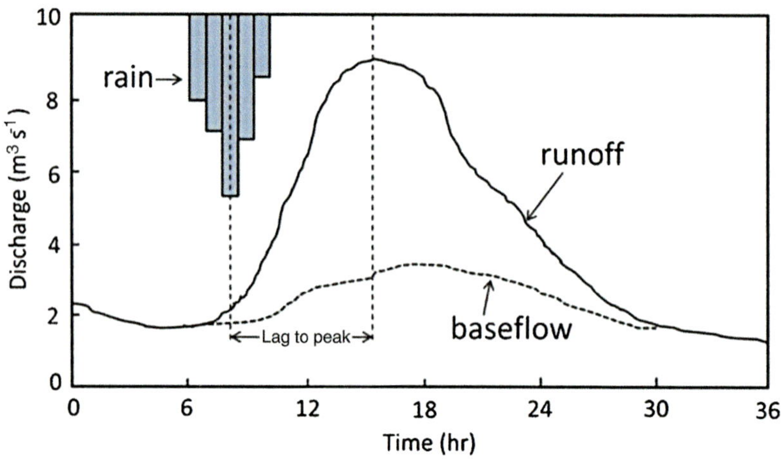

Fig. 2 A conceptual illustration of the runoff evolution due to precipitation

runoff (Horton overland flow) and (ii) the saturation excess runoff (Dunne or saturation overland flow) (Fig. 2).

Infiltration Excess Runoff

The classical model of surface runoff generation (Fig. 3) is by an infiltration excess mechanism in which rainfall intensity exceeds the local infiltration capacity of the soil for a sufficient period for any depression storage. In this way, the downslope flow is initiated at the soil surface. This will not occur where the permeability of the soil is high in comparison with expected rainfall intensities. This is not for the case that all the rainfall infiltrates before bringing the surface to saturation (the time to ponding). Such a basis of the infiltration excess was formulated in the 1930s by Robert Horton (Horton et al. 1940). It should be mentioned that the infiltration here is considered independent of overland flow dynamics resulting in weak coupling of the two processes. Horton overland flow is rare in vegetated humid region. It is common in areas devoid of vegetation such as semiarid rangelands and compacted soil. The concept in such a formation mechanism is soil infiltration capacity, $f = f(t)$, implying the maximal rate at which rainwater can be adsorbed by soil under given conditions, function f is potential infiltration.

Due to nonlinearity of flow in unsaturated soil, f decreases continuously throughout the rainfall period, and thus it behaves similar to the decay function. All rain water may go into the soil when the precipitation (p) is less than the potential infiltration. When p is greater than f, the water accumulation on the ground surface will be $q = p - f$, representing the potential of the process of water flow over the surface.

$$i(t) = min[p(t), f(t)] \tag{1}$$

where $i(t)$ is the actual rate of infiltration. The quantitative description of the process implies determining the moment t_i when the infiltration capacity f equals p. At this

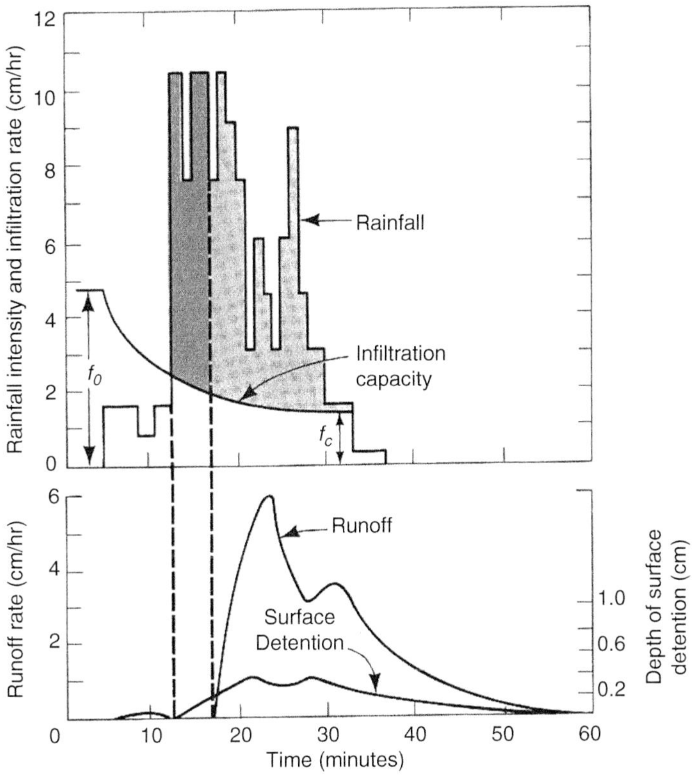

Fig. 3 Schematic chart of infiltration-excess runoff formation

time, the soil moisture reaches to its maximum value nearly equal to the soil porosity. Subsequently, $f = f(t-t_i)$ becomes less than p, resulting in surface runoff. Because the precipitation rate exceeds the infiltration capacity, there is excess precipitation available for surface runoff,

$$q(t) = \begin{cases} 0, & 0 < t < t_i \\ p - f(t - t_i), & t > t_i \end{cases} \tag{2}$$

$q(t)$ is the infiltration excess runoff rate (or intensity). Mathematically, the ponding time, t_i, is the moment when the flux boundary condition changes to a head boundary condition.

Saturation Excess Runoff

Interestingly, the surface runoff can occur in land surface with high permeability in the regions where the precipitation is always less than the infiltration capacity. This is especially true in the humid region of which groundwater table is high and the soils and vadose zone rocks are heterogeneous. Such a phenomenon is usually referred to saturation excess runoff. This formation uses the Dunne assumption that all precipitation enters the soil and runoff occurs due to the inability of soil to absorb any more water (Dunne 1978; Dunne et al. 1975). They reflect the limited ability of soil and underlying rocks to accumulate water, for example, in areas where a thin soil layer is underlain by low-permeability deposits or in areas with shallow groundwater table. When falling onto such areas, precipitation has no storage reserve, it will directly transform into surface runoff. Therefore, the Dunne overland flow generation is controlled, along with soil hydraulic properties, by two major factors including the geomorphology (e.g., the shape and the slope) of the catchment and its subsurface hydrology (Willgoose and Perera 2001).

The saturation excess runoff in humid regions with coarse-texture soils is generated by saturation from below or by a rising groundwater table or by discharging sporadic horizontal flow of water (through flow) within the soil layer. Many studies have shown that overland flow in areas with humid climate form within relatively small areas (as compared with the total watershed area) with higher water table. This commonly occur in hill slopes, river valleys, and swales. Such saturated areas, where saturation excess runoff is produced, are referred as variable source areas (VSAs). The reason is that these areas are limited to the close vicinity of the stream, but they can expand during the storm resulting in larger rates of runoff generation. Previous studies showed that up to two-thirds of direct runoff can originate from the VSAs occupying only 5–20% of the watershed area (Boughton 1993; Ogden and Watts 2000). In general, the peak rate of saturation excess runoff varies, but it is less than that of infiltration excess runoff, because only a portion of the drainage basin is contributing saturation overland flow. Flow velocity of the former is somewhat smaller than that of infiltration excess runoff, because saturation overland flow takes place on gentle vegetated surface.

General Observational Techniques of Surface Runoff

Runoff Plot Measurement

Surface runoff, or overland flow, is a fundamental process of interest in hydrology. Surface runoff generation can occur at multiple scales, ranging from small pools of excess water that propagate downhill to stream networks that drain large catchments. Accurate measurement of runoff quantity is vital to clarify the mechanisms and effects of overland flow and also indispensable to understand fundamental hydrological processes. As a prerequisite of watershed scale investigations, plot-scale studies are often conducted to evaluate the rainfall–runoff processes with better control over the controlling, under which circumstances, runoff plot experiments were used widely to study runoff and/or sediment losses from different field sites around the world (Negi et al. 1998; Sarkar and Dutta 2011; Sarkar et al. 2008; Singh et al. 1983). Most of the experimental set-ups used either rainfall simulators or inflow–outflow methods for evaluation of plot-scale hydrologic responses, which are also prerequisites to developing any regional hydrological model.

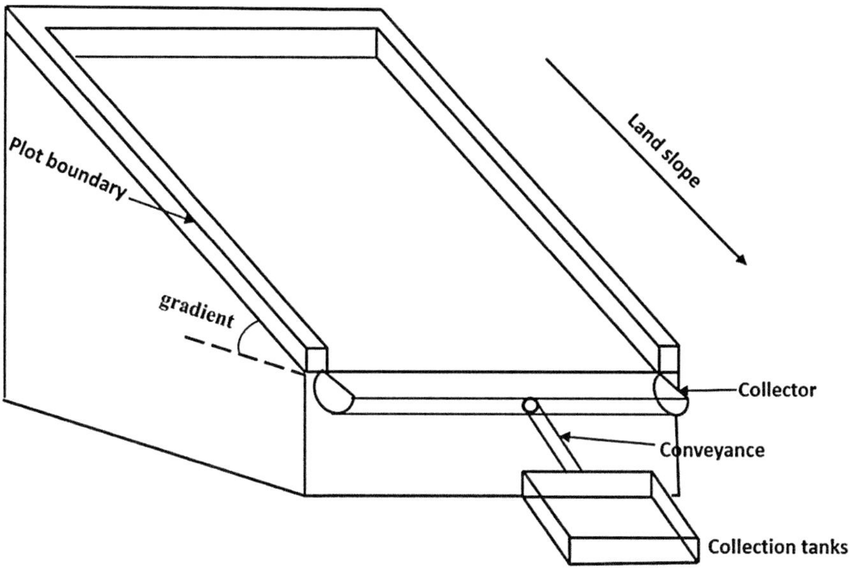

Fig. 4 Typical design of simple runoff plot layout

Outlined in general, following equipments are needed for a typical runoff plot (Mutchler 1963) (See in Fig. 4).

1. Boundaries around the plot to define the measured area.
2. Collect channel to catch and concentrate runoff from the plot.
3. Conveyance equipment to carry runoff to a collection tank.
4. Collection tanks to hold aliquot portions of runoff.

Other equipment is sometimes desired. It is helpful in analysis to have a runoff hydrograph. Hence, a rate-measuring flume can be placed between the collector unit and the sampling unit. Special heating equipment is needed if snow-melt runoff is evaluated. Principles of the key components equipped with the runoff plot can be explained in details as follows.

Plot Boundaries

As a general rule, runoff plots with planar surfaces off large stones, steep slope, or sag encourage the flow to occur. Plot area influences the amount of runoff that needs to be collected, stored, and measured after a rainfall event or series of rainfall events (Zhang et al. 2015). Several methods have been used to define runoff plot areas. Dikes in combination with terrace channels have been used generally on plots larger than one-fourth acre. Strips of 16-gage galvanized steel approximately 9 inches high by 6- to 12-feet long, with corrugations running across the small dimension, make excellent boundaries for cultivated plots. These are comparatively easy to install and maintain. Where the boundaries are permanently installed, smooth galvanized steel strips, 14-gage are preferred.

Runoff Collector

Collecting equipment of many different designs and materials has been used on the runoff stations around the world. The collector generally acts as a weir across the bottom of the plot and a channel for runoff to the sampling unit. Sheet-metal construction is preferred to concrete, because the collector elevation must be adjusted to the level of the plot as erosion occurs. An endplate made of heavy gage galvanized steel blocks off the plot end and furnishes a stable attachment for the trough of the collector. The endplate should extend at least 8 inches below the collector trough.

The collector trough acts as a channel for the runoff material. This trough (together with the endplate) is designed to reach across the entire width of narrow plots. For plots wider than 14 feet, it is best to concentrate the runoff before collecting it.

The major elements of collector trough design are depth, width, and bottom slope. Design depth can be figured two ways, depending on whether a measure flume is used or whether runoff is conducted directly to the sludge tank. If a flume is used, depth of the collector is controlled by the size of the approach channel required by the flume. In other words, the design is started by choosing the type and size flume necessary to handle maximum runoff. Thus, depth of the collector is equal to depth of the flume plus about 10% freeboard.

When only a conveyance pipe is used (no rate measurement), the collector depth is based on the pipe size needed to carry the runoff load. After the collector depth is calculated as discussed above, a freeboard of approximately 0.4 foot is added to the collector trough. This freeboard is needed primarily to form a notch across the plot end and may be changed to suit local design requirements.

Figure 5a shows an example. Although the design worked appropriately during many storms, it did not work in some other cases. Replacement by a design that concentrated flows vertically and produced supercritical flow in the channel leading to collection tanks (Fig. 5b, c) overcame the problem. However, there are a number of designs where flows are forced to concentrate on the eroding area so that they produce results that are open to question (Kinnell 2016).

Figure 6a shows an example where flow was forced to concentrate on the eroding surface and sedimentation occurred on the plot just upslope of a flume. The kite-shaped plot shown in Figure 5.1.1 in Kuhn et al. (2014) provides another example where flows are forced to concentrate on the eroding surface. Figure 6b shows schematics of other designs that have been used in experiments reported in literature (Strohmeier et al. 2016; Vaezi et al. 2008; Zhang et al. 2015) but produce results that are open to question. Ensuring that surface water flows freely over the whole of the eroding surface is essential in all rainfall erosion experiments no matter what the scale or whether the experiments are done in the field or in the laboratory.

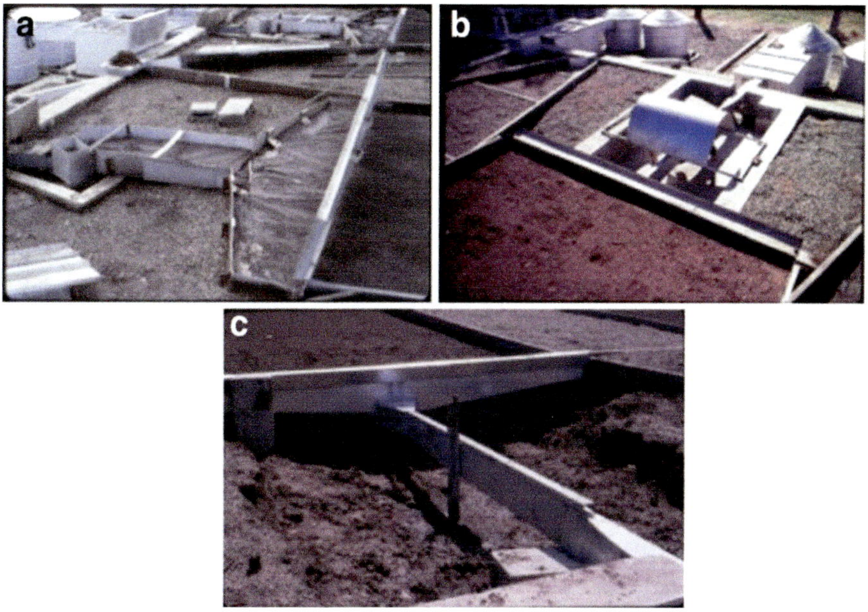

Fig. 5 Runoff collection equipment with 40 m long 2.6 m wide bare fallow plots at Gunnedah, New South Wales, Australia: (**a**) shows the design commonly used by the then Soil Conservation Service of New South Wales and (**b**) and (**c**) show the design that replaced it (Kinnell 2016)

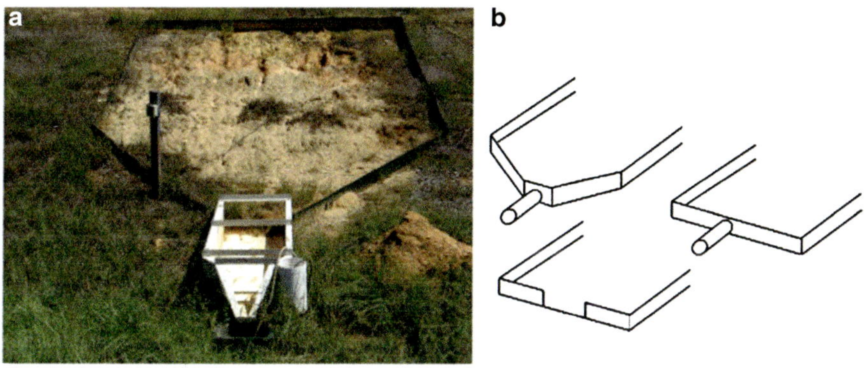

Fig. 6 Examples of designs of runoff collection systems that should not be used on runoff and soil loss plots (Kinnell 2016)

Collection Tanks

Tanks are used to store all the sludge and the aliquot of the soil loss-runoff mixture. Oval end stock water tanks make good sludge tanks and are commercially available in 2-foot height. Usually 1 mm of runoff on a square meter of plot will result in 1 L of water for collection and storage so that the runoff collection system should not overflow between the end of the plot and tanks designed to collect runoff (Kinnell

2016). Hence, the storage system should be made of inexpensive mild sheet steel, painted against rust and be tailored to the runoff producing capacity of the plots in the climate at the location being studied (Kinnell 2016). Some companies will make tanks 29 or 30 inches high on special order, which are preferable to a 2-foot height because of the higher storage capacity. Round tanks are recommended for aliquot storage; these usually have to be custom made.

The sludge tank unit has two major functions: (i) To retain all the heavy soil material and pass only a suspended sediment mixture to the divisor unit and (ii) to store sludge which will make up the bulk of the soil loss from the runoff plot. Turbulence in the sludge tank due to high entrance velocities from the runoff plot is reduced by placing two screens across the flow through the sludge tank, thus increasing deposition. The screens also keep trash from clogging up the divisor. The screens do not extend to the tank bottom and freeboard is allowed at the top. This is done to insure flow even though the screens become filled with trash. Floor space between the tank inlet and first screen is left for a can to catch low flows, so that the entire tank need not be cleaned after every rain shower. The 50-ton-per-acre maximum soil loss figure discussed under design criteria is used to calculate storage capacity of the sludge tank. A cubic foot of soil weighs from 60 to 100 pounds.

As plot size increases, the ability to store all runoff and soil loss becomes increasingly difficult. In some cases, a series of connected tanks is installed with the overflow from one tank feeding to another being subsampled by devices called multislot divisors. Outflow from the tank flows through a number of slots with only one being used to pass a known portion of the outflow to the next tank (Pinson et al. 2004). Deposition of coarse sediment usually takes place in the first tank so that the flow into the downstream tank usually contains only fine suspended load. Cascading to additional tanks with the outflows passing through multislot divisors provides the means of storing information on runoff and soil loss for large runoff events that would be impossible to obtain otherwise. The separation of coarse and fine loads also enables the runoff in the downstream tanks to be mixed and subsampled to obtain sediment concentrations that can be used to determine the sediment load in the tanks. However, in the first tank, that cannot be done, because the fast settling nature of the coarse sediment results in underestimation of the sediment concentration (Ciesiolka et al. 2006). Consequently, the most accurate method involves collecting the coarse sediment from the water in the first tank to enable it to be dried and weighed separately. The fine material in the water from the first tank can then be determined through subsampling the liquid once the coarse material has been removed. This approach was adopted by Kinnell (1983) who ensured that subsamples were taken while stirring the mixture after removal of coarse material. This resulted in variations in pairs of subsamples about their mean being frequently within $\pm 1\%$ and seldom exceeding $\pm 3\%$.

Apart from multislot devisors, some other devices have been developed to reduce the quantities of runoff and soil loss that have to be stored during one or more erosion events. With the Coshocton Wheel sampler (Brakensiek et al. 1979; Carter and Parsons 1967), flow from the runoff collector rotates a slot through the outflow collector with the speed of rotation controlled by flow falling on to vanes (Fig. 7a).

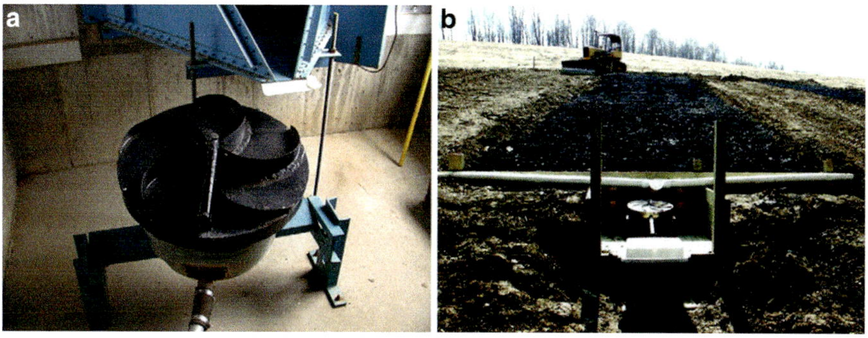

Fig. 7 (a) An early version of the Coshocton Wheel used by the USDA-ARS. (b) An example of a Coshocton Wheel installed on a plot to measure runoff and soil loss (Kinnell 2016)

Runoff measuring devices like H-flumes can cause deposition to occur upslope of Coshocton Wheels and care needs to be taken to ensure that such devices do not cause ponding and deposition on the eroding surface such as shown in Fig. 7a. A H-flume–Coshocton Wheel system (see Fig. 2 in Mutchler et al. (1988)) was used in the evaluation of the product of runoff rate and rain kinetic energy flux as an alternative to the EI30 index at Holly Springs, MS, USA (Kinnell 1995; Kinnell et al. 1994). Coarse material deposited upslope of the flume was measured separately from the material collected by the Coshocton Wheel.

Another approach to measuring runoff centers about the use of the tipping bucket method to measure runoff (Nehls et al. 2011; Yu et al. 1997; Zhao et al. 2001). Tipping buckets were initially developed as a meteorological device for measuring rainfall but have been expanded in size to measure runoff (Edwards et al. 1974; Khan and Ong 1997; Pillsbury et al. 1962). Sediment sampling is achieved in some cases by using slots to collect runoff and sediment as the bucket empties (Deasy et al. 2009; Silgram et al. 2010). Some designs are better suited to situations where only fine material is discharged with runoff.

As an example, in Fig. 8, runoff water was collected and quantified at the lower end of each plot throughout the growing season using a tipping-bucket runoff metering apparatus (Fig. 8). Buckets were calibrated (one tip represents 3 L of runoff) and maintained to provide precise measure of amount of runoff per tip. Numbers of tips were counted using mechanical runoff counters. Collection of samples in 3.79 L borosilicate glass bottles was carried out through a flow-restricted composite collection system (approximately 40 ml per tip was collected) (Antonious 2015).

Experiments Using Artificial Rainfall

Comprehensively speaking, a number of instruments have been used to quantify runoff. The most basic measurement method involves diverting flow to a barrel or similar structure (Dosskey et al. 2007; Hudson 1993; Meals and Braun 2006). Water quantity, chemistry, and sediment measurements can then be taken on the collected water. However, for specific application, designs of runoff plot varied according to the local climate and other conditions and can be conducted under two

Fig. 8 Surface runoff water collection using tipping buckets installed down the field slope. A gutter was installed across the lower end of each plot with 5% slope to direct runoff to the tipping buckets and collection bottles for runoff (Antonious 2015)

circumstances: in-situ measurement without any artificial water pumping as the above introduction and experiment simulations with water input by rainfall simulators or other sources. Below are several runoff plot examples applicable to different circumstances. Experiments using runoff plots under natural rainfall such as those used to develop the Universal Soil Loss Equation are still undertaken from time to time in various parts of the world but, as noted earlier, the USDA-Purdue rainfall simulator, or "rainulator," was developed by Meyer and McCune (1958) as a tool to conduct experiments to supplement the USLE natural rainfall database. Wischmeier and Mannering (1969) used the rainulator on 55 soils to examine the relationship of soil properties to erodibility. Each test consisted of three storms. The first was for 60 min on an initially dry soil. The second storm applied the next day was applied under what was considered to be a wet condition, and that was then followed by another storm on the same day on a surface that was considered to be in very wet condition. The different values of erodibility recorded for the different antecedent moisture conditions were then used in an equation that calculated the K for the climate in central USA (Romkens 1985).

There are many different rainfall simulator designs using a wide variety of nozzles reported in the literature (Iserloh et al. 2013a; Meyer 1994). The rainfall simulators that have been the development of WEPP and RHEM have used TeeJet 2HH-SS50WSQ nozzles. In WEPP interrill erosion model given in Flanagan et al. (Flanagan and Nearing 1995), there is a term (Fnozzle) to take account of the use of other nozzles. That provision is not documented well in many other places and consequently, the data on interrill erodibilities published without taking account of differences between the nozzles used and TeeJet 2HH-SS50WSQ nozzles (Mahmoodabadi and Cerdà 2013; Romero et al. 2007). Apart from Iserloh et al. (2013b), there appears to be little or no work reported on the effect of different rainfall simulators on sediment discharged from interrill areas. Figure 9 shows the

Fig. 9 The artificial simulated rainfall device (Mayerhofer et al. 2017)

artificial simulated rainfall device used in studying the effects of land use and land cover on surface runoff in the catchment area of the Alps.

Runoff Measurement by Curve Number (CN) Method

Theory

Runoff estimates are often needed for ungauged watersheds for engineering design of hydraulic structures, watershed the U.S. Department of Agriculture (USDA) – Soil Conservation Service (SCS) developed a method for estimating rainfed runoff volume based on measured total rainfall and direct runoff, and physical watershed features. This method is simple to use and requires basic descriptive inputs that are converted to numeric values for estimation of watershed direct-runoff volume. The curve number (CN) method is widely used by engineers and hydrologists as a simple watershed model, and as the runoff-estimating component in more complex, watershed models. The method depends on using measured watershed runoff and rainfall data to develop a CN value that reflects the CN value that should be developed from measured data.

The maximum potential retention (S) can be calculated from the CN value which is able to be determined in considering hydrological, soil property, land use and surface conditions, and soil moisture content before runoff occurs (Mishra and Singh 2013). However, the CN method does not consider rainfall intensity, and there are questions as to whether it is applicable for areas outside of the USA (Yamashita et al. 2006).

On the other hand, Chong and Green (1983) introduced the following Eq. (1) which combining with the SCS rainfall-runoff equation and the maximum potential retention (S) of a watershed in order to estimate the value of sorptivity. This equation shows that it is possible to estimate the volume of rainfall runoff from a watershed, if there is a relationship between sorptivity values and initial soil moisture contents

$$S = \frac{1}{2} R_i^{-1/2} K_{Sat}^{-1/2} [Sp(\theta)]^2 \tag{3}$$

where S is the maximum potential retention. $Sp(\theta)$ is soil sorptivity, K_{sat} is saturated soil hydraulic conductivity, and R_i: rainfall intensity. The term "sorptivity" was introduced by Philip (1957a) in his well-known two-term infiltration equation. As described by Philip, sorptivity, $Sp(\theta)$, is a measure of the uptake of water by soil without gravitational effects. According to the Philip two-term equation, this coefficient is one of the most important soil parameters governing the early portion of infiltration.

Thus the relationship between sorptivity values and soil moisture contents and estimated the maximum potential retention using Eq. (3) with rainfall intensity and saturated soil hydraulic conductivity can be clarified. Finally, the estimated surface runoff volumes for each rainfall event were calculated using sorptivity. On the other hand, surface runoff volumes using the CN method were also estimated in order to compare the surface runoff volumes.

Experiment Setup

A water reservoir to measure rainfall runoff volume from a catchment area was constructed in an experimental field with covered plastic film sheet to prevent percolation into soils. In order to measure the water level of this reservoir, a water pressure sensor with a data logger was installed at the bottom of the reservoir. At the same time, we measured the atmospheric pressure using another pressure sensor with a data logger so that we can get a water depth in the reservoir to calculate a difference value of both pressure sensors. Other meteorological data such as air temperature, relative humidity, wind speed, solar radiation, and rainfall were also collected in the experimental field (Fig. 10).

A disc tension permeameter can be fabricated in order to measure water infiltration in the soil, which is characterized by in situ saturated and unsaturated soil hydraulic properties. It is mainly used to provide estimates of sorptivity and the hydraulic conductivity of the soil near saturation. In order to clarify the relationship between sorptivity values and initial moisture contents in soils in the catchment area, we carried out an experiment using the disc tension permeameter near the catchment area located in ATRAC. The steps for measuring sorptivity are as follows.

Firstly, surface top soils of in 2–3 cm thickness are moved out and a metal cylinder of diameter 15 cm is vertically inserted into soils. The disc tension permeameter with 4 cm suction is installed on the cylinder. After starting infiltration into soils of the metal cylinder, accumulated infiltration amounts at each elapsed time are measured. Water to surface soil can also be supplied close to the cylinder in order to change soil moisture content of top soils so that sorptivity values under different soil moisture conditions could be measured.

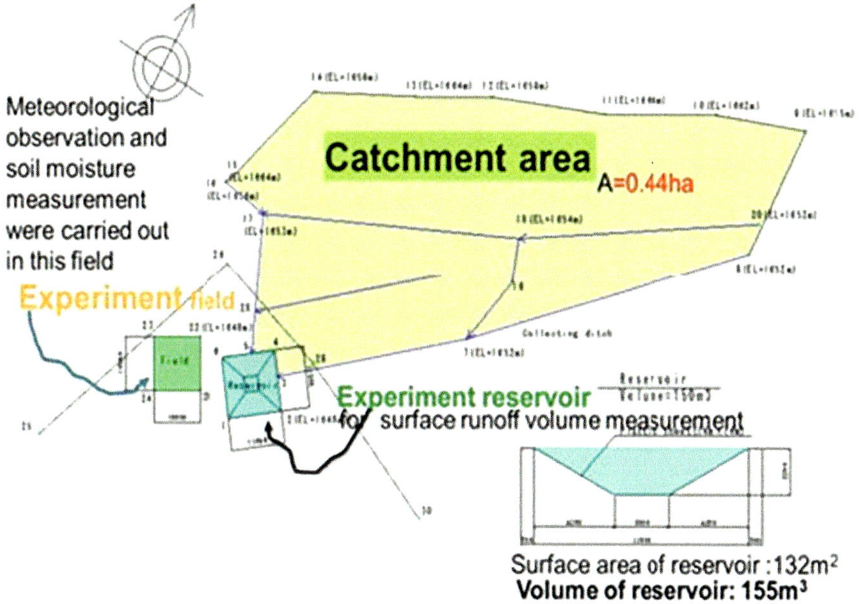

Fig. 10 Layout plan of experimental reservoir, catchment area, canals, and experiment field conducted by Watanabe et al. (2012)

Undisturbed soil cores were collected using a 100 cm^3 soil sampler to measure the moisture content of the soil surface close to the cylinder so that the relationship between sorptivity values and initial soil moisture content could be clarified for each sorptivity measurement.

Runoff Measurement by Isotopic Tracers

Theory

Hydrological responses of hillslopes are determined by a number of factors associated with parent geological material, topography, climate, and vegetation. The processes whereby rainfall becomes runoff continue to be difficult to quantify and conceptualize. Hydrograph separation with natural tracers or isotopes has become a popular method to gain comprehensive insights into runoff processes which can date back nearly 50 years (Hubert et al. 1969).

The simplest concept of hydrograph separation distinguishes between event and pre-event water. Over the hillslope, event water is water from rain that enters and flows through the system, and pre-event water is soil that is already stored in the system at the beginning of the event. If the two end-members have a distinct difference in their isotopic signature, the surface runoff hydrograph can be separated in their contributions based on a mass balance approach (Buttle 1998):

$$Q_t = Q_p + Q_e \tag{4}$$

$$C_t Q_t = C_p Q_p + C_e Q_e \tag{5}$$

$$F_p = \frac{(C_t - C_e)}{(C_p - C_e)} \tag{6}$$

where Q_t is the surface flow/streamflow; Q_p the contribution from pre-event water; Q_e the contribution of event water; C_t, $C_{p,}$ and C_e are the d values of surface runoff, pre-event water, and event water, respectively; and F_p is the fraction of pre-event water in the surface runoff. Abundance of stable water isotopes is based on the isotopic ratios ($^{18}O/^{16}O$ and $^{2}H/^{1}H$).

At the initial stage, Sklash et al. (1976) and Sklash and Farvolden (1979) provided the first, clear exposition of the main underlying assumptions implicit in the technique (initially four), which were later refined and extended to five underlying assumptions (Buttle 1994; Moore 1989):

1. The isotopic content of the event and the pre-event water are significantly different.
2. The event water maintains a constant isotopic signature in space and time or any variations can be accounted for.
3. The isotopic signature of the pre-event water is constant in space and time or any variations can be accounted for.
4. Contributions from the vadose zone must be negligible, or the isotopic signature of the soil water must be similar to that of groundwater.
5. Surface storage contributes minimally to the surface runoff.

For now, it is important to note that early isotopic hydrograph separation (IHS) work also assumed that the pre-event water could be described by a single isotopic value of water in the stream prior to the event; describing in essence, a single, integrated pre-event water signal that is assumed to be representative of the entire stored water that may contribute to surface runoff (Sklash and Farvolden 1979). A number of follow-on studies to the early two-component IHS used a multi-component approach to account for additional contributing end-members. In such cases, the standard mixing Eq. (7) and (8) were extended as follows:

$$Q_t = Q_1 + Q_2 + Q_3 + \cdots + Q_n \tag{7}$$

$$C_t Q_t = C_1 Q_1 + C_2 Q_2 + C_3 Q_3 + \cdots + C_n Q_n \tag{8}$$

where Q_n is the discharge of a particular runoff component and C_n the tracer concentration of a particular runoff component. In the case of three flow components, a second tracer or a measurement of one flow component was required. Most commonly, a stable isotope tracer was combined with a geochemical tracer (Wels et al. 1991), but sometimes a second stable isotope was used (Rice and Hornberger, 1998).

In the past, several improvements and modifications of the original hydrograph separation procedure were suggested (Harris et al. 1995; Mcdonnell et al. 1990). At the initial stage, hydrograph separation method to the hillslope runoff generation were usually applied to the simple partition of pre-event and event by adding an appropriate weighting technique (Burlando 1999; Mcdonnell et al. 1990).

Drawbacks related to a simplistic use of isotopes in runoff separation studies can be avoided by applying more sophisticated experimental and modeling approaches, ideally including fully distributed process-based multidimensional numerical modeling of the relevant hydrological processes. To prevent an excessive computational cost of the numerical solution of two- (2D) or three-dimensional (3D) governing equations of the hillslope scale flow and transport processes, it is possible to decouple the essentially 3D flow into one-dimensional (1D) vertical variably saturated flow and 1D lateral saturated flow along soil-bedrock interface (Fan and Bras 1998; Hilberts et al. 2007; Troch et al. 2002). Vogel et al. (2010) applied vertical one-dimensional dual-continuum model to describe soil water dynamics and stable isotope transport in a hillslope soil. In their analysis, the oxygen isotope was used as a natural tracer to study preferential flow effects at the site of interest. In the subsequent study, lateral component of rapid shallow subsurface flow at the same site, in addition to preferential vertical movement, was considered (Dusek et al. 2012).

Experiment Setup

Taking the studies conducted by Vogel et al. (2010) as examples. The experimental hillslope site Tomsovka is situated in the small mountain catchment Uhlirska, Jizera Mountains, Czech Republic (total area 1.78 km^2, average altitude 820 m above sea level, annual precipitation exceeding 1300 mm, average annual temperature 4.7°C).

1. Isotope sampling

   The $\delta^{18}O$ values were determined in (i) precipitation collected from the rain gauge (or from snowfall and snowpack samples during winter), (ii) surface and subsurface hillslope discharge collected from the experimental trench, and (iii) soil water extracted from the selected depths by suction cups.

   During the vegetation seasons, rainwater samples were collected at daily intervals. Because each sample represents cumulative rainfall for the period between the two samplings, the measured $\delta^{18}O$ value was averaged over the respective time interval. During the winter, weekly precipitation totals were measured by a storage gauge (rain–snow standpipe). The snow depth and snow density were measured by a snow sampling tube. Both the storage gauge water and the snowpack water were sampled (synchronously and separately) to determine the ^{18}O content. The monitoring of hillslope discharge by tipping bucket flux meters was discontinued due to freezing temperatures.

   Soil water was extracted by suction cups installed at the depths of 30 and 60 cm below the soil surface. At the beginning of the soil water extraction, the air was pumped out from the probe with a hand vacuum pump until a pressure head of about −500 cm was reached. A soil water sample was then collected in 2 to 4 d, depending on the soil moisture conditions. The extraction time was adjusted so as to obtain at least a 20 cm^3 water sample. The pressure head increase in the probe during the extraction was about 100 cm. The sampling was repeated at approximately monthly intervals over the period of interest.

The oxygen and hydrogen isotope ratios ($\delta^{18}O$ and δD) of all of the samples can be measured on a Picarro L2130-i liquid analyzer. The measured values of $\delta^{18}O$ and δD are expressed in parts as per mil (‰) of their deviations, with respect to Vienna Standard Mean Ocean Water (V-SMOW2).

2. Related factors measurement

Soil water pressure within the soil profile was measured using a set of automated tensiometers installed at three different depths below the soil surface. The discharge of shallow subsurface flow was measured by means of experimental trench. Water entering the trench was collected at the depth of about 75 cm below the soil surface into PVC pipes. The pipe discharge was measured by tipping bucket gauges, separately for two trench sections denoted as A and B (each 4 m long). The discharge rates QA and QB were measured continuously during vegetation seasons (from May to October). The hillslope length contributing to measured subsurface runoff was estimated to be about 25 m (Hrnčíř et al. 2010), although the geographic catchment divide is located approximately 130 m above the experimental trench, winding through a gently sloping plateau. The contributing hillslope length estimate was based on the comparison of hillslope discharge to the trench and observed catchment outlet discharge (assuming that hillslope subsurface flow represents a dominant part of the catchment response and is uniform across the catchment). For the modeling purposes, the hillslope microcatchments corresponding to trench sections A and B were assumed to have approximately same geometric and material properties (hillslope length, depth to bedrock, soil stratification, soil hydraulic properties, etc.).

From the point of view of the model, the ^{18}O signature in precipitation water represents an input signal, and the isotopic composition of the subsurface hillslope discharge an output signal. The model conveys our hypothesis about the processes controlling the input-output transformation. The transformation takes place in both the soil matrix and the preferential flow domains, each being characterized by different mixing patterns of "old" and "new" water (Fig. 11). The mixing is also affected by the interdomain exchange of soil water.

Runoff Measurement by Salts

Mean flow velocity (V_m) is one of the most important hydraulic variables in soil erosion modeling, since it is dependent upon flow discharge, slope gradient, topography, and surface condition (Zhang et al. 2002). It is used to calculate friction coefficient, runoff concentration time, and other hydraulic parameters such as stream power and unit stream power, which are used to simulate the processes of both detachment and sediment transport in the process-based erosion model (De Roo et al.

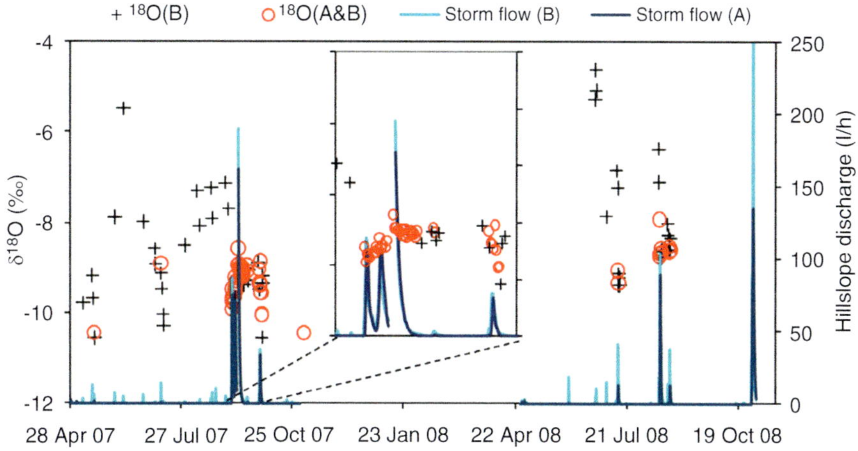

Fig. 11 Observed hillslope discharges and the corresponding ^{18}O contents. The crosses indicate the instances when only section B of the experimental trench contributed to the collected samples while no discharge was observed from section A. The circles indicate that both trench sections contributed to the samples. The period of 1 September to 4 October is enlarged to provide closer view of the major storm (the missing rising limb of the third peak is due to incomplete observation data) (Vogel et al. 2010)

2015; Zhang et al. 2010a). The measurement of shallow water flow often involves the use of a tracer. Tracers used have included dyes (Abrantes et al. 2018; Zhang et al. 2010a) and salts (electrolytes) (Planchon et al. 2005; Lei et al. 2005). Most of these methods necessarily involve the use of instrumentation to detect the tracer movement. Here we will focus on an improved method for shallow water flow velocity measurement with practical electrolyte inputs.

Theory

Salt solution in water flow is transported under influences of both convection and dispersion. Its transportation is influenced by many factors such as the flow rate, flow velocity, and the water quality. When the flow is reasonably assumed to be a one-dimensional and steady flow, its behavior is well defined and quantified by a partial differential equation (PDE).

The convectional and dispersion processes of salt a steady water flow are defined by Fick's law and the mass conservation law and is given by the differential equation for the one- D solute transport as:

$$hw\frac{\partial C}{\partial t} + hwu\frac{\partial C}{\partial x} = \frac{\partial}{\partial x}\left(hwD_H\frac{\partial C}{\partial x}\right) \tag{9}$$

where h is the depth of the water flow, m; w is the width of the flow, m; C is the electrolyte concentration, kg m^{-3}, a function of distance x and time t, proportional to the electrical conductivity of the solution; x is the coordinate down the slope, m; u is the flow velocity, m s^{-1}; t is time, s; and D_H is the hydrodynamic dispersion coefficient, m^2 s^{-1}.

When rainfall and infiltration are ignored, the flow rate is a constant, and the velocity of the laminar flow varies little, such that:

$$Q_0 = hwu = constant \tag{10}$$

$$u = constant \tag{11}$$

which leads to

$$hw = constant \tag{12}$$

where Q_0 is the flow rate, $m^3 s^{-1}$

Combining Eq. (12) with Eq. (9) yields:

$$\frac{\partial C}{\partial t} + \frac{\partial C}{\partial x} = \frac{\partial}{\partial x}\left(hw\frac{\partial C}{\partial x}\right) \tag{13}$$

When the upper boundary condition is assumed to be a pulse, the initial and boundary conditions for Eq. (13) are given as:

$$C(x,t) = C_0\delta(t) \quad x = 0 \tag{14}$$

$$C(x,t) = 0 \quad x = \infty \tag{15}$$

$$C(x,t) = 0 \quad t = 0 \tag{16}$$

The solution to Eq. (13) as a time-dependent function is given by (Lei et al. 2005) as:

$$C(x,t) = C_0 \frac{x}{2t\sqrt{\pi D_H t}} \exp\left(-\frac{(x-ut)^2}{4D_H t}\right) \tag{17}$$

This is the analytical solution to Eq. (13), with the pulse function as the upper boundary condition. The solution quantifies the transient transport of solutes in the flowing water, under a pulse input. Equation (17) is an error function. There are three important parameters, i.e., C_0, u, and D_H to be determined to specify the functional distribution of the transient transport process.

Experiment Setup

An experimental flume of 4 m long and 15 cm wide was used to simulate the water flow, in which the solute was transported (Lei et al. 2010). The experimental system is shown in Fig. 12. The system included a computer installed with specially designed software for control of salt solute injection and sensed data logging, an interface unit, electric conductivity sensors, a salt solute injector, the flume, and the

Fig. 12 Experimental equipment system (Lei et al. 2010)

water supply. The experiments involved a combination of three flow rates (Q = 12, 24, and 48 L min^{-1}) and three slope gradients (S = 4°, 8°, and 12°). The regulated water flow was introduced into the flume from the upper end. Once the flow was stabilized (within 1 min fluctuation), about 6 ml of highly saturated salt solution of KCl was injected at a location 1 m from the upper end of the flume, allowing some distance for the establishment of the steady flow.

The injection of the salt solution into the water flow was done through a computer-controlled electrical valve. The six sensors were located at 5, 30, 60, 90, 120, and 150 cm from the solute injector. The sensor 5 cm from the solution injection point was used to register the input signal. The electrical conductivity values measured at the six locations were logged into the computer through the specially designed data logger, controlled by the specially designed software.

When using this system for velocity measurement of shallow water flow in the laboratory by the method discussed above, these procedures are typically as following:

1. Wire up the system and set the flume at the required slope.
2. Place the electrical conductivity (EC) sensors at the designate locations, with one sensor located at 5 cm from the KCl solution injection point.
3. Put the prepared KCl solution into the injector container.
4. Introduce a specific water flow into the flume from its upper end, stabilize and measure it.
5. Start the software specially designed to start taking measurements.
6. The computer initiates data logging and injects salt solute into the stream via the interface unit.
7. Allow time for the solute to transport and pass through each sensor.
8. The EC values as function of time are recorded by the computer.
9. Stop data logging.
10. Run the analysis part of the software to fit the logged data from the first channel to determine the input boundary signal.

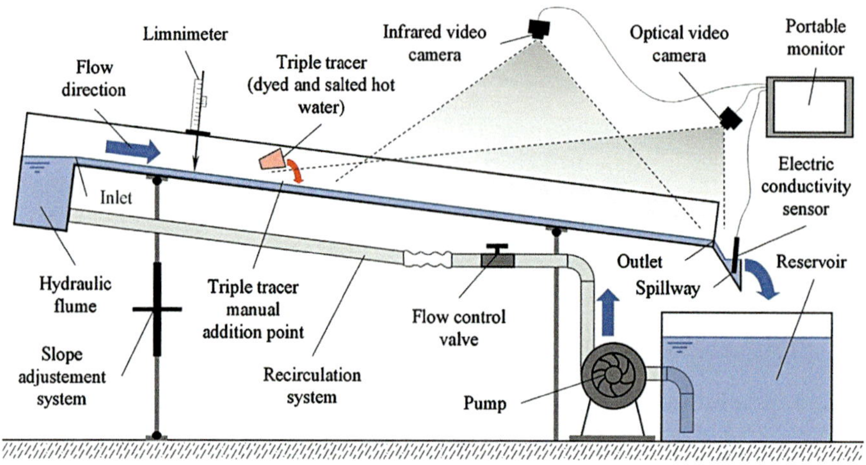

Fig. 13 Scheme (side view) of the laboratory setup used in the triple-tracer experiments (Abrantes et al. 2018)

11. Fit the logged data from the other channels with the integral equations to estimate the velocities at different locations.
12. Output the computed velocity values and other information required.

Similar experiments laboratory experiments were also conducted to compare the traditional dye and salt tracer techniques to the more recent thermal tracer technique for estimating shallow flow velocities and investigating the effects of a wide range of hydraulic conditions on the correction factor used to determine mean flow velocity (Abrantes et al. 2018) (Fig. 13).

Measurement of Elements Related to Surface Runoff

Soil Temperature and Humidity

In each of the above-mentioned runoff-monitoring plots, two 1.5 m deep wells will be constructed to monitor soil moisture and temperature synchronously. Soil moisture and soil temperature sensors will be installed in adjacent wells at different depths. Soil moisture can be determined by a frequency domain reflectometer (FDR) using a calibrated soil moisture sensor equipped with a Theta-probe. Volumetric soil moisture can be derived from changes in the soil dielectric constant, converted to a millivolt signal, with an accuracy of ±2%. Soil temperature was monitored using a thermal resistance sensor sensitive to temperature changes in the range of -40 to 50°C, with an overall system precision of ±0.02°C. The thermal resistance sensors were developed by the State Key Laboratory of Frozen Soil Engineering in Lanzhou, China, using Fluke 180 series digital multimeters (Fluke Co., USA). The sensors had been even successfully used in other projects on the

Fig. 14 Photograph of (**A**) 30 cm diameter (inner) and 60 cm diameter (outer) double-ring infiltrometer, (**B**) 15 cm diameter (inner) and 30 cm diameter (outer) double-ring infiltrometer, and (**C**) Mariotte siphon developed to maintain a constant inner head in the infiltration rings (Gregory et al. 2005)

Qinghai–Tibet Plateau over the past 20 years (Wu and Liu 2004; Wu et al. 2002). All of the soil temperature and moisture data can be collected automatically once every 30 min by a CR1000 data logger.

Soil Infiltration Rate Measurement

Too many methods have been applied to measure infiltration rate like the ring infiltration method, hydrological method, artificial rainfall method, among which the classical double-ring infiltrometer stands out (Mathieu and Pieltain 1998).

Double-ring infiltrometer usually consists of two concentric metal rings (Fig. 14). The rings are driven into the ground and filled with water. The outer ring helps to prevent diver-gent flow. The drop-in water level or volume in the inner ring is used to calculate an infiltration rate. The infiltration rate is the amount of water per surface area and time unit which penetrates the soils. The diameter of the inner ring should be approximately 50–70% of the diameter of the outer ring, with a minimum inner ring size of four inches. The infiltration velocity was measured from the beginning of the experiment until a stationary regime, the steady-state infiltration rate (SIR) was reached. The steady-state rate was assumed to be reached when three consecutive similar measurements were observed after 90 min. Taking into account the spatial variability, three replicates were performed in each zone with a distance of 5–10 m between each other. In all cases, the land slope was less than 5% (Neris et al. 2012).

Evapotranspiration Measurement

This weighing lysimeter was constructed in 1990 and put into operation in 1991. It is placed in the middle of a 1.0 Ü 106 m^2 cultivated field. The basic components of the lysimeter are illustrated in Fig. 15. Component I is a steel oil cylinder with a surface area of 3.14 m^2 and a soil profiles with depth of 4.5 m overlying 0.5 m of fine sand. The aboveground part is 0.05 m in height. The steel cylinder wouldn't be cut into the soil until the lysimeter was constructed. Therefore, the lysimeter was filled by undisturbed soil. A neutron probe access tube (IV) is installed in the column. The soil column rests on a sensitive weighing system (V) capable of measuring the total mass of approximately 35 Mg to the nearest 60 g. A Mariotte system (II) is

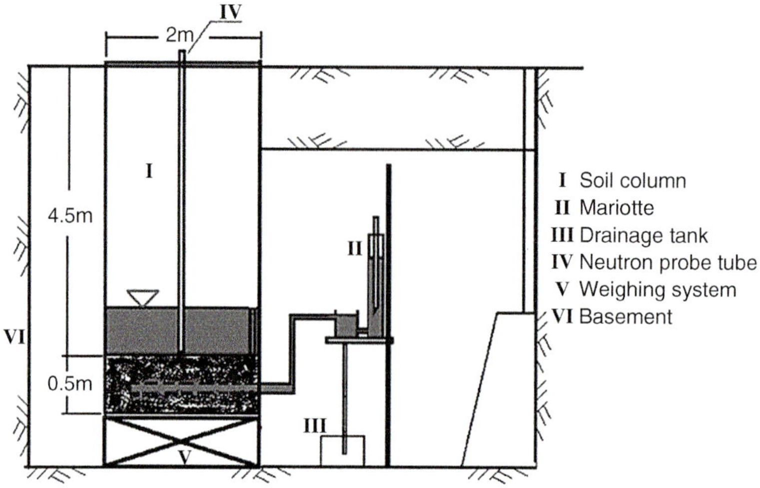

Fig. 15 The measured evapotranspiration in the lysimeter (Yucheng Comprehensive Experimental Station, Chinese Academy of Sciences, 1999) (Luo et al. 2003)

Fig. 16 A picture for the weighing lysimeter

connected to the soil column to control and record the water table inside, and measure the amount of water that is supplied to the soil column and/or leaks out of it. Gravity drainage is collected by a drainage tank (III). By recording the weight change of the soil column, water leakage from or water supply to the soil column, and the irrigation and/or rainfall amount, the total evapotranspiration, at certain time intervals, from the lysimeter can be obtained through a water balance approach. Generally, observations are made at 08:00 and 20:00 each day (Luo et al. 2003). The weighing system is calibrated every year. A picture for the in-situ weighing lysimeter is shown in Fig. 16.

General Modeling of Surface Runoff

Basic Concepts in Hydrological Models

Hydrological models have been developed to improve our understanding of surface runoff generated from complex watersheds, which should capture the essence of the physical controls of soil, vegetation, and topography on runoff production. Generally, there are three mechanisms generating surface runoff: (i) unsaturated surface runoff (Hortonian-type runoff), (ii) saturation-excess surface runoff, and (iii) return of subsurface storm flow, where the last is detectable in some cases already on the plot scale but becomes increasingly important when moving from the plot to the catchment scale and from the event to longer time scales. However, not all excess water generated by these mechanisms contributes to surface runoff because some is stored on the surface as depression storage (infiltrating after rain events) and detention storage (partly running off after events). Factors involved in the process of runoff, such as soil characteristics, vary extensively over small distances.

A modeling approach to simulate the physical processes of runoff would be ideal to investigate the effects of changes in a catchment on its generation, due to the spatial and temporal heterogeneity of the factors involved in runoff at catchment scale. The hydrological models generally integrate existing knowledge into a logical framework of relationships and rules. They can be used to be predictive tools for water resources management and to improve our understanding of environmental systems as a tool for hypothesis testing. Some models are more simplified than others but at the base of each model, there is a mathematical description that simplifies the factors that are being considered and that enables models to make quantitative predictions. The selection of a suitable model should depend on study objectives, additionally other factors such as data availability, money, and time should also be taken into account. Due to the differences in soil and climate characteristics of catchments, there is a growing awareness that catchments respond to rainfall in a variety of ways. Therefore, increasing the complexity of a model structure through emphasizing more on the physical basis of natural processes does not necessarily improve the model performance.

Classifications of the Model

The mathematical descriptions of hydrological models are simplifications of the actual processes of streamflow in nature. The models whether empirical, physical, or combinations of the two are therefore based on many assumptions. In general, surface runoff models can be classified from stochastic to deterministic models, from black-box or empirical to conceptual models, from lumped to physically based (white-box) distributed models, and from land surface to global hydrological models. In a stochastic modeling approach, randomness or uncertainty in the possible outcome of the model is permitted due to the uncertainty introduced by the input data. Besides, both the input and output variables of stochastic runoff models are described in terms of a probability density distribution. On the other hand, deterministic models focus on the simulation of the physical processes involved in the transition from precipitation to runoff. They can further be divided into conceptual- and physical-based runoff modeling approaches.

The predictions obtained from lumped modeling approaches are single values, whereas the distributed modeling approaches make spatially distributed predictions. Lumped modeling approaches consider a catchment to be one unit and a single average value representing the entire catchment is used for the variables in the model. However, the use of such simple models tends to generalize details of environmental processes, which may result in the loss of both spatial and temporal information. The distributed models make predictions that are distributed in space allowing to assess the effects of land use/cover changes in a catchment on the rate at which runoff is generated. Nevertheless, making a certain model more physical based implies that the input parameters are also increased and are more complicated to attain. Most parameters are obtained through it may introduce some error into the model when the parameters are processed contributes to the overall inaccuracy of a model.

Conceptual Models

The empirical models were simple transfer functions describing a linear relationship between rainfall and surface runoff like models of the Green and Ampt (1911), Philip or Horton (Horton et al. 1940) type, and the CS curve number. Watershed-scale models dealing with surface runoff and soil erosion from arable land. Assuming that surface sealing during heavy rainfall events dominates runoff generation on partly bare soils, they often stick to Hortonian-type surface runoff generation approaches. The simplest case of such runoff models are index models, which assume a certain constant fraction of the total rain falling in a catchment becomes surface runoff, or there is a constant loss rate from the total rain falling in a catchment before surface runoff occurs. These models are simple ways of obtaining approximate runoff estimates and are still widely used. Larger-scale models typically use Green and Ampt or Philip approaches assuming that there is the existence of sharp wetting front having a constant matric potential and the wetting zone is uniformly wetted with a constant hydraulic conductivity the initial soil moisture conditions. The US soil conservation Curve Number Method, developed by the U.S. Department of Agriculture and Natural Resources through the analysis of runoff volumes from small catchments in the USA, is another simple empirical method for estimating the amount of rainwater available for runoff in a catchment. This type of model has little data demanding and is easy to apply. However, limited quantitative information can be obtained on how the model parameters are developed and their application to conditions for which they are not developed may lead to questionable results.

Process-Based Models

Distributed hydrological models take account of the spatial variability of climate, soil, vegetation, and terrain, which are able to make a series of hydrological processes interconnected, such as snow accumulation and melt, soil moisture dynamics, runoff generation, recharge to groundwater, and evapotranspiration. Despite their complexity, the distributed hydrological models are very helpful to study the changes in hydrological processes caused by human activities, such as deforestation, urbanization, and forestation. This feature also offers the potential to improve hydrologic predictions since these elements are divided into smaller units that are more homogenous than the whole watershed. In physical-based modeling approaches, the characteristics and properties of nature processes are based on the laws of conservation of mass, energy, and momentum. Thus, such types of models are complicated and demanding in their data requirement (Dingman 2002). Due to the fast development of 3S (RS/GPS/GIS) technology, the distributed hydrological models have been well developed during the last decades. A wide range of physical-based rainfall-runoff models are available today, such as HBV, TOPMODEL, and the SHE. The representative semi-distributed hydrological model TOPMODEL, developed in 1971, describes runoff generation process including both saturation excess and infiltration excess runoff according to topographic index derived from digital elevation model (DEM). After TOPMODEL, distributed hydrological models such as SHE (System Hydrologic European) and SWAT (Soil and Water Assessment Tool) are fully distributed and contain more complex hydrological processes. Although the distributed hydrological models have more solid physical base compared to the lumped models, these models often require a large number of parameters to run them and most of the parameters required are obtained through calibration, making such approaches expensive and time-consuming. So selecting models depends on objectives, application, and availability of data (Table 1).

General Vegetation Controls on Surface Runoff Processes

Vegetation controls surface runoff by means of its canopy, roots, and litter components to reduce raindrops energy effectively and thus redistribute rainfall (Gyssels et al. 2005). Vegetation changes will change the properties of the land surface, thus affecting the slope runoff process and soil moisture infiltration process. Therefore, it is of great significance to reveal the mechanism of hydrological response under the condition of vegetation change.

Vegetation can affect surface runoff by intercepting rainfall, changing kinetic energy of raindrops, increasing surface roughness and infiltration, improving soil physical properties, and directly fixing soil by root (Ju et al. 2007; Xi et al. 2008; Gan et al. 2010). Vegetation also has a considerable influence on the shallow soil flow, which is manifested in the soil reinforcement of the root system, the soil anchorage

Table 1 Summary of the models commonly used surface runoff simulation

Model type	Model	Country	Reference
Lumped model	AWBM	Australia	Boughton and Chiew (2003)
	GR4J	Australia	Perrin et al. (2003)
	HBV	Sweden	Bergstrom (1995)
	HEC	USA	Feldman (1981)
	HSPF	USA	Crawford and Linsley (1966)
	HYDROLOG	Australia	Porter and McMahon (1971)
	IHACRES	Australia	Jakeman and Hornberger (1993)
	Sacramento	USA	Burnash et al. (1973)
	SIMHYD	Australia	Chiew et al. (2002)
	SWM	USA	Crawford (1962)
	Tank	Japan	Sugawara et al. (Sugawara et al. 1986)
	Xinanjiang	China	Zhao (1992)
	SRM	Nordic	Martinec (1975)
Physical model	CEQUEAU	Canada	Morin et al. (2002)
	HYDROTEL	USA	Fortin et al. (Fortin et al. 2001)
	IHDM	USA	Calver et al. (1987)
	MIKE-SHE	Denmark	Refsgaard and Storm (Refshaard and Storm 1995)
	SHE	Denmark	Abbott et al. (1986)
	SLURP	UK	Kite (1995)
	SWAT	USA	Neitsch et al. (2002)
	SWMM	USA	Rossman (2009)
	TOPMODEL	UK	Beven and Kirkby (1979)
	WATELOOD	Canada	Kouwen et al. (Kouwen et al. 2002)

in the root system, the regulation of soil moisture, the soil support and dome, the load-bearing effect, root wedge effect, and wind transmission (Barker 1995; Morgan and Rickson 1995; Nordin 1995). Figure 17 depicts this process.

Effects of Vegetation Components on Surface Runoff

Canopy Layer Regulate Surface Runoff by Intercepting Rainfall

The canopy layer of forest vegetation distributes the precipitation for the first time. Larger interception capacity reduces the effective precipitation that the precipitation reaches the ground, reduces the raindrop's falling speed, and prolongs the duration of the precipitation and runoff. When the vegetation canopy interception reaches the limit value, it will no longer affect the net rainfall and redistribute to the precipitation. This effect has continued.

The results of most of the related studies show that the canopy layer interception rates of different forest types are as follows: coniferous forest<broad-leaved forests, deciduous forest<evergreen forest, single-layer forest<stratified heterogeneous

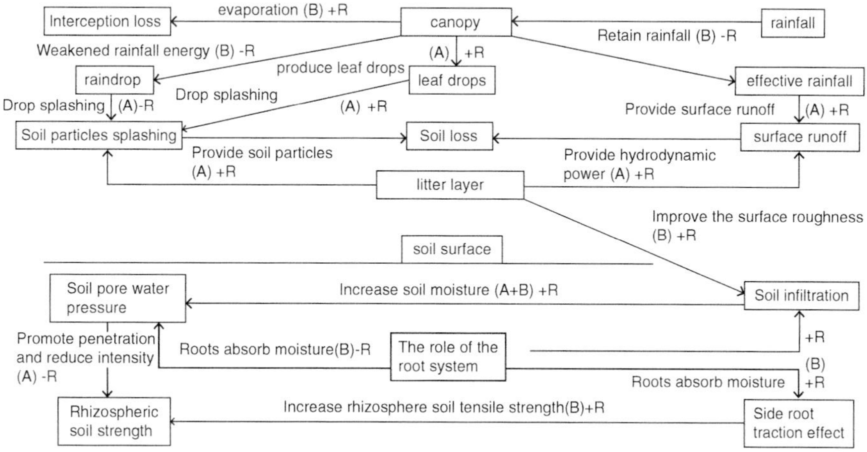

Fig. 17 A conceptual model of the comprehensive effects of the forest on slope protection. A: Adverse effect, B: Benificial effect, +R: accelerating action, -R: weakening action (Zhou 2000)

forest, and pure forest<mixed forest. Canopy density has a greater impact on canopy interception, and dense canopy offers higher interception rate (Tang 2012). The canopy layers of different forest stands have different effects on the kinetic energy of raindrops. Taking *Huashan pine* forests as an example, when the canopy height exceeds 7 m, the impact of the canopy layer on the kinetic energy of the raindrops becomes negligible (Lei 1997; Yu et al. 2006, Yu et al. 2010). Hierarchical vegetation communities protect soil better than monolayers and reduce water erosion.

Hydrological effects of forest canopy under different rainfall intensity are also different. When the rainfall intensity is small, the effect of branches and leaves of canopy on the accumulation of rainfall and raindrops is very prominent. The kinetic energy of raindrops under forest rain and the splashing effect on soil are also obvious (Xie et al. 1994; Wang and Zhang 1998; Zhou 2000). However, when rainfall intensity is larger, the canopy interception is more obvious.

Plant Roots Improve Soil Corrosion Resistance, Soil Permeability, and Soil Stabilization

Vegetation can effectively prevent soil erosion during runoff (Liu et al. 2010). The role of plant roots in soil erosion control is much larger than that of aerial parts, and its role cannot be ignored. The root system's entanglement and distribution in soil are the key factors affecting the degree of soil hydraulic properties, including the density, distribution, and branching characteristics of the root system (Shi et al. 2009).

Soil mechanics composition, structure, porosity, water content, and other indicators all affect the permeability of soil. Plant root system has a great influence on soil physical properties and therefore directly affects the soil infiltration capacity. Some researchers believe that the effect of plant roots on soil infiltration capacity is mainly reflected in the effect of the root system on the bonding of single soil, the dispersion of solidified compacted soil, and the decomposition of the root system itself to humus on the soil aggregate structure and pore condition (Zhu and Ren 1992).

Root growth can improve the friction between roots and soil, root system. At the same time, the root system itself has the ability of antishear and anti-pull. Combining these two points, the root system can play a role of stabilizing the soil. Plant roots are integrated with the soil during the growth process, developed and integrated into a soil-plant-atmosphere continuum, which plays a role in soil consolidation (Yang et al. 1996).

The Litter Layer Again Intercept Rainfall, Thus Prolonging Runoff Formation Time

The effect of litter layers on slope runoff is mainly manifested in two aspects. First, its water storage and holding capacity can prolong runoff and improve soil infiltration capacity; in addition, the presence of litter increases the effective surface roughness, which reduces runoff flow rate to prevent soil erosion (Zhang et al. 2005; Hu et al. 2008).

Effects of Vegetation Covers on Surface Runoff

Changes in vegetation coverage will change the underlying surface properties, which has a significant impact on the slope runoff process. Loss of vegetative cover may lead to the formation of soil seals that increase runoff during the early stages of seal development (Singer and Le Bissonnais 1998).

Quantitative studies under different environmental conditions have demonstrated the positive effect of vegetative cover in reducing water erosion at small scales, with increased soil infiltration accompanied by surface runoff diminished (linearly or exponentially) (Cerda 1999; Dunne et al. 1991; Francis and Thornes 1990; Muñoz-Robles et al. 2011; Quinton et al. 1997; Reid et al. 1999). Vacca et al. (2000) studied runoff in three areas under different land uses (abandoned grazing land, burned machia, and Eucalyptus sp.) and found that different amounts of runoff result from different land uses. The highest runoff was found under Eucalyptus sp. (135 mm), followed by abandoned grazing land (45.25 mm) and burned machia (30.45 mm). Reid et al. (1999) noted that the total runoff was significantly different among three types of land patches, the highest being from bare intercanopy patches, the intermediate being vegetated intercanopy patches, and the lowest being canopy patches. In another study, the decrease in canopy cover density as a result of overgrazing led to rapid runoff yield in rangelands (Oztas et al. 2003).

Seeger (2007) confirmed that increased vegetation coverage can effectively reduce runoff and sediment yield. Eshghizadeh et al. (2015) selected erosion sites in northeastern Iran as the target region and kept monitoring the soil erosion during the period of 2008 to 2015. After analyzing major natural factors affecting runoff and soil erosion in semiarid areas, he concluded that the vegetation coverage has a linear relationship with runoff (Fig. 18).

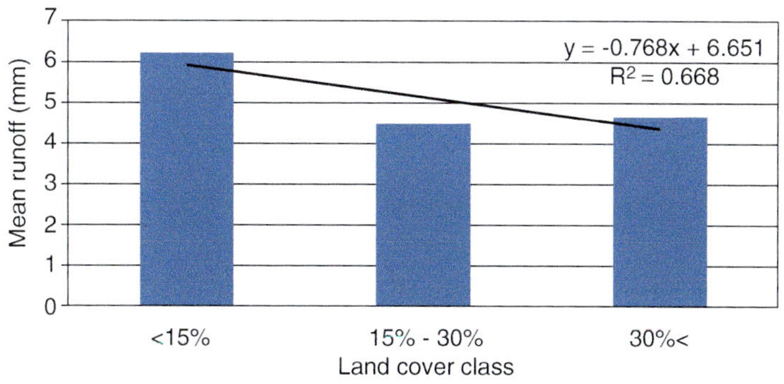

Fig. 18 Mean runoff in each class of land covers (canopy and litter) (Eshghizadeh et al. 2015)

Loss of vegetative cover as a result of human activities such as overgrazing and deforestation leads to the formation of soil seals (Singer and Le Bissonnais 1998) that increase the risk of runoff (Al-Jubeh 2006; Singer and Le Bissonnais 1998; Snyman and Du Preez 2005). Snyman and Preez (Snyman and Du Preez 2005) and Al-Jubeh (2006) observed that rangeland degradation usually leads to increased surface water runoff due to decreased plant cover, reduced aggregate stability, reduced soil fertility, and decreases in the soil water content in all soil layers. In another study, Merzer (2007) reported that bare plots produced significantly more runoff as compared to variety of vegetative plots.

Actually, vegetation can function to reduce runoff yield only if coverage has achieved to a certain threshold. The optimal vegetation coverage refers to the one which can make the soil loss less than the allowable soil erosion. Vegetation coverage has two different definitions of critical coverage and effective coverage. When the critical coverage is exceeded, the effect of vegetation coverage on surface runoff would not increase with the increase of coverage, in addition to the heavy rainfall (Li et al. 2005). When the rainfall intensity exceeds the critical threshold of rainfall, the effect of vegetation reduction would decrease, and the impact of coating on the runoff yield would be very small (Yao et al. 2011). Zhu et al. (2010) studied the effects of herbaceous vegetation cover on slope runoff erosion by using artificial simulated rainfall in the field. The results show that when vegetation coverage is 0% to ~60%, the increase of vegetation coverage can effectively reduce runoff yield and sediment yield. When the vegetation coverage is more than 80%, the influence of the increase of coverage on the runoff coefficient is weakened, and the influence of vegetation on runoff and sediment yield tends to be stable. Therefore, the critical vegetation coverage is 60–80%. Through the artificial simulated rainfall experiments, Zhao et al. (2015) found that the vegetation coverage would have a significant impact on soil moisture content and runoff. When the vegetation coverage reaches 18%, the time to stabilize seepage is shortened, and the runoff coefficient significantly decreases (Table 2). However, in the case of heavy rainfall, the runoff yield is still fast even though the coverage of the slope land reaches more than 50% in Loess Plateau.

Table 2 Runoff coefficient under different vegetation fractional coverage (Zhao et al. 2015)

Site number	Vegetation fractional coverage (%)	Runoff coefficient		Decreased runoff compared to bare soils (%)	
		5 min	10 min	5 min	10 min
2#	0	0.55	0.57	0	0
3#	11	0.37	0.48	32.73	15.79
4#	18	0.23	0.28	58.18	50.88
5#	43	0.29	0.37	47.27	35.09
6#	53	0.20	0.26	63.64	54.39

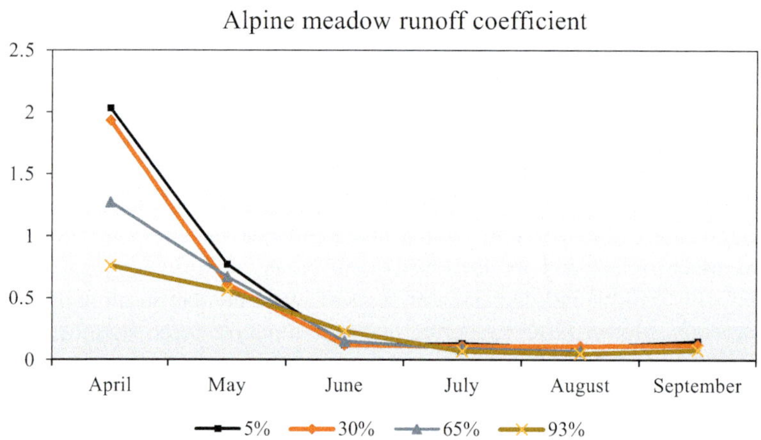

Fig. 19 Seasonal dynamics of surface runoff in alpine meadow under different vegetation coverage (Wang and Zhang 2016)

Fan (2014) studied the variation of runoff under different forest densities and grassland coverage. Both the forestland and the grassland significantly delayed the runoff time and effectively controlled the slope infiltration process. With the increase of vegetation coverage, stream intensity decreases. As the process of surface freezing and thawing profoundly affects the surface runoff process, the process of runoff generation in the cold area is inevitably different from that in the nonfrozen area. Wang and Zhang (2016) revealed the runoff generation mechanism of alpine meadow and alpine swamp in permafrost watershed of central Tibet Plateau from the observations in two catchments. Under different coverage conditions, seasonal dynamics in the runoff coefficients of alpine meadows and high-cold swamps can be seen in Figs. 19 and 20. Surface runoff process in the cold area is significantly different from the seasonal dynamic in the nonfrozen area because of the influence of surface freezing and thawing process.

In summary, depending on different vegetation coverage, runoff yield patterns varied. When vegetation has achieved to a certain threshold, the larger the vegetation coverage, the lesser the rainfall runoff will be. A common method for decreasing runoff is via stable and suitable vegetative cover (Chaplot and

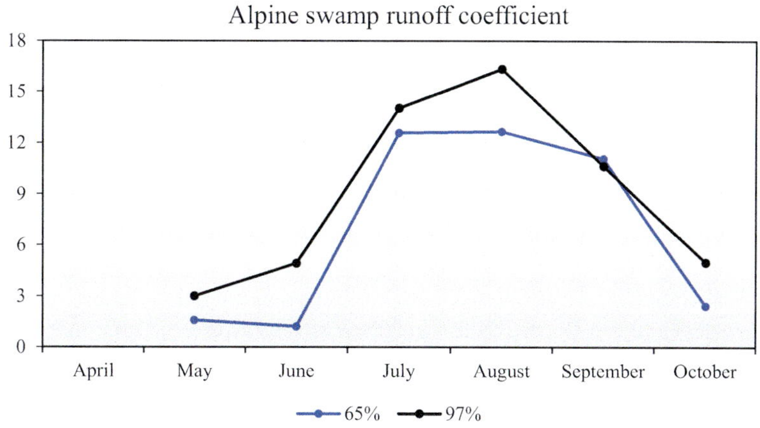

Fig. 20 Seasonal dynamics of surface runoff in alpine swamp under different vegetation coverage (Wang and Zhang 2016)

Le Bissonnais 2003; Dunjó et al. 2004; Kothyari et al. 2004; Mohammed 2005; Zhang et al. 2004).

Effects of Vegetation on Soil Water Infiltration

Infiltration is the process whereby water enters the soil and adds to the total soil moisture (Philip 1957a; Warrick 2003). Soil infiltration is an indispensable physical parameter to describe the soil characteristics and a critical process to link other hydrological components (precipitation, surface water, soil water, groundwater) during phase changes of water cycles. There are a number of factors that affect infiltration, including the vegetation, slope, rainfall regime, soil texture, soil structure, and so on (Leonard and Andrieux 1998). It is of great significance to conduct studies of relations between soil water infiltration process and vegetation changes to reveal hydrological response mechanism under the circumstances of changing vegetation.

Several models, including Green-Ampt equation (Green and Ampt 1911), Kostiakov equation (Kostiakov 1932), Hortan equation (Horton 1941), Philip equation (Philip 1957b), Holtan equation (Holtan 1961), and Smith equation (Smith 1971), can be used to simulate the processes of soil infiltration. The importance of vegetation cover in maintaining and improving soil stability and permeability is well known and has been discussed extensively (Branson et al. 1972; Colman 1953). The mechanism of how vegetation affects soil infiltration can be expressed as follows. One option is the cover may intercept raindrop energy and prevent surface sealing. Alternatively, vegetation increases surface roughness, which changes the flow-routing pattern and erosion processes (Li et al. 1992; Yu et al. 2006). Lastly, vegetation changes hydraulic properties of the underneath and surrounding soil by modifying the structure of the soil pore spaces as a result of the formation of the root system (Thompson et al. 2010a, b). Most studies revealed that soil hydraulic conductivity under vegetation would be 3–5 times higher than the bare soil (Bhark and Small 2003; Bromley et al. 1997; Dunkerley 2000; Titus et al. 2002).

Researches about the processes that generate vegetation-infiltration capacity feedbacks have been widely explored in drylands dating back to 1930s. However, at the initial stage, the theory of how vegetation affects infiltration is not well developed. For example, Johnson and Niederhof (Johnson and Niederhof 1941) failed to discover any simple relationship between vegetation cover density and infiltration capacity measured with infiltrometers, whereas Smith and Leopold (1942) documented large changes in infiltration with only modest changes in vegetation density in Pecos River watershed in New Mexico. Since then, detailed studies of the effect of plants on infiltration were conducted by many others (Duley and Domingo 1949; Dyksterhuis and Schmutz 1947; Woodward 1943). Experiment by Marston (1952) in the Davis County Experimental Watershed demonstrated that vegetation cover of 65% or more significantly increased infiltration. Lyford and Qashu (1969) measured constant rates of infiltration with a double-ring cylinder infiltrometer (with 11 and 30 cm diameter rings) at different distances from three desert shrub in a 45 cm deep sandy loam. Each plant stem was surrounded by a topographic mound (of unreported height) with a diameter approximately equal to that of the crown cover. The mounds were covered with annual weeds. The results show a strong lateral gradient in infiltration capacity (Fig. 21). Johnson and Gordon (1988) also demonstrated an approximated doubling of 30-min average infiltration rates beneath shurubs, as compared with grassy areas between the shrubs.

A complete review of the relationship of vegetation to infiltration in semiarid regions is summarized by Branson et al. (1981). On hillslopes of natural length and roughness, vegetation plays an important role in decreasing the average velocity of flow, increasing its residence time, and allowing significant post-storm infiltration to decrease runoff volumes (Dunne and Dietrich 1980) and created a simple water balance model to predict soil infiltration based on 4 years field observations.

In recent years, numerous studies have also been carried out, focusing on the mechanism of soil infiltration influenced by vegetation changes. McLeod et al. (2006) researched the soil water regimes of a Brown Chromosol on the Northern Tablelands of NSW, Australia, under three pasture types and noted that the vigorous phalaris plus white clover pasture yielded the greatest potential for water storage. Yi and Mingan (2007) showed that when other conditions are the same, the higher the vegetation coverage and the larger the initial and steady infiltration rates, the greater the recharge coefficient of precipitation infiltration. The infiltration rates and cumulative infiltration amount with different coverage can be expressed as power function relations with time. Wang et al. (2008) studied the influence of vegetation on infiltration and redistribution patterns with the aim of identifying tools for rebuilding desert ecosystems and suggested that vegetation had a significant effect on infiltration and redistribution patterns in stabilized sand dunes. Schwartz et al. (2010) studied soil water redistribution under sweep tillage and in untilled control plots and found that tillage with a sweep of 0.07–0.1 m significantly reduced net water storage at soil depths above 0.3 m but did not affect the water content at depths $\geq$0.2 m.

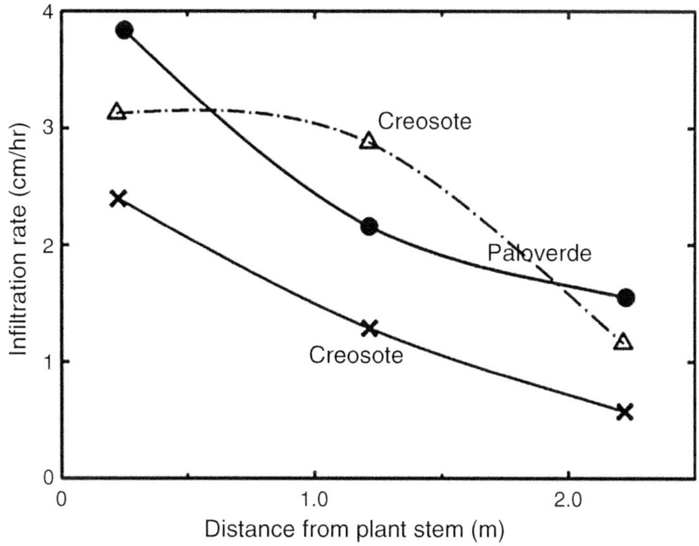

Fig. 21 Variation of infiltration capacity with distance from the stems of desert shrubs after (Lyford and Qashu 1969)

Numerous mathematical simulations of the infiltration process have also been conducted. Zhou et al. (2015) compared the simulation results of Horton, Philip, Kostiakov, and Green-Ampt model with in-situ measurements. Results revealed Hortan equations described the infiltration characteristics much better than the others. By artificial rainfall, Zhao and Wu (2004) concluded that Smith-Parlange equation can better capture the slope infiltration processes than Hortan and Kostiakov equations.

In summary, with the increase of vegetation coverage, there exists an obvious increase of soil infiltration whereas when the soil moisture content, rainfall amount, and rainfall density have achieved to a large extent, the impact degree of vegetation coverage decreased.

Observation of Vegetation Impacts on Surface Runoff: Precipitation Interception

Concept of Precipitation Interception

Precipitation interception by vegetation is an important fraction of water cycles, which is collected and temporarily stored on vegetation before being evaporated. Many studies have shown that interception losses are of major importance in influencing the water yield of forested areas relative to deforested area and other vegetative cover area such as shrubs, crops, and grass (Crockford and Richardson 2000). Better understanding of the interception process will allow reliable estimate

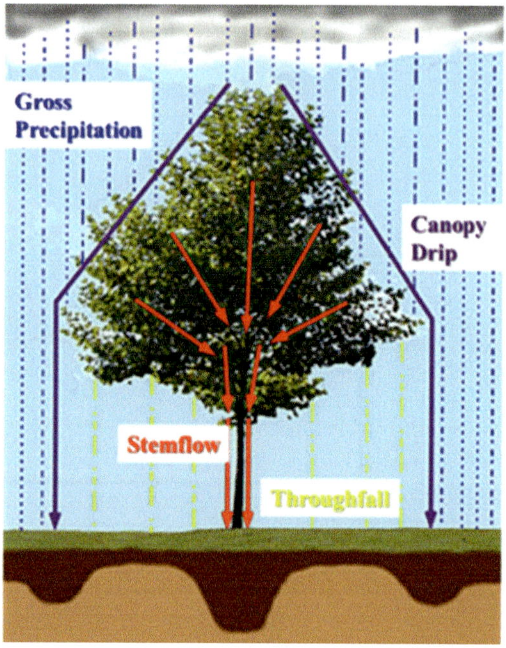

Fig. 22 Modification of falling precipitation by vegetation. The relative quantity of precipitation entering the soil is indicated in dark brown (Pidwirny 2006)

of its impact on surface runoff. Vegetation type, ground cover, and climate condition affect the amount of precipitation that reaches the ground surface.

Precipitation, i.e., gross precipitation (P) that input from above the canopy, can be partitioned into three fractions (Fig. 22): (i) that remains on the vegetation and is evaporated after or during rainfall (interception, I); (ii) that which flows to the ground via trunks or stems (stemflow, SF); and (iii) that which may or may not contact the canopy and which falls to the ground between the various components of the vegetation (throughfall, TF). The mass balance of partitioning of rainfall is generally expressed as:

$$I = P - TF - SF \tag{18}$$

Throughfall and stemflow are the two hydrological processes responsible for the transfer of precipitation and solutes from a vegetative canopy to the soil (Levia and Frost 2003), which typically account for 70–90% of the incident gross precipitation in temperate forests.

Precipitation Interception Effect on Surface Runoff Redistributing

Canopy interception of precipitation plays an important role in modulating the hydrologic process and water budget. Canopy interception indirectly influences the distribution ratio of runoff and infiltration by changing the amount of rainfall that reaches the ground, delaying the time to runoff and decreasing the flow volume and flow rate (Love et al. 2010; Zhang et al. 2016). It is usually described as interception rate (IR), which is the ratio of canopy interception divided by total rainfall at certain time intervals (Gavazzi et al. 2016). Interception rates are influenced by meteorological conditions such as wind speed and direction, evaporation rate, rainfall rate and duration, and canopy structure.

Influence of Vegetation Species

In the early twentieth century, Horton (Horton 1919) presented one of the first detailed analyses of rainfall interception across species and found that conifers intercepted more rainfall than hardwoods, but only hardwoods exhibited seasonal differences. Rates have exhibited high variability between and across species, the spruce-fir-hemlock forest type has been shown to have the highest canopy IR followed by pines and then hardwoods (Helvey and Patric 1965). Higher streamflow rates in beech (*Fagus sp.*) versus spruce (*Picea sp.*) dominated watersheds. Global estimates and spatial differences in the distribution of interception over different ecosystems showed interception loss was responsible for the evaporation of 13% of the total incoming rainfall over broadleaf evergreen forests, 19% in broadleaf deciduous forests, and 22% in needleleaf forests (Miralles et al. 2010). It was also found that interception was higher in dryer regions, indicating greater impacts on water resources (Soto-Schönherr and Iroumé 2016). However, unless under a large density, the effect of rainfall interception in grass is not as obvious as in other types of plants because mostly grass is low and short, whose leaves are small and soft and cannot hold much water (Zhang et al. 2016). Interception rate of shrub-grass layer in Liupan mountain region was about 1.8–12.6% (Liu et al. 1994). The clearing of native forest and establishment of grass on a small catchment in southwest of Western Australia resulted in a large streamflow increase (30% rainfall yr^{-1}), which was brought about by a decrease in transpiration and interception loss (Ruprecht and Schofield 1989). Greater amounts of vegetation and accumulated litter generate greater resistance to runoff, which decreases with decreasing grassland coverage (Zhang et al. 2016).

Influence of Meteorological Conditions

In study on evergreen broad leaves forest in east of coastal China, Peng et al. (2014) suggested that lag time of stemflow was about two times longer than throughfall and canopy interception contributed the rainfall redistribution up to 25–31%. The percentage of rainfall intercepted decreased as rainfall rate increased. The storage capacity of foliage was between 0.5 to 1.8 mm per storm. Interception rates averaged 25% during heavy rains of long duration but could be as high as 100% during light rainfall events when total rainfall did not exceed the storage capacity (Horton 1919). Peng's results (see in Fig. 23) support the well-established exponential decay relationship between the gross precipitation and the interception percentage after the canopy is saturated (Peng et al. 2014). Even the intercepted snow during the snow accumulation season could be removed from the canopy via three pathways: sublimation, mechanical removal (sliding leading to mass release), and meltwater

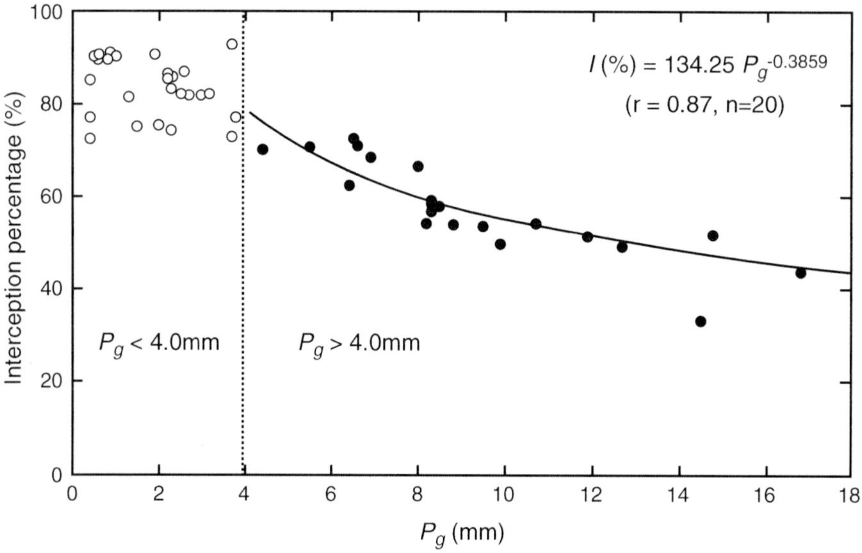

Fig. 23 The relationship between gross precipitation (P_g) and interception percentage (I) (Peng et al. 2014)

drip, which results in reduced snow water equivalent quantities and delayed lag time to surface runoff (Suzuki et al. 2008). During the snow accumulation season, a portion of the snowfall is intercepted by the forest canopy and sublimates or melts. Snowfall interception is important for understanding the water and energy balance of snowy watersheds (Suzuki et al. 2008).

Moreover, the decade-long data of canopy rainfall interception, study found that interannual variability of canopy interception is higher than reported in short-term studies (Gavazzi et al. 2016). It indicates uncertainties of rainfall interception should also be considered on surface runoff research and accurately predicting the hydrologic impacts of management and climate change.

Measurement for Interception

Interception cannot be measured directly but can be calculated through observation of gross precipitation, throughfall, and stemflow as in Eq. (18).

Gross Precipitation Measurement

As interception is the difference between gross and net rainfall, generally expressed as percentage of gross rainfall, accuracy in measurement of both parameters is critical (Crockford and Richardson 2000). Problems in measurement of gross rainfall are commonly ignored comparing with the measurement of net rainfall (throughfall and stemflow). The horizontal angle between the rain gauge and the top of the nearest trees was smaller than 45°, so little disturbance on gross rainfall measurement was caused by its surrounding environment (Asdak et al. 1998). However, sites for forest studies seldom have well-positioned rainfall gauges (Dohnal et al. 2014; Peng et al. 2014; Zimmermann et al. 2010). Aerodynamic effects both on canopy level gauges and on gauges in forest clearings are likely to lead to systematic errors in the measurement of rainfall, i.e., gauges can either underestimate or overestimate owing to wind-induced loss (Crockford and Richardson 2000; Robinson et al. 2004) or receive water blown from the canopy if placed close enough to the canopy, respectively. Wind-induced underestimation of precipitation is more likely with small drop-size rainfall. Larger drops have more momentum and are less likely to be blown away from the collector. Accurate measurement of precipitation is difficult when the study area is within a larger forest and gauges have to be placed above the canopy or in gaps in the canopy; the latter can result in overestimation of precipitation as a result of blow-in from the nearby higher tree crowns and the former can experience wind loss (Crockford and Richardson 2000). Therefore, the information of precise position of rain gauge in relation to the experimental plot, and where the gauge is noted as being in a clearing, the size of the clearing and the location of the gauge are important, because small errors in precipitation measurement can lead to large errors in the estimation of interception.

Throughfall Measurement

The accurate estimation of throughfall is generally difficult among the whole measurement of interception. The great error will be happened due to the spatial canopy heterogeneity of high forest, shrubbery, and even some grass ecosystem that may cause spatial variability of throughfall. Therefore, the search for robust sampling schemes, from size of devices (Crockford and Richardson 2000) to sampling design (Zimmermann et al. 2010) has attracted some interest in recent decades.

Collector Devices

Usually, the variability problem can be overcome by increasing sampling area of interception observation by using large plastic-sheet (Calder and Rosier 1976) or mostly trough-type net rainfall collectors (see in Fig. 24) (Fan et al. 2014; Herbst et al. 2006; Peng et al. 2014; Silva and Okumura 1996). Moreover, sampling strategy is also developed for accuracy purpose. (Zimmermann et al. 2010) studied throughfall by different sampling designs and variable sample supports (see in Fig. 25). Large number of collectors will be required to estimate mean throughfall, particularly for small events for a relative error limit of 20%. A larger enough sampling area is also needed for estimating throughfall of an uneven density canopy, i.e., more collectors are required. In order to control the error limit of 5%, several to dozens, even up to163 of rain gauges (Llorens and Domingo 2007), were designed to throughfall collection according to variable canopy characteristic (Czarnowski and Olszewski 1970), seasonal variation (Helvey and Patric 1965), and the area of sampling sites. There is even equation to determine the number of collectors required to estimate throughfall volume according to characteristic of canopy heterogeneity (Carlyle-Moses et al. 2004; Puckett 1991).

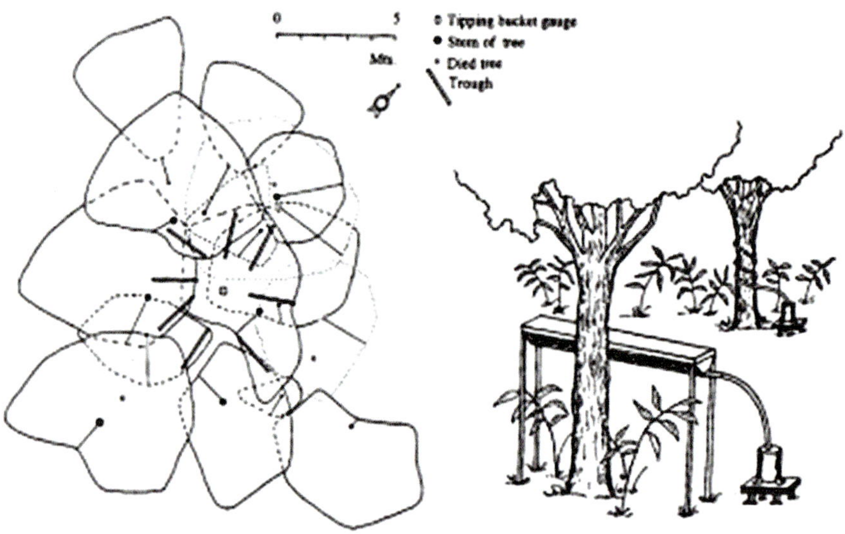

Fig. 24 Vertical crown projection map and layout of instruments of trough-type collector and stemflow collector in a natural oak forest (Silva and Okumura 1996)

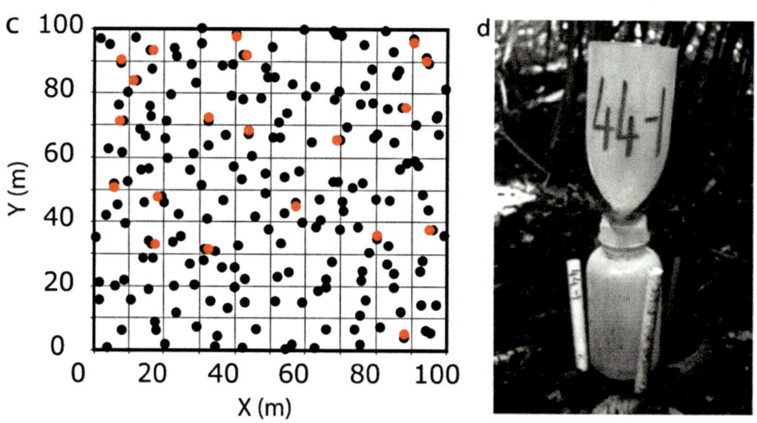

Fig. 25 (**a**) The close-up view of the 1 ha sampling area shows the location of sampling points (dots). Different colors of dots indicate collector placement according to different sampling strategies, that is, black dots refer to locations that were chosen according to a design-based sampling strategy, whereas red dots mark locations selected by a model-based strategy. (**b**) The photo shows one of the funnel-type samplers (Zimmermann et al. 2010)

Sampling Location Design

In addition to the difference of collectors, sampling location and design will also influence the result. Random sampling is the method that gauges are randomly located within a fixed plot or relocated periodically (Link et al. 2004). It will provide even coverage of the entire plot while capturing the full range of variation. Systematic sampling is the method that gauges are evenly distributed in regular grids or equidistance (Herbst et al. 2006; Xu et al. 2010). There is no significant difference between two methods according to comparative studies (Holwerda et al. 2006; Zimmermann et al. 2010). The random relocation of small, funnel-type collectors during the sampling period, subject to some assumptions, improves the efficiency of sampling estimates by increasing the spatial coverage. However, for the common problems of estimating throughfall for particular events, there is no consensus on the best strategy, but there is recognition that estimates of this variable are subject to considerable sampling error, and that these errors will propagate when the estimates are used in subsequent modeling (Zimmermann et al. 2010). Study showed that interception rate had a range between 3% and 25% among 23 subplots of lowland tropical forest canopy even in the same rainfall intensity (Manfroi et al. 2006). Therefore, location design will only reflect the rainfall interception of sampling plot scale. The estimation of spatial interception amount will be limited if the size and location of the observed plot cannot represent spatial feature of whole canopy.

Stemflow

Accurate measurement of stemflow is even more difficult, but in some areas, stemflow is a very small proportion of net precipitation and sometimes is not measured. In some forests, however, stemflow can be large enough to warrant special care (Crockford and Richardson 2000; Zhang et al. 2007). There is no standard protocol to adequately sample stemflow volume, likely the result of the diverse vegetation cover from which stemflow is collected and the differing objectives among stemflow studies (Levia and Frost 2003). Stemflow, however, is usually captured by circumferential collars which are formed by a small tire wrapped around the tree, attached with galvanized iron nails filled by polyurethane foam and finally sealed with neutral silicone sealant (Crockford and Richardson 2000; Dohnal et al. 2014; Dunkerley 2000). Then the stemflow is diverted to collection bin or tipping-bucket gauge to measure (see in Figs. 24 and 26). Following Hanchi and Rapp (Hanchi and Rapp 1997), the tree-level stemflow was upscaled to the stand-level stemflow for both forest stands using Eq. (19).

$$S_f = \sum_{i=l}^{n} \frac{S_n \cdot m}{A \cdot 10^4} \tag{19}$$

where S_f is the upscaled stemflow depth (mm) for a specified stand area of A (m^2), n is the number of diameter at breast height classes, and S_n is the average stemflow volume (ml) collected from m trees in the diameter at breast height classes.

Experiment on Grassland and Meadow

The above methods are usually used on study of trees and shrubs. Theoretically, those methods can be used in study on rainfall interception of grassland. However, it is difficult to design the experiment to measure the throughfall and stemflow of

Fig. 26 Stemflow collector consisting of plastic tubing and a collection bottle. Left figure for shrubs (Li et al. 2008) and right figure for trees (Germer et al. 2010)

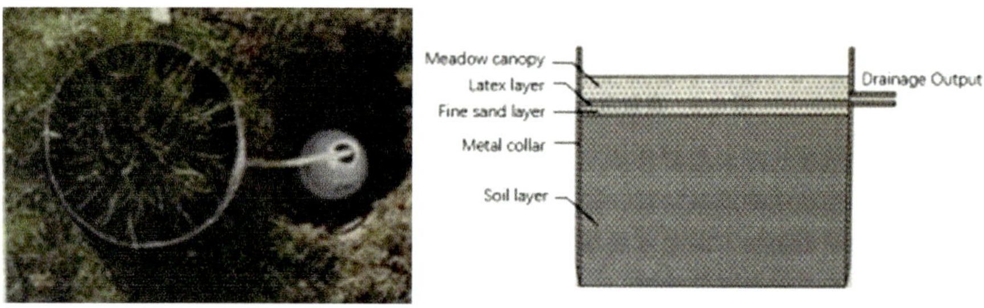

Fig. 27 Rainfall interception observation device for meadow (Wang and Zhang 2016)

dwarf and creeping growth grass. There are not recognized common experiment methods for the interception of grass and meadow study. One way is measuring throughfall and stemflow as one part together, using water balance to calculate the interception amount by the difference of gross precipitation and drainage output. According to Corbett's study (Corbett 1968), an improved device (see in Fig. 27) was designed and used for hydrological study in typical Swamp meadow and Alpine meadow in Tibetan Plateau (Li et al. 2012). A certain size metal collar was driven into soil to seal the soil surface. A layer of fine sand was sifted onto the soil to provide a uniform, sloping surface for drainage. Then soil surface was sealed with the neoprene latex emulsion. A drain carried off the throughfall and stemflow when precipitation happens.

Other researchers observed the interception directly by measuring the rainfall storage capacity. This method is based on measurement of water retained by vegetal entities and the up-scaling of these measurements using the determination of the surface of canopy feature (Li et al. 2009; Liu et al. 2015). The leaves and stem were randomly selected and taken back to laboratory, weighted to obtain fresh weight. Then they were maintained in the water for certain times (Llorens and Gallart 2000), reweighed to obtain wet weight when no more droplets. Water storage capacity, i.e., interception, was calculated by the difference of weight before and after soaking.

Interception Model

The various components of the rainfall interception process have been measured and modeled for many vegetation types. Apart from allowing extrapolation of measurement results in space and time, interception models also provide insights into the mechanisms of the interception process (Muzylo et al. 2009). Empirically derived relationships between interception loss and gross rainfall highly depended on species character of the interception process, which cannot be confidently applied where conditions are distinctly different. Conceptual model, i.e., physically based model which describes interception as a process, has been paid more attention due to its robustness in physics and less need for calibration of empirical coefficients (Cui and Jia 2014). To predict interception losses using readily available meteorological variables, researchers have developed over 15 physically based interception models.

Almost all the physically based models have been derived from two basic interception models: Rutter-type model that deals with the interception of rainfall drops using probability distribution and Gash-type model that deals with the redistribution of rainfall volume using a mass balance equation. The Rutter model represents the interception process by a running water balance of rainfall input, storage, and output in the form of drainage and evaporation (Rutter et al. 1971). The Gash model represents rainfall input as a series of discrete storms that are separated by intervals sufficiently long for the canopy and stems to dry completely, during which three phases can be distinguished – canopy wetting-up, saturation, and drying (Gash 1979). Subsequent two kinds of models have been improved. Muzylo et al. (2009) compared these models and found the original and revised Gash's analytical models to be the most commonly employed.

The input rainfall values of temporal resolution is from 10 min (Massman 1983) to daily (Mulder 1985) or mostly hourly (Cui and Jia 2014; Gash 1979; Rutter et al. 1971). The total number of model parameters varies considerably from model to model. In the case of the single-layer models, the number varies from 2 to 7, and in the case of the multilayer models, from 8 to 16 for stands comprising two layers (cf. Table 1 in Muzylo et al. (2009)). The models include parameter representing the storage of intercepted rain in the canopy or, to be more specific, the threshold amount of rain that can be stored in the canopy. Whereas, the analogous designations of this storage threshold are misleading, because its exact definition may differ significantly between models. Models also include a parameter describing the structure of the canopy. The bulk of the Rutter- and Gash-type models employ the free throughfall coefficient to partition the throughfall.

Satellite remote sensing observations can provide land surface information with high spatial and temporal resolution. Unfortunately, estimate of interception loss at regional scale using remote sensing information is not straight forward. Empirically based models have been applied to remote sensing observations for regional estimate of interception evaporation (Cui and Jia 2014). Nowadays, remote sensing can provide many variables or inputs needed by the physically based sophisticated models, for instance LAI, fractional vegetation cover (FVC). More and more attempts have been made in applying these satellite observed parameters to interception models, since there are still a few critical parameters which cannot be directly derived from remote sensing observations, e.g., canopy storage capacity, trunk storage capacity, and the proportion of the rainfall diverted to stemflow.

Observation of Vegetation Impacts on Surface Runoff: Root System

Surface runoff initiation is the result of rainfall intensity exceeding soil infiltrability, thereby leading to temporary saturated conditions at/near the soil surface (Cantón et al. 2011). Therefore, soil infiltrability should be regarded as the first scale of analysis. Soil structure in the top soil layers is particularly critical as it dominates water transport in the saturated and near-saturated range (Creswell et al. 1992; Bronick and Lal 2005). Plant roots are key drivers of soil structure enhancing aggregate formation and stability and improving soil shear strength (Haynes and Beare 1997; Kavdır and Smucker 2005; Gyssels et al. 2005; Baets et al. 2008). Root-induced macropores are of particular importance for runoff mitigation due to their large diameters and high connectivity, enhancing rapid rainfall infiltration and percolation to deeper soil layers (Cresswell and Kirkegaard 1995; Ghestem et al. 2011; Horn and Smucker 2005; Suwardji and Eberbach 1998).

Plant effects on soil structure are species dependent. For annual plants root diameter was shown to be an important trait for root induced pore formation, particularly in soil with high mechanical resistance (Bengough 2012). Legume species with coarse root axes created substantially more macropores compared to fine rooted species (Bodner et al. 2013).

Measurements of Root Parameters

The parameters commonly used to express root growth and distributions are number, weight, surface, volume, diameter, length, and the number of root tips (Bohm, 1979). The manner in which the root data is best expressed should be considered before starting an experiment. Not only the research aim but also the time and labor needed for the determination of the root parameter which has been chosen must be considered.

(1) Root number

Counting root number is the common procedure when working with the core-break method, profile wall methods, and glass wall methods. Root numbers counted in the different soil layers generally give a good impression of the rooting density in a soil profile. In studied where the entire undamaged root system of a single plant can be extracted, the number of main and lateral roots is counted to provide an estimate of their total length. Although root number is not an ideal parameter, as long and short roots are regarded and counted as equal units, high correlations with other root parameters are possible (Melhuish 1968).

(2) Root weight

Root weigh is the most commonly used parameter for studies of root growth in response to environment. Generally, the washed roots are dried and then their weight is determined. If root weight is to be determined in an ecological study, dry weight should be preferred. To determine dry weight, the washed and cleaned roots are dried in an oven at 105°C for about 10–20 h depending on the amount of roots. The drying can also be done at 60–75°C which will take longer but can have the advantage that this lower temperature prevents roots from being pulverized (Schuurman and Goedewaagen 1965).

Root weight is a good parameter for characterizing the total mass of roots in a soil. In all cases in which the productivity of underground terrestrial vegetation is to be determined, root dry weight should be the criterion for evaluation. This includes also all research in which the contribution of the roots for the amount of humus in the soil is to be studied. Root weight also can be regarded as a fundamental measure of photosynthate storage in a plant. The known shoot-root relations based mainly on shoot and root weight.

(3) Root surface

Surface area has seldom been determined in ecological research, although this parameter seems one of the best when experiments on water-or nutrient uptake are made. The most direct method is to determine the average diameter of a large number of individual roots and to measure the total root length per sample. From these data, root surface can be calculated easily (Geisler and Maarufi 1975; Adepetu and Akapa 1977). The main drawback to direct measurement is that the operator can decide only by inspection if the measured roots are still alive.

Wilde and Voigt (1949) proposed a method to determine the total root surface by titration. A simple method for determining relative root surface area is a gravimetric method developed by Carley and Watson (1966). Comparative studies made by Carley and Watson (1966) show that no significant differences in relative root surface areas exist between the titration method of Wilde and Voigt (1949) and this gravimetric method. The results obtained with the titration method have a higher accuracy, but the gravimetric method can be suitable in routine studies due to its greater speed and simplicity. A drawback of the gravimetric method is that fine roots can rope together and behave like thicker roots (Pearson 1974).

(4) Root volume

Root volume can be calculated by measuring root diameter and the root length. Such calculations, however, have seldom been done in practice. In general, for measuring root volume, the water displacement technique is preferred. The measurement is done in a special container with an overflow spout. This container is filled with water until it overflows from the spout. Then fresh-washed roots which have been carefully dried with a soft cloth are immersed and the overflow water volume is measured in a graduated cylinder. The main reason for this limited use is certainly that root volume measurements from species with few large roots can be equal to species with large amounts of small fibrous roots. So the volume data alone are of limited value for most research problems in root ecology. Data from volumetric measurements in general should be used only to supplement other parameters.

(5) Root Diameter

In studies with annual plants, root diameter in most cases has been measured only for calculating root surface or root volume. But knowledge of the diameter of roots can give important information on the relationship between the pore size in a soil and the potential of root penetration (Wiersum 1957). The diameter is measured directly on freshly washed root samples with the aid of a microscope fitted with an ocular micrometer. Before starting the measurements, the roots are placed for some hours in water because many roots dry irregularly. Furthermore, it must be taken into account that there can be considerable diurnal variations in root diameter. On dry, sunny days, the diameter of roots can shrink to about 60% of its maximum size (Huck et al. 1970). Predominantly, in tree root studies, research workers confronted with the problem to differentiate between roots of various diameters. It is often convenient to divide roots into six classes with different root diameters in Table 3 (Böhm 1979). This classification must be regarded as arbitrary, and it would be wrong to identify root diameter with any particular kind of root function.

(6) Root Length

For direct measurements, the wet roots are placed in a flat glass dish containing a small amount of water. Graph paper is placed under the dish. The roots are straightened with forceps so that they do not overlap and are held in position by a glass plate. The lengths of the given roots or root segments are then estimated to the nearest millimeter by eye inspection or by the use of a magnifying glass. Obtaining root-length data by these direct measurements is tedious and time-consuming, and the procedure is recommended only for estimating the length of

Table 3 Classes of roots with different diameters

Root diameter (mm)	Classes of roots
<0.5	Very fine
0.5–2	Fine
2–5	Small
5–10	Medium
10–20	Large
>20	Very large

single root, not as a routine method in ecological research. Instead of tedious direct measurements, root length can be calculated more rapidly by counting the intersection method. The total number of intersections between the roots and the vertical and horizontal lines of a grid on the glass observation windows were counted. Comparisons of estimated intersection data with the measured actual root length show a linear relation between the number of intersections and the actual root length.

Independently of this practical line intersection method, Newman (1966) developed a theory that root length can be estimated by the Eq. (20):

$$R = \pi AN/2H \tag{20}$$

where R is the total length of roots in a field of area A and N are the number of intersections between the roots and random straight lines of total length H. Newman's method has been modified and improved by several research workers (Torssell et al. 1968; Evans 1970; Marsh 1971; Tennant 1975). The main change is that for the area over which the roots are spread, any convenient size of grid system can be used.

Practical tests done by Head (1966), Reicosky et al. (1970), and Tennant (1975) indicate that there are only small differences in the precision between the data from the direct measurements and those obtained from the intersection method. In no case has the standard deviation been more than $\pm 10\%$ of the true lengths, and in most cases, it has been much less.

(7) Root Tips

For determination, the cleaned roots are placed in a flat dish where they are kept under water. Here they are arranged so that they do not overlap. Then the number of live root tips is counted using a stereomicroscope. The distinction between live and dead root tips can be made by evaluating their morphology and color. Root tips are generally assumed to be alive if they are turgid, and white to light brown in color (Weller 1971). The main lamination seems to be that the counting procedure is very time-consuming and therefore not well adapted for routine studies.

(8) Shoot–Root Relations

The common parameter for evaluating the relations between above and below ground growth of plants is the shoot:root ratio. It is a measure of the distribution of dry weight between shoot and root system of the plants.

Many of the results of all this research indicate that the common assumption of a persistent tendency towards a positive correlation between shoots and roots are not generally valid. Ecological conditions can change the shoot:root ratio considerably when plants are growing under filed conditions. Nevertheless, studying shoot–root relations should be a central research filed for root ecologists. Adequate information is essential for analyzing or simulating growth pattern of whole plants (Mayaki et al. 1976; Sivakumar et al. 1977) and for estimating primary productivity of ecosystems (Lieth and Whittaker 1975). However, when considering the efficiency of a root system for plant growth, root distribution in the soil profile, rather than total root weight, seems the more important factor.

With the development of technology, image analyses systems provide a quick determination of various root morphological parameters (root length, average diameter, surface area, and so on). Two image analyses programs using different measuring algorithms are often used, one is a commercial package WinRHIZO and the other is a freeware ROOTEDGE (Himmelbauer et al. 2004). ROOTEDGE is a computer program running under DOS and based on the edge chord algorithm (Ewing and Kaspar 1995). The program operates with binary (black-white) images in uncompressed tagged file (TIFF) format: root objects are assumed to be black and the background white. Root length, projected area, and average root diameter were determined. The commercial software package WinRHIZO 4.1 works under Windows and uses a skeletonization method for measuring the root parameters (Arsenault et al. 1995; Bauhus and Messier 1999; Smit et al. 1994). The program operates with 256 levels grayscale images in TIFF file format, which are further converted into threshold (binary) and skeleton images. Threshold images are used by the system for evaluation of root diameter, while root length is measured on the so-called skeleton images (Bauhus and Messier 1999). Measurements involved total root length, average root diameter, projected surface area, plus length and area measurements as a function of different root diameter classes. Additionally, the program detects overlapping parts and takes them into account when calculating root parameters.

Plant Root–Hydraulic Conductivity Relations

Soil Hydraulic Conductivity

A number of procedures have been developed for estimating soil hydraulic conductivity from tension infiltrometer data. These include methods by White and Perroux (1989), Smettem and Clothier (1989), and Reynolds and Elrick (1991). These methods vary greatly in their capabilities, procedures, assumptions, and complexity. The procedure developed by Reynolds and Elrick (1991) have been widely used (Yu et al. 2016). According to Reynolds and Elrick (1991), the soil hydraulic conductivity (K_h) was given:

$$K(h)=K_s \exp\left(\alpha^* h\right) \tag{21}$$

h is the pressure head (cm), K_s is the saturated hydraulic conductivity (cm h^{-1}), and α (cm^{-1}) is the sorptive number. So the K_s can be calculated by the Gardner's exponential model (Gardner 1958). Parameters (K_s and α) were obtained by fitting the function of $K(h)$ to the measured K_h vs. h data pairs using a nonlinear fitting procedure with a Levenberg Marquardt optimization algorithm.

For the different species and positions, the soil hydraulic conductivity at different supply pressure heads measured by tension infiltrometry are different. Table 4 shows

Table 4 Soil hydraulic conductivity (Kh) at different pressure heads (h = −15, −10, −3, −1 cm) measured by tension infiltrometry (Yu et al. 2016)

	K_{15} (cm h^{-1})		K_{10} (cm h^{-1})		K_3 (cm h^{-1})		K_1 (cm h^{-1})	
Species	Row	Interrow	Row	Interrow	Row	Interrow	Row	Interrow
V. sativa	0.24	0.22	0.39	0.40	4.26	2.98	13.13	9.24
Mixture 1	0.15	0.05	0.27	0.19	2.46	2.13	7.46	5.94
M. officinalis	0.19	0.17	0.40	0.26	5.60	1.45	23.36	3.02
L. sativus	0.31	0.21	0.60	0.37	5.73	2.46	20.32	6.18
Mixture 2	0.26	0.16	0.49	0.28	4.47	1.87	15.32	3.94
T. lexandrinum	0.32	0.23	0.64	0.39	5.23	2.2	14.36	4.78
S. alba	0.30	0.24	0.52	0.43	1.99	2.72	3.16	6.28
F. esculentum	0.29	0.23	0.51	0.40	4.00	2.15	10.66	4.50
L. sitatissimum	0.33	0.24	0.64	0.41	5.29	3.03	15.08	8.38
P. tanacetifolia	0.24	0.21	0.48	0.40	3.48	3.62	8.46	10.42
R. sativus	0.20	0.20	0.38	0.34	3.44	2.27	11.35	5.73
S. cereale	0.26	0.19	0.47	0.33	4.15	2.69	12.25	7.37

the soil hydraulic conductivity at different supply pressure heads measured by tension infiltrometry for the different species and positions in Australia (Yu et al. 2016). Statistical analysis revealed a significant interaction between species and position ($p < 0.001$). Species clusters for the row and interrow position was given in Fig. 28 (Yu et al. 2016).

Species with significantly higher K_h in the row compared to the interrow were *Melilotus officinalis*, *L. sativus*, Mixture 2, *Trifolium alexandrinum*, *L. usitatissimum,* and *Secale cereale*. Among the species, the differentiation in K_h was higher in the row (CV = 0.37) compared to the interrow (CV = 0.30). In the row, the legumes as well as *L. usitatissimum* commonly induced an average high K_h, while the brassicas together with *S. cereale* and Mixture 1 had low values. In the interrow, there was no evident distinction in K_h between plant families. Some legume species with high K_h in the row had a rather low value in the interrow such as *M. officinalis* or *T. alexandrinum* as well as the legume based Mixture 2. In the interrow, *Phacelia tanacetifolia* and *V. sativa* had the highest K_h, while *L. usitatissimum* showed high K_h in both the row and interrow.

The two parameters of the exponential Gardner model to describe the entire K_h curve are derived and shown in Table 5. Both parameters showed a significant interaction of species × position and had higher values in the row than in the interrow. Among species, *V. sativa*, *M. officinalis,* and *L. sativus* had both very high K_s and α values, while *Sinapis alba* and Mixture 1 were at the lower end in both parameters. For *Raphanus sativus,* the value of α was in very low range, while they had an intermediate K_s.

Plant Root–Hydraulic Conductivity Relations

Based on the parameters characterizing the root systems and the soil hydraulic conductivity function, which root trait provides the best predictor for hydraulic

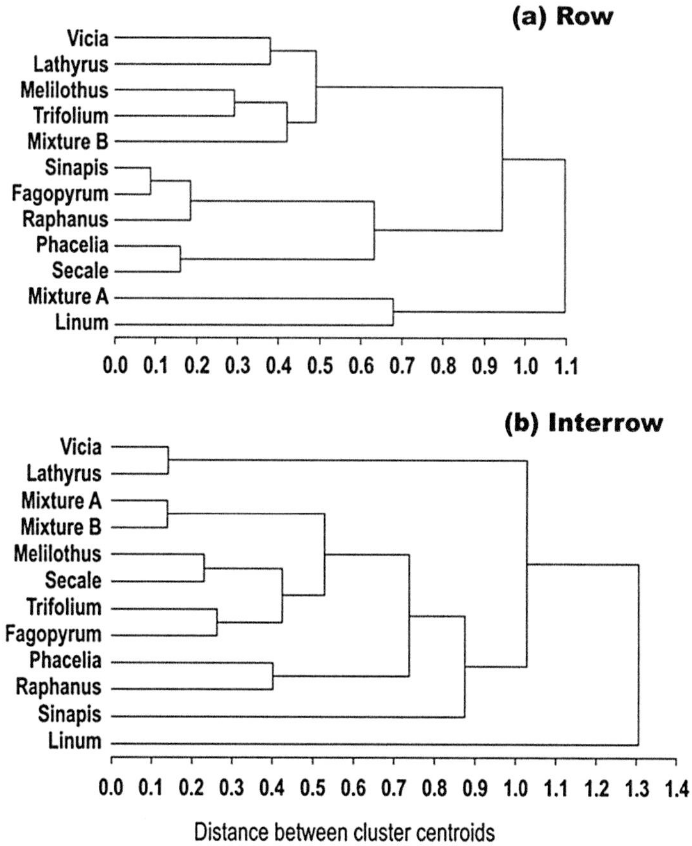

Fig. 28 Dendrogramm showing similar root system types among cover crop species based on the cluster analysis using root traits for row (**a**) and interrow (**b**) (Yu et al. 2016)

conductivity was analyzed by Yu et al. (2016). As shown in Fig. 29, in the row, specific root length (SRL) was the best root predictor for K_s and α, being negatively related to both parameters, i.e., species with a predominant root biomass investment in coarse root length enhanced saturated hydraulic conductivity while showing a quick decrease of the K_h function toward lower pressure heads. In the interrow, the best root predictor for the Gardner parameters was the median root radius (MRR). Interestingly, at this position, MRR showed a negative relation to the hydraulic conductivity descriptors, i.e., species with their root volume allocated to coarser axes here tended to reduce K_s while showing a lower decrease of K_h toward smaller pores. The coarse root might enlarge the soil porosity and a non-Darcy behavior (The relationship between infiltration rate and head loss rate is not proportional.) occurs.

Table 5 Parameters of Gardner's soil hydraulic conductivity function (*Ks* is saturated hydraulic conductivity, α is the sorptive number) as influenced by cover crop species and measurement position (row vs. interrow) (Yu et al. 2016)

	K_s (cm h^{-1})	α (cm^{-1})
Species (S)	0.160	0.111
Position (P)	<0.001	<0.001
S _ P	0.037	0.004
Position		
Row	23.22	0.49
Interrow	10.08	0.26
s.e.d.	2.55	0.02
Species		
V. sativa	19.93	0.40
Mixture 1	11.91	0.30
M. officinalis	25.72	0.45
L. sativus	24.82	0.42
Mixture 2	17.66	0.41
T. alexandrinum	15.43	0.41
S. alba	6.71	0.25
F. esculentum	11.92	0.38
L. usitatissimum	19.19	0.41
P. tanacetifolia	14.84	0.34
R. sativus	15.14	0.35
S. cereale	16.57	0.39
s.e.d.	5.97	0.07

Root Impact on Surface Runoff Mitigation

The changes in soil hydraulic conductivity driven by morphologically different root systems of cover crops have a significant impact on surface runoff. Yu et al. (2016) analyzed the potential impact of root-induced modifications of soil hydraulic conductivity on surface runoff, which was estimated by HYDRUS 2D. Figure 30 shows the soil hydraulic property parameters calculated to run the HYDRUS simulation. The Gardner exponential model estimates higher K_h values in the wet range while decreasing quickly toward the dry range. The Mualem–Van Genuchten hydraulic conductivity function on the contrary quickly falls below the saturated conductivity while the function then decreases more smoothly toward lower pressure heads. The cross-over between the average hydraulic conductivity functions based on K_s and Gardner's vs. Mualem–Van Genuchten's α is at a pressure head value of about $h = 0.1$ m. Below this cross-over point ($h = 0.1$ m), the tension infiltrometer measurement points in the near-saturated range are nearer to the Mualem–Van Genuchten function, while the tension infiltrometer measurement points in the near-saturated range are in better agreement with the Gardner model at higher pressure heads.

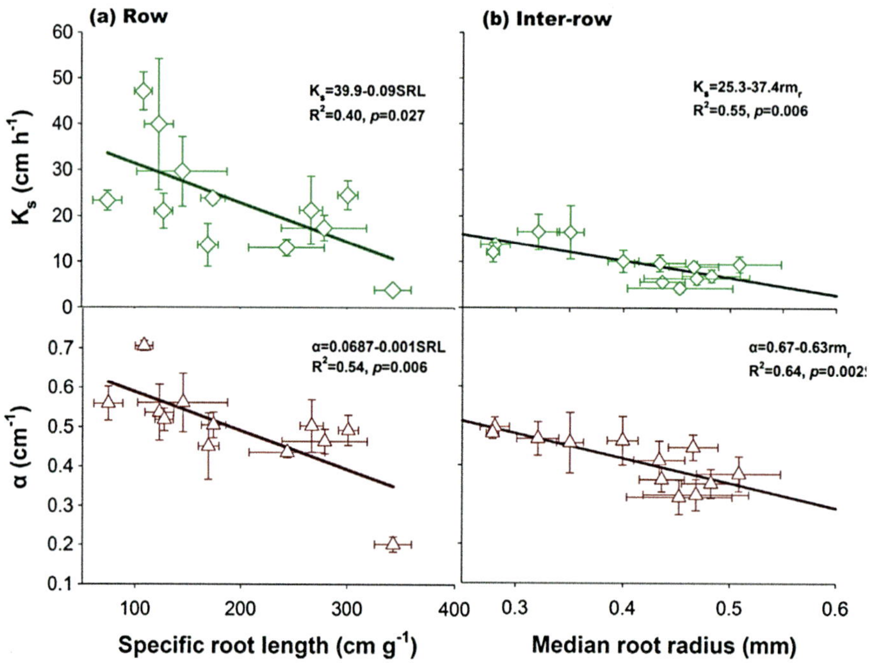

Fig. 29 Relationships between best root predictor and parameters of Gardner's exponential soil hydraulic conductivity function (**a**) in the row and (**b**) in the interrow (Yu et al. 2016)

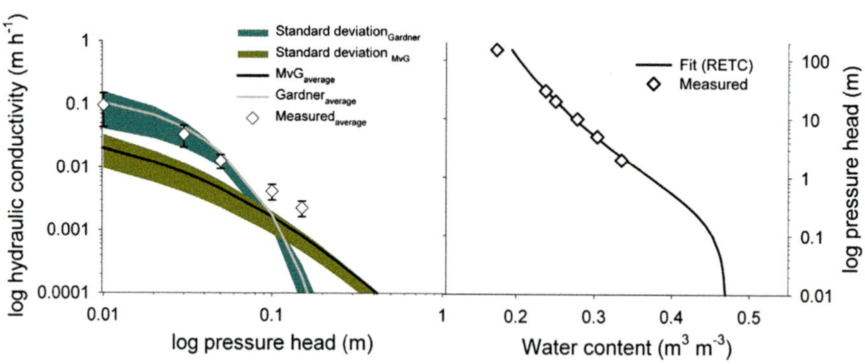

Fig. 30 Soil hydraulic functions used for runoff simulation with HYDRUS 2D. Left figure: Example of a representative hydraulic conductivity curve (mean of all measurements and standard deviation) using the Gardner and Mualem–Van Genuchten functions compared to point measurements. Right figure: Van Genuchten retention curve estimated from pressure plate data using RETC (Yu et al. 2016)

Figure 31 shows the runoff simulation result estimated by HYDRUS 2D when using the distinct row and interrow hydraulic properties corresponding to the different cover crop species. In each scenario, the cumulative rainfall was 20 mm over 72 h, with scenario 1 representing high intensity rainfall (intensity = 20 mm h^{-1}) on a dry soil,

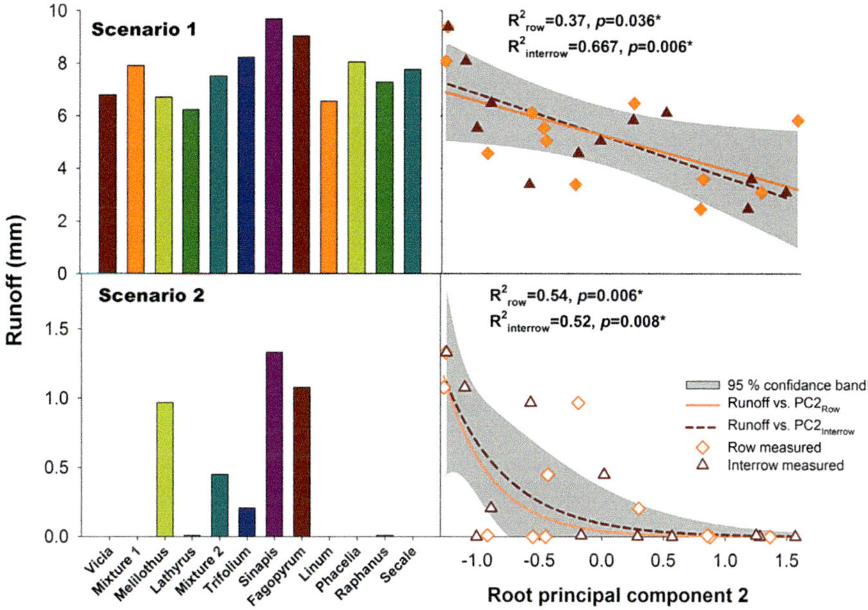

Fig. 31 Simulated runoff for different cover crops and two rainfall scenarios (left), relation between runoff and best root predictor (right). The root principal component 2 is a multivariate descriptor for axes thickness containing root diameter, specific root length, and median root radius. Scenario 1 represents a short, high intensity rainfall on initially dry soil; scenario 2 simulates rainfall over longer duration with low intensity on initially wet soil. In both scenarios, the cumulative rainfall amount is 20 mm (Yu et al. 2016)

while scenario 2 was a more extended rainfall (maximum intensity = 5 mm h^{-1}) on an initially wet soil. As expected, the intense rainfall scenario resulted in higher runoff losses and for all species, while under reduced rainfall intensity, soil infiltrability under most cover crops was high enough to take up all incoming precipitations. The low hydraulic conductivity of soil planted with *S. alba* showed highest runoff with nearly 50% of total rainfall lost. *L. sativus* on the contrary exhibited the lowest runoff losses with only 31% of incoming rainfall. In case of high intensity rainfall, the root effect on K_s is predominant explaining 79% of variability in the estimated amount of runoff. If root principal component 2 (The root principal component 2 is a multivariate descriptor for axes thickness containing root diameter, specific root length, and median root radius.) as a comprehensive descriptor of traits related to root axes thickness (SRL, MRR, RV) with a positive value describing root systems with high root volume of predominantly coarse axes was used, the amount of runoff was reduced by such rooting types which is expressed by the average lower runoff under legumes. However, the surface runoff of *L. usitatissimum* with a highly dense root system is efficiently mitigated. Under less intensive rainfall, only five species had runoff higher 1% of incoming rainfall with peak values for *S. alba*, *Fagopyrum esculentum*, and *M. officinalis*. Here the limiting factor appeared to be related to a instead of K_s, where the interrow showed the strongest relation ($R^2 = 0.60$, $p = 0.03$) with higher α values constraining infiltrability. The relation to root axes thickness expressed by the composite principal component descriptor was similar to the high intensity scenario but following an exponential trend.

Outlooks

The runoff generation and flow concentration are the vital processes of terrestrial ecosystem during the water cycles. This process includes the water movement in the soil, the subsurface runoff, slope surface runoff, and the runoff generation in watershed. The formation of surface runoff is complex and highly variable on spatial and temporal scales, which is always the root of uncertainty and the theoretical bottleneck of hydrological analysis and accurate predictions as well as simulations.

1. To explore the impact of vegetation on the mechanism of surface runoff formation, it is necessary to carry out the in situ observations of runoff process on slope and watershed scales, respectively under different vegetation conditions. So far, such observations have only been done in typical vegetation basins, which lead to the ambiguous mechanism of runoff formation on basin or regional scales. Therefore, the establishment of surface runoff observation network in different ecological system is not only the ultimate approach to further understanding of mechanism of surface runoff formation but also a basis for constructions of high precision hydrological models.
2. The development of distributed hydrological models based on physical mechanism, to some extent, has played an important role in the accurate simulation of surface runoff. However, it is undeniable that the data processing of ecological factors tends to be simple, taking a common case of vegetation to be parameterized as input in most of the models without any consideration to apply merits of land surface process models into hydrological models. Therefore, the combination of land surface model and distributed hydrological model will greatly improve the accuracy and effectiveness of simulation results. This is also the main developing direction of watershed hydrological models in the future.
3. For the ecological process, most of the studies can be based on individual organisms, community, or transect. However, most of the surface runoff processes were identified based on the scales of sample plots, slope, catchment area, or small watershed. Therefore, the slope surface, water catchment unit, or subbasin become the ideal scales to link ecology and hydrology. However, how to integrate the ecological process and the runoff process on different scales or on larger scale (such as watershed or regional scale)? How to couple the surface runoff process and the ecological process under the background climate change?

The lack of the feasible observation technique and simulation method has become the obstacles on the way and also one of the greatest challenges for the development and application of eco-hydrology.

Acknowledgments This research was financially supported by the National Natural Science Foundation of China (41430748 and 41661144025).

References

M.B. Abbott, J.C. Bathurst, J.A. Cunge, P.E. O'Connell, J. Rasmussen, An introduction to the European Hydrological System – Systeme Hydrologique Europeen, "SHE", 1: history and philosophy of a physically-based, distributed modelling system. J. Hydrol. **87**(1), 45–59 (1986)

J.R.C.B. Abrantes, R.B. Moruzzi, A. Silveira, J.L.M.P.D. Lima, Comparison of thermal, salt and dye tracing to estimate shallow flow velocities: novel triple-tracer approach. J. Hydrol. **557**, 362–377 (2018)

J.A. Adepetu, L.K. Akapa, Root growth and nutrient uptake characteristics of some cowpea varieties. Agron. J. **69**(6), 940–943 (1977)

O. Al-Jubeh, Studies of natural vegetation characteristics at different Environment and range improvement practices at southern West Bank, M. Sc thesis, Hebron University, 2006

G.F. Antonious, Reducing herbicide residues from agricultural runoff and seepage water. (2015)

J.L. Arsenault, S. Poulcur, C. Messier, R. Guay, WinRHIZO™, a root-measuring system with a unique overlap correction method. Hortscience Am. Soc. Horticult. Sci. **30**(4), 906 (1995)

C. Asdak, P.G. Jarvis, P.V. Gardingen, A. Fraser, Rainfall interception loss in unlogged and logged forest areas of Central Kalimantan, Indonesia. J. Hydrol. **206**(3), 237–244 (1998)

S.D. Baets, J. Poesen, B. Reubens, K. Wemans, J.D. Baerdemaeker, B. Muys, Root tensile strength and root distribution of typical Mediterranean plant species and their contribution to soil shear strength. Plant Soil **305**(1–2), 207–226 (2008)

D.H. Barker, *Vegetation and Slopes: Stabilization, Protection and Ecology* (Thomas Telford, London, 1995)

J. Bauhus, C. Messier, Evaluation of fine root length and diameter measurements obtained using RHIZO image analysis. Agron. J. **91**(1), 142–147 (1999)

A.G. Bengough, Water dynamics of the root zone: rhizosphere biophysics and its control on soil hydrology. Vadose Zone J. **11**(2), 460–460 (2012)

S. Bergstrom, *Computer Models of Watershed Hydrology.* (1995), pp. 443–476

K.J. Beven, M.J. Kirkby, Taylor & Francis Online: A physically based, variable contributing area model of basin hydrology/Un modèle à base physique de zone d'appel variable de l'hydrologie du bassin versant. Hydrol. Sci. Bull. **24**(1), 43–69 (1979)

E.W. Bhark, E.E. Small, Association between plant canopies and the spatial patterns of infiltration in shrubland and grassland of the Chihuahuan Desert, New Mexico. Ecosystems **6**(2), 0185–0196 (2003)

G. Bodner, L. Daniel, N. Alireza, S. Monika, M. Karl, K. Hans-Peter, A statistical approach to root system classification. Front. Plant Sci. **4**(4), 292–292 (2013)

W. Böhm, *Methods of Studying Root Systems* (Springer, Berlin/Heidelberg/New York, 1979), pp. 1–3

W. Boughton, A hydrograph-based model for estimating the water yield of ungauged catchments. Proc. Hydrol. Water Resour. Symp. Newcastle Inst. Engs. Aust. Nat. Conf. Publ. **93**(14), 317–324 (1993)

W. Boughton, F. Chiew, Calibrations of the AWBM for use on ungauged catchments. Am. Soc. Civ. Eng. **14**(6), 562–568 (2003)

D.L. Brakensiek, H.B. Osborn, W.J. Rawls, *Field Manual for Research in Agricultural Hydrology.* (1979)

F. Branson, G. Gifford, J. Owen, Rangeland hydrology. Range Science Series No. 1, Soc. Range Manage., Denver CO. (1972)

F. Branson, G. Gifford, K. Renard, R. Hadley, Rangeland Hydrology. Range Sciences Series No. 1, October 1972, 1981. Soc. Range Manage., Denver CO. Kendall (ed.), (Hunt Publishing Co., Dubuque, 1981)
J. Bromley, J. Brouwer, A. Barker, S. Gaze, C. Valentine, The role of surface water redistribution in an area of patterned vegetation in a semi-arid environment, south-west Niger. J. Hydrol. **198**, 1–29 (1997)
C.J. Bronick, R. Lal, Soil structure and management: a review. Geoderma **124**(1–2), 3–22 (2005)
P. Burlando, Hydrograph separation of runoff components based on measuring hydraulic state variables, tracer experiments, and weighting methods. Paper presented at integrated methods in catchment hydrology: tracer, remote sensing and new hydrometric techniques: proceedings of an international symposium held during IUGG 99, the XXII General Assembly of the International Union of Geodesy and Geophysics, Birmingham, IAHS, 18–30 July 1999
R.J.C. Burnash, R.L. Ferral, R.A. Mcguire, *A generalized streamflow simulation system – conceptual modeling for digital computers*. (1973)
J.M. Buttle, Isotope hydrograph separations and rapid delivery of pre-event water from drainage basins. Prog. Phys. Geogr. **18**(1), 16–41 (1994)
J.M. Buttle, in *Chapter 1 – Fundamentals of Small Catchment Hydrology*, Isotope Tracers in Catchment Hydrology (1998)
I.R. Calder, P.T.W. Rosier, The design of large plastic-sheet net-rainfall gauges. J. Hydrol. **30**(4), 403–405 (1976)
A. Calver, W.L. Wood, V.P. Singh, The Institute of Hydrology distributed model (1987)
Y. Cantón, A. Solébenet, V.J. De, C. Boixfayos, A. Calvocases, C. Asensio, J. Puigdefábregas, A review of runoff generation and soil erosion across scales in semiarid south-eastern Spain. J. Arid Environ. **75**(12), 1254–1261 (2011)
H.E. Carley, R.D. Watson, A new gravimetric method for estimating root-surface areas. Soil Sci. **102**(5), 289–291 (1966)
D.E. Carlyle-Moses, J.S.F. Laureano, A.G. Price, Throughfall and throughfall spatial variability in Madrean oak forest communities of northeastern Mexico. J. Hydrol. **297**(1–4), 124–135 (2004)
C.E. Carter, D.A. Parsons, Field tests on the Coshocton-type wheel runoff sampler, 0133–0135 (1967)
A. Cerda, Parent material and vegetation affect soil erosion in eastern Spain. Soil Sci. Soc. Am. J. **63**(2), 362–368 (1999)
V.A. Chaplot, Y. Le Bissonnais, Runoff features for interrill erosion at different rainfall intensities, slope lengths, and gradients in an agricultural loessial hillslope. Soil Sci. Soc. Am. J. **67**(3), 844–851 (2003)
F.H.S. Chiew, M.C. Peel, A.W. Western, V.P. Singh, D. Frevert, Application and testing of the simple rainfall-runoff model. SIMHYD, 335–367 (2002)
S.K. Chong, R.E. Green, Sorptivity measurement and its application. (1983)
C.A.A. Ciesiolka, B. Yu, C.W. Rose, H. Ghadiri, D. Lang, C. Rosewell, Improvement in soil loss estimation in USLE type experiments. J. Soil Water Conserv. **61**(4), 223–229 (2006)
E.A. Colman, Vegetation and watershed management (1953)
E.S. Corbett, Rainfall interception by annual grass and chaparral; losses compared (1968)
N.H. Crawford, The synthesis of continuous streamflow hydrographs on a digital computer (1962)
N.H. Crawford, R.E. Linsley, Digital simulation in hydrology: Stanford watershed model IV, Evapotranspiration, 39 (1966)
H.P. Cresswell, J.A. Kirkegaard, Subsoil amelioration by plant-roots – the process and the evidence. Soil Res. **33**(2), 221–239 (1995)
H.P. Creswell, D.E. Smiles, J. Williams, Soil structure, soil hydraulic properties and the soil water balance. Aust. J. Soil Res. **30**(3), 265–283 (1992)
R.H. Crockford, D.P. Richardson, Partitioning of rainfall into throughfall, stemflow and interception: effect of forest type, ground cover and climate. Hydrol. Process. **14**(16–17), 2903–2920 (2000)
Y. Cui, L. Jia, A modified Gash model for estimating rainfall interception loss of forest using remote sensing observations at regional scale. Water **6**(4), 993–1012 (2014)
M.S. Czarnowski, J.L. Olszewski, Number and spacing of rainfall-gauges in a deciduous forest stand. Oikos **21**(1), 48–51 (1970)

A.P.J. De Roo, R.J.E. Offermans, N.H.D.T. Cremers, LISEM: a single-event, physically based hydrological and soil erosion model for drainage basins. II: sensitivity analysis, validation and application. Hydrol. Process. **10**(8), 1119–1126 (2015)

C. Deasy, J.N. Quinton, M. Silgram, A.P. Bailey, B. Jackson, C.J. Stevens, Mitigation options for sediment and phosphorus loss from winter-sown Arable Crops. J. Environ. Qual. **38**(38), 2121–2130 (2009)

S.L. Dingman, Physical hydrology. Phys. Hydrol. **11**(1), 112–120 (2002)

M. Dohnal, T. Černý, J. Votrubová, M. Tesař, Rainfall interception and spatial variability of throughfall in spruce stand. J. Hydrol. Hydromech. **62**(4), 277–284 (2014)

M.G. Dosskey, K.D. Hoagland, J.R. Brandle, Change in filter strip performance over ten years. J. Soil Water Conserv. **62**(1), 21–32 (2007)

F. Duley, C.E. Domingo, Effect of grass on intake of water (1949)

G. Dunjó, G. Pardini, M. Gispert, The role of land use–land cover on runoff generation and sediment yield at a microplot scale, in a small Mediterranean catchment. J. Arid Environ. **57**(2), 239–256 (2004)

D. Dunkerley, Hydrologic effects of dryland shrubs: defining the spatial extent of modified soil water uptake rates at an Australian desert site. J. Arid Environ. **45**(2), 159–172 (2000)

T. Dunne, Field studies of hillslope flow processes, in *Hillslope Hydrology*, ed. by M. J. Kirkby (Ed), (Wiley, New York, 1978), pp. 227–293

T. Dunne, W.E. Dietrich, Experimental study of Horton overland flow on tropical hillslopes; 1, Soil conditions, infiltration and frequency of runoff. Z. Geomorphol. **2**, 40–80 (1980)

T. Dunne, T.R. Moore, C.H. Taylor, Recognition and prediction of runoff-producing zones in humid regions. Hydrol. Sci. Bull, 20, 305–327 (1975)

T. Dunne, W. Zhang, B.F. Aubry, Effects of rainfall, vegetation, and microtopography on infiltration and runoff. Water Resour. Res. **27**(9), 2271–2285 (1991)

J. Dusek, T. Vogel, M. Dohnal, H.H. Gerke, Combining dual-continuum approach with diffusion wave model to include a preferential flow component in hillslope scale modeling of shallow subsurface runoff. Adv. Water Resour. **44**(44), 113–125 (2012)

E. Dyksterhuis, E. Schmutz, Natural mulches or "litter" of grasslands: with kinds and amounts on a southern prairie. Ecology **28**(2), 163–179 (1947)

I.J. Edwards, W.D. Jackon, P.M. Fleming, Tipping bucket gauges for measuring run-off from experimental plots. Agric. Meteorol. **13**(2), 189–201 (1974)

M. Eshghizadeh, A. Talebi, M.T. Dastorani, H. Azimzadeh, Effect of natural land covers on runoff and soil loss at the hill-slope scale. Glob. J. Environ. Sci. Manag. **2**, 125–134 (2015)

P.S. Evans, Root growth of Lolium perenne L. 1. Effect of plant age, seed weight, and nutrient concentration on root weight, length and number of apices. N. Z. J. Bot. **8**, 544–556 (1970)

R.P. Ewing, T.C. Kaspar, Accurate perimeter and length measurement using an edge chord algorithm. J. Comput.-Assist. Microsc. **7**, 91–100 (1995)

D.X. Fan, Study on slope soil erosion response and model simulation in Beijing mountainous area, 166 (2014)

Y. Fan, R.L. Bras, Analytical solutions to hillslope subsurface storm flow and saturation overland flow. Water Resour. Res. **34**(4), 921–928 (1998)

J. Fan, K.T. Oestergaard, A. Guyot, D.A. Lockington, Measuring and modeling rainfall interception losses by a native Banksia woodland and an exotic pine plantation in subtropical coastal Australia. J. Hydrol. **515**(13), 156–165 (2014)

A.D. Feldman, Hec models for water resources system simulation: theory and experience. Adv. Hydrosci. **12**, 297–423 (1981)

D. Flanagan, M. Nearing, USDA-Water Erosion Prediction Project: hillslope profile and watershed model documentationRep., NSERL report (1995)

J.P. Fortin, R. Turcotte, S. Massicotte, R. Moussa, J. Fitzback, J.P. Villeneuve, Distributed watershed model compatible with remote sensing and GIS data. II: application to Chaudière watershed. J. Hydrol. Eng. **6**(2), 91–99 (2001)

C. Francis, J. Thornes, Runoff hydrographs from three Mediterranean vegetation cover types, vegetation and erosion. Process. Environ., 363–384 (1990)

Z. Gan, J. Ye, Q. Zhou, Z. Zhou, Z. Shangguan, Effects of grass vegetations on the processes of soil erosion over slope lands in simulated rainfalls. Acta Ecol. Sin. **30**(9), 2387–2396 (2010)

W.R. Gardner, Some steady-state solutions of the unsaturated moisture flow equation with application to evaporation from a water table. Soil Sci. **85**(4), 228–232 (1958)

J. Gash, An analytical model of rainfall interception by forests. Q. J. R. Meteorol. Soc. **105**(443), 43–55 (1979)

M.J. Gavazzi, G. Sun, S.G. McNulty, E.A. Treasure, M.G. Wightman, Canopy rainfall interception measured over ten years in a coastal plain loblolly pine (Pinus taeda L.) plantation. Trans. ASABE **59**(2), 601–610 (2016)

G. Geisler, D. Maarufi, Investigations on the importance of the root systems of cultivated plants, 1: the influence of the soil water content and nitrogen manuring on plant growth, root morphology, transpiration and nitrogen absorption (1975)

S. Germer, L. Werther, H. Elsenbeer, Have we underestimated stemflow? Lessons from an open tropical rainforest. J. Hydrol. **395**(3–4), 169–179 (2010)

M. Ghestem, R.C. Sidle, A. Stokes, The influence of plant root systems on subsurface flow: implications for slope stability. Bioscience **61**(11), 869–879 (2011)

W.H. Green, G. Ampt, Studies on soil phsyics. J. Agric. Sci. **4**(1), 1–24 (1911)

J.H. Gregory, M.D. Dukes, G.L. Miller, P.H. Jones, Analysis of double-ring infiltration techniques and development of a simple automatic water delivery system, Appl. Turfgrass Sci., 2 (2005)

G. Gyssels, J. Poesen, E. Bochet, Y. Li, Impact of plant roots on the resistance of soils to erosion by water: a review. Prog. Phys. Geogr. **29**(2), 189–217 (2005)

A. Hanchi, M. Rapp, Stemflow determination in forest stands. For. Ecol. Manag. **97**(3), 231–235 (1997)

D.M. Harris, J.J. Mcdonnell, A. Rodhe, Hydrograph separation using continuous open system isotope mixing. Water Resour. Res. **31**(1), 157–171 (1995)

R.J. Haynes, M.H. Beare, Influence of six crop species on aggregate stability and some labile organic matter fractions. Soil Biol. Biochem. **29**(11), 1647–1653 (1997)

G.C. Head, Estimating seasonal changes in the quantity of white unsuberized root on fruit trees. J Pomol Hortic Sci **41**(2), 197–206 (1966)

J. Helvey, J.H. Patric, Canopy and litter interception of rainfall by hardwoods of eastern United States. Water Resour. Res. **1**(2), 193–206 (1965)

M. Herbst, J.M. Roberts, P.T.W. Rosier, D.J. Gowing, Measuring and modelling the rainfall interception loss by hedgerows in southern England. Agric. For. Meteorol. **141**(2–4), 244–256 (2006)

A.G.J. Hilberts, P.A. Troch, C. Paniconi, J. Boll, Low-dimensional modeling of hillslope subsurface flow: Relationship between rainfall, recharge, and unsaturated storage dynamics. Water Resour. Res. **43**(3), 305–310 (2007)

L.M. Himmelbauer, Loiskandl, Kastanek, Estimating length, average diameter and surface area of roots using two different image analyses systems: new challenges for rhizosphere research at the entrance of the 21st century. Plant Soil (2004)

H.N. Holtan, Concept for infiltration estimates in watershed engineering (1961)

F. Holwerda, F.N. Scatena, L.A. Bruijnzeel, Throughfall in a Puerto Rican lower montane rain forest: a comparison of sampling strategies. J. Hydrol. **327**(3), 592–602 (2006)

R. Horn, A. Smucker, Structure formation and its consequences for gas and water transport in unsaturated arable and forest soils. Soil Tillage Res. **82**(1), 5–14 (2005)

R.E. Horton, Rainfall interception. Mon. Weather Rev. **47**(9), 603–623 (1919)

R.E. Horton, An approach toward a physical interpretation of infiltration-capacity. Soil Sci. Soc. Am. J. **5**(C), 399–417 (1941)

R. Horton, R. Horton, R.R. Horton, R.E. Horton, R. Horton, R.E. Horton, An approach towards physical interpretation of infiltration capacity (1940)

M. Hrnčíř, M. Šanda, A. Kulasová, M. Císlerová, Runoff formation in a small catchment at hillslope and catchment scales. Hydrol. Process. **24**(16), 2248–2256 (2010)

S.P. Hu, Y.U. Xinxiao, Y.U.E. Yongjie, Hydrological effects of forest litters and soil in Baihua Mountain. J. Soil Water Conserv. **22**(1), 146–150 (2008)

P. Hubert, E. Marin, M. Meybeck, P. Olive, E. Siwertz, Aspects Hydrologiques et Sedimentologigues de la Crue Exceptionnelle de la Dranse du Chablais du 22 Septembre 1968 (1969)

M.G. Huck, B. Klepper, H.M. Taylor, Diurnal variations in root diameter. Plant Physiol. **45**(4), 529 (1970)

N. Hudson, Field measurement of soil erosion and runoff, Food & Agriculture Org (1993)

T. Iserloh, J.B. Ries, J. Arnáez, C. Boix-Fayos, V. Butzen, A. Cerdà, M.T. Echeverría, J. Fernández-Gálvez, W. Fister, C. Geißler, European small portable rainfall simulators: a comparison of rainfall characteristics. Catena **110**(11), 100–112 (2013a)

T. Iserloh, J.B. Ries, A. Cerdà, M.T. Echeverría, W. Fister, C. Geißler, N.J. Kuhn, F.J. León, P. Peters, M. Schindewolf, Comparative measurements with seven rainfall simulators on uniform bare fallow land. Z. Geomorphol. **57**(3), 11–26 (2013b)

A.J. Jakeman, G.M. Hornberger, How much complexity is warranted in a rainfall-runoff model? Water Resour. Res. **29**(8), 2637–2649 (1993)

C. Johnson, N. Gordon, Runoff and erosion from rainfall simulator plots on sagebrush rangeland. Trans. ASAE **31**(2), 421–427 (1988)

W. Johnson, C. Niederhof, Some relationships of plant cover to run-off, erosion, and infiltration on granitic soils. J. For. **39**(10), 854–858 (1941)

T.J. Ju, L.I.U. Puling, X.U. Xuexuan, W. Shuanquan, S.H.I. Xinhe, Experimental study on runoff and sediment process in primary-lands in loess hilly regions under different rainfall conditions. J. Sediment Res. **4**, 65–71 (2007)

Y. Kavdır, A.J.M. Smucker, Soil aggregate sequestration of cover crop root and shoot-derived nitrogen. Plant Soil **272**(1–2), 263–276 (2005)

A.A.H. Khan, C.K. Ong, Design and calibration of tipping bucket system for field runoff and sediment quantification. Paper presented at J. Soil Water Conserv. (1997)

P.I.A. Kinnell, The effect of kinetic-energy of excess rainfall on soil loss from non-vegetated plots. Aust. J. Soil Res. **21**(4), 445–453 (1983)

P. Kinnell, The IXEA index: an index with the capacity to give more direct consideration of hydrology in the USLE modelling environment. J. Soil Water Conserv. **50**, 507–512 (1995)

P.I.A. Kinnell, A review of the design and operation of runoff and soil loss plots. Catena **145**, 257–265 (2016)

Kinnell, P. I. A., K. C. Mcgregor, and C. J. Rosewell (1994), The IXEA index as an alternative to the EI30 erosivity index, 37(5), 1449-1456.

G. Kite, in *The SLURP Model*, ed. By Singh. Computer Models of Watershed Hydrology (1995)

A.N. Kostiakov, On the dynamics of the coefficient of water-percolation in soils and on the necessity of studying it from a dynamic point of view for purposes of amelioration, Trans. 6th Cong. International. Soil Science, Russian Part A, 17–21 (1932)

B. Kothyari, P. Verma, B. Joshi, U. Kothyari, Rainfall–runoff-soil and nutrient loss relationships for plot size areas of Bhetagad watershed in Central Himalaya, India. J. Hydrol. **293**(1), 137–150 (2004)

N. Kouwen, S.F. Mousavi, V.P. Singh, D.K. Frevert, WATFLOOD/SPL9 hydrological model & flood forecasting system. Mathematical Models of Large Watershed Hydrology (2002)

N.J. Kuhn, P. Greenwood, W. Fister, Use of field experiments in soil erosion research, 175–200 (2014)

R. Lei, Effect of a larix principis-rupprechtii forestplantation on soil in middle zone of south-facingslope of the Qinling mountains. Sci Silvae Sin. (1997)

T. Lei, W. Xia, J. Zhao, Z. Liu, Q. Zhang, Method for measuring velocity of shallow water flow for soil erosion with an electrolyte tracer. J. Hydrol. **301**(1–4), 139–145 (2005)

T.W. Lei, C. Ruiyuan, J. Zhao, X.N. Shi, L. Lin, An improved method for shallow water flow velocity measurement with practical electrolyte inputs. J. Hydrol. **390**(1–2), 45–56 (2010)

J. Leonard, P. Andrieux, Infiltration characteristics of soils in Mediterranean vineyards in Southern France. Catena **32**(3), 209–223 (1998)

D.F. Levia, E.E. Frost, A review and evaluation of stemflow literature in the hydrologic and biogeochemical cycles of forested and agricultural ecosystems. (Elsevier B.V., 2003), pp. 1–29

Y. Li, X. Xu, X. Zhu, Preliminary-study on mechanism of plant-roots to increase soil anti-scouribility on the Loess Plateau. Sci. China Ser. B: Chem. **35**(9), 1085–1092 (1992)

M. Li, W. Yao, Z. Li, Progress of the effect of grassland vegetation for conserving soil and water on loess plateau. Adv. Earth Sci. **20**(1), 74–80 (2005)

X.Y. Li, L.Y. Liu, S.Y. Gao, et al., Stemflow in three shrubs and its effect on soil water enhancement in semiarid loess region of China[J]. Agric. For. Meteorol. **148**(10), 1501–1507 (2008)

C. Li, D.X. Ren, G.X. Wang, H.U. Hong-Chang, L.I. Tai-Bing, G.S. Liu, Analysis of artificial precipitation interception over two meadow species on Qinghai-Tibet Plateau. Adv. Water Sci. **20**(6), 769–774 (2009)

C.J. Li, G.X. Wang, T.X. Mao, Rainfall interception by swamp meadow and alpine meadow and its seasonal distribution pattern on the Qinghai-Tibet Plateau, China. Adv. Mater. Res. **610–613**, 2675–2682 (2012)

H. Lieth, R.H. Whittaker, *Primary Productivity of the Biosphere* (Springer-Verlag, New York, 1975), p. 274

T.E. Link, M. Unsworth, D. Marks, The dynamics of rainfall interception by a seasonal temperate rainforest. Agric. For. Meteorol. **124**(3), 171–191 (2004)

X. Liu, Q. Wu, H. Zhao, The vertical interception function of forest vegetation and soil and water conservation. Res. Soil Water Conserv. **1**(3), 8–13 (1994)

H. Liu, W. Cao, X. Wang, Response of water and sediment to land use in watershed of Losses plateau. Agric. Res. Arid Areas. **28**(2), 237–242 (2010)

Y. Liu, Q. Wang, T. Yang, J. Lv, G. Zhao, Study on interception characteristics of different plants. J. Soil Water Conserv, **29**(3), (2015)

P. Llorens, F. Domingo, Rainfall partitioning by vegetation under Mediterranean conditions. A review of studies in Europe. J. Hydrol. **335**(1–2), 37–54 (2007)

P. Llorens, F. Gallart, A simplified method for forest water storage capacity measurement. J. Hydrol. **240**(1), 131–144 (2000)

D. Love, S. Uhlenbrook, G. Corzo-Perez, S. Twomlow, P. van der Zaag, Rainfall–interception–evaporation–runoff relationships in a semi-arid catchment, northern Limpopo basin, Zimbabwe. Hydrol. Sci. J. **55**(5), 687–703 (2010)

Y. Luo, Z. Ouyang, G. Yuan, D. Tang, X. Xie, Evaluation of macroscopic root water uptake models using lysimeter data. Trans. ASAE **46**(3), 625 (2003)

F.P. Lyford, H.K. Qashu, Infiltration rates as affected by desert vegetation. Water Resour. Res. **5**(6), 1373–1376 (1969)

M. Mahmoodabadi, A. Cerdà, WEPP calibration for improved predictions of interrill erosion in semi-arid to arid environments. Geoderma **204–205**(4), 75–83 (2013)

O.J. Manfroi, K. Kuraji, M. Suzuki, N. Tanaka, T. Kume, M. Nakagawa, T.O. Kumagai, T. Nakashizuka, Comparison of conventionally observed interception evaporation in a 100-m 2 subplot with that estimated in a 4-ha area of the same Bornean lowland tropical forest. J. Hydrol. **329**(1–2), 329–349 (2006)

B.A. Marsh, Measurement of length in random arrangements of lines. J. Appl. Ecol. **8**(1), 265–267 (1971)

R.B. Marston, Ground cover requirements for summer storm runoff control on aspen sites in northern Utah. J. For. **50**(4), 303–307 (1952)

J. Martinec, Snowmelt-runoff model for stream flow forecasts. Iwa Publish. **6**, 145–154 (1975)

W.J. Massman, The derivation and validation of a new model for the interception of rainfall by forests. Agric. Meteorol. **28**(3), 261–286 (1983)

C.M. Mathieu, F.O. Pieltain, Analyse physique des sols (1998)

W.C. Mayaki, I.D. Teare, L.R. Stone, Top and root growth of irrigated and nonirrigated soybeans. Crop Sci. **16**(1), 92–94 (1976)

C. Mayerhofer, G. Meißl, K. Klebinder, B. Kohl, G. Markart, Comparison of the results of a small-plot and a large-plot rainfall simulator – effects of land use and land cover on surface runoff in Alpine catchments. Catena **156**, 184–196 (2017)

J.J. Mcdonnell, M. Bonell, M.K. Stewart, A.J. Pearce, Deuterium variations in storm rainfall: implications for stream hydrograph separation. Water Resour. Res. **26**(3), 455–458 (1990)

M. McLeod, D. MacLeod, H. Daniel, The effect of degradation of phalaris+ white clover pasture on soil water regimes of a Brown Chromosol on the Northern Tablelands of NSW, Australia. Agric. Water Manag. **82**(3), 318–342 (2006)

D.W. Meals, D.C. Braun, Demonstration of methods to reduce E. coli runoff from dairy manure application sites. J. Environ. Qual. **35**(4), 1088–1100 (2006)

F.M. Melhuish, A precise technique for measurement of roots and root distribution in soils. Ann. Bot. **32**(125), 15–22 (1968)

T. Merzer, The effects of different vegetative cover on the local hydrological balance of a semi arid afforestation. Albert Katz International School for Desert Studies (2007)

L. Meyer, Rainfall simulators for soil erosion research (1994)

L.D. Meyer, D.L. Mccune, Rainfall simulator for runoff plots (1958)

D.G. Miralles, J.H. Gash, T.R.H. Holmes, R.A.M. de Jeu, A.J. Dolman, Global canopy interception from satellite observations. J. Geophys. Res. **115**(D16), D16122 (2010)

S.K. Mishra, V.P. Singh, Soil conservation service curve number (SCS-CN) methodology. Springer Science & Business Media (2013)

A. Mohammed, Rangeland condition at southern West Bank. Hebron Univ. Res. J. **2**, 42–54 (2005)

R.D. Moore, Tracing runoff sources with deuterium and oxygen-88 during spring melt in a headwater catchment, southern Laurentians, Quebec. J. Hydrol. **112**(1), 135–148 (1989)

R.R.C. Morgan, R.J. Rickson, *Slope Stabilization and Erosion Control. A Bioengineering Approach* (E & EN Spon, London, 1995)

G. Morin, V.P. Singh, D.K. Frevert, CEQUEAU hydrological model. Cequeau Hydrological Model (2002), pp. 507–576

J.P.M. Mulder, *Simulating Interception Loss Using Standard Meteorological Data* (Springer Netherlands, 1985), pp. 177–196

C. Muñoz-Robles, N. Reid, M. Tighe, S.V. Briggs, B. Wilson, Soil hydrological and erosional responses in patches and inter-patches in vegetation states in semi-arid Australia. Geoderma **160**(3), 524–534 (2011)

C.K. Mutchler, Runoff plot design and installation for soil erosion studies (1963)

C.K. Mutchler, C.E. Murphree, K.C. Mcgregor, R. Lal, Laboratory and field plots for soil erosion studies. Soil Erosion Res. Methods **9**, 36 (1988)

A. Muzylo, P. Llorens, F. Valente, J.J. Keizer, F. Domingo, J.H.C. Gash, A review of rainfall interception modelling. J. Hydrol. **370**(1), 191–206 (2009)

G. Negi, H. Rikhari, S. Garkoti, The hydrology of three high-altitude forests in Central Himalaya, India: a reconnaissance study. Hydrol. Process. **12**(2), 343–350 (1998)

T. Nehls, Y.N. Rim, G. Wessolek, Technical note on measuring run-off dynamics from pavements using a new device: the weighable tipping bucket. Hydrol. Earth Syst. Sci. **15**(15), 1379 (2011)

S.L. Neitsch, J.G. Arnold, J.R. Kiniry, J.R. Williams, K.W. King, Soil and water assessment tool theoretical documentation: version 2000, GSWRL Report 02-01, BRC Report 02-05, TR-191. College Station, Texas: Texas Water Resources Institute (2002)

J. Neris, C. Jiménez, J. Fuentes, G. Morillas, M. Tejedor, Vegetation and land-use effects on soil properties and water infiltration of Andisols in Tenerife (Canary Islands, Spain). Catena **98**(6), 55–62 (2012)

E.I. Newman, A Method of Estimating the Total Length of Root in a Sample. J. Appl. Ecol. **3**(1), 139–145 (1966)

A.R. Nordin, Eco-Engineering Practices in Malaysia. Ph.D. Thesis, University of New castle Upon Tyne, Glasgow, 1995

F.L. Ogden, B.A. Watts, Saturated area formation on nonconvergent hillslope topography with shallow soils: a numerical investigation. Water Resour. Res. **36**(7), 1795–1804 (2000)

T. Oztas, A. Koc, B. Comakli, Changes in vegetation and soil properties along a slope on overgrazed and eroded rangelands. J. Arid Environ. **55**(1), 93–100 (2003)

R.W. Pearson, Significance of rooting pattern to crop production and some problems of root research, Proc. Plant Root Environ. (1974)

H. Peng et al., Canopy interception by a spruce forest in the upper reach of Heihe River basin, Northwestern China. Hydrol. Process. **28**(4), 1734–1741 (2014)

C. Perrin, C. Michel, V. Andréassian, Improvement of a parsimonious model for streamflow simulation. J. Hydrol. **279**(1–4), 275–289 (2003)

J.R. Philip, The theory of infiltration: 1. The infiltration equation and its solution. Soil Sci. **83**(5), 345–358 (1957a)

J.R. Philip, The theory of infiltration: 4. Sorptivity and algebraic infiltration equations. Soil Sci. **84**(3), 257–264 (1957b)

M. Pidwirny, *Interception, Stemflow, Canopy Drip, and Throughfall Fundamentals of Physical Geography*, 2nd edn. (2006)

A.F. Pillsbury, R.E. Pelishek, J.F. Osborn, T.E. Szuszkiewicz, Effects of vegetation manipulation on the disposition of precipitation on chaparral-covered watersheds. J. Geophys. Res. **67**(2), 695–702 (1962)

W.T. Pinson, D.C. Yoder, J.R. Buchanan, W.C. Wright, J.B. Wilkerson, Design and evaluation of an improved flow divider for sampling runoff plots. Appl. Eng. Agric. **20**(4), 433 (2004)

O. Planchon, N. Silvera, R. Gimenez, D. Favis-Mortlock, J. Wainwright, Y.L. Bissonnais, G. Govers, An automated salt-tracing gauge for flow-velocity measurement. Earth Surf. Process. Landf. **30**(7), 833–844 (2005)

J.W. Porter, T.A. Mcmahon, A model for the simulation of streamflow data from climatic records. J. Hydrol. **13**, 297–324 (1971)

L.J. Puckett, Spatial variability and collector requirements for sampling throughfal. Can. J. For. Res. **21**(11), 1581–1588 (1991)

J.N. Quinton, G. Edwards, R. Morgan, The influence of vegetation species and plant properties on runoff and soil erosion: results from a rainfall simulation study in south east Spain. Soil Use Manag. **13**(3), 143–148 (1997)

J.C. Refshaard, B. Storm, MIKE SHE, in *Computer Models of Watershed Hydrology* (1995)

D.C. Reicosky, R.J. Millington, D.B. Peters, A comparison of three methods for estimating root length. Agron. J. **62**(4), 451–453 (1970)

K.D. Reid, B.P. Wilcox, D.D. Breshears, L. MacDonald, Runoff and erosion in a piñon–juniper woodland influence of vegetation patches. Soil Sci. Soc. Am. J. **63**(6), 1869–1879 (1999)

W.D. Reynolds, D.E. Elrick, Determination of hydraulic conductivity using a tension infiltrometer. J. Infect. Dis. **163**(4), 921–922 (1991)

M. Robinson, S.J. Grant, J.A. Hudson, Measuring rainfall to a forest canopy: an assessment of the performance of canopy level raingauges. Hydrol. Earth Syst. Sci. **8**(3), 327–333 (2004)

C.C. Romero, L. Stroosnijder, G.A. Baigorria, Interrill and rill erodibility in the northern Andean highlands. Catena **70**(2), 105–113 (2007)

M.J.M. Romkens, The soil erodibility factor: a perspective (1985)

L.A. Rossman, *Storm Water Management Model User's Manual* (2009)

V.G. Rumynin, *Surface Runoff Generation, Vertical Infiltration and Subsurface Lateral Flow* (Springer International Publishing, 2015)

J.K. Ruprecht, N.J. Schofield, Analysis of streamflow generation following deforestation in south-west Western Australia. J. Hydrol. **105**(1–2), 1–17 (1989)

A.J. Rutter, K.A. Kershaw, P.C. Robins, A.J. Morton, A predictive model of rainfall interception in forests, 1. Derivation of the model from observations in a plantation of Corsican pine. Agric. Meteorol. **9**(71), 367–384 (1971)

R. Sarkar, S. Dutta, Field investigation and modeling of rapid subsurface stormflow through preferential pathways in a vegetated hillslope of northeast India. J. Hydrol. Eng. **17**(2), 333–341 (2011)

R. Sarkar, S. Dutta, S. Panigrahy, Characterizing overland flow on a preferential infiltration dominated hillslope: case study. J. Hydrol. Eng. **13**(7), 563–569 (2008)
J.J. Schuurman, M.A.J. Goedewaagen, Methods for the examination of root systems and roots (1965)
R. Schwartz, R. Baumhardt, S. Evett, Tillage effects on soil water redistribution and bare soil evaporation throughout a season. Soil Tillage Res. **110**(2), 221–229 (2010)
M. Seeger, Uncertainty of factors determining runoff and erosion processes as quantified by rainfall simulations. Catena **71**(1), 56–67 (2007)
Z.J. Shi, Y.H. Wang, L.H. Xu, X. Wei, P.T. Yu, G. Hao, L.Y. Zhang, Hydrological functions of litter layer of typical forest types in the Liupan Mountains of Ningxia, northwestern China. J. Beijing For. Univ. **31**(1), 91–99 (2009)
M. Silgram, D.R. Jackson, A. Bailey, J. Quinton, C. Stevens, Hillslope scale surface runoff, sediment and nutrient losses associated with tramline wheelings. Earth Surf. Process. Landf. **35**(6), 699–706 (2010)
I.C. Silva, T. Okumura, Throughfall, stemflow and interception loss in a mixed white oak forest (Quercus serrata Thunb.). J. For. Res. **1**(3), 123–129 (1996)
M.J. Singer, Y. Le Bissonnais, Importance of surface sealing in the erosion of some soils from a Mediterranean climate. Geomorphology **24**(1), 79–85 (1998)
J.S. Singh, A.N. Pandey, P.C. Pathak, A hypothesis to account for the major pathway of soil loss from Himalaya. Environ. Conserv. **10**(4), 343–345 (1983)
M.V.K. Sivakumar, H.M. Taylor, R.H. Shaw, Top and root relations of field-grown soybeans. Agron. J. **69**(3), 470–473 (1977)
M.G. Sklash, R.N. Farvolden, The role of groundwater in storm runoff. J. Hydrol. **43**(1), 45–65 (1979)
M.G. Sklash, R.N. Farvolden, P. Fritz, Erratum: A conceptual model of watershed response to rainfall develop. Can. J. Earth Sci. **63**(13), 1016–1020 (1976)
K.R.J. Smettem, B.E. Clothier, Measuring unsaturated sorptivity and hydraulic conductivity using multiple disc permeameters. Eur. J. Soil Sci. **40**(3), 563–568 (1989)
A.L. Smit, J.F.C.M. Sprangers, P.W. Sablik, J. Groenwold, Automated measurement of root length with a three-dimensional high-resolution scanner and image analysis. Plant Soil **158**(1), 145–149 (1994)
R. Smith, Infiltration envelope-results from a theoretical infiltrometer. Paper presented at Transactions-American Geophysical Union, Amer Geophysical Union 2000 Florida Ave NW, Washington, DC 20009 (1971)
H.L. Smith, L.B. Leopold, Infiltration studies in the Pecos River watershed, New Mexico and Texas. Soil Sci. **53**(3), 195–204 (1942)
H.d. Snyman, C. Du Preez, Rangeland degradation in a semi-arid South Africa – II: influence on soil quality. J. Arid Environ. **60**(3), 483–507 (2005)
S. Soto-Schönherr, A. Iroumé, How much water do Chilean forests use? A review of interception losses in forest plot studies. Hydrol. Process. **30**(25), 4674–4686 (2016)
S. Strohmeier, G. Laaha, H. Holzmann, A. Klik, Magnitude and occurrence probability of soil loss: a risk analytical approach for the plot scale for two sites in Lower Austria. Land Degrad. Dev. **27**(1), 43–51 (2016)
M. Sugawara, I. Watanabe, E. Ozaki, Y. Katsuyama, Tank model programs for personal computer and the way to use (second report). Rev. Res. Disaster Prev. **113**, 1–200 (1986)
P. Suwardji, P.L. Eberbach, Seasonal changes of physical properties of an Oxic Paleustalf (Red Kandosol) after 16 years of direct drilling or conventional cultivation. Soil Tillage Res. **49**(1–2), 65–77 (1998)
K. Suzuki, Y. Kodama, T. Yamazaki, K. Kosugi, Y. Nakai, Snow accumulation on evergreen needle-leaved and deciduous broad-leaved trees. Boreal Environ. Res. **13**(5), 403–416 (2008)
Y. Tang, Research on soil erosion characteristics and vegetation cover and management factor C value of slope farmland in purple hilly region, 79 (2012)

D. Tennant, A test of a modified line intersect method of estimating root length. J. Ecol. **63**(3), 995–1001 (1975)

S. Thompson, C. Harman, P. Heine, G. Katul, Vegetation-infiltration relationships across climatic and soil type gradients. J. Geophys. Res. Biogeosci. **115**(G2), 1–12 (2010a)

S.E. Thompson, G.G. Katul, A. Porporato, Role of microtopography in rainfall-runoff partitioning: an analysis using idealized geometry. Water Resour. Res. **46**(7), W07520 (2010b)

J.H. Titus, R.S. Nowak, S.D. Smith, Soil resource heterogeneity in the Mojave Desert. J. Arid Environ. **52**(3), 269–292 (2002)

B. Torssell, J.E. Begg, C.W. Rose, G.F. Byrne, Stand morphology of Townsville lucerne (Stylosanthes humilis): seasonal growth and root development. Aust. J. Exp. Agric. **8**(34), 533–543 (1968)

P. Troch, E.V. Loon, A. Hilberts, Analytical solutions to a hillslope-storage kinematic wave equation for subsurface flow. Adv. Water Resour. **25**(6), 637–649 (2002)

A. Vacca, S. Loddo, G. Ollesch, R. Puddu, G. Serra, D. Tomasi, A. Aru, Measurement of runoff and soil erosion in three areas under different land use in Sardinia (Italy). Catena **40**(1), 69–92 (2000)

A.R. Vaezi, S.H.R. Sadeghi, H.A. Bahrami, M.H. Mahdian, Modeling the USLE K-factor for calcareous soils in northwestern Iran. Geomorphology **97**(3–4), 414–423 (2008)

T. Vogel, M. Sanda, J. Dusek, M. Dohnal, J. Votrubova, Using oxygen-18 to study the role of preferential flow in the formation of hillslope runoff. Vadose Zone J. **9**(2), 252–259 (2010)

L.X. Wang, Z.Q. Zhang, Advances in the study of ecohydrological effects from vegetation changes, World Forestry Research (6), 10 (1998)

G. Wang, Y.E.A. Zhang, *Ecohydrology in Cold Regions: Theory and Practice* (Science Press, Beijing, 2016)

X.P. Wang, Y. Cui, Y.X. Pan, X.R. Li, Z. Yu, M. Young, Effects of rainfall characteristics on infiltration and redistribution patterns in revegetation-stabilized desert ecosystems. J. Hydrol. **358**(1), 134–143 (2008)

A.W. Warrick, *Soil Water Dynamics* (Oxford University Press, New York, 2003)

F. Watanabe, Y. Kobayashi, S. Suzuki, T. Hotta, S. Takahashi Estimating the volume of surface runoff from in situ measured soil sorptivity, 沙漠研究: 日本沙漠学会誌, 22 (2012)

F. Weller, A method for studying the distribution of absorbing roots of fruit trees. Exp. Agric. **7**(4), 351–362 (1971)

C. Wels, R.J. Cornett, B.D. Lazerte, Hydrograph separation: a comparison of geochemical and isotopic tracers. J. Hydrol. **122**(1–4), 253–274 (1991)

I. White, K.M. Perroux, Estimation of unsaturated hydraulic conductivity from field sorptivity measurements. Soil Sci. Soc. Am. J. **53**(2), 324–329 (1989)

L.K. Wiersum, The relationship of the size and structural rigidity of pores to their penetration by roots. Plant Soil **9**(1), 75–85 (1957)

S.A. Wilde, G.K. Voigt, Absorption-transpiration quotient of nursery stock. J. For. **47**(8), 643–645 (1949)

G. Willgoose, H. Perera, A simple model of saturation excess runoff generation based on geomorphology, steady state soil moisture. Water Resour. Res. **37**(1), 147–156 (2001)

W.H. Wischmeier, J.V. Mannering, Relation of soil properties to its erodibility. Proc. Soil Sci. Soc. Am. **33**(1), 131–137 (1969)

L. Woodward, Infiltration-capacities of some plant-soil complexes on Utah range watershed-lands. Eos. Trans. AGU **24**(2), 468–475 (1943)

Q. Wu, Y. Liu, Ground temperature monitoring and its recent change in Qinghai–Tibet Plateau. Cold Reg. Sci. Technol. **38**(2–3), 85–92 (2004)

Q.B. Wu, B. Shi, Y.Z. Liu, Research on the interaction of permafrost and highway along Qinghai-Tiber highway. (2002)

H.Y. Xi, F. Qi, C. Zhongqiang, C. Yufei, S.I. Jianhua, S.U. Yonghong, G.U.O. Rui, Permeability characteristics of soils and their dependence on soil conditions in Ejina Oasis. J. Glaciol. Geocryol. **30**(6), 976–982 (2008)

M. Xie, H. Zhang, Y. Wang, Study on the influence of soil and water conservation measures on the speciality of run-off production and its flowing and on the starting conditions of debris flow occured by hydrodynamics. J. Beijing For. Univ. (1994)

L.H. Xu, Z.J. Shi, Y.H. Wang, W. Xiong, P.T. Yu, Canopy interception characteristics of main vegetation types in Liupan Mountains of China. Chin. J. Appl. Ecol. **21**(10), 2487–2493 (2010)

M. Yamashita, A. Ichikawa, F. Satoh, H. Shibata, Quantitative approach and problems of river hydrological simulation models. Jpn. J. Limnol. **67**(3), 267–280 (2006)

Y. Yang, Y. Mo, Z. Wang, Z. Liao, J. Deng, X. Zhang, Experimental study on anti-water erosion and shear strength of soil-root composite. J. China Agric. Univ. (1996)

W.-Y. Yao, P.-Q. Xiao, Z.-Z. Shen, C.-X. Yang, Responses of runoff process and threshold of sediment generation for different vegetation-covered plot. Shuili Xuebao (J. Hydraul. Eng.) **42**(12), 1438–1444 (2011)

L. Yi, S. Minǵan, Experimental study on influence factors of rainfall and infiltration under artificial grassland coverage. Trans. Chin. Soc. Agric. Eng. **2007**(3) (2007)

B. Yu, C.W. Rose, K.J. Coughlan, B. Fentie, Plot-scale rainfall-runoff characteristics and modeling at six sites in Australia and South East Asia. Trans. ASAE **40**(5), 1295–1303 (1997)

X. Yu, X. Zhang, J. Li, M. Zhang, Y. Xie, Effects of vegetation cover and precipitation on the process of sediment produced by erosion in a small watershed of loess region. Acta Ecol. Sin. **26**(1), 1–8 (2006)

X.X. Yu, Y.Q. Geng, L.L. Niu, Y.J. Yue, Effect of sampling scale on spatial variability of forest soil nutrient in typical watershed with Miyun Chaoguan west watershed for example. Sci. Silvae Sin. **46**(10), 162–166 (2010)

Y. Yu, W. Loiskandl, H.P. Kaul, M. Himmelbauer, W. Wei, L. Chen, G. Bodner, Estimation of runoff mitigation by morphologically different cover crop root systems. J. Hydrol. **538**, 667–676 (2016)

G.H. Zhang, B.Y. Liu, M.A. Nearing, C.H. Huang, K.L. Zhang, Soil detachment by shallow flow. Trans. ASAE Am. Soc. Agric. Eng. **45**(2), 351–357 (2002)

B. Zhang, Y.-s. Yang, H. Zepp, Effect of vegetation restoration on soil and water erosion and nutrient losses of a severely eroded clayey Plinthudult in southeastern China. Catena **57**(1), 77–90 (2004)

Z.M. Zhang, Y.U. Xin-Xiao, J.Z. Niu, L.U. Shao-Wei, W.F. Sun, X.P. Liu, Y. Zhang, Ecohydrological functions of litter on different forest stands. J. Soil Water Conserv. **19**(3), 139–143 (2005)

X. Zhang, L. Luo, W. Jing, S. Wang, R. Wang, Z. Che, Study on the distribution effect of canopy interception of Picea crassifolia forest in Qilian Mountains. J. Mt. Sc. (2007)

G.H. Zhang, R.T. Luo, Y. Cao, R.C. Shen, X.C. Zhang, Correction factor to dye-measured flow velocity under varying water and sediment discharges. J. Hydrol. **389**(1–2), 205–213 (2010a)

G.H. Zhang, M.K. Tang, X.C. Zhang, Temporal variation in soil detachment under different land uses in the Loess Plateau of China. Earth Surf. Process. Landf. **34**(9), 1302–1309 (2010b)

L. Zhang, J. Wang, Z. Bai, C. Lv, Effects of vegetation on runoff and soil erosion on reclaimed land in an opencast coal-mine dump in a loess area. Catena **128**, 44–53 (2015)

X. Zhang, W. Zhao, Y. Liu, X. Fang, Q. Feng, The relationships between grasslands and soil moisture on the Loess Plateau of China: a review. Catena **145**, 56–67 (2016)

R.J. Zhao, The Xinanjiang model applied in China. J. Hydrol. **135**(1–4), 371–381 (1992)

X.N. Zhao, F.Q. Wu, Developments and reviews of soil infiltration research. J. Northwest For. Univ. **19**(1), 42–45 (2004)

S.L. Zhao, E.C. Dorsey, S.C. Gupta, J.F. Moncrief, D.R. Huggins, Automated water sampling and flow measuring devices for runoff and subsurface drainage. J. Soil Water Conserv. **56**(4), 299–306 (2001)

M. Zhao, W. Yao, J. Wang, P. Xiao, Experimental study on influence of vegetation coverage to soil infiltration and runoff-producing of the Loess Plateau region. Soil Water Conserv. China **6**, 3 (2015)

Y. Zhou, Vegetation and erosion control: exploration on basic principle of slope engineering. Chin. J. Appl. Ecol. **11**(2), 297 (2000)

X. Zhou, Y. Wang, Q. Liu, Adaptability of four infiltration formulas on fluvo-aquic. Soil J. Irrig. Drain. **34**(8), 5 (2015)

X. Zhu, M. Ren, Formation process and regulation measures of Loess Plateau in China. Soil Water Conserv. China **119**(2), 8 (1992)

B. Zhu, Z. Li, P. Li, Z. You, Effect of grass coverage on sediment yield of rain on slope. Acta Pedol. Sin. **47**(3), 401–407 (2010)

B. Zimmermann, A. Zimmermann, R.M. Lark, H. Elsenbeer, Sampling procedures for throughfall monitoring: a simulation study. Water Resour. Res. **46**(1), W01503 (2010)

Subsurface Flow

Guangrong Hu and Xiaoyan Li

Contents

Abstract

Subsurface flow refers to any flow below the surface of the ground which includes low flow (base flow) and quick flow (subsurface stormflow). Subsurface flow has been attracting attention as an important topic of research in recent years because of its crucial role in water cycle calculation, flood prediction, slope stability, nutrient recycling, and soil–water–vegetation exchange processes. In early subsurface flow research, trenches (or pits) combined with hydrometric approaches are the main observation techniques. In recent decades, great progress has been made on the source, pathway, and residence time of subsurface flow due to the application of tracer and geophysical techniques. However, there is yet no broad consensus on the responsible mechanisms that explain runoff processes in the

G. Hu · X. Li (✉)
State Key Laboratory of Earth Surface Processes and Resource Ecology, Faculty of Geographical Science, Beijing Normal University, Beijing, China

School of Natural Resources, Faculty of Geographical Science, Beijing Normal University, Beijing, China
e-mail: 201631190007@mail.bnu.edu.cn; xyli@bnu.edu.cn

X. Li, H. Vereecken (eds.), *Observation and Measurement of Ecohydrological Processes*, Ecohydrology, https://doi.org/10.1007/978-3-662-48297-1_9

ground. This chapter seeks to provide an overview of current research on subsurface flow. We first give a brief description of the basic concept of subsurface flow and base flow and then focus on the flow regimes, controlling factors, and monitor and model of subsurface stormflow. In the end of this chapter, future research directions are proposed.

Keywords
Subsurface flow · Subsurface stormflow · Preferential flow · Base flow · Trenching · Tracer · Geophysical methods

Terminology

In traditional hydrology, flow is divided into surface flow and subsurface flow, where subsurface flow can be divided into quick flow and low flow. In general, the quick flow refers to subsurface stormflow, and low flow refers to base flow (Fig. 1). Subsurface flow refers to any flow below the surface of the ground (WMO 2012). The flow can be saturated or unsaturated, and the flow direction can be lateral or vertical (Dunne 1990; Lin and Zhou 2008). It is part of the infiltrated effective precipitation that circulates in the ground and reaches the surface through drain channels. Subsurface flow is an important source of streamflow, contributed streamflow together with glacier runoff and overland flow (Winter 2007). The exerted role of subsurface flow has been shown to be of key importance for substantially maintaining low flow (base flow) during the rainless period, quickly responding, and contributing to quick flow (subsurface stormflow) during storm periods. The total volume of subsurface flow is the sum of base flow and subsurface stormflow (Winter 2007; Ghasemizade and Schirmer 2013).

Compared with subsurface stormflow, base flow is often viewed as a “dull” or “stable” flow. Hence, the spatial and temporal scale of the study base flow is usually larger (or longer) than the subsurface stormflow. Current research on base flow

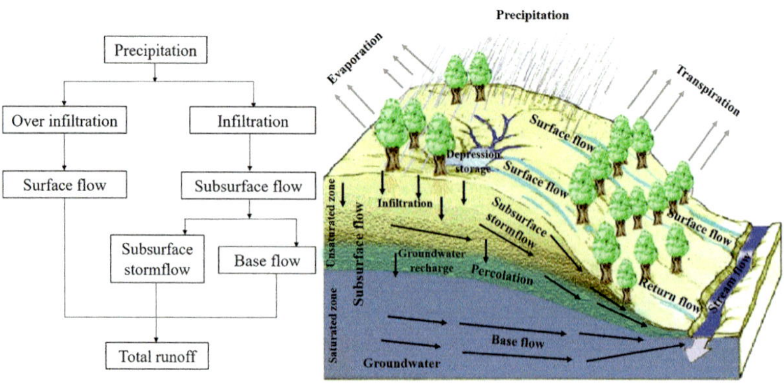

Fig. 1 Subsurface stormflow and base flow in hillslopes. (Modified from Tarboton 2003)

mainly focused on base flow separation methods, regression model, and influencing factors at watershed scale or year scale. However, research on subsurface stormflow is mainly focused on rainfall–runoff processes, transport mechanisms at hillslope/pedon scale, or single precipitation event scale. In the following, we will give a brief description of the base flow, then focus on rainfall–runoff processes, and mainly discuss current research of subsurface stormflow.

Base Flow

Base flow is also known as low flow, groundwater flow, percolation flow, underrun, seepage flow, and sustained flow in the hydrological literature. Base flow can be considered as subsurface flow derived from deep percolation of infiltrated water that enters the permanent saturated groundwater flow system and discharges into the river channel especially during the prolonged rainless period (Freeze 1972; Frohlich et al. 1994). It is not the direct consequence of the rainfall event but is considered as the outflow of the groundwater reservoir feeding the river (Ghasemizade and Schirmer 2013). The base flow rate is relatively stable, generally taking the lowest flow rate of multi-year as the average value.

Base flow is an important part of assessing water resources in a watershed, especially in arid and semiarid areas. In the dry season, the base flow is the most important hydrological feature of the watershed. These streamflow characteristics can be used by watershed planners and regulators to determine water use allocations, assimilative capacities of streams, and aquatic habitat needs. It is essential that watershed planners and regulators have access to accurate and easily obtainable base flow (low flow) characteristics to make informed decisions (Stuckey 2006). At present, the research on base flow is mainly focused on separation methods, regression model, and influencing factors at watershed scale or year scale. Few studies have investigated the mechanisms which generate streamflow during inter-storm/seasonal base flow periods (Payn et al. 2012). In recent years, problems of droughts have focused attention on base flow periods and the processes sustaining water resources for both human consumption and ecosystem needs during dry spells (Lehner et al. 2006). Future research is needed to understand how base flows sustain water supplies and aquatic ecosystems. We believe that better understanding of the interacting controls on low-flow generation mechanisms can lead to better management of limited water resources (Ghasemizade and Schirmer 2013).

Subsurface Stormflow

Subsurface stormflow is also known as subsurface runoff, interflow, throughflow, lateral flow, transient groundwater, fast subsurface flow, or soil water flow in the hydrological literature. It was first recognized by Engler (1919) and first quantitatively studied by Hursh and Brater (1941). As an important component of subsurface flow, subsurface stormflow most describes runoff generation processes that discharges into the transient, near-channel wetlands that develop during the storm period and reaches the channel without entering the groundwater zone (Whipkey 1965; Genereux and Hooper 1998). Subsurface stormflow is considered to be a dominant runoff-producing mechanism in many upland environments, which determines the hydrological responses of headwater catchments to rainstorms. Subsurface stormflow has a slower drainage capacity than overland flow but a faster one than groundwater flow. It's mainly responsible for quickly responding of streamflow during a precipitation or snowmelt event. It can lead to unexpected fast drainage of soils and transmits water to contribute to storm flow, thus delaying or even preventing the saturation of the soil and generation of saturation overland flow and determining the hydrological responses of headwater catchments to rainstorms (Weiler et al. 2005; Tromp-van Meerveld and McDonnell 2006a, b; Kienzler 2007). The essential conditions for the generation of subsurface stormflow include steep slopes and highly permeable surface soils underlain by an impermeable or semipermeable layer (Turton and Miller 1992). It often moves laterally in a downslope direction through the water-impeding layer or soil–bedrock interface. Subsurface stormflow mainly occurs in humid catchments, where the terrain is steep, the surface soil is permeable, and an impermeable soil layer or soil–bedrock interface exists near the surface (Whipkey 1965; Sloan and Moore 1984; Turton and Miller 1992). In drier climate and gentler terrains, subsurface stormflow might occur only under extreme conditions, such as high-intensity rainfall or high antecedent soil moisture (Wilcox et al. 1997). It plays a crucial role in water cycle calculation, flood prediction, nutrient recycling, sediment transport, and soil–water–vegetation exchange processes (Weiler et al. 2005; Fox and Wilson 2010; McIntosh et al. 2017).

Flow Regimes at Different Scales

Subsurface stormflow describes all runoff generation processes close to the soil surface that result in a stream channel hydrograph response during a precipitation event. These processes in subsurface soil may be coupled directly to flow in preferential pathways like macropores and layers or areas with high permeability. The main flow regimes in subsurface soil may be subdivided into homogeneous matrix flow (uniform flow) and preferential flow (nonuniform flow) (Green and Ampt 1911; Weiler et al. 2005). Both types of flow often occur simultaneously but have considerably different consequences for water flow and chemical leaching.

Homogeneous matrix flow (uniform flow or piston flow) refers to flow with the same velocity magnitude and direction at every point; the velocity pattern within a cross section does not change. The movement of water and solutes through the soil is relatively slow and even, while all pore spaces are sampled, obeying the convective dispersion theory, which assumes that water follows an average flow path through soil (Atkinson 1978). It has a laminar hydraulic regime, conforms to Darcy's law, and leads to stable wetting fronts that are parallel to the soil surface. Uniform flow includes downslope and vertical components and might occur under both saturated and unsaturated conditions.

Preferential flow (nonuniform flow) mainly refers to the uneven and often rapid movement of water and solutes through porous media, typically soil, characterized by regions of enhanced flux such that a small fraction of media participates in most of the flow, allowing much faster transport of a range of contaminants including pesticides, nutrients, trace metals, and manurial pathogens. In few cases, preferential flow can also occur in highly uniform soils in the form of unstable "fingers," especially in coarse, single-grained, and water-repellent soils (Jarvis et al. 2016). Preferential flow and transport occur in the unsaturated zone when the vertical velocities in a small fraction of the total pore volume are much faster than the rates of lateral equilibration of water pressures and/or solute concentrations with the surrounding water-filled pores and lead to irregular wetting fronts (Beven and Germann 1982). It reflects the transport mechanism of soil water moving from homogeneous to heterogeneous; it has been acknowledged as a significant process contributing to stormflow generation in headwater catchments. Preferential flow in the vadose zone occurs on different spatial and temporal scales under a wide variety of conditions and includes several types. Each of these types of preferential flow is caused by different physical mechanisms. It comprises all phenomena, where water and solutes move along certain pathways while bypassing a fraction of the porous matrix. This physical non-equilibrium phenomenon can occur on three different scales: pore scale, Darcian scale (Fig. 2), and areal scale (Fig. 3).

On the pore scale, we can observe several types of macropore flow. Macropore flow is used to indicate preferential flow in continuous root channels, earthworm burrows, fissures, cracks, or irregular interaggregate voids created by tillage implements. Water bypasses the denser and less permeable soil matrix by using the pathway of least resistance through macropores, leading to a non-equilibrium with

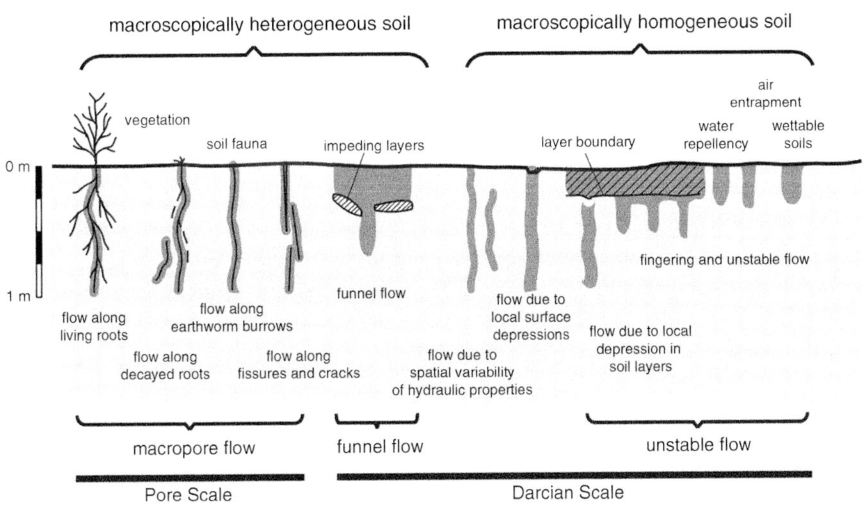

Fig. 2 Schematic showing different preferential flow mechanisms observed on pore and Darcian scales (Hendrickx and Markus 2001)

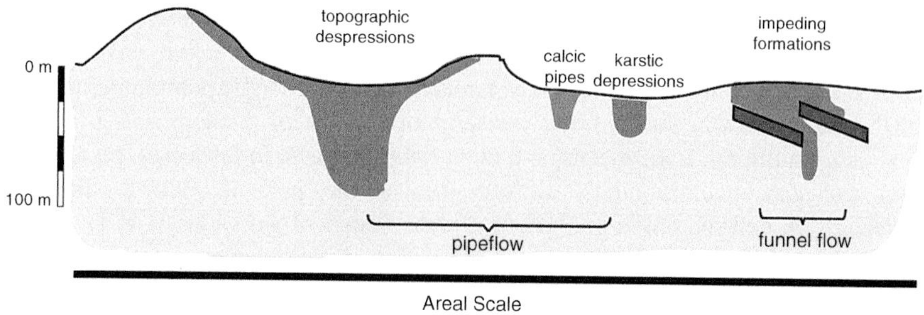

Fig. 3 Schematic showing different preferential flow mechanisms observed on the areal scale (Hendrickx and Markus 2001)

the soil matrix. The flow paths vary in type from individual pores to a highly connected pore network (Gerke 2006). It occurs predominantly in fine-textured soils or media with a pronounced structure. According to different formation reasons, we can divide macropore flow into crack flow and burrow flow. Crack flow refers to preferential flow along continuous cracks through an unsaturated soil profile. Cracking occurs during drying of certain duration in soils with significant clay content. Burrow flow refers to flow through channels created by soil fauna when runoff occurs (Zehe and Fluhler 2001; Allaire et al. 2009). Crack flow and burrow flow together form macropore flow.

On the Darcian scale, preferential flow can be divided into funnel flow, finger flow, and unstable flow due to the spatial variability of hydraulic properties, local surface depressions, local depressions in soil layers, water repellency, and presence of large stones. Each of these types of preferential flow is caused by different physical mechanisms (Hendrickx and Markus 2001). Funnel flow refers to a phenomenon that flow caused by redirection of water at soil textural boundaries. Funnel flow soils are usually characterized by their squat hopper geometry. It is a special case of preferential flow at the Darcian scale. It occurs when water again moves along the pathway of least resistance and can be redirected through a series of less permeable layers embedded in the soil profile. If the underlying region is coarser, finger flow might also occur. Finger flow occurs when infiltrating water accumulates at the interface between two soil layers, usually in sandy soils, with a coarser layer underlying a finer layer, single-grained soils, or water-repellent soils (Starr et al. 1978). The water enters the subjacent layer through fingers (preferential flow paths) rather than uniformly through the entire layer (Rezanezhad et al. 2006). Finger flow might sometimes occur under the following conditions: infiltration of ponded water with compression of air ahead of the wetting front, water-repellent soils, and continuous nonponding infiltration. Unstable flow refers to unstable wetting fronts that start out as horizontal wetting fronts and, under certain conditions, break into fingers or preferential flow paths as the front moves downward, similar to rain running off a sheet of glass and breaking into streams. The occurrence of preferential flow has always been reported in conjunction with pronounced soil structures such as earthworm burrows, root channels, or fractures. Recent numerical analyses, however, have shown that preferential flow can also occur in macroscopically homogeneous soils with spatially variable hydraulic properties but no pronounced macropore structure (Kung 1990; Hendrickx and Markus 2001).

On the areal scale, pipe flow is the main type of preferential flow. Funnel flow can also occur on the areal scale (Fig. 3). Pipe flow is a type of liquid flow within a closed conduit. Pipe flow might occur due to topographic depressions, calcic pipes, or karstic depressions.

The concept of preferential flow is now a hot theme in hillslope hydrology. In recent years, the hydrologic community has come to a consensus that subsurface flow is generally dominated by preferential flow of various types (Jones 2010; Lin 2010). It is a fundamentally important soil hydrologic process that controls a variety of soil physical, chemical, and biological functions.

Controlling Factors of Subsurface Stormflow

Subsurface stormflow is considered to be ubiquitous, especially in steep and humid catchments. However, the transport process of subsurface stormflow is very complex due to the spatial heterogeneity of the catchment and the spatial–temporal variability of precipitation distribution. Even in small catchments or hillslopes, the formation and transport of subsurface stormflow are often different because of the heterogeneity of the soil, vegetation, and topography. Many studies suggested that the occurrence and intensity of subsurface stormflow are mainly affected by a number of factors such as rainfall characteristics (amount, intensity, and duration of rainfall), soil characteristics (soil texture, soil thickness, soil structure, antecedent soil moisture, and fissures created by swell-shrink or freeze-thaw), biological characteristics (vegetation cover, plant root, animal burrow, and land use patterns), topography (slope, surface, and subsurface topography), and land use patterns.

Rainfall Characteristics

Rainfall characteristics (amount, intensity, and duration of rainfall) constitute a dominant control of the runoff response. Furthermore, the temporal distribution of input and patterns of evapotranspiration strongly influence the soil moisture, which represents a key factor driving hydrological processes (Jost et al. 2005; McNamara et al. 2005). Many rapid hydrological processes, such as runoff and preferential flow, are not continuous but are triggered by thresholds. The occurrence of subsurface stormflow is linked to the timing, intensity, and magnitude of rainfall events (Beven and Germann 1982). The temporal dynamics of preferential flow triggering due to between-storm and within-storm rainfall variability has been previously explored via numerical simulation approaches using synthetic rainfall time series. Many studies documented that there is a positive but nonlinearity relationship between rainfall and subsurface stormflow. Fu (2012) studied characteristics of rainfall–runoff processes in a small coastal granite catchment in the Pearl River Basin in southern China; the results showed that a large proportion of subsurface stormflow was observed only when the precipitation amount or intensity was sufficiently large, the primary direction of rainwater movement at plot scale was vertical, and rainwater was consumed primarily by evapotranspiration and infiltration into the bedrock. Tromp-van Meerveld and McDonnell (2006a) analyzed subsurface stormflow in response to 147 rainstorms at a trenched hillslope in the Panola Mountain Research Watershed between February 1996 and May 1998. They found a clear threshold response of subsurface stormflow to total storm precipitation. For storms smaller than the precipitation threshold of 55 mm, little subsurface stormflow was observed. For events exceeding the threshold, the subsurface stormflow increased by almost two orders of magnitude compared with subsurface stormflow from storms smaller than the threshold.

Furthermore, an increase in the rainfall intensity enhances macropore flow because the soil water pressure is closer to saturation during rainfall and larger macropores conduct water. Flow in a macropore of any given size is triggered only when the pressure potential exceeds its water-entry pressure. Therefore, in addition to the rain intensity, the initial soil moisture, duration of rainfall, and saturated hydraulic conductivity of the matrix determine when macropore flow occurs. Anderson (2009) found that the subsurface stormflow velocity was most closely related to the 1-hour rainfall intensity. However, some studies showed that there is little correlation between rainfall intensity and subsurface stormflow rates (Kienzler and Naef 2008). It seems that the effect of the rainfall intensity is strongly connected to other factors and can be both highly relevant and subordinate when buffered by other controls.

Soil Characteristics

The physical properties and depth of the soil are very important controls of subsurface stormflow. The soil texture can affect the subsurface stormflow direction of the soil, while the thickness of the soil layer can affect the difficulty of subsurface stormflow generation. In sandy and coarse-textured soils (with predominant sand and stones), vertical flow usually dominates. In clays and fine-textured soils, there is resistance to vertical flow, and sometimes lateral or shallow subsurface stormflow occurs quickly. If the soil layer is thick, subsurface stormflow generation needs more precipitation to infiltrate the soil, and the response might be delayed. If the soil layer is thin, it's easy for soil water to accumulate on the impermeable layer and generate subsurface stormflow.

The soil architecture is also an extremely important influencing factor of subsurface stormflow. Heterogeneous soil structure leads to complex interactions among different hydrologic processes, soil moisture dynamics, and subsurface flow and solute transport (Li et al. 2018). Fissures, cracks, or channels less likely occur, nor are important, in coarse-textured soil. In fine-textured or layered soils, fissures, cracks, or channels are more likely to occur, providing possible routes for flow and largely replacing textural voids as the main paths for unsaturated and saturated flow. Noguchi (1999) found that individual macropores that make up preferential flow are typically short but can be connected through various mechanisms over relatively long hillslope distances. Vervoort (1999) reported that preferential flow in soil profiles is related to soil structural differences (e.g., macropore flow and fractional flow) or textural differences (e.g., fingering flow and funnel flow).

The soil moisture plays an important role in the initiation and connectivity of subsurface stormflow. Many researches show that hillslope discharge was distinctly threshold-like with a nonlinear response to the soil moisture (Detty and McGuire 2010; McGuire and McDonnell 2010). Penna (2011) studied the role of soil moisture on the threshold runoff response in a small headwater catchment, and the result showed that with further increasing wetness, a moisture threshold was exceeded, resulting in a marked increase of streamflow and likely the triggering of transient subsurface stormflow on the hillslopes as suggested by the abrupt increase in runoff coefficients above the 45% soil moisture threshold and the much larger increase in runoff depth with increasing precipitation. A connection was likely established between the riparian area and hillslopes, which became hydrologically active zones. Newman (1998) found that when soils were at or near saturation (> 33% volumetric water content), a very large volume of subsurface stormflow is produced. Grayson (1997) studied the effect of the temporal variation of soil moisture on the spatial patterns of soil moisture. In their study, two spatial patterns of soil moisture were proposed. One is the lateral flow-dominated spatial pattern, and the other is the vertical flow-dominated spatial pattern. They found that the transition between two patterns likely occurs in the range of relative soil saturation of 0.6–0.8. McNamara (2005) found that the soil moisture status is related to the hillslope hydraulic connectivity: low rates of water input to the soil during winter allow dry-soil regions to persist at the soil–bedrock interface, which act as barriers to lateral flow. Once the dry-soil flow barriers are wetted, the whole-slope hydraulic connectivity is established; subsequently, lateral flow occurs, and upland soils are in direct connection with the near-stream soil moisture. Recent study showed that field capacity can be indirectly determined to indicate the potential of subsurface stormflow and quickly response to streamflow. When soil water content was above the thresholds, soil-free water content (the soil water content beyond the field capacity) started to increase. This indicated that the subsurface stormflow started to occur (Lai et al. 2018).

Furthermore, soil cracks; fissures created by seasonal drying/wetting, swelling/shrinking, and freezing/thawing; and irregular interaggregate voids form macropores in the soil due to the rainfall intensity and rates of evapotranspiration and determine the moisture budget of the soil and hillslope and affect the transport of subsurface stormflow paths and flow regimes.

Biological Characteristics

Biological characteristics belong to the most important factors in the process of soil formation and mainly include plant roots and macrofauna, which have an important effect on the soil structure, formation of macropores, and subsurface stormflow.

The structure is central to soil functioning because it controls water, gas, and nutrient fluxes and storage and therefore influences the activity and growth of living organisms. In return, plants, fauna, and microbes will affect the structure in complex ways. Vegetation affects the soil structural form and stability on different scales through various direct and indirect mechanisms. At the pore scale, roots could change soil pores size and number and stabilize soil aggregates, which determine soil water storage capacity. On the Darcian scale, roots form macropores which enhance preferential flow transport. Roots of species form macropores, even in stagnic soil horizons that ameliorate poorly structured soils and enhance infiltration and drainage and ultimately aeration. Due to the enhanced infiltration of the soil, roots form macropores, which favor preferential flow transport. On the areal scale, roots could change soil structure and porosity, which affect subsurface stormflow volume and pathway. The anchorage of roots and exudation of cementing material stabilize the soil structure. Finally, as a source of C, roots and plant residues provide a food source to the microflora and fauna, which contribute to structure formation and stabilization. In return, plant-induced changes of the structure will affect plant growth, mostly by modifying the root physical environment and water and nutrient cycles (Angers and Caron 1998). Mitchell (1995) conducted an experiment to study the effect of root systems on preferential flow in swelling soil under two cropping systems. The two crops were alfalfa (*Medicago sativa, L.*) and wheat (*Triticum turgidum, L.*), respectively, which provided sharply contrasting root systems; wheat possesses fine fibrous roots, while alfalfa has a taproot system. The result showed that decaying roots of alfalfa produce stable macropores, while wheat roots and earthworm (Lumbricus terrestris) channels do not produce such stable macropores.

Surface and Subsurface Topography Characteristics

McGuire et al. (2005) examined surface topographic controls on residence time for seven catchments (0.085–62.4 km^2) that represent diverse geologic and geomorphic conditions in the western Cascade Mountains of Oregon. The results illustrate that, compared with the area, surface topography, internal form, and structure of the basin define the first-order control on subsurface stormflow residence time. Many studies in the past assumed that hillslope flow is controlled by surface topography. However, Freer (1997) found that the flow path directions were dominant by subsurface topography. They analyze the distribution of topographic index patterns through digital terrain analysis for both surface and subsurface topography and monitor subsurface stormflow by artificial trench. The result showed that the subsurface topography (bedrock topography) is distinctly different from surficial topography and the subsurface topography (bedrock topography) has a considerable influence on local hydrological gradients and therefore the dominant flow path directions. Buttle and McDonald (2002) found that a saturated layer with thin soil may quickly develop above the lithic bedrock on forested slopes on the Canadian Shield. Haga (2005) demonstrated that saturated subsurface stormflow above the soil–bedrock interface is dominant subsurface runoff in a small catchment in the Mizugaki Research Watershed. Tromp-van Meerveld and McDonnell (2006a) proposed the fill and spill hypothesis to explain the effect of the bedrock topography on the hillslope-scale connectivity of saturated subsurface areas. These findings indicate subsurface topography of the impeding layers may play a more important role than surface topography on storm water partitioning and redistribution in moderately sloping catchments (Du et al. 2016).

In addition, the position of the measurement site on a hillslope is also a major factor determining the amount of subsurface stormflow observed. The distance from the divide is related to the amount of flow accumulated from the upslope, and the slope gradient helps to determine the rate of flow. Both the slope-profile convexity or concavity and the convergence or divergence of flow lines in plan also influence the flow rates. These topographic factors interact with the soil factors already discussed; the two sets are related via the soil catena concept and produce complex areal patterns of subsurface and overland flow.

Land Use Pattern Characteristics

Land use can change the infiltration and soil hydraulic conductivity by altering soil surface conditions and modifying the soil structure; it affects the total macroporosity, size distribution of large pores, and also their continuity (Jarvis 2007). Different land use systems and management practices affect not only the types and abundance of macropores but also the potential for subsurface stormflow. For example, the long-term use of organic amendments in arable farming produces organic-rich topsoils that can become water repellent under dry conditions and therefore are prone to fingering (Vanderborght et al. 2001). Undisturbed soils have a much higher infiltration capacity than soils in agricultural land. Heavy machinery used in agriculture exacerbates soil compaction, particularly at depth, via the development of plow pans. These compacted, dense layers have a significantly higher bulk density and lower total soil porosity than the soil directly above or below and frequently become preferential zones of runoff generation (Burt and Slattery 2005).

Monitoring of Subsurface Stormflow

Subsurface stormflow has been observed to substantially contribute to hillslope and catchment runoff. However, the direct observation of the occurrence and distribution of subsurface lateral flow on hillslopes has been difficult because of the complexity of nonuniform flow and the lack of appropriate means to directly observe flow pathways and their dynamic changes. Trenching, tracer, and geophysical techniques are three primary methods for the characterization and detection of subsurface stormflow networks that have been reported in the literature.

Trenching

In early subsurface stormflow research, physically based direct field observations, such as trenches (or pits), combined with hydrometric data were used extensively for hillslope investigations and subsurface stormflow measurements. Trenching is a method used to investigate lateral subsurface stormflow processes, where soil is excavated from the surface to a certain depth or to bedrock; the lateral flow is monitored in a cross section of the hillslope (Whipkey 1965). This conventional method is used extensively in hillslope investigations and is the only direct means of measuring subsurface stormflow. It is one of the most representative methods and might be the most intuitive and accurate method to study hillslope runoff processes. Many other methods need to be verified with trenching measurements. To understand the rainfall–runoff transformation, much advancement has been made through the use of trenches on experimental hillslopes (Freer et al. 2002). Trench studies can be used to identify subsurface stormflow sources, groundwater–stream water relations, and the role of macropores and preferential flow in subsurface stormflow generation (Tromp-van Meerveld and McDonnell 2006b). Trench studies have also demonstrated that a variety of subsurface stormflow pathways are involved and important in controlling and transporting hillslope contributions to streams. Graham (2010) used an instrumented (trenched) 0.5 ha hillslope in the southern tier of New York State, USA, to provide new insights on how variable source areas and associated flow pathways form and combine to connect rainfall with downstream water flow across a hillslope. The results showed that near-surface runoff in the upper 10 cm of the soil profile significantly contributes to the event and pre-event water (37%–62%) during storm events with wet antecedent conditions and large rainfall amounts. However, deeper subsurface water from below the fragipan layer significantly contributes (33–71%) to the total discharge from the hillslope during events with dry antecedent conditions. Thus, increasing subsurface saturation decreases the vertical percolation of water through the fragipan and the depth to the lateral subsurface stormflow occurrence; lateral flow increasingly occurs atop instead of below the fragipan layer.

Tracer Methods

In the past few decades, tracers have provided the best new insights into the age, origin, and pathway of subsurface stormflow transport. They have been widely applied in hydrological studies because of their ability to provide more detail about some of the underlying processes that control the chemical and physical behavior of elements and compounds in the natural environment. The ability to study widespread effects has generally made naturally occurring tracers more useful and more environmentally accepted than artificially introduced tracers and also more applicable to a broader range of geochemical problems (Gibson et al. 2005).

Tracer methods mainly include isotope, geochemical, and dye tracers. Isotopic tracing techniques are of particular interest in hydrological studies. They were introduced into catchment hydrology research in the 1960s as tools complementary to conventional hydrologic methods to address questions with respect to where water goes when it rains, what pathways it takes to the stream, and how long the water resides in the catchment (McDonnell 2003). Stable isotope tracers mainly refer to oxygen-18 (^{18}O) and deuterium (^{2}H or D). Hydrogen/oxygen isotopes and mixing models are often used to separate stormflow into its event and pre-event components as time source components to determine the temporal origin of subsurface stormflow. The identification of these runoff components is achieved through the use of stable isotopes; thus, these models are also referred to as isotope hydrograph separations (Inamdar 2011). Pre-event water is the main source producing the stream hydrograph peak instead of Hortonian overland flow that was first truly proven through isotope hydrograph separations. Compared with isotope tracers, geochemical tracers are utilized for hydrograph separations from geographic sources. These mixing models can be referred to as geochemical hydrograph separations (Genereux and Hooper 1998). In the 1990s, Christophersen and Hooper (1992) introduced a multivariate statistical analysis method and dramatically increased the interest in using geochemical tracers to identify subsurface stormflow sources. They used a suite of tracers (e.g., cations, anions, and silica) for mixing models, which is referred to as end-member mixing analysis (EMMA). Subsequently, chemical tracers and nonconservative isotope tracers became an effective tool for the identification of subsurface stormflow paths and resident times because of their "active" characteristics and composition changes due to the reaction between water and soil. The use of geochemical and isotopic tracers has proven to be a valuable aid in field research integrating process understanding on larger spatial scales (Buttle 1994). Moreover, tracer data, when combined with simple hydrological flow models, provide a means of estimating subsurface stormflow sources, pathways, and residence times (McGuire et al. 2002; Weiler and Naef 2003; McDonnell 2003).

Dyes are important tracers for the investigation of subsurface stormflow movement. Dye tracers have provided clues about the hydrological cycle and subsurface stormflow and transport processes. They are important tools to assess the flow pathways of such contaminants with respect to groundwater contamination often originating in the vadose zone. Dye tracers, followed by destructive sampling, have long been used by hydrologists and soil scientists to identify vertical subsurface stormflow paths and flow networks on the pedon scale. However, this method also has great limitations. The experimental procedure is complicated. It is often difficult to observe clear staining traces in the soil after long-distance transport of the dye solution when applying dye tracers to large-scale water transport studies. Therefore, it is necessary to find a better stain to trace the soil flow.

Geophysical Methods

Traditional methods have largely been destructive or based on the interpolation between point-based measurements (Sidle et al. 2001; Freera et al. 2004). As a result, transient phenomena or the spatially varying occurrence of subsurface preferential flow could not be easily and precisely determined. In recent years, with the increasing use of nondestructive microscopic imaging techniques in the laboratory (such as X-ray computed tomography, CT; and nuclear magnetic resonance, NMR), widespread use of noninvasive geophysical and remote sensing methods (such as ground-penetrating radar, GPR; electromagnetic induction, EMI; and electrical resistivity tomography, ERT), and growing arrays of sensor networks for in situ real-time monitoring (such as soil moisture and temperature monitoring networks), new opportunities emerged to better document and understand the preferential flow occurrence and its networks. Only based on such technological breakthroughs, we can determine detailed network-like flow and transport dynamics for new types of hydrologic models (Lin 2010).

On the pore and core scales, CT has been widely used to directly characterize the quantity, morphology, and continuity of macropores and related vertical preferential flow networks (Luo et al. 2008). Three-dimensional visualization and quantification of soil macropores have been performed by synthesizing a series of two-dimensional CT images (Pierret et al. 2002). Sequential scans with and without a tracer allow the determination of active flow networks. Pierret (1999) applied medical CT to isolate the macropore domains from the soil matrix and monitored the solute transport within an intact soil column. Luo (2008) showed that connected macropore networks could extend throughout soil columns with a diameter of 10 cm and length of

30 cm. However, the CT methods are limited to laboratory studies using intact soil cores or columns.

On the Darcian to areal scales, geophysical methods, such as GPR (Holden 2004), EMI (Zhu et al. 2010), and ERT (Ward et al. 2010), have been used to observe subsurface stormflow networks in situ. The GPR refers to a range of electromagnetic techniques designed primarily for the location of objects or interfaces buried beneath the Earth's surface or located within a visually opaque structure (Daniels 1996). Electromagnetic waves are transmitted to detect changes of physical properties within the shallow subsurface; GPR has become a popular geophysical tool for the detection of subsurface features (Poisson et al. 2009). It can be applied either directly or indirectly to map lateral preferential flow in situ. Direct GPR is an effective method to monitor real-time subsurface stormflow dynamics. Indirect GPR can be used to identify subsurface cavities (such as soil pipes and macropores). Lateral preferential flow pathways can be inferred from the hydrological connectivity of lateral subsurface pipes or macropores (Guo et al. 2014). Overmeeren (1997) tested GPR against capacitance probes and showed that this technology accurately measures the volumetric subsurface water content without excavation or other disturbances. Yoder (2001) used both EMI and GPR to identify locations in an agricultural field, where soil and bedrock features were conducive to promoting lateral preferential flow, although measurements of actual water fluxes or stores were not carried out. Gormally (2011) used GPR technology to detect and map the three-dimensional structure of subsurface lateral macropore networks at a Mid-Atlantic riparian wetland field study site. The results showed that the temporal and spatial variations of the stream discharge and nitrate flux are due to preferential flow mechanisms. Depth-weighted soil electrical conductivity (EC) measurements obtained using various EMI meters have been widely used to map different soil and hydrologic properties (James et al. 2003; Sherlock and McDonnell 2003; Robinson et al. 2009). Zhu (2010) demonstrated that repeated EMI surveys are useful in capturing the dynamics of soil moisture change and related subsurface stormflow paths in the landscape. The results indicated a significantly higher EC in areas close to subsurface stormflow paths on an agricultural hillslope, especially during wetter periods. Ward (2010) used ERT in a combined tracer and monitoring approach to identify subsurface stormflow paths in the hyporheic zone in a first-order stream meander in the Leading Ridge Watershed in central Pennsylvania. They applied a salt tracer to stream water and used ERT to monitor fluxes through an abandoned stream bend. The above-cited studies show the potential of geophysical techniques in probing the subsurface and parameterizing subsurface stormflow networks. Further work is needed to both identify new geophysical methods and determine how spatial and temporal resolution limitations of EMI and GPR can be surmounted to pinpoint subsurface stormflow networks.

Great achievements have been made in hydrology due to the application of tracer and geophysical techniques. However, each method has advantages and disadvantages. It is necessary to clearly research objectives and design experiments together with the advantages and disadvantages of various methods. In general, trenching is the only direct means of measuring subsurface stormflow. Many other methods need to be verified with trenching measurements. However, trenching has the disadvantage of (1) destroying the soil architecture, interrupting the hydrograph, distorting the net of the hydraulic potential on the slope, and altering the water flux and (2) only collecting saturated subsurface stormflow. Tracer-based approaches have the advantage of providing more process-based information about temporal and geographic sources of runoff which do not disturb the hydrologic system (Weiler and Naef 2003). But affected by different instruments and operating methods, there is a high variability in test results. Recently, geophysical technology has been widely used for its nondestructive and easy-to-measure features. Geophysical technology also can measure 2D or 3D network of soil architecture or subsurface stormflow. However, the technology still has the following disadvantages: first, it requires special expertise and interpretation often difficult, and, second, instruments are usually very expensive. Finally, instrument parameters and results need to destructive sampling test (Allaire et al. 2009).

Model Simulation

Models are convenient and useful tools in subsurface stormflow quantitative simulation and hypothesis formulation (Weill et al. 2011). Subsurface stormflow models mainly include the following types: Darcy–Richards model, two-domain model, dual-porosity model, kinematic wave model, two-phase model, stochastic model, and numerical model.

The Darcy–Richards model is based on the continuum principle of soil water movement (also called equilibrium principle) and Darcy's law. The Darcy–Richards model studies the subsurface stormflow from a microscopic point of view and cannot directly determine the flow discharge in the soil profile at the outlet section of the slope. Based on the degree of simplification of the Richards equation in the solution process, it can be divided into one-dimensional, two-dimensional, and three-dimensional models. The Darcy–Richards model has a high precision; however, it is difficult to predict and forecast the flood of rainstorms due to the complexity of the solution process (Sloan and Moore 1984).

The two-domain model assumes that the soil consists of two domains, one representing the soil matrix and the other one representing the macropores in the soil. Depending on the different transports of water and solutes in the two domains and the exchange of water and solutes between the two domains, the two-domain model can be divided into the following types: dual-porosity model, kinematic wave model, and boundary layer flow theory. The dual-porosity model considers the macropore domain and matrix domain as pore systems, each of which is considered to be a homogeneous medium with the properties of individual flow and solute transport. It assumes that both water flow and solute transport can be described by two equations, which are coupled using a term characterizing the exchange of fluid or solutes between the two pore regions. The system consists of two flow velocities, two pressure heads, and two water contents and solute concentrations (Gerke and Genuchten 1993). The kinematic wave theory was first discussed by Lighthill and Whitham and has been widely applied to a variety of hydrological processes such as overland flow, channel flow, base flow, unsaturated flow, macropore flow, furrow flow, and preferential flow. It is a simplified description of complex flow processes, mathematically expressed by the law of conservation of mass through the continuity equation and a flux–concentration relationship. The theory includes the theory of continuous wave formation, theory of shock wave formation, and theory of continuous wave formation from shock waves (Singh 2001).

The two-phase model was first discussed by Hosang (1993). The model divides the generation of preferential flow into two stages. The first phase represents times of redistribution and low infiltration, which does not generate preferential flow. Times of heavy infiltration, inducing preferential flow, are handled by the second phase. The Richards equation is applied to both phases as the governing flow equation, and preferential flow is allowed by increasing the hydraulic conductivity of the soil during the second phase. The two-phase model is simple and easy to understand.

The stochastic model uses the stochastic mathematical method to describe the transport of preferential flow. Because macropores are randomly distributed in the soil, the stochastic model can be used to better describe the transport of water and solutes in the soils with macropores. In the numerical model, the flow paths through which the water and solutes flow are essentially in vertical direction, with gravity being the main driving force. The various flow paths are considered to be connected to each other or grouped together. The pore size groupings are established by introducing a piecewise linear conductivity function. It consists of a number of lines tangent to the conductivity function in the "Darcy flow region," with the slope of each consecutive tangent line always being an integer multiple of the preceding line. Macropores are simulated by adding one or more pore groupings above the region, where Darcy's Law is valid (Steenhuis et al. 1990).

Connecting Subsurface Stormflow with Ecohydrology Processes

There is an increasing recognition of connecting aboveground ecohydrology and belowground subsurface stormflow in recent years (Li et al. 2009). The ecosystem structure influences the spatial and temporal dynamics of water flowing across the terrain surface and subsurface. Hydrology influences the rates of transport, deposition, and recycling of inorganic materials, which, in turn, influence the spatial and temporal dynamics of ecosystem processes. Thus, detailed studies of subsurface stormflow can help us to better understand ecohydrological processes (Pcovich 2005).

The discharge, infiltration, and residence times of subsurface stormflow are all influenced by the ecosystem. Aboveground vegetation and surface flow networks appear to have a close connection to belowground root and subsurface stormflow networks. Vegetation can be considered as a key link connecting aboveground and belowground. The actual pathways from rainfall to streamflow usually involve a combination of surface and subsurface stormflow, and many of them are related to biological factors including the canopy, stem, and root. Aboveground, vegetation canopies can redistribute incident precipitation and affect the vertical and horizontal spatial distribution of water within the plant community by interception, stemflow, and throughfall. Belowground, plant root systems create networks of subsurface stormflow and thus can alter subsurface stormflow and nutrient transport paths in the vadose zone. Subsurface stormflow can occur in channels formed by dead or decaying roots, channels formed by decayed roots that are newly occupied by living roots, and channels formed around live roots (Ghestem et al. 2011). The subsurface stormflow volume, flow paths, and residence times can affect the sediment transport, nutrient cycling, and biological productivity and ecosystem dynamics. Li (2009) confirmed that root channels are the preferred pathways for most stemflow water moved into the soil and stemflow is conducive to concentrate and store water in the deeper layers of the soil profiles, creating favorable soil water conditions for plants to survive droughts under arid conditions.

Recently, a study on global warming, plant growth seasons, and terrestrial carbon sinks has highlighted the importance of the terrain and deeper subsurface geophysical structure in controlling plant water availability (Hu et al. 2010). Hu (2010) used observations of stable hydrogen isotopes (δD) of snow, rain, and extracted xylem water from three dominant tree species. The results showed that the trees relied heavily on snowmelt, even late into the growing season. By coupling the model, they found that the annual forest carbon uptake highly depends on snow water; the annual gross primary productivity was estimated to be 67%, 77%, and 71%, respectively, depending on snowmelt. Their isotope results suggest that rain events typically hydrate only the upper soils and rarely reach depths between 30 and 35 cm; however, all three dominant tree species access water deeper than 35 cm. They suggested that the lower reliance on summer rain could be attributed to two reasons: (1) the trees were tapping into a deeper and more consistent water source during the growing season; and (2) the rain events were too small to penetrate deeply into the soils. They then suggested that global warming induces longer growing seasons, which could lead to less carbon sequestration in a subalpine forest because of less available snowmelt resources. Precipitation patterns and temperatures have been affected by global climate change. Changes in climate factors, such as precipitation patterns and temperature, will in turn affect the runoff patterns (surface and subsurface flow), the plant water use sources, and the dynamics of vegetation. This provides new insights into the mutual feedback mechanism of the subsurface stormflow and ecohydrological processes.

Some studies suggested that the patterns in crop development are often directly related to lateral and vertical changes in soil texture, causing changes of the available water and resource supply for plant growth. Hence, better detection, delineation, and quantification of the subsurface variability are needed (Rudolph et al. 2015). Rudolph (2015) used multi-receiver EMI data and large-scale multi-temporal and multispectral satellite imagery in conjunction with selected ERT transects and conventional soil sampling to investigate the lateral and vertical changes of soil properties that influence the spatial and temporal variability of crop performance in arable fields [satellite-derived leaf area index (LAI) patterns]. For the first time, the LAI patterns with better crop performance in large-scale multi-temporal satellite imagery were interpreted by proximal soil sensor data; a higher water storage capacity and increased correlations between the large-offset EC data and subsoil clay content and soil profile depth imply that the subsoil is mainly responsible for better crop development in drought periods along the buried paleo-river structure; and stagnant water in the subsoil indicates that this paleo-river structure still plays an important role in subsurface hydrology.

Future Research Directions

Subsurface flow research aims to address questions about the flow source, pathways, resident times, dominant control factors, and deciphering runoff mechanisms across diverse hillslopes and catchments. Current research and challenge of subsurface flow are mainly focus on (1) subsurface flow (preferential flow) initiation condition, (2) observation methods for the detection and quantification of preferential flow, (3) fist-order control of flow generation, (4) thresholds and nonlinearity of the subsurface flow process, (5) transit time distribution and pathway, and (6) interactions among hydraulic conductivity distributions, subsurface topography, soil architecture, and subsurface flow transport processes.

Up until recently, it has been difficult to derive general hydrologic principles from single research studies within intensively studied small basins. Future research of subsurface flow will need to focus on the following aspects: first, we need to continue process-based field experimental observations to better understand subsurface flow transport processes; second, we need to combine tracers, geophysical or other new techniques (such as GPR, CT, or laser spectrometer technologies), to obtain high-precision and systematic field datasets in time and space, especially from different types of catchments and environments, to better characterize the soil architecture and hydrological process; third, we need to extract commonalities from multiple different well-instrumented sites to define the first-order controls of subsurface stormflow generation, flow paths, and residence time; and fourth, we need more stringent tests of models; assess the uncertainty of the model; capture the mechanisms of subsurface stormflow transport processes, transport paths, resident times, key thresholds, hysteresis, and other nonlinearity behaviors through high-precision and systematic field datasets to test old hypotheses; formulate new hydrologic theories; and provide feedback for catchment water resource management and prediction.

Acknowledgments The study was financially supported by the National Natural Science Foundation of China (NSFC 41730854 and 91425301) and the PCSIRT (IRT_15R06), and projects were supported by the State Key Laboratory of Earth Surface Processes and Resource Ecology and by Ten Thousand Talent Program for leading young scientist.

References

S.E. Allaire, S. Roulier, A.J. Cessna, Quantifying preferential flow in soils: a review of different techniques. J. Hydrol. **378**(1–2), 179–204 (2009)

S.P. Anderson, W.E. Dietrich, D.R. Montgomery, R. Torres, M.E. Conrad, K. Loague, Subsurface flow paths in a steep, unchanneled catchment. Water Resour. Res. **33**(12), 2637–2653 (1997)

A.E. Anderson, M. Weiler, Y. Alila, R.O. Hudson, Subsurface flow velocities in a hillslope with lateral preferential flow. Water Resour. Res. **45**, W11407 (2009)

D.A. Angers, J. Caron, Plant-induced changes in soil structure: processes and feedbacks. Biogeochemistry **42**(1–2), 55–72 (1998)

T.C. Atkinson, Techniques for measuring subsurface flow on hillslopes, in Hillslope Hydrology, ed. by M. J. Kirkby (Wiley, Hoboken, 1978), pp. 73–78

K. Beven, P. Germann, Macropores and water-flow in soils. Water Resour. Res. **18**(5), 1311–1325 (1982)

T.P. Burt, M.C. Slattery, in *Land Use and Land Cover Effects on Runoff Processes: Agricultural Effects*, ed. by M. G. Anderson (Wiley, Hoboken, 2005), pp. 1805–1811

J.M. Buttle, Isotope hydrograph separations and rapid delivery of pre-event water from drainage basins. Prog. Phys. Geogr. **18**(1), 16–41 (1994)

J.M. Buttle, D.J. McDonald, Coupled vertical and lateral preferential flow on a forested slope. Water Resour. Res. **38**(5), 18–1–18–16 (2002)

N. Christophersen, R.P. Hooper, Multivariate analysis of stream water chemical data: the use of principal components analysis for the end-member mixing problem. Water Resour. Res. **28**(1), 99–107 (1992)

D.J. Daniels, *Ground Penetrating Radar*, 2nd edn. (Institution of Electrical Engineers, London, 1996)

J.M. Detty, K.J. McGuire, Threshold changes in storm runoff generation at a till-mantled headwater catchment. Water Resour. Res. **46**, W07525 (2010)

E.H. Du, C.R. Jackson, J. Klaus, J.J. McDonnell, N.A. Griffiths, M.F. Williamson, J.L. Greco, M. Bitew, Interflow dynamics on a low relief forested hillslope: lots of fill, little spill. J. Hydrol. 534, 648–658 (2016)

T. Dunne, Hydrology mechanics, and geomorphic implications of erosion by subsurface flow. Geol. Soc. Am. **252**, 1–28 (1990)

A. Engler, Untersuchungen über den Einfluss des Waldes auf den Stand der Gewässer. Kommissionsverlag von Beer and Cie, Zürich, (1919), pp. 626. in Wieler et al. 2005.

G.A. Fox, G.V. Wilson, The role of subsurface flow in hillslope and stream bank erosion: a review. Soil Sci. Soc. Am. J. **74**(3), 717–733 (2010)

J. Freer, J.J. Mcdonnell, K.J. Beven, D. Brammer, D. Burns, R.P. Hooper, C. Kendal, Topographic controls on subsurface storm flow at the hillslope scale for two hydrologically distinct small catchments. Hydrol. Process. **11**, 1347–1352 (1997)

J. Freer, J.J. McDonnell, K.J. Beven, N.E. Peters, D.A. Burns, R.P. Hooper, B. Aulenbachand, C. Kendall, The role of bedrock topography on subsurface storm flow. Water Resour. Res. **38** (12), 5–1–5–16 (2002)

J.E. Freera, H. McMillanb, J.J. McDonnell, K.J. Beven, Constraining dynamic TOPMODEL responses for imprecise water table information using fuzzy rule based performance measures. J. Hydrol. **291**, 254–277 (2004)

R.A. Freeze, Role of subsurface flow in generating surface runoff 2, upstream source areas. Water Resour. Res. **8**(5), 1272–1283 (1972)

K. Frohlich, W. Frohlich, H. Wittenberg, Determination of ground-water recharge by baseflow separation – regional analysis in Northeast China. 2nd international conference on flow regimes from international experimental and network data 221, 69–75 (1994)

C. Fu, J. Chen, S. Zeng, Observation and analysis of rainfall-runoff characteristics in a coastal granite catchment in southern China. J. Hydrol. Eng. **17**(1), 138–149 (2012)

D.P. Genereux, R.P. Hooper, Oxygen and hydrogen isotopes in rainfall-runoff studies, in *Isotope Tracers in Catchment Hydrology*, ed. by C. Kendall, J. J. McDonnell (Elsevier, Amsterdam, 1998), pp. 319–346

H.H. Gerke, Preferential flow descriptions for structured soils. Soil Sci. Plant Nutr. **169**(3), 382–400 (2006)

H.H. Gerke, M.T. Genuchten, A dual-porosity model for simulating the preferential movement of water and solutes in structured porous media. Water Resour. Res. **29**(2), 305–319 (1993)

M. Ghasemizade, M. Schirmer, Subsurface flow contribution in the hydrological cycle: lessons learned and challenges ahead-a review. Environ. Earth Sci. **69**(2), 707–718 (2013)

M. Ghestem, R.C. Sidle, A. Stokes, The influence of plant root systems on subsurface flow: implications for slope stability. Bioscience **61**(11), 869–879 (2011)
J.J. Gibson, T.W.D. Edwards, S.J. Birks, N.A. St Amour, W.M. Buhay, P. McEachern, B.B. Wolfe, D.L. Peters, Progress in isotope tracer hydrology in Canada. Hydrol. Process. **19**(1), 303–327 (2005)
K.H. Gormally, M.S. McIntosh, A.N. Mucciardi, G.W. McCarty, Ground-penetrating radar detection and three-dimensional mapping of lateral macropores: II. Riparian application. Soil Sci. Soc. Am. J. **75**(4), 1236 (2011)
C.B. Graham, R.A. Woods, J.J. McDonnell, Hillslope threshold response to rainfall: (1) a field based forensic approach. J. Hydrol. **393**(1–2), 65–76 (2010)
R.B. Grayson, A.W. Western, F. Chiew, G. Bloschl, Preferred states in spatial soil moisture patterns: local and nonlocal controls. Water Resour. Res. **33**(12), 2897–2908 (1997)
W.H. Green, G.A. Ampt, Studies on soil physics. I. The flow of water and air through soils. J. Agric. Sci. **4**, 1–24 (1911)
L. Guo, J. Chen, H. Lin, Subsurface lateral preferential flow network revealed by time-lapse ground-penetrating radar in a hillslope. Water Resour. Res. **50**(12), 9127–9147 (2014)
H. Haga, Y. Matsumoto, J. Matsutani, M. Fujita, K. Nishida, Y. Sakamoto, Flow paths, rainfall properties, and antecedent soil moisture controlling lags to peak discharge in a granitic unchanneled catchment. Water Resour. Res. **41**, W12410 (2005)
M.H. Hendrickx, F. Markus, Uniform and preferential flow mechanisms in the vadose zone, in *Conceptual Models of Flow and Transport in the Fractured Vadose Zone*, ed. by P.A. Hsieh (National Academy Press, Washington, 2001), pp. 149–198
J. Holden, Hydrological connectivity of soil pipes determined by ground-penetrating radar tracer detection. Earth Surf. Process. Landf. **29**(4), 437–442 (2004)
J. Hosang, Modeling preferential flow of water in soils – a two phase approach for field condition. Geoderma **58**, 149–163 (1993)
J. Hu, D.J. Moore, S.P. Burns, R.K. Monson, Longer growing seasons lead to less carbon sequestration by a subalpine forest. Glob. Chang. Biol. **16**(2), 771–783 (2010)
C.R. Hursh, E.F. Brater, Separating storm-hydrographs from small drainage areas into surface and subsurface flow. Transactions of the American Geophysical Union. 22, 863–871 (1941)
S. Inamdar, in *The Use of Geochemical Mixing Models to Derive Runoff Sources and Hydrologic Flow Paths*, ed. by D. F. Levia (Springer, Dordrecht/Heidelberg/London/New York, 2011), pp. 163–181
I.T. James, T.W. Waine, R.I. Bradley, J.C. Taylor, R.J. Godwin, Determination of soil type boundaries using electromagnetic induction scanning techniques. Biosyst. Eng. **86**(4), 421–430 (2003)
N.J. Jarvis, A review of non-equilibrium water flow and solute transport in soil macropores: principles, controlling factors and consequences for water quality. Eur. J. Soil Sci. **58**(3), 523–546 (2007)
N. Jarvis, J. Koestel, M. Larsbo, Understanding preferential flow in the vadose zone: recent advances and future prospects. Vadose Zone J. **15**(12), 1–11 (2016)
J.A. Jones, Soil piping and catchment response. Hydrol. Process. **24**(12SI), 1548–1566 (2010)
G. Jost, G. Heuvelink, A. Papritz, Analysing the space-time distribution of soil water storage of a forest ecosystem using spatio-temporal kriging. Geoderma **128**(3–4), 258–273 (2005)
P.M. Kienzler, Experimental study of subsurface stormflow formation combining tracer, hydrometric and geophysical techniques. Albert Ludwig University, Germany (2007)
P.M. Kienzler, F. Naef, Subsurface storm flow formation at different hillslopes and implications for the 'old water paradox'. Hydrol. Process. **22**(1), 104–116 (2008)
K.J.S. Kung, Preferential flow in a sandy vadose zone. 1. Field observations. Geoderma **46**, 51–58 (1990)
X.M. Lai, Z.W. Zhou, Q. Zhua, K.H. Liao, Comparing the spatio-temporal variations of soil water content and soil free water content at the hillslope scale. Catena **160**, 366–375 (2018)
B. Lehner, P. Doll, J. Alcamo, T. Henrichs, F. Kaspar, Estimating the impact of global change on flood and drought risks in europe: a continental, integrated analysis. Clim. Chang. **75**(3), 273–299 (2006)
X.Y. Li, Z. Yang, Y. Li, H. Lin, Connecting ecohydrology and hydropedology in desert shrubs: stemflow as a source of preferential flow in soils. Hydrol. Earth Syst. Sci. **13**(7), 1133–1144 (2009)

X.Y. Li, H. Lin, H.H. Gerke, Frontiers in hydropedology: interdisciplinary research from soil architecture to the critical zone. Vadose Zone J. (2018). https://doi.org/10.2136/vzj2018.03.0045
H. Lin, Linking principles of soil formation and flow regimes. J. Hydrol. **393**(1–2SI), 3–19 (2010)
H. Lin, X. Zhou, Evidence of subsurface preferential flow using soil hydrologic monitoring in the Shale Hills catchment. Eur. J. Soil Sci. **59**(1), 34–49 (2008)
L. Luo, H. Lin, P. Halleck, Quantifying soil structure and preferential flow in intact soil using x-ray computed tomography. Soil Sci. Soc. Am. J. **72**(4), 1058 (2008)
J.J. McDonnell, Where does water go when it rains? Moving beyond the variable source area concept of rainfall-runoff response. Hydrol. Process. **17**(9), 1869–1875 (2003)
K.J. McGuire, J.J. McDonnell, Hydrological connectivity of hillslopes and streams: characteristic time scales and nonlinearities. Water Resour. Res. **46**, W10543 (2010)
K.J. McGuire, D.R. DeWalle, W.J. Gburek, Evaluation of mean residence time in subsurface waters using oxygen-18 fluctuations during drought conditions in the mid-Appalachians. J. Hydrol. **261**, 132–149 (2002)
K.J. McGuire, J.J. McDonnell, M. Weiler, C. Kendall, B.L. McGlynn, J.M. Welker, J. Seibert, The role of topography on catchment-scale water residence time. Water Resour. Res. **41**, 1-14 (2005)
J.C. McIntosh, C. Schaumberg, J. Perdrial, A. Harpold, A.V. Ortega, C. Rasmussen, D. Vinson, X. Zapata-Rios, P.D. Brooks, T. Meixner, J. Pelletier, L. Derry, J. Chorover, Geochemical evolution of the critical zone across variable time scales informs concentration-discharge relationships: Jemez River Basin Critical Zone Observatory. Water Resour. Res. **53**, 4169–4196 (2017)
J.P. McNamara, D. Chandler, M. Seyfried, S. Achet, Soil moisture states, lateral flow, and streamflow generation in a semi-arid, snowmelt-driven catchment. Hydrol. Process. **19**(20), 4023–4038 (2005)
A.R. Mitchell, T.R. Ellsworth, B.D. Meek, Effect of root systems on preferential flow in swelling soil. Commun. Soil Sci. Plant Anal. **26**(15–16), 2655–2666 (1995)
B.D. Newman, A.R. Campbell, B.P. Wilcox, Lateral subsurface flow pathways in a semiarid ponderosa pine hillslope. Water Resour. Res. **34**(12), 3485–3496 (1998)
S. Noguchi, Y. Tsuboyamaa, R.C. Sidle, I. Hosoda, Morphological characteristics of macropores and the distribution of preferential flow pathways in a forested slope segment. Soil Sci. Soc. Am. J. **63**(5), 1413–1423 (1999)
R.A. Overmeeren, S.V. Sariowan, J.C. Gehrels, Ground penetrating radar for determining volumetric soil water content; results of comparative measurements at two test sites. J. Hydrol. **197**(1), 316–338 (1997)
R.A. Payn, M.N. Gooseff, B.L. McGlynn, K.E. Bencala, S.M. Wondzell, Exploring changes in the spatial distribution of stream baseflow generation during a seasonal recession. Water Resour. Res. **48**, W04519 (2012)
A. Pcovich, in *Ecosystem Processes*, ed. by M. G. Anderson (Wiley, Hoboken, 2005), pp. 1535–1537
D. Penna, H.J. Tromp-Van Meerveld, A. Gobbi, M. Borga, G.D. Fontana, The influence of soil moisture on threshold runoff generation processes in an alpine headwater catchment. Hydrol. Earth Syst. Sci. **15**, 689–702 (2011)
A. Pierret, Y. Capowiez, C.J. Moran, A. Kretzschmar, X-ray computed tomography to quantify tree rooting spatial distributions. Geoderma **90**(3–4), 307–326 (1999)
A. Pierret, Y. Capowiez, L. Belzunces, C.J. Moran, 3D reconstruction and quantification of macropores using X-ray computed tomography and image analysis. Geoderma **106**(3–4), 247–271 (2002)
J. Poisson, M. Chouteau, M. Aubertin, D. Campos, Geophysical experiments to image the shallow internal structure and the moisture distribution of a mine waste rock pile. J. Appl. Geophys. **67**(2), 179–192 (2009)
F. Rezanezhad, H.J. Vogel, K. Roth, Experimental study of fingered flow through initially dry sand. Hydrol. Earth Syst. Sci. Discuss. **3**(4), 2595–2620 (2006)
D.A. Robinson, I. Lebron, B. Kocar, K. Phan, M. Sampson, N. Crook, S. Fendorf, Time-lapse geophysical imaging of soil moisture dynamics in tropical deltaic soils: an aid to interpreting hydrological and geochemical processes. Water Resour. Res. **45**, W00D32 (2009)
S. Rudolph, J. Kruka, C. Hebela, M. Alia, M. Herbsta, C. Montzkaa, S. Pätzoldb, D.A. Robinsonc, H. Vereeckena, L. Weihermüllera, Linking satellite derived LAI patterns with subsoil

heterogeneity using large-scale ground-based electromagnetic induction measurements. Geoderma **241–242**, 262–271 (2015)
M.D. Sherlock, J.J. McDonnell, A new tool for hillslope hydrologists: spatially distributed groundwater level and soilwater content measured using electromagnetic induction. Hydrol. Process. **17**(10), 1965–1977 (2003)
R.C. Sidle, S. Noguchi, Y. Tsuboyama, K. Laursen, A conceptual model of preferential flow systems in forested hillslopes: evidence of self-organization. Hydrol. Process. **15**(10), 1675–1692 (2001)
V.P. Singh, Kinematic wave modelling in water resources: a historical perspective. Hydrol. Process. **15**(4), 671–706 (2001)
P.G. Sloan, I.D. Moore, Modeling subsurface stormflow on steeply sloping forested watersheds. Water Resour. Res. **20**(12), 1815–1822 (1984)
J.L. Starr, H.C. Deroo, C.R. Frink, J.Y. Parlange, Leaching characteristics of a layered field soil. Soil Sci. Soc. Am. J. **42**(3), 386–391 (1978)
T.S. Steenhuis, J.Y. Parlange, M.S. Andreini, A numerical-model for preferential solute movement in structured soils. Geoderma **46**(1–3), 193–208 (1990)
M. H. Stuckey, Low flow, base flow, and mean flow regression equations for Pennsylvania streams. US Geological Survey Scientific Investigations Report 2006-5130 (2006)
D. G. Tarboton (ed.), *Rainfall-Runoff Processes* (Utah State University, Lohan, 2003)
H.J. Tromp-van Meerveld, J.J. McDonnell, Threshold relations in subsurface stormflow: 1. A 147-storm analysis of the Panola hillslope. Water Resour. Res. **42**, W02410 (2006a)
H.J. Tromp-van Meerveld, J.J. McDonnell, Threshold relations in subsurface stormflow: 2. The fill and spill hypothesis. Water Resour. Res. **42**, W024112 (2006b)
D.J. Turton, E.L. Miller, Subsurface flow responses of a small forested catchment in the Ouachita mountains. Hydrol. Process. **6**, 111–125 (1992)
J. Vanderborght, M. Vanclooster, A. Timmerman, P. Seuntjens, D. Mallants, D.J. Kim, D. Jacques, L. Hubrechts, C. Gonzalez, J. Feyen, J. Diels, J. Deckers, Overview of inert tracer experiments in key Belgian soil types: Relation between transport and soil morphological and hydraulic properties. Water Resour. Res. **37**(12), 2873–2888 (2001)
R.W. Vervoort, D.E. Radcliffe, L.T. West, Soil structure development and preferential solute flow. Water Resour. Res. **35**(4), 913–928 (1999)
A.S. Ward, M.N. Gooseff, K. Singha, Imaging hyporheic zone solute transport using electrical resistivity. Hydrol. Process. **24**(7), 948–953 (2010)
M. Weiler and F. Naef, An experimental tracer study of the role of macropores infiltration in grassland soils. Hydrol. Process. **17**, 477 (2003)
M. Weiler, J. J McDonnell, I. Tromp-Van Meerveld, T. Uchida, in *Subsurface Stormflow*, ed. by M. G. Anderson (Wiley, Hoboken, 2005), pp. 1720–1729
S. Weill, A. Mazzia, M. Putti, C. Paniconi, Coupling water flow and solute transport into a physically-based surface-subsurface hydrological model. Adv. Water Resour. **34**(1), 128–136 (2011)
R.Z. Whipkey, Subsurface stormflow from forested slopes. Int. Assoc. Sci. Hydrol. Bull. **10**(2), 74–85 (1965)
B.P. Wilcox, B.D. Newman, D. Brandes, D.W. Davenport, K. Reid, Runoff from a semiarid ponderosa pine hillslope in New Mexico. Water Resour. Res. **33**(10), 2301–2314 (1997)
T.C. Winter, The role of ground water in generating streamflow in headwater areas and in maintaining base flow. J. Am. Water Resour. Assoc. **43**(1), 15–25 (2007)
WMO Rep, International Glossary of Hydrology, World Meteorological Organization, Geneva (2012), p. 336
R.E. Yoder, R.S. Freeland, J.T. Ammons, L.L. Leonard, Mapping agricultural fields with GPR and EMI to identify offsite movement of agrochemicals. J. Appl. Geophys. **47**(3–4), 251–259 (2001)
E. Zehe, H. Fluhler, Preferential transport of isoproturon at a plot scale and a field scale tile-drained site. J. Hydrol. **247**(1–2), 100–115 (2001)
Q. Zhu, H. Lin, J. Doolittle, Repeated electromagnetic induction surveys for improved soil mapping in an agricultural landscape. Soil Sci. Soc. Am. J. **74**(5), 1763 (2010)

Photosynthesis (NPP, NEP, Respiration)

Xufeng Wang, Haibo Wang, Xin Li, and Youhua Ran

Contents

Abstract

Carbon exchange between terrestrial ecosystems and environment is paid great attention in recent decades, because it can regulate the atmospheric carbon dioxide concentration. Photosynthesis is the key process in the carbon cycle.

X. Wang (✉) · H. Wang · Y. Ran
Heihe Remote Sensing Experimental Research Station, Northwest Institute of Eco-Environment and Resources, Chinese Academy of Sciences, Lanzhou, China
e-mail: wangxufeng@lzb.ac.cn; whb@lzb.ac.cn; ranyh@lzb.ac.cn

X. Li
Institute of Tibetan Plateau Research, Chinese Academy of Sciences, Beijing, China

CAS Center for Excellence in Tibetan Plateau Earth Sciences, Chinese Academy of Sciences, Beijing, China
e-mail: xinli@itpcas.ac.cn

X. Li, H. Vereecken (eds.), *Observation and Measurement of Ecohydrological Processes*, Ecohydrology, https://doi.org/10.1007/978-3-662-48297-1_10

GPP, NPP, and NEP are key variables in carbon cycle study. Thus, accurate estimation of these carbon fluxes is important for understanding the interactions between terrestrial ecosystems and atmosphere. In this chapter, we introduce measuring methods of these carbon fluxes at field scale and estimating models of these carbon fluxes based on remote sensing data at regional or global scale. The processes and key questions in these methods or models are specifically analyzed.

Keywords
Photosynthesis · Remote sensing · Gross primary production · Carbon cycle · Light using efficiency model

Introduction

Carbon exchange between terrestrial ecosystems and environment is paid great attention in recent decades, because it can regulate the carbon dioxide concentration in the atmosphere. In recent 50 years, 25–30% of human fossil-released CO_2 was absorbed by terrestrial ecosystems (Le Quere et al. 2009). Terrestrial ecosystems absorb CO_2 from the atmosphere through vegetation photosynthesis and release CO_2 to the atmosphere through vegetation and soil respiration. Gross primary production (GPP) is the total amount of carbon fixed in the process of photosynthesis by plants in an ecosystem in a given length of time. Net primary production (NPP) is defined as GPP minus the amount of carbon respired by plants themselves in autotrophic respiration. It is the net production of organic carbon by plants in an ecosystem usually measured over a given period (usually a year) (Kirschbaum et al. 2001). Net ecosystem production (NEP) is GPP reduced by losses resulting from the respiration of the plants (autotrophic respiration) and respiration of litter and soil microbe (heterotrophic respiration) (Kirschbaum et al. 2001).

To better understand the role of terrestrial ecosystem in the global carbon cycle, GPP, NPP, NEP, and respiration of terrestrial ecosystem need to be accurately estimated. The widely used methods in estimating regional terrestrial carbon fluxes are (1) remote sensing-based light use efficiency models, such as MOD17 algorithm (Running et al. 2000), VPM (Xiao et al. 2004), C-FIX (Veroustraete et al. 2002), and EC-LUE (Yuan et al. 2010); (2) process-based ecosystem models, such as Biome-BGC (Running and Gower 1991), LPJ (Sitch et al. 2003), and CLM (Oleson et al. 2010); and (3) scale-up from field observation. Remote sensing-based light use efficiency models are efficient tools to estimate regional carbon fluxes, because they need few parameters and forcing data and have simple mathematic formulas. Process-based ecosystem models contain detailed carbon transfer in ecosystems, but they need numerous parameters and huge forcing data, which may result in great uncertainties in the results of the regional simulation. Estimating regional carbon fluxes from field observations, such as forest inventories and eddy covariance carbon flux measurement, is one of the typical bottom-up methods. The disadvantage of this method is that the sparse observing points cannot present the spatial variation of carbon fluxes and field work is time- and money consuming.

This chapter will mainly introduce how to accurately get regional carbon fluxes, such as GPP, NEP, and respiration, using remote sensing-based methods. In section "Introduction," we gave some general background of this chapter. In section "Field Measurements at Point Scale," we will introduce how to measure GPP, NEP, and reparation at leaf and canopy scale in the field, which can be used to validate and calibrate remote sensing-based light use efficiency model. In section "Remote Sensing-Based GPP and NEP Estimation at Regional Scale," we will introduce the widely used remote sensing models, how to get the parameters of the models, and how to use them at the regional scale. In section "Summary," we will give a summary to this chapter.

Field Measurements at Point Scale

Leaf Scale Measurement of Photosynthesis

Infrared Gas Analyzer Measurement

Leaves are photosynthetic organs that absorb light and convert the photon energy of light to chemical energy for use in CO_2 assimilation. Photosynthesis is a process used by plants and other organisms to convert light energy into chemical energy, which is the primary physiological process that drives plant growth and crop productivity and influences many other processes. Photosynthesis and respiration of terrestrial plants are two opposite but interdependent metabolic pathways (Raghavendra and Padmasree 2003). They are main forms of gas exchange between leaf and atmosphere and important components of carbon-water exchange between terrestrial ecosystems and atmosphere.

The infrared gas analyzer (IRGA) is one of the most frequently used methods to measure gas exchange between leaf and atmosphere. The IRGA estimated the gas density by using the different absorption efficiencies in CO_2 and H_2O to infrared ray, which measures the decrements of gas when it passed through the infrared rays. The open portable IRGA system Li-6400/XT from LI-COR Inc. (LI-COR Biosciences 2011) is a widely used analyzer for photosynthesis measurements, which measures photosynthesis and transpiration based on the differences of CO_2 and H_2O in an air stream that flows through the leaf cuvette. The LI-6400 improves upon traditional open systems by having the gas analyzers in the sensor head. As an open system, the gas is pumped into the analyzer, and then it is divided into two paths, 25% of the gas flows into the referenced cuvette, while the other part enters the sample cuvette, which estimated the gas flow rates by the gas analyzer in the sample chamber. The IRGA measures the absorption rates of IR light in a certain wavelength region. This absorption is dependent on the molecule density of the molecule species absorbed in this wavelength region. The photosynthesis rate is calculated from the differences of CO_2 densities between the reference cuvette and the sample cuvette; considering the other variables, such as gas flow rates and leaf area indexes, the system can compute the leaf-level CO_2 net assimilation rates (photosynthesis rate and respiration rate), transpiration rate, stomatal conductance, intercellular CO_2 concentration, etc.

The LI-6400 provides mechanisms for modifying the incoming air CO_2 and H_2O concentrations. There are chemical tubes for scrubbing CO_2 and H_2O, and air can be diverted through these tubes in any proportion desired. The 6400-01 CO_2 Mixer option is specified as having an upper range of 2000 μmol mol^{-1}, and the lower limit of control is typically 40 or 50 μmol mol^{-1}. This range can be adjusted by varying the pump speed (LI-COR Biosciences 2011).

With the LI-6400 analyzer, the carbon assimilation rates (A) in relation to intercellular CO_2 concentration (A-Ci response curves) can be measured, which can provide mechanistic information on the underlying biochemical limitations of carbon assimilation that can vary by genotype (Stinziano et al. 2017), and can be easily assimilated into vegetation models (Farquhar et al. 1980). The A-Ci curves can also provide estimates of V_{cmax} and J_{max}, along with other valuable information on carbon fluxes in the plant (Sharkey et al. 2007). As the development of technique, the new instrumentation LI-6800 can obtain measures of V_{cmax} and J_{max} in ~5 min, or possibly even faster, which will allow for more rapid screening of crops, mutants, and populations of plants in natural environments, bringing gas exchange into the phenomic era (Stinziano et al. 2017).

Leaf-Level Gas Exchange Modeling

Leaf gas exchange rates vary depending upon environmental conditions such as light, temperature, CO_2 concentration, and humidity. Advances in photosynthesis physiology and modeling studies improved our understanding of environmental responses of leaf-level gas exchange rates. The steady-state biochemical model of CO_2 assimilation considers photosynthetic metabolism as a composite of several processes, namely, RuBP carboxylation, electron transport, and regeneration. The net CO_2 assimilation is a balance between the carboxylation of ribulose-1,5-bisphosphate (RuBP) catalyzed by ribulose-1,5-bisphosphate carboxylase/oxygenase (Rubisco) and the release of CO_2 by photorespiration and mitochondrial respiration.

The model of Farquhar et al. (1980) ("Farquhar model") has the most widespread use over the past three decades in simulation studies of the response of C_3 photosynthesis to environmental change because the model is mechanistic and capable of describing the underlying processes that affect photosynthesis (Sharkey et al. 2007). Many studies demonstrated that the Farquhar model accurately describes leaf-level photosynthesis and that its parameters have intuitive ecological meaning and have been incorporated into a number of ecosystem models as a mechanistic basis describing the photosynthetic response to acclimation across a range of scales (Hanson et al. 2004; Thornton et al. 2002). The carbon assimilation rates are limited by the minimization of three processes: (1) the maximum ribulose-1,5-bisphosphate carboxylase/oxygenase (Rubisco) activity with respect to carboxylation (Rubisco limitation); (2) the electron transport rate, limiting the regeneration of ribulose-1,5-bisphosphate (RuBP limitation); and (3) the regeneration of RuBP controlled by the rate of triose phosphate utilization (TPU) (TPU limitation) .

The CO_2 response curves were fitted based on the biochemically based model of CO_2 assimilation originally published by Farquhar et al. (1980). According to the biochemical model, the net CO_2 assimilation rate (P_N[μmol (CO_2) m^{-2} s^{-1}]) is limited by three processes:

$$P_N = \min\{P_c, P_j, P_p\} \tag{1}$$

where P_N [μmol m^{-2} s^{-1}] is the net photosynthesis rate, P_c [μmol m^{-2} s^{-1}] is the Rubisco carboxylation-limited photosynthesis rate, P_j[μmol m^{-2} s^{-1}] is the RuBP regeneration-limited photosynthesis rate, and $P_p = 3\text{TPU}$ is the rate at which triose phosphate utilization (TPU) is limited. The first limiting process is

$$P_c = \frac{(C_c - \Gamma_*)\cdot V_{c\max}}{C_c + K_c\cdot(1 + O/K_o)} - R \tag{2}$$

where $V_{c\max}$[μmol m^{-2} s^{-1}] is the maximum rate of Rubisco activity [μmol mol^{-1}], Γ^* [μmol mol^{-1}] is the photosynthetic CO_2 compensation point, K_c [μmol mol^{-1}] is the Michaelis-Menten constant of Rubisco for CO_2, K_o [mmol mol^{-1}] is the Michaelis-Menten constant of Rubisco for O_2, C_c and O are the partial pressures of CO_2 and O_2 at the Rubisco reaction rates (Pa), and R_D [μmol m^{-2} s^{-1}] is the rate of mitochondrial respiration. The second limiting process is

$$P_j = \frac{(C_c - \Gamma_*)\cdot J}{4\cdot C_c + 8\cdot\Gamma_*} - R \tag{3}$$

where J [μmol m^{-2} s^{-1}] is the potential electron transport rate at a specific light intensity. The light dependency of J is given by a non-rectangular hyperbolic function (Farquhar and Wong 1984):

$$J = \frac{\alpha\cdot\text{PAR} + J_{\max} - \sqrt{(\alpha\cdot\text{PAR} + J_{\max})^2 - 4\cdot\theta\cdot J_{\max}\cdot\alpha\cdot\text{PAR}}}{2\cdot\theta} \tag{4}$$

where $J_{\max}$ [μmol m^{-2} s^{-1}] is the maximum electron transport rate, α [mol(electrons) mol(photons)$^{-1}$] is the energy conversion efficiency for the incident light, and θ is the curvature (dimensionless) of the light response of J. Because α and θ do not vary much for C_3 species, α was fixed at 0.3 [mol(electrons) mol(photons)$^{-1}$] according to Long et al., and an average C_3 photosynthetic quantum yield of 0.093, leaf absorptance of 0.8, and θ of 0.90 were adopted.

The concentration of CO_2 inside the chloroplast (C_c) can be estimated using mesophyll conductance (g_m) (Sharkey et al. 2007):

$$C_c = C_i - \frac{P_N}{g_m} \tag{5}$$

where C_i is the partial pressure (Pa) of CO_2 in the intercellular air space and g_m is the mesophyll conductance [μmol(CO_2) m^{-2} s^{-1} Pa^{-1}].

The IRGA systems can validate the leaf-level net CO_2 assimilation rate and inversion of photosynthesis-related parameters. Gas exchange measurements (net photosynthesis and stomatal conductance to water) of maize were carried out by using a portable IRGA LI-6400 with a 6 cm^2 leaf chamber, from LI-COR Inc. (Lincoln, Nebraska, USA). Leaf

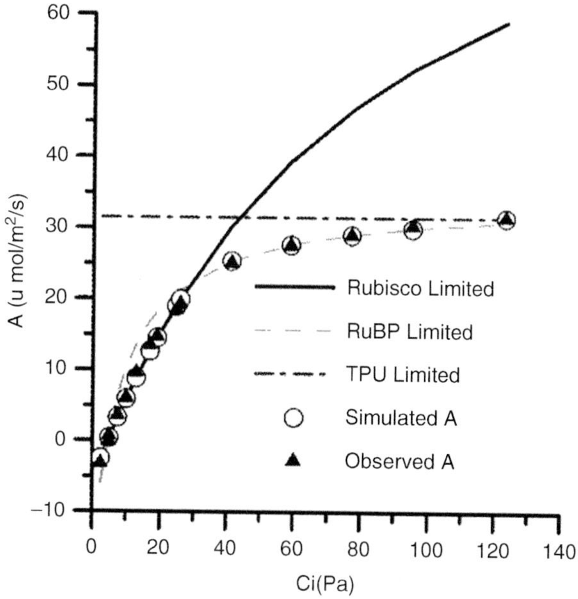

Fig. 1 Prediction of the net photosynthetic assimilation (P_N) of wheat as a function of the CO_2 partial pressure (C_i) using Farquhar model

photosynthesis rate was measured at a specific leaf position in the canopy, i.e., the uppermost fully unfolded leaves were selected for leaf photosynthesis measurements. Light intensity, CO_2 concentration, and temperature are controlled in the leaf chamber. The CO_2 response of leaf photosynthesis is measured at CO_2 concentration levels between 20 and 2000 μmol mol^{-1}. The starting CO_2 concentration was 400 μmol mol^{-1}, followed by concentrations of 300, 250, 200, 150, 100, 50, 400, 500, 600, 800, 1000, 1200, 1400, 1600, 1800, and 2000 μmol mol^{-1}. The air temperature in the leaf chamber was maintained at 25°C. The CO_2 response curves were determined at 1500 μmol m^{-2} s^{-1} of incident Photosynthetically Active Radiation (PAR). Figure 1 showed the result of net CO_2 assimilation rate estimation by Farquhar model. We used the model to estimate five critical parameters of the photosynthesis model (g_m [μmol m^{-2} s^{-1} Pa^{-1}], V_{cmax} [μmol(CO_2) m^{-2} s^{-1}], R_D[μmol (CO_2) m^{-2} s^{-1}], J_{max} [μmol m^{-2} s^{-1}], and TPU [μmol(CO_2) m^{-2} s^{-1}]). The predicted net photosynthetic assimilation (A) matches the observed values well.

Canopy and Ecosystem Photosynthesis Measurement

Eddy Covariance Techniques

Canopy photosynthesis is an integration of various physical, chemical, and biological processes extending from molecular-, cellular-, and organ-level processes to turbulent transport. The eddy covariance method is a micrometeorological technique for direct measurements of the transport of gases, heat, and momentum between the earth's surface and the atmosphere (Aubinet et al. 2012; Burba 2013). It typically provides a direct and sequential estimate of carbon dioxide (CO_2) flux absorbed/released by the ecosystem from/to the atmosphere with fine temporal resolution (Hikosaka et al. 2016).

Through the net ecosystem exchange (NEE) partitioning technique, the NEE can be partitioned into GPP and ecosystem respiration (ER) (Baldocchi 2014). Over the past decades, the EC technique has been widely applied to measure NEE of various biome types throughout the world, and a large amount of carbon flux data has been accumulated (Baldocchi 2014; Baldocchi et al. 2001). Presently, over 600 eddy covariance stations are in operation in over 120 countries (Burba et al. 2013). Growing confidence in long-term and quasi-continuous eddy flux measurements by the broader community led to the establishment of regional flux networks in North and South America (AmeriFlux, LBA, and Fluxnet-Canada), Europe (EuroFlux), Australia (OzFlux), Asia (ChinaFlux and AsiaFlux), and the global network, FLUXNET (Baldocchi 2003 2014). The eddy covariance method has become popular because it provides a direct measure of the flux density of the atmosphere-ecosystem interface, without disturbance of the vegetation and the soil (Baldocchi 2014).

The atmosphere contains turbulent motions of upward- and downward-moving air that transport trace gases such as CO_2. The EC technique samples these turbulent motions to determine the net difference of material moving across the canopy-atmosphere interface. The flux from the area of interest can be represented as a covariance between measurements of vertical velocity, the upward and downward movements, and the concentration of the entity of interest. The vertical flux (F) is equal to a mean product of air density (ρ_a), vertical wind speed (ω), and the dry mole fraction (s) of the gas of interest (Aubinet et al. 2012; Burba 2013):

$$F = \overline{\rho_a\ \omega\ s} \approx \overline{\rho_a \omega' s'} \tag{6}$$

where ρ_a is known from air temperature, humidity, and barometric pressure, ω is measured by a high-speed three-dimensional sonic anemometer, and s is measured by a high-speed gas analyzer (e.g., CO_2/H_2O, CH_4, etc.). In Eq. (6), the overbars denote time averaging, and primes represent fluctuations from the mean. The right part of the equation is the result of a derivation using Reynolds decomposition, where an instantaneous value for each member is presented as a product of an hourly or half-hourly mean and an instantaneous deviation from the mean (Baldocchi 2003). As a result, the hourly or half-hourly fluxes, or emission rates, are computed using instantaneous data, usually recorded 10–20 times per second.

The instrumentation of a basic eddy covariance station includes a three-dimensional sonic anemometer and a gas analyzer (Fig. 2a), which measures three-dimensional wind speed (u, v, w), the concentration of the gas of interest (e.g., CO_2 or CH_4), and water vapor simultaneously. The stations can additionally be equipped with instrumentation to measure weather variables (e.g., mean air temperature, relative humidity, wind speed, direction, precipitation amounts, etc.). These variables, measured at the same location and at the same time, are used to help interpret the flux data and to fill in potential gaps in data. The three-dimensional sonic anemometer (such as CSAT3 from Campbell Scientific Inc.) can obtain the instantaneous value of three-dimensional wind speed (u, v, w) continuously, while the gas analyzer (such as the

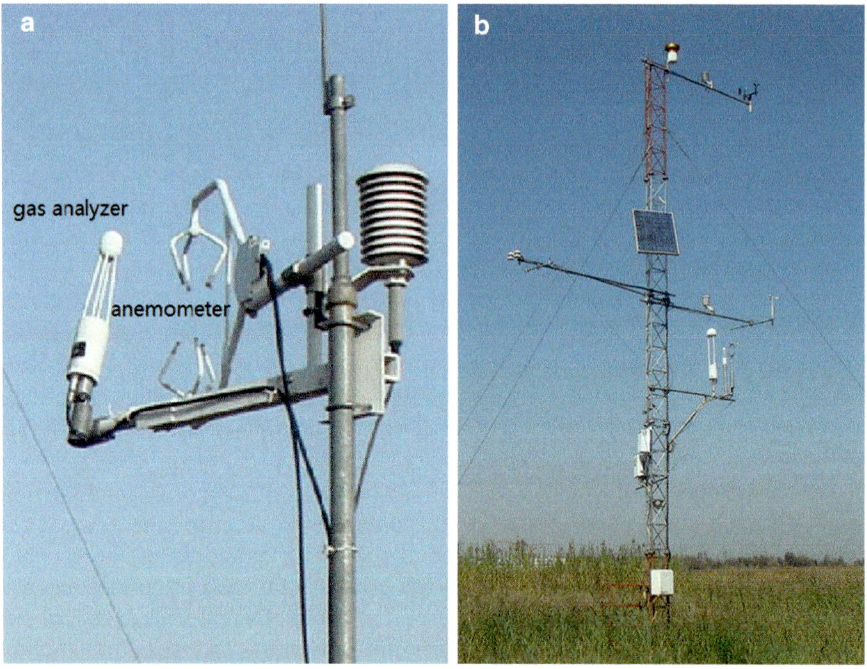

Fig. 2 (**a**) Standard instrumentation on the eddy covariance tower; (**b**) flux tower measuring emissions of CO_2, CH_4, and H_2O

open-path gas analyzer LI-7500/7500A and close-path analyzer products LI-6262 and LI-7000 from LI-COR Inc.) is a high-speed/high-precision infrared (~4.26 m and 2.59 m) gas analyzer, which measures densities of CO_2 and water vapor. In conjunction with sonic anemometer, the fluxes of CO_2 and H_2O can be determined. The LI-7700 open-path CH_4 analyzer uses a single-mode tunable near-infrared laser source, in the absorption band of methane near 1.6 μm and a wavelength modulation spectroscopy (WMS) technique to determine CH_4 concentration (Fig. 2b).

Flux Data Process Procedures

Processing eddy covariance data requires flexible software that fits a wide range of experimental designs and sampling conditions. Multiple automated packages are available for data collection, and recently, several packages have become available for comprehensive data processing with advanced options and user-friendly interface (Burba et al. 2013) such as the EdiRe software developed at the University of Edinburgh (http://www.geos.ed.ac.uk/homes/jbm/micromet/EdiRe/), EddyPro software developed by the LI-COR (http://www.licor.com/env/products/eddy_covariance/software.html), and the TK2/TK3 software developed by the University of Bayreuth (http://www.bayceer.uni-bayreuth.de/mm/en/software/software/software_dl.php). Mauder et al. compared several eddy covariance software, such as the TK2, EC Pack, EddyPro, EdiRe, eth-flux, and TUDD; the overall comparison indicated a good although not yet perfect agreement among the different software within 5–10% difference for 30-min CO_2 flux values.

The general processing procedures of the raw 10 Hz flux data included spike detection, lag correction of H_2O/CO_2 relative to the vertical wind component, sonic virtual temperature correction, coordinate rotation (2-D rotation), corrections for density fluctuation (Webb-Pearman-Leuning correction, WPL correction), and frequency response correction (Xu et al. 2013). The EC data are subsequently averaged over 30 min periods. The processed observation data quality was divided into three classes according to the quality assessment method proposed by Foken and Wichura, class 1 (level 0: $\Delta st < 30$ and $ITC < 30$), class 2 (level 1: $\Delta st < 100$ and $ITC < 100$), and class 3 (level 2: $\Delta st > 100$ and $ITC > 100$), representing high-, medium-, and low-quality data, respectively. Class 1 was selected for the analysis.

Despite the recent developments in the EC technique and the availability of instruments with low power consumption, there are always some common problems in the data acquisition, such as power breaks, damages to instruments, and also human actions like vandalism or robbery. These problems thus create gaps in the measurements. It is found that about 20–60% of the data was rejected by the different quality filters applied (Papale and Valentini 2010). Gap filling is a process that is sometimes unavoidable when daily to annual integrals are needed (Aubinet et al. 2012). Currently, many investigators have developed and implemented their own gap-filling techniques (Desai et al. 2005; Gove and Hollinger 2006; Hui et al. 2004; Moffat et al. 2007; Papale et al. 2006; Reichstein et al. 2005; Richardson et al. 2006; Schwalm et al. 2007; Stauch and Jarvis 2006) which are based on a wide range of approaches, including interpolation, probabilistic filling, look-up tables, nonlinear regression, artificial neural networks, and process-based models in a data assimilation system. Some commonly used gap-filling methods include the mean diurnal variation method (MDV), look-up table (LUT), artificial neural networks (ANNs), and nonlinear regression method (Falge et al. 2001). The mean diurnal variation method (MDV) is an interpolation technique that is based on the temporal autocorrelation of the fluxes . The method of implementation is easy, but the accuracy and performances are lower with respect to the other methods. The look-up table (LUT) is an empirical method and easy to implement, where the missing values are replaced with the average of valid measurements occurring under similar meteorological conditions. The artificial neural networks (ANNs) are purely empirical nonlinear regression models with a medium level of implementation difficulties. Similar to the LUT method, also in the ANN, it is important to select as input the appropriate and relevant environmental variables that drive the flux variability. The nonlinear regression method is based on parameterized nonlinear equations which express semiempirical relationships between the flux and environmental variables, often temperature and light for CO_2 fluxes. Falge et al. compared several gap-filling methods and found that they produced similar results and did not introduce any particular methodological bias, while Moffat et al. (2007) showed all the gap-filling techniques give on average good results when the gaps are shorter than 10 days and the relevant meteorological drivers available. Thus, the uncertainty about the selection of the gap-filling methods is small when the gap length is not too long and the dataset itself is of good quality.

The decision about which method to select should be based on different considerations. The most important consideration should be the availability of meteorological drivers. A different method has different needs of drivers. If there is no meteorological data available, the MDV method is often the only choice, and the uncertainty will be large because the MDV is an empirical method that does not require drivers and can be in theory applied to fill all the variables when temporal autocorrelation is expected (Aubinet et al. 2012). When using the LUT method, the drivers should be selected according to the site characteristics and take into consideration the environmental variables that are more important in gap filling. While ANN method requires a gap-free driver dataset, which needs to gap-fill the meteorological variables firstly. If there are not enough drivers available, the other method (e.g., MDV) should be used. Besides, the quality and the representativeness of the training dataset play also an important role in the ANN method (Aubinet et al. 2012). For the nonlinear regression method, since it is semiempirical, it is applicable only when the functions linking meteorological variables and fluxes are well known and consolidated. Since there is a strong link between gap-filling quality and uncertainty, it is important to choose the proper method carefully. As the performances of the ANN method are relatively good, it is used as the standard in the European database.

The EC approach can only determine the NEE. Through NEE partitioning approaches (Gitta et al. 2010; Reichstein et al. 2005), the NEE can be routinely partitioned into GPP and ER. Various flux partitioning methods are available and have been previously compared using measured or modeled data from single or multiple sites (Ankurr et al. 2008; Desai et al. 2008; Gitta et al. 2010; Hagen et al. 2006; Reichstein et al. 2005; Stoy et al. 2006; Yi et al. 2004). The common partitioning methods include the following procedures. Firstly, the NEE was divided into daytime and nighttime according to the downward shortwave radiation (Rg). During the nighttime, NEE is equal to ER. The nighttime ER was modeled with the air temperature-dependent exponential model, such as the van't Hoff equation. After this function was calibrated using the nighttime air temperature and nocturnal NEE, the equation can be used to estimate daytime ER with daytime air temperature as input. GPP was then calculated as the sum of the daytime net ecosystem exchange (NEE) and ecosystem respiration (ER). According to the differences of the form of models such as driving variables, the cost function used to estimate parameters, and the use of nighttime, daytime, or all eddy covariance data used for model parameterization (Moffat et al. 2007), the flux partitioning approaches can be classified into two types, those that use only filtered nighttime data to directly measure ER (Reichstein et al. 2005) and those that exploit both day- and nighttime data or only daytime data, using light response curves to estimate ER (Aubinet et al. 2012). According to the comparisons of these approaches by Stoy et al. (2006) and Gitta et al. (2010), the two types of methods showed generally good agreement, although some are prone to bias (Desai et al. 2008). There are large variations of carbon exchanges among different ecosystems in arid region.

We used flux and micrometeorological observations from a flux observation network across an arid and semiarid region, Northwest China. Our study area is in a typical inland river basin, Heihe River Basin (HRB), which is the second largest inland river

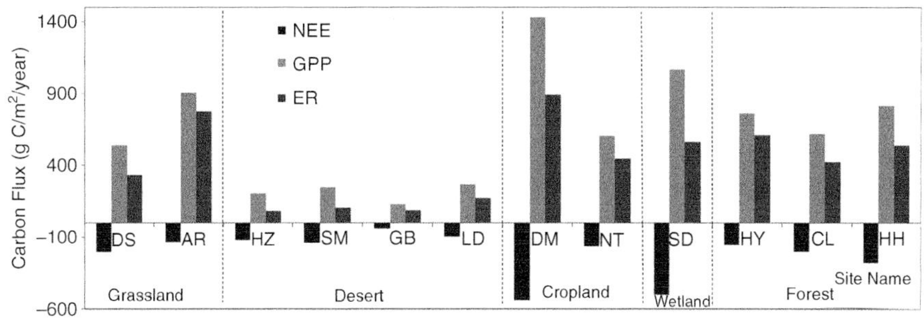

Fig. 3 Variations of annual cumulative NEE, GPP, and ER in different ecosystems during the observation period in HRB

basin and is representative of all of the inland river basins. To improve the observability of hydrological and ecological processes, we constructed 12 EC flux sites to measure fluxes of different ecosystems in HRB, covering 2 grassland sites, 4 desert sites, 2 cropland sites, 3 forest sites, and 1 wetland site, which almost covered the major ecoregions and typical ecosystem types in the arid region of an inland river basin.

Figure 3 illustrates the annual GPP, ER, and NEE for each site. As a whole, NEE of Heihe terrestrial ecosystems ranged from 41.77 to −537.23 g C/m^2/year. Most terrestrial ecosystems in HRB were large carbon sinks, with an average of −213.52 g C/m^2/year. The cropland ecosystems of maize showed the largest carbon sinks, with an NEE of −537.23 g C/m^2/year, followed by the wetland ecosystem, with an NEE of −501.73 g C/m^2/year. The lowest carbon sink was the Gobi Desert ecosystem, with an NEE of −41.77 g C/m^2/year. The GPP of HRB terrestrial ecosystems ranged widely, from 128.01 to 1430.20 g C/m^2/year.

From Fig. 3, we can found a large variations of annual carbon fluxes among different sites and vegetation types. The maize cropland ecosystem and wetland ecosystem had the highest carbon assimilation strengths and carbon sink strengths, with the annual carbon assimilation more than 1000 g C/m^2/year and carbon uptake strengths more than 500 g C/m^2/year. In comparison, the carbon assimilation desert ecosystem is very small, and the annual accumulated GPP was less than 300 g C/m^2/year in the four desert ecosystems; however, the annual carbon exchanges in these sites were large, some even with rivaling rates of carbon uptake in some of the grassland and forest ecosystems. As concerns, the annual Re, the fluctuations of Re, is very similar with that of GPP.

Remote Sensing-Based GPP and NEP Estimation at Regional Scale

Light Use Efficiency Model

Terrestrial ecosystems are machines supplied with energy from an external source, usually the sun (Monteith 1972; Monteith and Moss 1977). The absorbed energy is used to produce dry matters and heat ecosystems through the physical and chemical process. Only the solar radiation within the spectral range from 400 to 700 nm, which is called photosynthetically active radiation (PAR), can be absorbed by chlorophyll and converted to chemical energy in photosynthesis. In the perspective of energy conversion, the production of terrestrial ecosystems can be predicted from absorbed energy. Therefore, the GPP can be expressed as the product of PAR as follows (Hilker et al. 2008):

$$\mathrm{GPP} = \mathrm{PAR} \times \mathrm{FPAR} \times \varepsilon \quad (7)$$

where FPAR is the fraction of PAR absorbed by the canopy. ε is the photosynthetic efficiency which determines dry matter production by per unit PAR. Most frequently used units of GPP, NPP, and NEP are listed in Table 1.

Fraction of PAR (FPAR)

FPAR is one of the inputs of LUE models, which is retrieved from remote sensing data. The incident PAR is reflected, transmitted, and absorbed by the canopy, and only the absorbed PAR is effective in photosynthesis. FPAR is the ratio of canopy-absorbed PAR to incident PAR at top of the canopy. Detailed radiative transfer in the canopy is described by canopy radiative transfer models, such as PROSAIL (Jacquemoud et al. 2009). Physical models of FPAR estimation are very complicated and time-consuming. Empirical models of FPAR are witnessed high accuracy and widely used in LUE models.

Lots of linear and nonlinear empirical approaches are developed based on vegetation indices for different vegetation types. Numerous studies have shown a close relationship between FPAR and the top of the canopy reflectance measurements in the visible and near-infrared region (Daughtry et al. 1983; Tucker 1979). Normalized difference vegetation index (NDVI) and enhanced vegetation index (EVI) are most frequently used in FPAR empirical models. NDVI and EVI can be calculated from the reflectance of red, infrared, and blue band according to the following formula:

$$\mathrm{NDVI} = \frac{(\rho_{\mathrm{nir}} - \rho_{\mathrm{red}})}{(\rho_{\mathrm{nir}} + \rho_{\mathrm{red}})} \quad (8)$$

Table 1 Conversion between units of GPP, NPP, and NEP

1 Mg (megagram) = 1000 kg = 1 t (tonne)
1 Gg (gigagram) = 106 kg = 1 kt
1 Tg (teragram) = 109 kg = 1 Mt
1 Pg (petagram) = 1012 kg = 1 Gt
1 kg/m^2 = 10 t/ha
1 molC = 12 gC
1 tC/ha = 8.33 molC/m^2
1 molC/m^2 = 120 kgC/ha
1 molCO_2/m^2 = 440 kgCO_2/ha

$$\mathrm{EVI} = \mathrm{G}\frac{(\rho_{\mathrm{nir}} - \rho_{\mathrm{red}})}{(\rho_{\mathrm{nir}} + (C1 \cdot \rho_{\mathrm{red}} - C2 \cdot \rho_{\mathrm{blue}}) + L)} \tag{9}$$

where ρ_{blue}, ρ_{red}, and ρ_{nir} are reflectance at blue, red, and infrared band, $G = 2.5$, $C1 = 6$, $C2 = 7.5$, and $L = 1$. And FPAR can be estimated using the following empirical functions:

$$\mathrm{FPAR} = 1.24 \times \mathrm{NDVI} - 0.168 \tag{10}$$

$$\mathrm{FPAR}_{\mathrm{chl}} = a \times \mathrm{EVI} \tag{11}$$

where $\mathrm{FPAR}_{\mathrm{chl}}$ is the fraction of PAR absorbed by chlorophyll and a is an empirical parameter. The parameters in FPAR empirical models usually change with site, sensor, climate, etc. For a specific region, it is better to calibrate these empirical parameters before estimating GPP if there is sufficient FPAR observation. Moderate-resolution imaging spectroradiometer (MODIS), advanced very-high-resolution radiometer (AVHRR), and Satellite Pour l'Observation de la Terre VEGETATION (SPOT-VEGETATION) are the most widely used sensors in LUE models.

Light Use Efficiency ϵ

Light use efficiency ε is the rate of dry matter producing by per unit PAR for the canopy. It is stressed by environmental factors and can be expressed as follows:

$$\varepsilon = \varepsilon_{\max} \cdot f(T, W, \ldots) \tag{12}$$

where $\varepsilon_{\max}$ is maximum light use efficiency without any environmental stress for a canopy. $f(T, W, \ldots)$ is environmental factors' stress on canopy photosynthesis, such as temperature, water, phenology, and so on.

The maximum light use efficiency ($\varepsilon_{\max}$) is a key parameter in remote sensing-based light use efficiency model. It varies with plant species, photosynthesis pathways (C3/C4), leaf nitrogen content, canopy structure, human management, and environmental factor variation. For global modeling, $\varepsilon_{\max}$ is usually set to a fixed value for each plant functional type. For example, the MOD17 algorithm provides a $\varepsilon_{\max}$ look-up table for different vegetation types, which is determined from process-based vegetation model Biome-BGC. For a specific site, a lot of studies witnessed $\varepsilon_{\max}$ should be calibrated before using LUE model to predict GPP, because the default $\varepsilon_{\max}$ may result in bias in the results.

Meteorological Forcing

Spatial gridded meteorological forcing datasets are needed in the regional/global simulation of LUE models, including PAR, temperature, water condition, etc. Temperature and water condition data is usually converted to scalar between 0 and 1 using $f(T, W, \ldots)$ in Eq. (12) to limit the maximum light use efficiency. These meteorological forcing datasets could be interpolated from meteorological stations' observation and the output of climate models or retrieved from time continued remote sensing data.

Time continuous PAR can be directly obtained from networks of surface measuring of photosynthetic photon flux density (PPFD) or shortwave radiation. But these measurements are spatially discrete. When forcing the LUE models at regional or global scale, the discrete measurements should be interpolated into spatial continuous data with spatial resolution matching with other inputs of LUE models. Spatial continuous PAR also can be obtained from outputs of climate models (such as WRF) or land data assimilation systems (such as GLDAS). Shortwave irradiance can be retrieved from the top of the atmosphere solar radiance using optical modeling, and PAR is defined as a fraction of shortwave radiation.

Temperature has a great impact on photosynthesis and control activity of the enzyme in photosynthesis. Extreme high or low temperature could limit the ε close to zero. Therefore, the accurate temperature should be used in LUE models. Spatial continuous temperature could be obtained by interpolating weather station measurement, climate model outputs, land data assimilation system, and reanalysis datasets.

Water is one of the reactants in photosynthesis, and water supply will limit the production of photosynthesis. LUE models choose different variables to express water limitation on photosynthesis, such as soil water content (SWC), vapor pressure deficit (VPD), or land surface water index (LSWI). Soil water content is difficult to obtain. Microwave remote sensing can be used to retrieve soil moisture, but this approach contains great uncertainties. VPD can be calculated from temperature and relative humidity observation. VPD also is provided by outputs of climate models, land data assimilation systems, or reanalysis datasets. LSWI is an index calculated from shortwave infrared and near-infrared band to show the water condition of the land surface, and it is easy to obtain from remote sensing data.

Widely Used Light Use Efficiency Models

In this section, we will introduce six widely used LUE models which were developed in the recent two decades.

1. **MODIS GPP Algorithm**

MODIS GPP algorithm was developed by Running in 2000 (Running et al. 2000). The main functions of this model are listed as follows:

$$\mathrm{GPP} = \mathrm{PAR} \times \mathrm{FPAR} \times \varepsilon_{\max} \times f(T\mathrm{min}) \times f(\mathrm{VPD}) \tag{13}$$

$$f(T_{\min}) = \begin{cases} 0 & T\mathrm{min} < T\mathrm{min}_{\min} \\ \dfrac{T\mathrm{min} - T\mathrm{min}_{\min}}{T\mathrm{min}_{\max} - T\mathrm{min}_{\min}} & T\mathrm{min}_{\min} < T\mathrm{min} < T\mathrm{min} \\ 1 & T\mathrm{min} > T\mathrm{min}_{\max} \end{cases} \tag{14}$$

$$f(\mathrm{VPD}) = \begin{cases} 0 & \mathrm{VPD} > \mathrm{VPD}_{\max} \\ \dfrac{\mathrm{VPD}_{\max}\text{-}\mathrm{VPD}}{\mathrm{VPD}_{\max}\text{-}\mathrm{VPD}_{\min}} & \mathrm{VPD}_{\min} < \mathrm{VPD} < \mathrm{VPD}_{\max} \\ 1 & \mathrm{VPD} < \mathrm{VPD}_{\min} \end{cases} \tag{15}$$

Table 2 Parameter look-up table in MODIS GPP algorithm

	ε_0 [gC/MJ APAR]	$T\text{min}_{\text{min}}$ [°C]	$T\text{min}_{\text{max}}$ [°C]	VPD_{max} [Pa]	VPD_{min} [Pa]
ENF	1.008	−8.00	8.31	2500	650
EBF	1.159	−8.00	9.09	3900	1100
DNF	1.103	−8.00	10.44	3100	650
DBF	1.044	−8.00	7.94	2500	650
MF	1.116	−8.00	8.5	2500	650
WL	0.8	−8.00	11.39	3100	930
Wgrass	0.768	−8.00	11.39	3100	650
Cshrub	0.888	−8.00	8.61	3100	650
Oshrub	0.774	−8.00	8.8	3600	650
Grass	0.68	−8.00	12.02	3500	650
Crop	0.68	−8.00	12.02	4100	650

ENF evergreen needleleaf forest, *EBF* evergreen broadleaf forest, *DNF* deciduous needleleaf forest, *DBF* deciduous broadleaf forest, *MF* mixed forest, *WL* grassy woodland, *Wgrass* wooded grassland, *Cshrub* closed shrubland, *Oshrub* open shrubland, *Grass* grassland, *Crop* cropland

where GPP [gC/m^2/d] is gross primary production, PAR[MJ] is photosynthetically active radiation, FPAR is fraction of absorbed PAR, ε_{max} [gC/MJ APAR] is maximum light use efficiency, and *f*(*T*min) and *f*(VPD) are scalars of minimum temperature Tmin [°C] and VPD [Pa] on photosynthesis. This algorithm is used to produce MODIS GPP product MOD17. ε_{max}, $T\text{min}_{\text{min}}$ [°C], $T\text{min}_{\text{max}}$ [°C], VPDmax [Pa], and VPDmin [Pa] are parameters based on IGBP plant functional types (see Table 2).

2. **VPM Model**

The vegetation photosynthesis model (VPM) was developed by Xiao based on the conceptual partitioning of chlorophyll and non-photosynthetically active vegetation within a canopy. In the VPM, GPP is calculated using the equation as follows:

$$\text{GPP} = \text{PAR} \times \text{FPAR}_{\text{chl}} \times \varepsilon_{\text{max}} \times f(T) \times f(W) \times f(P) \tag{16}$$

$$\text{FPAR}_{\text{chl}} = a \times \text{EVIFPAR}_{\text{chl}} = a \times EVI \tag{17}$$

$$f(T) = \frac{(T - T_{\text{min}})(T - T_{\text{max}})}{(T - T_{\text{min}})(T - T_{\text{max}}) - (T - T_{\text{opt}})^2} \tag{18}$$

$$f(W) = \frac{1 + \text{LSWI}}{1 + \text{LSWI}_{\text{max}}} \tag{19}$$

$$f(P) = \begin{cases} \dfrac{1 + \text{LSWI}}{2}, & \text{during emergence to leaf full expansion} \\ 1, & \text{after full leaf expension} \end{cases} \tag{20}$$

$$f(P) = \begin{cases} \frac{1 + \text{LSWI}}{2} & \text{during emergence to leaf full expansion} \\ 1 & \text{after full leaf expension} \end{cases}$$ where

$FPAR_{chl}$ is the fraction of PAR absorbed by chlorophyll and estimated from EVI by a linear function and a is an empirical parameter with default value equals to 1. $f(T)$, $f(W)$, and $f(P)$ are scalars of temperature, water, and phenology to photosynthesis. $f(T)$ is calculated from mean air temperature by Eq. (18); T_{min}, T_{max}, and T_{opt} are empirical parameters. $f(T)$ is maximum when mean air temperature is close to T_{opt} and is minimum when mean air temperature is greater than T_{max} or less than T_{min}. $f(W)$ is calculated from remote sensed LSWI, and $LSWI_{max}$ is maximum LSWI during growing season. $f(P)$ is calculated depending on life expectancy of leaves. Therefore, $f(P)$ is estimated at two different phases for deciduous vegetation (see Eq. 20), and $f(P)$ is set to 1 for evergreen vegetation. The VPM model has very high accuracy in GPP estimation (Li et al. 2007; Wang et al. 2012; Xiao et al. 2005; Yan et al. 2009), because EVI was witnessed has a higher correlation with GPP than NDVI.

3. **EC-LUE Model**

Eddy covariance-light use efficiency (EC-LUE) model is developed by Yuan, et al. in 2007. The main equations of EC-LUE model are given as below:

$$\text{GPP} = \text{PAR} \times \text{FPAR} \times \varepsilon_{max} \times \min(f(T), f(W)) \quad (21)$$

$$f(W) = \frac{\text{LE}}{\text{HS} + \text{LE}} \quad (22)$$

where $f(T)$ uses the same formula with VPM. LE is latent heat flux, and HS is sensible heat flux. In EC-LUE model, the photosynthesis dependence on water condition is expressed as the ratio between LE and sum of HS and LE.

4. **VI Model**

Vegetation index model (VI) was developed by Wu et al. (2009a, 2010). Its equations are given as below:

$$\text{GPP} = \text{PAR} \times \text{EVI} \times \text{EVI} \times a \quad (23)$$

where a is scalar which converts PAR to GPP. Wu found that GPP has higher correlation with EVI × EVI than EVI or NDVI alone. Thus, in this model, GPP is directly estimated from EVI, PAR, and empirical parameter a. This model was validated in a lot of sites. But when used in a regional scale, uncertainties in parameter a are sensitive to the result.

5. **TG Model**

In order to develop a fully remote sensing-based GPP model, temperature and greenness (TG) model was presented by Sims et al. (2008). The main formulas are listed below:

$$\mathrm{GPP} = m \times \mathrm{ScaledEVI} \times \mathrm{ScaledLST} \tag{24}$$

$$\mathrm{ScaledEVI} = \mathrm{EVI} - 0.1 \tag{25}$$

$$\mathrm{ScaledLST} = \min\left[\frac{\mathrm{LST}}{30}, (2.5 - (0.05 \times \mathrm{LST}))\right] \tag{26}$$

When EVI is lower than 0.1, set ScaledEVI as 0. If LST is lower than 0°C or greater than 50°C, set ScaledLST as 0. m (gC/m2/d) is the parameter to convert ScaledEVI and ScaledLST to GPP, which changes with vegetation type and climate condition. How to get spatially distributed m is vital to this application at regional scale. Another problem is that spatial continuous LST retrieving is difficult, because of contamination of cloud.

Parameter Estimation of LUE Models

Before using LUE models to simulate GPP, NPP, or NEP, parameters of LUE models should be calibrated if there are sufficient observing datasets. Numerous studies reported that great bias exists in outputs of LUE models with default parameter values at some specific sites or regions.

A Bayesian method Markov chain Monte Carlo (MCMC) is used to derive the model parameters. According to Bayesian theory, posterior probability density functions (PDFs) of model parameters c can be obtained from prior knowledge of parameters and information generated by comparison of simulated and observed variables. It can be expressed as follows (Mosegaard and Sambridge 2002):

$$p(c|Z) = \frac{p(Z|c)p(c)}{p(Z)} \tag{27}$$

where $p(c|Z)$ is the posterior PDF of model parameters. $p(Z|c)$ is the conditional PDF of observed data; it is also called the likelihood function of parameter c. $p(c)$ is the prior PDF of parameter c. $p(Z)$ is the PDF of observed data. To carry out the parameter estimation, ranges and initial value for model parameters were specified first according to literature. Then the Metropolis-Hastings (M-H) algorithm was used to determine the parameters (Hastings 1970; Metropolis et al. 1953). In parameter selection process, there are two steps repeated: a proposing step and a moving step (Wu et al. 2009b; Xu et al. 2006).

In proposing step, the new proposing parameter was generated from previously accepted parameters with a random walk (Δc). Δc is calculated using a random number (r) between 0 and 1, the parameter range ($c_{\max} - c_{\min}$), and a step length factor (s) (Wu et al. 2009b).

$$c^{\text{new}} = c^{\text{old}} + \frac{(r - 0.5)}{s}(c_{\max} - c_{\min}) \tag{28}$$

where $c^{\text{new}}$and c^{old} are new parameter vector and previously accepted parameter vector.

In moving step, c^{new} is tested using M-H rule to determine whether to accept the parameter vector or not. The M-H rule is a probability to accept the proposed parameters, which is derived from likelihood functions of proposed parameters relative to the parameters accepted last time (Wu et al. 2009b). Likelihood function is usually expressed as follows:

$$p(Z|\, c) \propto \exp\left\{ - \sum_i \frac{1}{2\delta_i^2} \sum_{t \in \text{obs}(Z_i)} (e_i(t))^2 \right\} \tag{29}$$

where $e_i(t)$ is the difference between simulated value and observed value. In the case of this study, $e_i(t)$ is the difference between EC-observed GPP and model-simulated GPP. δ_i^2 is the measurement error variance of dataset. In M-H algorithm, the acceptance probability is calculated according to the ratio $R = \min\left\{1, \frac{p(c^{\text{new}})p\left(c^{\text{old}} \middle| c^{\text{new}}\right)}{p(c^{\text{old}})p\left(c^{\text{new}} \middle| c^{\text{old}}\right)}\right\}$. In every moving step, acceptance probability is compared with a uniform random number U [0,1]. Only when R is greater than U, c^{new} will be accepted.

Take VPM model as an example, VPM model totally contains six parameters, and they are $\varepsilon_{\max}$, a, $T_{\min}$, $T_{\max}$, T_{opt}, and $\text{LSWI}_{\max}$. $\text{LSWI}_{\max}$ is the maximum value of LSWI during growing season, and a usually is set to 1. Therefore, only $\varepsilon_{\max}$, $T_{\min}$, $T_{\max}$, and T_{opt} are parameters that need to be calibrated for a specific site. Using MCMC method, we calibrated parameters in VPM model, and the posterior distributions of $T_{\min}$, T_{opt}, $T_{\max}$, and $\varepsilon_{\max}$ (LUEmax in Fig. 4 are estimated (see Fig. 4). This calibrated parameter can greatly improve GPP estimating accuracy for a specific region. Parameter uncertainties have a great impact on GPP result (see Fig. 5); in this example, the $\varepsilon_{\max}$ is (LUEmax in Fig. 5) is the most sensitive parameter and has the greatest impact on GPP.

Comparison of Remote Sensing-Based GPP Model

There are many remote sensing-based GPP models that were developed in recent several decades. So the comparison of these models is valuable to help researchers choose the

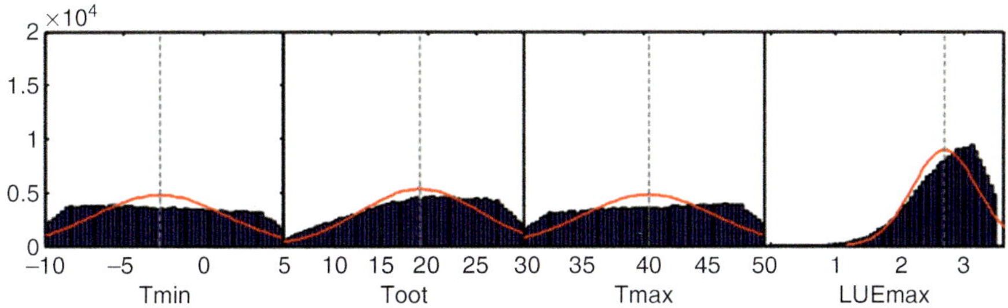

Fig. 4 Posterior distribution of parameters in LUE model (Blue bars are posterior distribution of parameter; red line is fitted normal distribution; and dashed lines are average of posterior distribution of parameter)

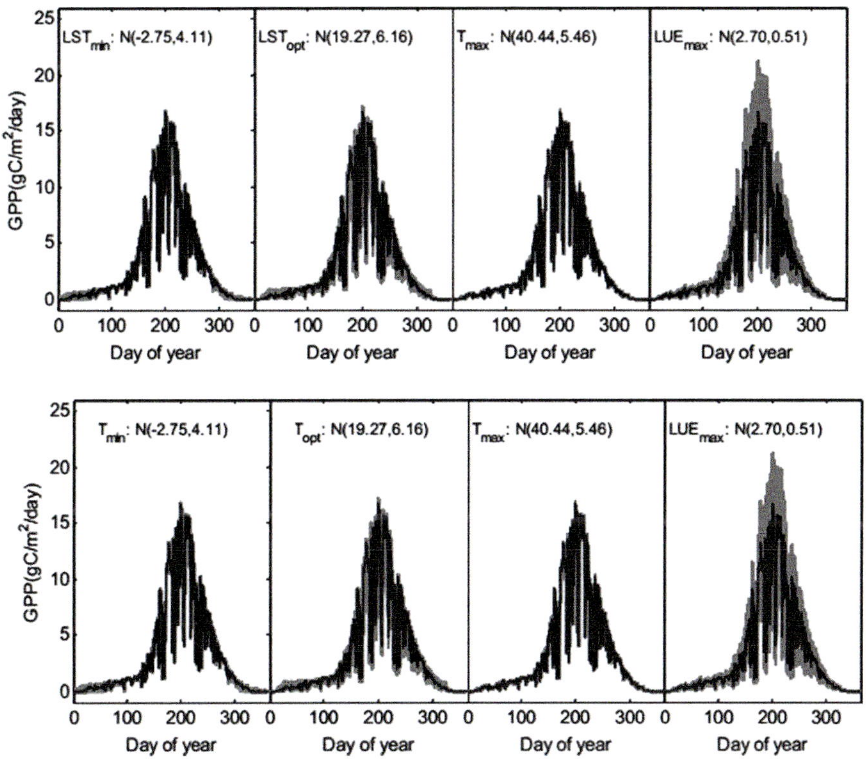

Fig. 5 Uncertainties in GPP simulation resulted from parameters in LUE models (Black line is GPP simulated with average of posterior distribution of parameter; gray background is GPP range with posterior distribution of parameter)

right model for a specific area or application. Here, we compared the performance of MODIS GPP algorithm (GPP_MOD), VPM model (GPP_VPM), EC-LUE model (GPP_EC-LUE), VI model (GPP_VI), and TG (GPP_TG) model at an alpine meadow site (A'rou station) in the year 2009 in the upstream of the Heihe River Basin. The results of these models are shown in Fig. 6; all the models capture the seasonal dynamic very well after calibration. Predicted GPP and EC-observed GPP are very close (see Table 3 and Fig. 7). R^2 of all these models are very high; MODIS GPP algorithm and VPM model show a little bit higher R^2 than other models (see Table 3).

GPP Estimation of Heihe River Basin

In order to further understand the application of LUE model, here we give an example of MODIS GPP algorithm application in the Heihe River Basin. Heihe River Basin is

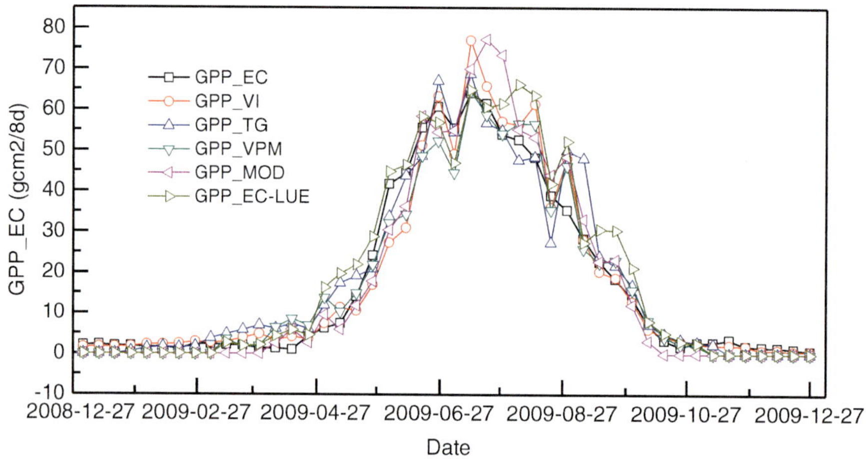

Fig. 6 Plot of GPP observed by EC and GPP predicted by models at A'rou station in 2009

Table 3 Comparison between EC-observed GPP and model-predicted GPP

		GPP_MOD	GPP_VPM	GPP_EC-LUE	GPP_TG	GPP_VI	GPP_EC
WY	R^2	0.96	0.96	0.95	0.94	0.95	
	a	1.11	0.95	1.06	0.98	1.02	
	b	−1.81	0.35	0.95	1.41	−0.006	
	GPP	806.1	778.2	899.6	848.8	816.4	804.2
GS	R^2	0.89	0.89	0.88	0.84	0.88	
	a	1.10	0.86	0.84	0.85	1.06	
	b	−1.52	4.55	11.75	7.12	−2.27	
	GPP	791.8	732.4	859.5	780.8	749.3	747.0

WY whole year, *GS* growing season (May–September), a and b are slope and intercept for linear regression, GPP unit, $gC/m^2/year$

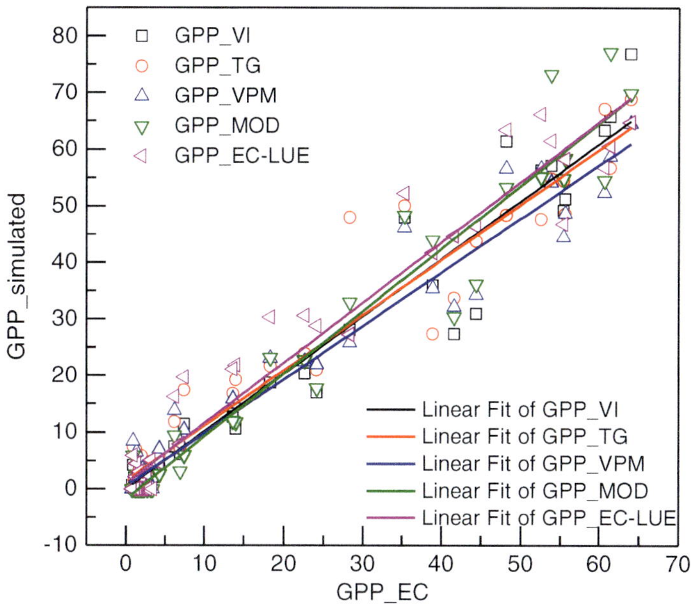

Fig. 7 Scatterplot of GPP observed by EC and GPP predicted by models at A'rou station in year 2009

the second largest inland river basin in China, and it is located in the northwest arid region of China. We have installed three EC towers to measure fluxes of the alpine grassland, evergreen needleleaf forest, and cropland, which are typical ecosystems in this river basin. We also collected another EC tower carbon flux data of cropland. Figure 8 shows the location of these sites and vegetation distribution of the Heihe River Basin. Barren or sparsely vegetated lands cover 80.7% of the area of Heihe River Basin, grassland covers 13.9%, cropland covers 2.3%, and other land cover types are smaller than 1%. Information of EC sites is listed in Table 4. MODIS GPP algorithm was validated at these EC towers in the Heihe River Basin. The results of this model before (MOD17A2) and after (MODIS-PSN) calibration were compared.

From Figs. 9 and 10 is indicated that the MODIS GPP product, which uses the default parameters' value in the simulation, greatly underestimated the GPP at the four stations. At AR station and DYK station, it was about one-half of GPP observed by EC and one-fifth of GPP observed at YK and LZ stations. The RMSE was up to 74gC/m^2/8d (see Table 5). In order to reduce uncertainties from model parameters and inputs, we use field-observed meteorological data as input and calibrate the model parameters. We use all the observed data to calibrate the model parameters, because the available data time series is very short. When local observed meteorological data and calibrated ε_0 were used as inputs of MODIS GPP algorithm (MODIS-PSN), GPP was correctly simulated at the four stations. The GPP predicted by MODIS-PSN was consistent with observed GPP. The slope of linear regression and RMSE were improved (see Table 4).

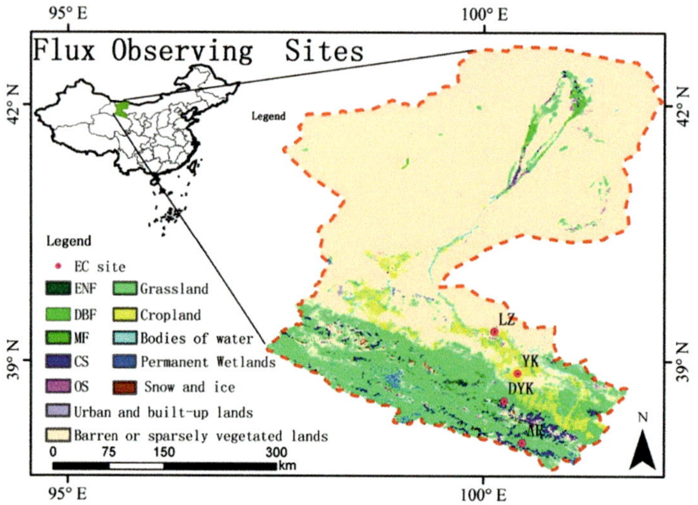

Fig. 8 Locations of the EC stations in the Heihe River Basin (*ENF* evergreen needleleaf forest, *DBF* deciduous broadleaf forest, *MF* mixed forest, *CS* closed shrublands, *OS* open shrublands)

Table 4 Information of observation sites in the Heihe River Basin

Station name	Vegetation type	Observing period
LZ	Cropland (maize)	2008 (Jun–Sep), 2009 (Jun–Sep)
YK	Cropland (maize)	2008–2009
DYK	Evergreen needleleaf forest	2010
AR	Grassland	2008–2009

Figure 11 shows standard MODIS GPP product and GPP estimated in this study in Heihe River Basin. Generally, standard MODIS GPP product underestimated the GPP of this region. Maximum yearly integrated GPP of standard MODIS GPP product was 804 gC/m^2/year in 2008 and 802 gC/m^2/year in 2009, but the maximum value of GPP estimated in this study was 1648 gC/m^2/year in 2008 and 1631 gC/m^2/year in 2009 (see Table 6). In the Heihe River Basin, the total GPP from standard MODIS GPP was 5.74*10^{12} gC/year in 2008 and 5.72*10^{12} gC/year in 2009, while that estimated from this study is 1.42*10^{13} gC/year in 2008 and 1.37*10^{13}gC/year in 2009 (see Table 6).

Scale Mismatch Between Remote Sensing Pixel and Eddy Covariance Footprint

Validating of remote sensing-based LUE models with in situ observation is very important, and a lot of studies have been conducted. One of the most widely validated LUE models is MODIS GPP algorithm, because MODIS GPP product

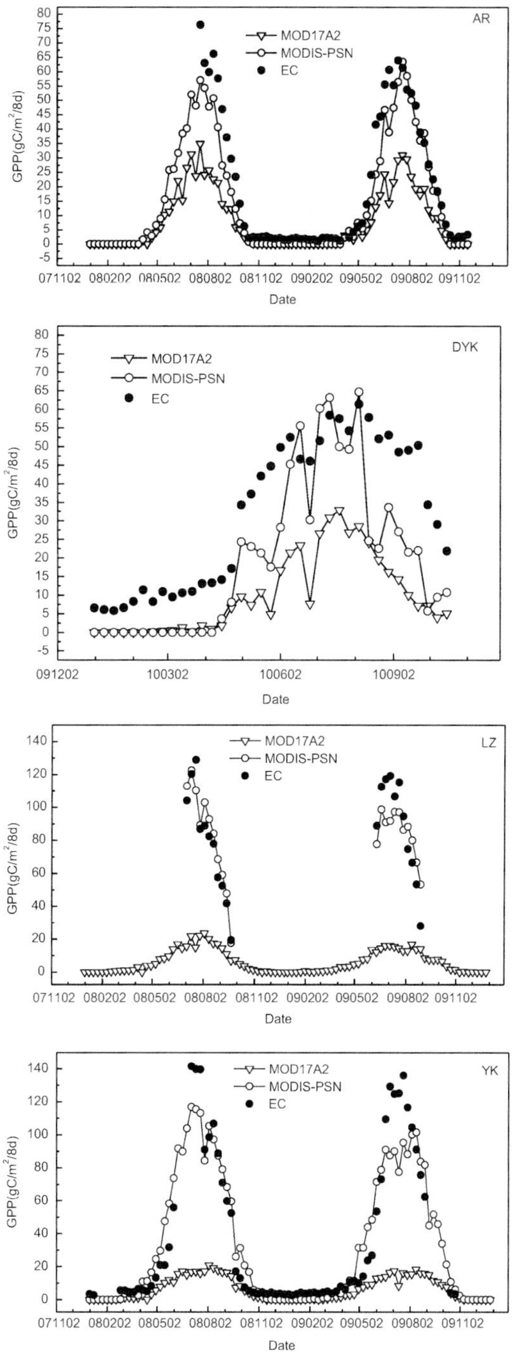

Fig. 9 Comparison of GPP from MODIS GPP product (MOD17A2), predicted by MODIS-PSN (MODIS_PSN) and observed by EC (GPP_EC) at AR station, DYK station, LZ station, and YK station

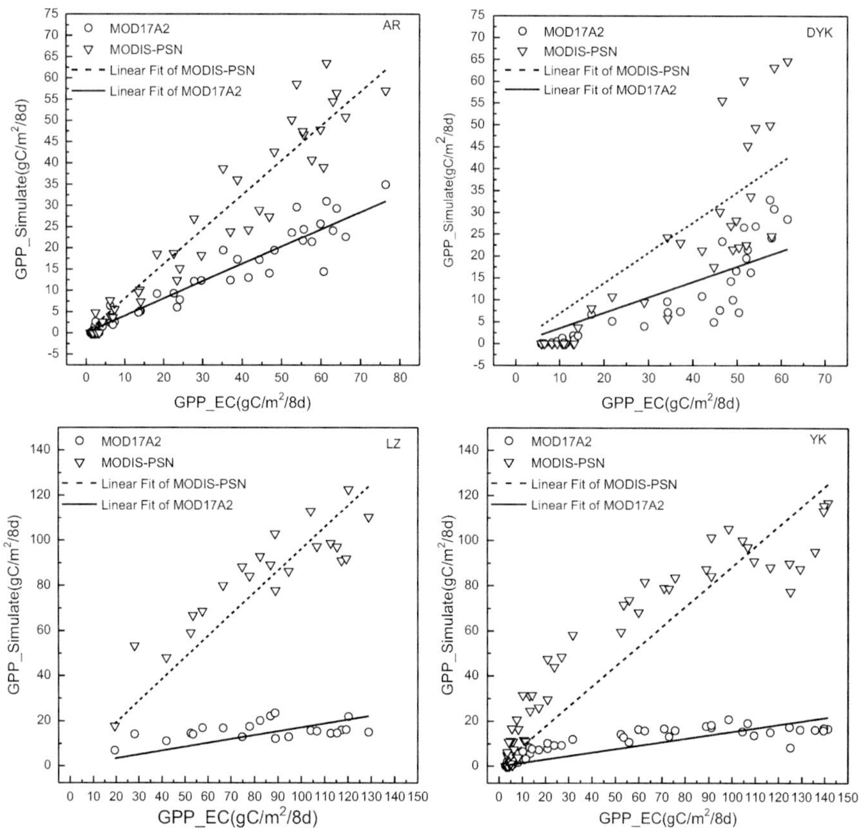

Fig. 10 Scatterplot of MODIS GPP product and GPP predicted by MODIS-PSN and observed by EC at AR station, DYK station, LZ station, and YK station. Intercept of linear fit was fixed equal to zero. Slope of linear fit was listed in Table 4

Table 5 Slope in linear fit and RSME for MODIS GPP product and GPP estimated in this study at each site

	MOD17A2		MODIS-PSN	
Name	Slope	RMSE	Slope	RMSE
AR	0.41	19	0.86	8.2
DYK	0.35	25	0.69	16.1
LZ	0.17	74	0.96	15.9
YK	0.15	50	0.88	13

produced is widely used in carbon cycle study. Mostly, LUE models were validated by comparing the quantity and seasonal dynamic of simulated GPP with eddy covariance-obtained GPP (Zhang et al. 2008). In most validation studies, LUE GPP of one pixel or the average of 3 by 3 pixels where the eddy covariance tower

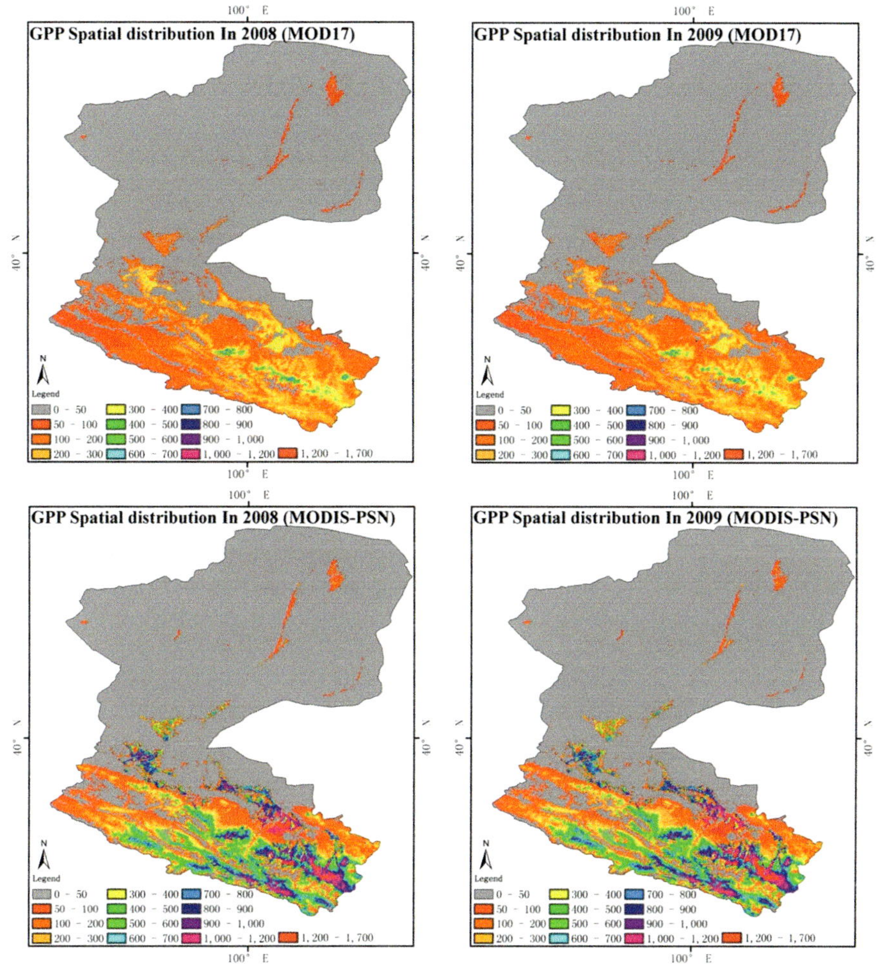

Fig. 11 Spatial distribution of GPP in the Heihe River Basin estimated by MODIS GPP algorithm before (MOD17) and after (MODIS-PSN) parameter calibration in year 2008 and 2009

Table 6 Total GPP in Heihe River Basin

	MOD17		MODIS-PSN	
Year	sum (gC/year)	max (gC/m2/year)	sum (gC/year)	max (gC/m2/year)
2008	$5.74*10^{12}$	804	$1.42*10^{13}$	1648
2009	$5.72*10^{12}$	802	$1.37*10^{13}$	1631

is located in comparing with GPP obtained from eddy covariance (Plummer 2006; Wang et al. 2013). For more accurate validation, the footprint of eddy covariance should be estimated, and remotely sensed GPP within the footprint should be used to compare with in situ-observed GPP.

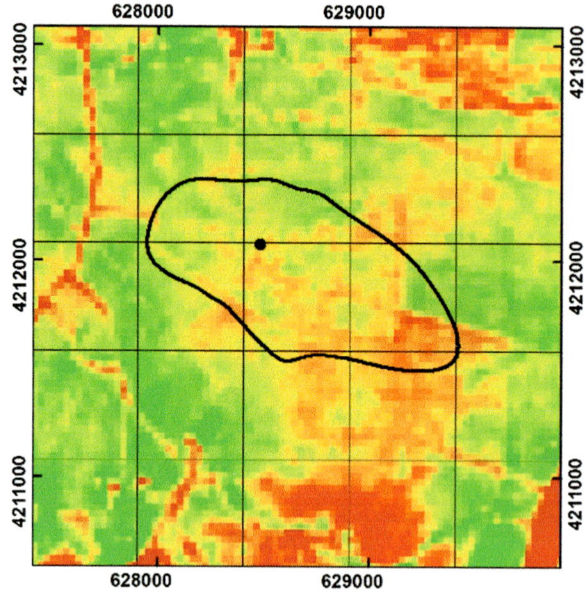

Fig. 12 Footprint climatology calculated by FFP model at a EC tower in the Heihe River Basin. The background image is Landsat 8 OLI image; black grids are boundaries of MODIS 1 km image

Footprint shows the flux source area of eddy covariance, which changes with the height of EC equipped, wind speed, wind direction and vegetation height of the sites. Therefore, the footprint is changing with time. In order to get the long-term footprint of a specific EC tower, fine time step (usually half hour) footprints were weighted by carbon flux to get an average footprint for a year, which is called footprint climatology (Chen et al. 2009). Some open-source footprint models can be downloaded from the Internet, such as Flux Footprint Prediction (FFP) (Kljun et al. 2015) and flux source area model (FSAM) (Schmid 1994). Figure 12 is an example of footprint of eddy covariance measurement at Daman station in the Heihe River Basin.

Solar-Induced Fluorescence (SIF) in GPP Estimation

Chlorophyll pigments absorb photons for photosynthesis process. Some of the photons are reemitted at a longer wavelength, in a process called chlorophyll fluorescence. Based on this process, GPP can be estimated using fluorescence. The fluorescence can be measured by high spectral resolution spectrometers, for example, the solar-induced chlorophyll fluorescence (SIF) that is emitted during absorption of solar radiation by chlorophyll.

With the development of remote sensing technique, it has become possible to measure solar-induced fluorescence (SIF) from space by satellites, such as Greenhouse Gases Observing Satellite (GOSAT) (Frankenberg et al. 2012), Global Ozone Monitoring Mission Experiment-2 (GOME-2) (Joiner et al. 2014), and Orbiting Carbon Observatory-2 (OCO-2) (Hammerling et al. 2012). In the recent decade, numerous studies show SIF was highly correlated with GPP observed by EC (Verma et al. 2017; Wagle et al. 2016; Yang et al. 2015). This research field is developing fast and will improve the GPP monitoring in the next decade.

Summary

Photosynthesis is the key process in the carbon cycle. GPP, NPP, and NEP are key variables in carbon cycle study. Accurate estimation of these variables at regional and global scale will improve our understanding mechanism of the carbon cycle in terrestrial ecosystem and interaction between terrestrial ecosystem and atmosphere in the background of climate change. That can help human make right strategy to face the climate change. In this chapter, we systematically reviewed GPP, NPP, and NEP measuring methods in the field and retrieving models based on remote sensing. Field measurements can capture the season dynamic of carbon fluxes very well and get accurate carbon flux quantity estimation, but it is difficult to upscale these carbon flux measurements to regional or global scale, and it is expensive to build and maintain flux net. Remote sensing based carbon flux estimation models is effective way to get the carbon fluxes at a regional or global scale, but parameters and forcing data of these models could result in great uncertainties in carbon flux estimation. Therefore, integrating the carbon fluxes of the two methods by data assimilation methods can get most accurate carbon fluxes at the regional or global scale. Meanwhile, some new technology will contribute to the estimation of carbon fluxes, such as SIF sensor boarded on satellite.

References

D. Ankurr et al., Cross-site evaluation of eddy covariance GPP and RE decomposition techniques. Agric. For. Meteorol. **148**(6), 821–838 (2008)

M. Aubinet, Vesala, T., & Papale, D. (Eds.). Eddy covariance: a practical guide to measurement and data analysis. Springer Dordrecht Heidelberg London New York. (2012).

D.D. Baldocchi, Assessing the eddy covariance technique for evaluating carbon dioxide exchange rates of ecosystems: past, present and future. Glob. Chang. Biol. **9**(4), 479–492 (2003)

D. Baldocchi, Measuring fluxes of trace gases and energy between ecosystems and the atmosphere–the state and future of the eddy covariance method. Glob. Chang. Biol. **20**(12), 3600–3609 (2014)

D. Baldocchi et al., FLUXNET: a new tool to study the temporal and spatial variability of ecosystem–scale carbon dioxide, water vapor, and energy flux densities. Bull. Am. Meteorol. Soc. **82**(11), 2415–2434 (2001)

G. Burba, *Eddy Covariance Method for Scientific, Industrial, Agricultural and Regulatory Applications: A Field Book on Measuring Ecosystem Gas Exchange and Areal Emission Rates* (LI-Cor Biosciences, Lincoln, 2013)

G. Burba, R. Madsen, K. Feese, Eddy covariance method for CO2 emission measurements in CCUS applications: principles, instrumentation and software. Energy Procedia **40**, 329–336 (2013)

B. Chen et al., Assessing tower flux footprint climatology and scaling between remotely sensed and Eddy covariance measurements. Bound.-Layer Meteorol. **130**, 30 (2009)

C. Daughtry, K. Gallo, M.E. Bauer, Spectral estimates of solar radiation intercepted by corn canopies. Agron. J. **75**(3), 527–531 (1983)

A.R. Desai, P.V. Bolstad, B.D. Cook, K.J. Davis, E.V. Carey, Comparing net ecosystem exchange of carbon dioxide between an old-growth and mature forest in the upper Midwest, USA. Agric. For. Meteorol. **128**(1), 33–55 (2005)

A.R. Desai et al., Cross-site evaluation of eddy covariance GPP and RE decomposition techniques. Agric. For. Meteorol. **148**(6), 821–838 (2008)

E. Falge et al., Gap filling strategies for defensible annual sums of net ecosystem exchange. Agric. For. Meteorol. **107**(1), 43–69 (2001)

G.D. Farquhar, V.S. von von Caemmerer, J. A. Berry, A biochemical model of photosynthetic CO2 assimilation in leaves of C3 species. Planta. **149**(1), 78–90 (1980)

G. Farquhar, S. Wong, An empirical model of stomatal conductance. Funct. Plant Biol. **11**(3), 191–210 (1984)

C. Frankenberg, C.O. Dell, L. Guanter, J. McDuffie, Remote sensing of near-infrared chlorophyll fluorescence from space in scattering atmospheres: implications for its retrieval and interferences with atmospheric CO_2 retrievals. Atmos. Meas. Tech. **5**(8), 14 (2012)

L. Gitta et al., Separation of net ecosystem exchange into assimilation and respiration using a light response curve approach: critical issues and global evaluation. Glob. Chang. Biol. **16**(1), 187–208 (2010)

J.H. Gove, D.Y. Hollinger, Application of a dual unscented Kalman filter for simultaneous state and parameter estimation in problems of surface-atmosphere exchange. J. Geophys. Res. Atmos. **111**(D8), 21 PP (2006)

S.C. Hagen et al., Statistical uncertainty of eddy flux–based estimates of gross ecosystem carbon exchange at Howland Forest, Maine. J. Geophys. Res. Atmos. **111**(D8), 811–830 (2006)

D.M. Hammerling, A.M. Michalak, S.R. Kawa, Mapping of CO2 at high spatiotemporal resolution using satellite observations: Global distributions from OCO-2[J]. Journal of Geophysical Research: Atmospheres, 2012, 117(D6).

P.J. Hanson et al., Oak forest carbon and water simulations: model intercomparisons and evaluations against independent data. Ecol. Monogr. **74**(3), 443–489 (2004)

W.K. Hastings, Monte Carlo sampling methods using Markov chains and their applications. Biometrika **57**(1), 97–109 (1970)

K. Hikosaka, U. Niinemets, N.P. Anten, *Canopy Photosynthesis: From Basics to Applications* (Springer, Dordrecht, 2016)

T. Hilker, N.C. Coops, M.A. Wulder, T.A. Black, R.D. Guy, The use of remote sensing in light use efficiency based models of gross primary production: a review of current status and future requirements. Sci. Total Environ. **404**(2), 411–423 (2008)

D. Hui et al., Gap-filling missing data in eddy covariance measurements using multiple imputation (MI) for annual estimations. Agric. For. Meteorol. **121**(1), 93–111 (2004)

S. Jacquemoud et al., PROSPECT+ SAIL models: a review of use for vegetation characterization. Remote Sens. Environ. **113**, S56–S66 (2009)

J. Joiner et al., The seasonal cycle of satellite chlorophyll fluorescence observations and its relationship to vegetation phenology and ecosystem atmosphere carbon exchange. Remote Sens. Environ. **152**, 16 (2014)

M. Kirschbaum, D. Eamus, R. Gifford, S. Roxburgh, P. Sands, *Definitions of Some Ecological Terms Commonly Used in Carbon Accounting* (Cooperative Research Centre for Carbon Accounting, Canberra, 2001), pp. 2–5

N. Kljun, P. Calanca, M.W. Rotach, H.P. Schmid, A simple two-dimensional parameterisation for Flux Footprint Prediction (FFP). Geosci. Model Dev. **8**, 18 (2015)

C. Le Quere et al., Trends in the sources and sinks of carbon dioxide. Nat. Geosci. **2**(12), 831–836 (2009)

Z. Li et al., Modeling gross primary production of alpine ecosystems in the Tibetan Plateau using MODIS images and climate data. Remote Sens. Environ. **107**(3), 510–519 (2007)

LI-COR Biosciences Using the LI-6400 / LI-6400XT Portable Photosynthesis System, 10th edition. (LI–COR Biosciences Inc.: Lincoln, 2011) Available at https://www.licor.com/env/support/product?p=6400XT [Verified 26 October 2017]

N. Metropolis, A.W. Rosenbluth, M.N. Rosenbluth, A.H. Teller, E. Teller, Equation of state calculations by fast computing machines. J. Chem. Phys. **21**, 1087 (1953)

A.M. Moffat et al., Comprehensive comparison of gap-filling techniques for eddy covariance net carbon fluxes. Agric. For. Meteorol. **147**(3), 209–232 (2007)

J. Monteith, Solar radiation and productivity in tropical ecosystems. J. Appl. Ecol. **9**(3), 747–766 (1972)

J.L. Monteith, C. Moss, Climate and the efficiency of crop production in Britain [and discussion]. Philos. Trans. R. Soc. Lond. B: Biolo. Sci. **281**(980), 277–294 (1977)

K. Mosegaard, M. Sambridge, Monte Carlo analysis of inverse problems. Inverse Prob. **18**(3), R29 (2002)

K.W. Oleson, D.M. Lawrence, B. Gordon, M.G. Flanner,E. Kluzek, J. Peter, , ... C.L. Heald, Technical description of version 4.0 of the Community Land Model (CLM). (2010)

D. Papale, R. Valentini, A new assessment of European forests carbon exchanges by eddy fluxes and artificial neural network spatialization. Glob. Chang. Biol. **9**(4), 525–535 (2010)

D. Papale et al., Towards a standardized processing of net ecosystem exchange measured with eddy covariance technique: algorithms and uncertainty estimation. Biogeosciences **3**(4), 571–583 (2006)

S. Plummer, On validation of the MODIS gross primary production product. IEEE Trans. Geosci. Remote Sens. **44**, 10 (2006)

A.S. Raghavendra, K. Padmasree, Beneficial interactions of mitochondrial metabolism with photosynthetic carbon assimilation. Trends Plant Sci. **8**(11), 546–553 (2003)

M. Reichstein et al., On the separation of net ecosystem exchange into assimilation and ecosystem respiration: review and improved algorithm. Glob. Chang. Biol. **11**(9), 1424–1439 (2005)

A.D. Richardson et al., Comparing simple respiration models for eddy flux and dynamic chamber data. Agric. For. Meteorol. **141**(2), 219–234 (2006)

S.W. Running, S.T. Gower, FOREST-BGC, a general model of forest ecosystem processes for regional applications. II. Dynamic carbon allocation and nitrogen budgets. Tree Physiol. **9**(1-2), 147–160 (1991)

S.W. Running, P.E. Thornton, R. Nemani, J.M. Glassy, Global terrestrial gross and net primary productivity from the earth observing system. Methods Ecosyst. Sci. **3**, 44–45 (2000)

H.P. Schmid, Source areas for scalar and scalar fluxes. Bound.-Layer Meteorol. **67**, 25 (1994)

C.R. Schwalm, T.A. Black, K. Morgenstern, E.R. Humphreys, A method for deriving net primary productivity and component respiratory fluxes from tower-based eddy covariance data: a case study using a 17-year data record from a Douglas-fir chronosequence. Glob. Chang. Biol. **13**(2), 370–385 (2007)

T.D. Sharkey, C.J. Bernacchi, G.D. Farquhar, E.L. Singsaas, Fitting photosynthetic carbon dioxide response curves for C3 leaves. Plant Cell Environ. **30**(9), 1035–1040 (2007)

D.A. Sims, A.F. Rahman, V.D. Cordova, B.Z. El-Masri, D.D. Baldocchi, P.V. Bolstad, L.B. Flanagan, A.H. Goldstein, D.Y. Hollinger, L. Misson, R.K. Monson, W.C. Oechel, H.P. Schmid, S.C. Wofsy, L.K. Xu, A new model of gross primary productivity for North American ecosystems based solely on the enhanced vegetation index and land surface temperature from MODIS. Remote Sens. Environ. **112**, 14 (2008)

S. Sitch et al., Evaluation of ecosystem dynamics, plant geography and terrestrial carbon cycling in the LPJ dynamic global vegetation model. Glob. Chang. Biol. **9**(2), 161–185 (2003)

V. Stauch, A. Jarvis, A semi-parametric gap-filling model for eddy covariance CO2 flux time series data. Glob. Chang. Biol. **12**(9), 1707–1716 (2006)

J.R. Stinziano et al., The rapid A-Ci response: photosynthesis in the phenomic era. Plant Cell Environ. **40**, 1256–1262 (2017)

P.C. Stoy et al., An evaluation of models for partitioning eddy covariance-measured net ecosystem exchange into photosynthesis and respiration. Agric. For. Meteorol. **141**(1), 2–18 (2006)

P. Thornton et al., Modeling and measuring the effects of disturbance history and climate on carbon and water budgets in evergreen needleleaf forests. Agric. For. Meteorol. **113**(1), 185–222 (2002)

C.J. Tucker, Red and photographic infrared linear combinations for monitoring vegetation. Remote Sens. Environ. **8**(2), 127–150 (1979)

M. Verma et al., Effect of environmental conditions on the relationship between solar-induced fluorescence and gross primary productivity at an OzFlux grassland site. J. Geophys. Res. Biogeo. **122**, 17 (2017)

F. Veroustraete, H. Sabbe, H. Eerens, Estimation of carbon mass fluxes over Europe using the C-Fix model and Euroflux data. Remote Sens. Environ. **83**(3), 376–399 (2002)

P. Wagle, Y. Zhang, C. Jin, X. Xiao, Comparison of solar-induced chlorophyll fluorescence, light-use efficiency, and process-based GPP models in maize. Ecol. Appl. **26**(4), 11 (2016)

X. Wang et al., Vegetation primary production estimation at maize and alpine meadow over the Heihe River Basin, China. Int. J. Appl. Earth Obs. Geoinf. **17**, 94–101 (2012)

X. Wang et al., Validation of MODIS-GPP product at 10 flux sites in northern China. Int. J. Remote Sens. **34**(2), 587–599 (2013)

C.Y. Wu, Z. Niu, J. Tang, W.J. Huang, Remote estimation of gross primary production in wheat using chlorophyll-related vegetation indices. Agric. For. Meteorol. **149**, 7 (2009a)

X. Wu et al., Conditional inversion to estimate parameters from eddy-flux observations. J. Plant Ecol. **2**(2), 55–68 (2009b)

C.Y. Wu, X.Z. Han, J.S. Ni, Z. Niu, W. Huang, Estimation of gross primary production in wheat from in situ measurements. Int. J. Appl. Earth Obs. Geoinf. **12**, 7 (2010)

X. Xiao et al., Satellite-based modeling of gross primary production in an evergreen needleleaf forest. Remote Sens. Environ. **89**(4), 519–534 (2004)

X. Xiao, Q. Zhang, D. Hollinger, J. Aber, B. Moore, Modeling gross primary production of an evergreen needleleaf forest using MODIS and climate data. Ecol. Appl. **15**(3), 954–969 (2005)

T. Xu, L. White, D. Hui, Y. Luo, Probabilistic inversion of a terrestrial ecosystem model: analysis of uncertainty in parameter estimation and model prediction. Glob. Biogeochem. Cycles. **20**(2) (2006)

Z. Xu et al., Intercomparison of surface energy flux measurement systems used during the HiWATER-MUSOEXE. J. Geophys. Res. Atmos. **118**(23) (2013)

H. Yan et al., Modeling gross primary productivity for winter wheat–maize double cropping system using MODIS time series and CO 2 eddy flux tower data. Agric. Ecosyst. Environ. **129**(4), 391–400 (2009)

Yang X, Tang J, Mustard J.F, et al. Solar-induced chlorophyll fluorescence that correlates with canopy photosynthesis on diurnal and seasonal scales in a temperate deciduous forest[J]. Geophysical Research Letters. 42(8): 2977–2987 (2015)

C. Yi et al., A nonparametric method for separating photosynthesis and respiration components in CO 2 flux measurements. Geophys. Res. Lett. **31**(17), 123–138 (2004)

W. Yuan et al., Global estimates of evapotranspiration and gross primary production based on MODIS and global meteorology data. Remote Sens. Environ. **114**(7), 1416–1431 (2010)

Y.Q. Zhang, Q. Yu, J. Jiang, Y.H. Tan, Calibration of Terra/MODIS gross primary production over an irrigated cropland on the North China Plain and an alpine meadow on the Tibetan Plateau. Glob. Chang. Biol. **14**, 10 (2008)

Leaf Area Index: Advances in Ground-Based Measurement

Yonghua Qu

Contents

Abstract

Leaf area index (LAI) is a primary parameter for vegetation structure and is one of the important products from remote sensing data. Ground-based LAI estimation of LAI is the important activity for global satellite product validation. In this chapter, the main methods on LAI measurement were reviewed, and the emphasis was put on the new advanced method on LAI measurement. We present two new instruments (LAINet and LAISmart) designed by Beijing Normal University which use either modern communication network or mobile computing platform to obtain LAI with high efficiency and low cost. LAINet is an instrument constructed on the base of wireless sensor network, and the principle of LAINet is capturing sunlight transmittance using a series of wireless sensors in different

Y. Qu (✉)
State Key Laboratory of Remote Sensing Science, Beijing Normal University and Institute of Remote Sensing and Digital Earth of Chinese Academy of Sciences, Beijing, China

Beijing Engineering Research Center for Global Land Remote Sensing Products, Institute of Remote Sensing Science and Engineering, Faculty of Geographical Science, Beijing Normal University, Beijing, China
e-mail: qyh@bnu.edu.cn

X. Li, H. Vereecken (eds.), *Observation and Measurement of Ecohydrological Processes*, Ecohydrology, https://doi.org/10.1007/978-3-662-48297-1_11

sun zenith angles. And LAI is estimated from the sensed transmittances. Borrowing the wireless communication technique, the measured data can be transferred to remote computer server, thus, LAINet can reduce the cost of field data collection. LAISmart is a mobile application deployed on the smartphone, and the LAI is calculated by the classification of captured image. By integration of capturing images and real time computing of smartphone, LAISmart provide automatic measurement method compared with the traditional digital hemispherical photography method. In the end of this chapter, the prospect of the methods on the LAI ground-based measurement is summarized, and it is pointed out that the integration of passive and active optical signal to produce low cost and light weight and thus affordable and portable device may be a promising tendency.

Keywords
Leaf area index · Ground-based method · Wireless sensor network · Mobile smartphone · LAINet · LAISmart

Introduction

Leaf area index (LAI) is defined as half of the leaf surface area per horizontal ground area unit and is an important parameter for ecological and hydrological modeling (Chen and Black 1991). Estimation of the LAI using remote sensing data has been an important method for obtaining LAI data at regional and global scales. LAI ground measurement methods are classified as either direct or indirect (Garrigues et al. 2008). The direct method is considered to be the most reliable for LAI ground truth. However, limited to its damaging to vegetation and its low efficiency, the direct method can only be used to obtain the ground LAI reference values, which are used to calibrate other indirect measuring instruments.

In the methods of indirect measurement, there are also two types of subdivision depending on the raw data obtained methods, i.e., imaging and non-imaging method.

The basic principle of the imaging method is to take photographs of a canopy at single or multiple shooting angles. The captured images are then segmented into leaf and background, which is known as gap classification, to obtain the gap fraction of the canopy at single or multiple angles. The LAI can then be computed on the basis of a gap fraction model (Nilson 1971).

For the non-imaging method, LAI is measured from canopy transmittance under specified conditions (Bréda 2003). For example, the LAI-2000 or LAI-2200 (Li-Cor, Lincoln, Nebraska, USA) should be measured in diffuse light. Furthermore, the approximately real-time total downward solar radiation must be measured in open spaces or outside of the canopy. Then the ratio of the two measurements in five angles can be used to calculate the canopy transmittance (Arias et al. 2007). Unlike the LAI-2000, which uses diffuse light, DEMON (CSIRO, Canberra, Australia) repeatedly measures direct radiation (Whitford et al. 1995). The sensor needs to move linearly over a certain distance underneath the canopy during measurement, and the measurements must be repeated several times in a day to collect data across a range of zenith angles.

However, the above two types of indirect methods faced a challenge nowadays on the application of remote sensing and ecological aspect, especially in the case where they are used in large space region and high frequent temporal scale. It was hoped to obtain data automatically by observing data from different measurement instruments and interconnecting them with a wireless network (Schirrmann et al. 2015). Other researchers have indicated that a simple and replicable method for obtaining a ground sample of vegetation parameters is important for ecological observations and for the application of remote sensing data (Morais et al. 2008).

In the field of earth observations, a new method based on wireless sensor network technology, which is called the Earth Observation Sensor Web (EOSW), has gained increasing attention (Di et al. 2010). The EOSW has been preliminarily applied in many research fields such as sea ice, soil moisture, and environmental monitoring. Currently, in terms of vegetation wireless sensor network monitoring, a significant amount of work has been performed to measure the environmental parameters (such as temperature and moisture) of vegetation growth (Rundel et al. 2009). However, few studies have evaluated the automatic measurement system of vegetation LAI. Compared to the newly emerging EOSW technique, over the past 20 years, ground-based LAI measurements have developed relatively slowly. In the section "WSN-Based Methods: LAINet" of this chapter, we will present the prototype of using the wireless sensor network technology to automatically measure the vegetation LAI. In this section, LAI is measured through a series of light sensitive non-imaging sensors, which are connected by a wireless network.

As mentioned above, besides the non-imaging sensors in measuring LAI, another tendency of indirect method is to estimate LAI from camera images. Diverse types of imaging devices are available such as professional single-lens reflex (SLR) cameras, common digital cameras, and pre-1970 film cameras. Among imaging sensor-based methods, a new development that uses a smartphone camera sensor rather than a general-purpose digital camera has emerged recently (Fuentes et al. 2012; Qu et al. 2016, 2017). Building on the high performance-price ratio and multi-sensor integration of the smartphone, this approach has attracted much attention to measuring LAI using smartphone camera sensors. As a result, in the section "Mobile Smartphone-Based Method: LAISmart" of this chapter, we will introduce the advance using smartphone to measure vegetation LAI.

WSN-Based Methods: LAINet

Methodology (Qu et al. 2014)

Overview of the WSN System

LAINet is an automatic device that obtains multi-angle transmittance of the vegetation canopy during the day following sun's movement, based on wireless sensor network technology, and then calculates the vegetation LAI from multi-angle transmittance using a vegetation gap probability model (Qu et al. 2014). The framework of this system is illustrated in Fig. 1. This wireless sensor network consists of three

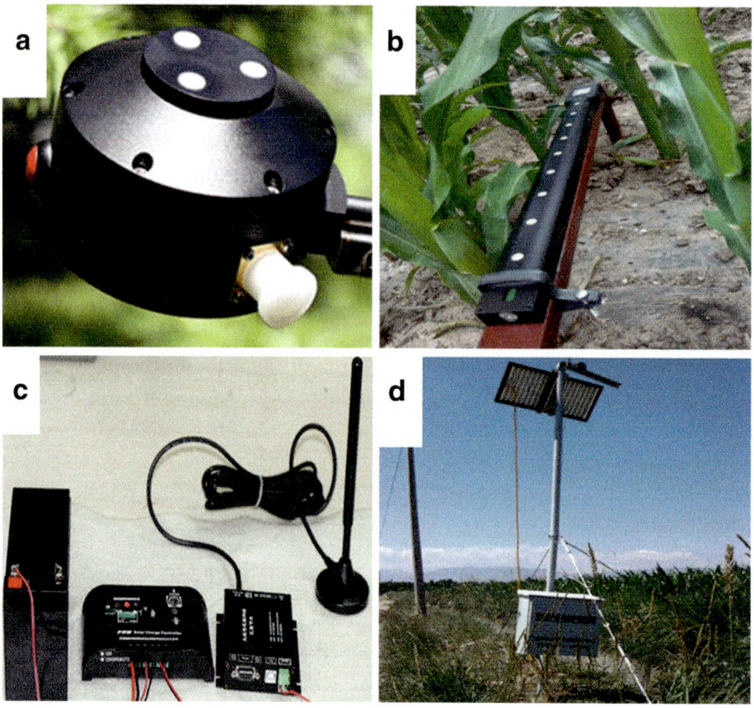

Fig. 1 The LAINet nodes, including (**a**) a node above the canopy, (**b**) a linear node with 9 light sensors under the canopy, (**c**) the sink node (right part) integrated with GRPS model together with the solar power control(left and middle part). All the components in (**c**) are deployed in the instrument container in the bottom of (**d**) and the above node in (**a**) is also deployed on the top of the support of (**d**)

different node types. The first node type is a circle node with three light sensors that are deployed above the canopy to receive the total downward solar radiation (Fig. 1a). Next, linear node types are used that have nine light sensors under the canopy and that receive the transmitted radiation (Fig. 1b). Finally, sink nodes are used to collect and transmit data to a remote data server through the General Packet Radio Service (GPRS) network (Fig. 1c). Communication between the nodes is implemented with a Zigbee network, which is based on the standard IEEE802.15.4 low-power field network protocol. The solar power components (the left and middle part in Fig. 1c) are deployed in the instrument container which is fixed on the upright support as illustrated in Fig. 1d, and the above node is also fixed on the top of horizontal support in Fig. 1d. The collected data will be transferred to the remote data server through a GPRS module (the right part in Fig. 1c).

If only a few measurements in a limited space are needed and the measurements are not in time series, LAINet is not as advantageous as traditional LAI measurement instruments. However, if continuous long time series observations are carried out at a large regional scale, LAINet has the following advantages:

1. It can reduce the time required to access the experimental site. LAINet only requires to enter the experimental site during instrument deployment and, therefore, can minimize plant disturbances by the experimenter, which is not possible with current traditional instruments.
2. Multiple measurement nodes are automatically synchronized, which ensures that the measured data of each node are obtained simultaneously and avoids the data uncertainty caused by differences in measurement time.
3. Wireless communication among measurement nodes requires no wiring. Communication both among the nodes under the canopy and among the nodes under and above the canopy is completed via Zigbee wireless networks. As a result, there is no need to lay and connect cables in the open area or under the canopy to facilitate instrument deployment.
4. The field data are remotely transmitted to the data center via public wireless networks and are automatically saved on the database server. Operator labor is not required to collect field data.

The Principle of LAINet

The data used to calculate LAI included the total solar radiation above the canopy and the transmitted radiation below the canopy. The canopy transmittance is obtained by calculating the ratio of the radiation transmitted below the canopy to the total incident radiation above the canopy at different solar angles throughout the day. These calculations are performed as following:

1. The solar elevation angle h is calculated based on the longitude, latitude, and local measurement time.
2. The start and end of the measurements above and below the canopy are read, and the overlapping time of the two nodes is calculated. Because the two types of nodes could work at different times, a uniform interpolation is applied to the observed sensor values within an overlapping time interval of 10 min. Therefore, the observations above and below the canopy are obtained at each time. These observations are denoted as E_0 and E_1, respectively.
3. The ratio of the observations below and above the canopy is calculated at each time to obtain the canopy transmittance at different solar elevation angles as Eq. (1),

$$T(h) = \frac{E_1}{E_0} \tag{1}$$

LAI Calculations

The gap probability model for planted canopies is used based on the Poisson distribution (Lang 1987; Haverd et al. 2012),

$$T(h) = e^{-\int_0^{\frac{\pi}{2}} A(\theta,h)\,\widehat{g}(\theta) LAI d\theta} \tag{2}$$

In Eq. (2), θ represents the leaf inclination angle; h represents the solar elevation angle; $\hat{g}(\theta)$ is the distribution function of the leaf inclination angle; and $A(\theta,h)$ is a trigonometric function as following,

$$A(\theta,h) = \begin{cases} \cos\theta & \theta \le h \\ \cos\theta[1 + 2(\tan\theta_0 - \theta_0)/\pi] & h < \theta < \frac{\pi}{2} \end{cases} \quad (3)$$

In Eq. (3), $\theta_0 = \text{acos}(\tan h \cot\theta)$. We combined Eq. (2) and Eq. (3) and calculated the logarithms on both sides of Eq. (2) to obtain Eq. (4), as following,

$$P(h) = -\ln(T(h)) = \int_0^{\frac{\pi}{2}} A(\theta,h)\widehat{g}(\theta)LAId\theta \quad (4)$$

Solving Eq. (4) requires a continuous observation of $P(h)$. However, in practical observations, we can only obtain discrete transmittance values. Therefore, it was difficult to solve the integral Eq. (4) directly. However, Eq. (4) can also be written in a discrete form combing the term $A(\theta,h)$ and $\widehat{g}(\theta)$ into one term $K(\theta,h)$:

$$K(\theta,h) = A(\theta,h)\widehat{g}(\theta) \quad (5)$$

Then, the leaf inclination angle is divided into M + 1 interval with equal intervals within the range of $[\theta, \pi/2]$. The length of each interval is $\Delta\theta$. The average leaf inclination angle of the interval midpoint is θ_i, with $i \in [0, M]$. The LAI within the interval of each leaf inclination angle is L_i. Therefore, the Eq. (4) can be written in a discrete form, as shown in Eq. (6).

$$P(h_j) = \sum_{i=1}^{M} K(\theta_i, h_i)L_i, j \in [1,N] \quad (6)$$

where N is the number of solar elevation angles.

When $M < N$, then L_i can be obtained by solving the equation. Next, the canopy LAI can be obtained using Eq. (7),

$$LAI = \sum_{i}^{M} L_i \quad (7)$$

Before calculating the LAI, information regarding the leaf inclination angle is needed. According to Geol (1988), the leaf inclination angle distribution patterns can be classified into six types. Based on previous results and field data e.g., for the corn crop type, we assumed that the probability density function of the vegetation leaf inclination angle is vertical as Eq. (8)

$$\widehat{g}(\theta) = e^{0.583\theta - 1} \quad (8)$$

where the mathematical expectation of θ is 63.2°.

Case Study

Experiment

The field experiments were part of the Heihe Watershed Allied Telemetry Experimental Research (HiWATER) project (Li et al. 2013), which is a multi-disciplinary and integrated remote sensing experiment evaluating ecological and hydrological processes in the Heihe River basin. This area is located in the Yingke irrigation districts (100°40′E, 38°80′N). The observations were made between June 25 and August 24 in the year of 2012.

We deployed nine sink nodes in the LAINet area within a range of 4 × 4 km (4 × 4 1 km pixels with a spatial location corresponding to MODIS data) as shown in Fig. 2. Each sink node was connected with 3–9 measurement nodes using the Zigbee network. As a result, a total of 42 measurement nodes under the canopy and 3 above the canopy were established.

In Fig. 2, the sink nodes are powered by solar energy, and their antenna systems are fixed on 2.5 m supports to ensure that the signals are not sheltered by the canopy (which had a maximum height of 2.2 m). The three nodes above the canopy and their corresponding convergent antennas are deployed on the upper side of the same

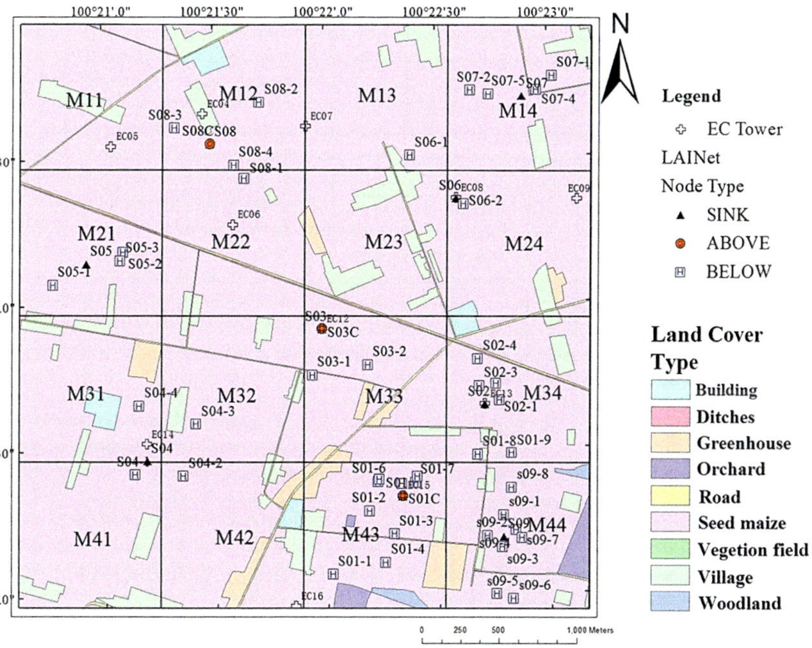

Fig. 2 Instrument nodes deployment map in the experimental area. There are 9 sink nodes (S01-S09) and 42 measurement nodes under the canopy (S0i-j: i stands for the sink node number and j stands for the measurement node number below the sink nodes), 3 measurement nodes for total radiation above the canopy (S01C, S02C and S03C) and 15 eddy covariance matrix towers (EC01-EC15) with 4 × 4 MODIS pixels (Mij, i, j = 1, 2, 3, 4)

supports to ensure that the nodes above the canopy were not sheltered by other objects. The nodes under the canopy, which are powered by rechargeable lithium batteries, are positioned upward and perpendicularly to the corn ridges. Each node under the canopy is supported by a 20 cm brace to ensure that the node was not immersed in water during agricultural irrigation. All measurement and sink nodes are placed between 200 and 500 m apart. Data communication between the measurement and sink nodes is conducted directly or with a multi-hop mechanism. Therefore, the measured node data are easily transmitted by newly formed routes when the data were not transmitted directly to the sink node.

In the course of the experiment, the sensors were set to operate from the 09:00 to 18:00 h (the minimum value of the local solar elevation angle ranged from 26° to 32° and the maximum value ranged from 67° to 81°). During this period, the sampling interval of the sensor is 15 min, with a sampling duration of 5 s. Therefore, during each 15-min sampling interval, the working duration is only 5 s. In addition, the sensor automatically shifts into sleep mode, which greatly reduces the instrument's energy consumption and extend its working duration.

To evaluate the LAINet measurement results, we use the LAI-2000 to measure the vegetation LAI at locations in the experimental area that are near some of the nodes. To maintain the vegetation's stable growth around the LAINet site and to avoid human disturbance, the measured points of the LAI-2000 are placed at approximately 50–100 m from the LAINet node. The selection of the LAI-2000 sampling locations was based on the vegetation growth at the LAINet node to ensure that the growth statuses at the two types of measurement sites are alike. In the LAI-2000 measurement scheme, an area of $5 \times 5\ m^2$ is used to collect between 7 and 9 readings in a diagonal pattern. Next, the average value of the readings is used to calculate the LAI of the measured points. The measurement revisit period for the LAI-2000 is 5 days before July 23 and 10 days after that time.

Results

For every node deployed in the field from June 25 to August 24, the observation period lasted approximately 60 days, but due to the unexpected obstacles, the maximum valid observation days were approximately 30 and most of the nodes captured LAI frequency as 2 or 3 days (Fig. 3). As a result, to compare the measured results from the two different instruments, we use two adjacent instruments (within 100 m). The time gaps between the measurements are 1 or 2 days. The results of these comparisons between the measured LAINet and LAI-2000 values are presented in Fig. 3.

As shown in Fig. 3, the coefficients of determination (R^2) of the results are from 0.27 to 0.97, with an average R^2 of 0.42. These results indicate that the differences between the data measured with the two different instruments varied by site. The highest relevant difference locates at the EC06 site. At this site, the data were collected from June 25th (DOY = 177) to August 11th (DOY = 224) with the LAINet values measured between 1.6 and 3.4. The lowest relevant difference was observed at the EC14 site, where the greatest difference between the two instruments was observed when the LAI values were lower than 1.5 or greater than 3.0.

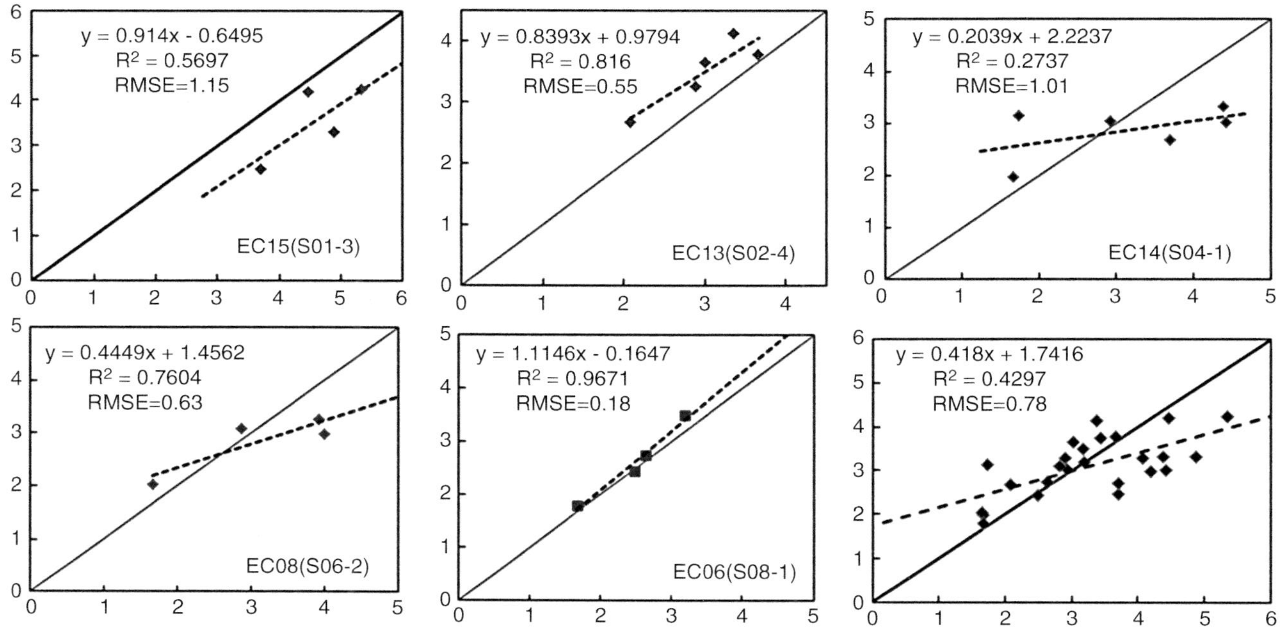

Fig. 3 Scatter plots of the measured LAI values for the LAINet and LAI-2000 observations in the location of eddy covariance matrix towers corresponding to the the ECx from subplot (**a**–**f**)

In Fig. 3, when the LAI is lower than 3.5, the LAI-2000 estimates are greater than the LAINet estimates. Conversely, when the LAI values are higher than 3.5, the observed LAI-2000 values are lower than the LAINet values. Unfortunately, the LAI destructives value for crops are not obtained in this study. Although the current measurements do not clarify which instrument is more accurate, many researchers have indicated that the LAI-2000 estimates are lower than the true LAI values. This finding is most notable when the LAI is relatively large, in which case the observed LAI-2000 values are much lower than their true values. For example, by comparing the canopy LAI of broad-leaf and coniferous forests, Martens et al. (1993) determined that the Li-Cor light quantum meter estimates (which used the direct light method as our method in this study) were greater than the LAI-2000 estimates. Compared with the direct measurement method, the Li-Cor light quantum meter method is the most accurate. Other researchers have reported similar results. For example, Mason et al. discovered that the ratio of direct measurement values to the LAI-2000 values was approximately 1.5 (Mason et al. 2012). We have similar findings. When the LAI is relatively large (higher than 3.5), the average ratio of the observed LAINet value (with the direct light principle) to the LAI-2000 value is approximately 1.3. This result indicates that the measured LAINet value is potentially closer to the true LAI value of the vegetation. However, this conclusion needs to be validated with true ground measurements in further work.

Mobile Smartphone-Based Method: LAISmart

Methodology (Qu et al. 2016)

Overview of the LAISmart System

LAISmart is composed of two smartphones connected by a support system and communicating by a wireless network. The frontend terminal serves as a data collection terminal, and the backend terminal serves as a user operation console (Fig. 4).

Currently, both the backend and frontend smart terminals are run on the Android operation system, and their hardware configuration must include a

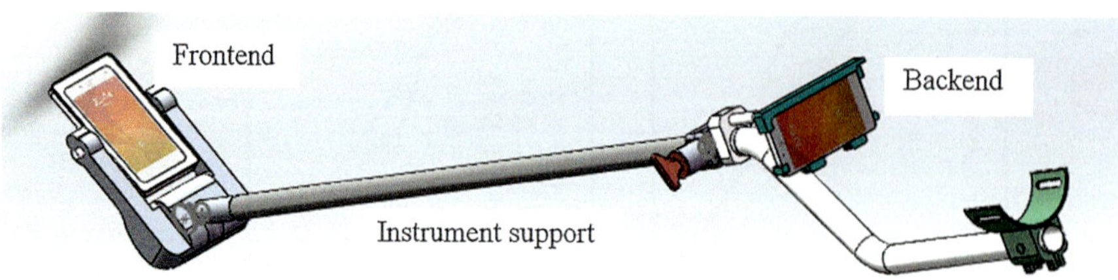

Fig. 4 Diagram of the LAISmart illustrating the main hardware modules

Global Positioning System (GPS) sensor, a gyro sensor, an imaging sensor, and a Wireless Fidelity (WIFI) sensor. The smart terminals of LAISmart are configured by two mobile phones with a resolution of 13 megapixels and a maximum FOV of 70°.

When operating LAISmart, the user can adjust the shooting angle of the frontend terminal by adjusting the direction of support system. In addition, the frontend angle value together with its GPS location and the preview image can be displayed at the backend screen through the wireless network, which is established by the hotspot used for the backend terminal. Before the measurement begins, the operator can set the photographing direction of the imaging sensor according to the vegetation height. For tall trees or short but continuous crops, the imaging sensor is recommended to be shot upward. In such situations, only vegetation and sky are in the FOV, whereas for short but discrete canopy, the operator can shoot downward with the sensor above the canopy. In such situations, the vegetation and soil are in the FOV.

Calculation of Canopy Gap Fraction

Extraction of the gap fraction from the images is actually achieved by automatic segmentation of vegetation and non-vegetation in the digital photographs. In LAISmart, the Otsu algorithm (Otsu 1979) is used to extract the gap fraction of the image automatically. The principle of the Otsu algorithm is to search for a segmentation threshold that maximizes the between-class variance of the segmented binary images.

However, owing to the effects of shooting conditions (e.g., direction of imaging sensor and ratio of sky light), the choice of the image segmentation feature for the foreground (vegetation) and background (sky or soil) varies in actual measuring situations. In fact, it is impractical to maintain one feature that can automatically segment images in a variety of conditions. Therefore, based on the original RGB image, the LAISmart system provides two candidate features. The first is greenness index (GI) (Booth et al. 2005),

$$GI = \frac{2^{*}G - R - B}{2^{*}G + R + B} \tag{9}$$

where R, G, B is the red, green, and blue band intensity in a color image, respectively. The second option is to use the blue band intensity value (B) directly (Welles 1990).

GI is effective for distinguishing vegetation from soil in downward photography or upward if the leaf vegetation has good transmittance, whereas the B feature can effectively distinguish vegetation and sky in upward photography with strong scattered light. The OpenCV library (http://opencv.org) was used in LAISmart to implement the Otsu algorithm.

Retrieval of LAI from Gap Fraction

On the condition that the leaves are randomly distributed and their sizes are significantly smaller than that of the canopy, the relationship between canopy gap fraction and LAI can be expressed as the Beer gap fraction model (Nilson 1971):

$$P(\theta) = e^{-G(\theta)LAI/\cos\theta}. \quad (10)$$

Then, LAI can be obtained as

$$LAI = -\ln P(\theta)\cos(\theta)/G(\theta), \quad (11)$$

where $P(\theta)$ represents the canopy gap fraction at the view zenith of θ, and $G(\theta)$ is the projection of the unit foliage area on the plane perpendicular to the view angle.

According to Eq. (11), knowledge of $G(\theta)$ is needed for retrieving LAI. Research results from Goudriaan (1988) indicated that the distribution function of the leaf angle could be approximated by a spherical distribution, and in such situations, the average projected area ratio of leaves $G(\theta)$ is 0.5 independent of the viewing angle of the sensor. Although arguments have been presented on this assumption (Pisek et al. 2013), many researchers have produced acceptable LAI by using this model when the leaf distribution function is unavailable (Chen and Black 1992; Chen 1996). Here, we take this assumption as our basis for calculating LAI. Thus,

$$LAI = -2\ln P(\theta)\cos(\theta), \quad (12)$$

where θ can be read from the gyro attitude sensor of the frontend smart terminal, and the canopy gap fraction $P(\theta)$ is equal to the proportion of background pixels in the image in the observation direction θ.

Once the desired photographs are captured correctly, LAI is determined by simply pressing the button on the backend terminal to trigger the real-time computing software.

Case Study

Experiment

Eight datasets were used covering four vegetation types in four provinces in China from 2015 to 2016 (Table 1). Two types of plot protocol were designed in the experiments according to the shape of the plot. For datasets S1, S3, and S4 in the province of Beijing, we deployed the sampling points along a line profile. When conducting measurement, the operator carried LAISmart waking along one profile in the plot and taking photographs at intervals of about four meters. The average length of a profile was about 30–40 m; thus, seven to nine photographs were taken for each profile. For the other datasets, the sampling points were located in a quadrat in 30 m × 30 m. In each quadrat, there were seven to nine LAISmart sampling points surrounded the center of the quadrat.

Depending on the vegetation height, two types of operation mode of LAISmart are used, as shown in Fig. 5. For tall forests (Qu et al, 2017) or short but continuous vegetation, the upward-looking mode is used in which the blue band is selected to segment vegetation and sky pixels. For short and discrete vegetation, the downward-

Table 1 List of experiment sites and the operation mode of LAISmart

Dataset	Province	Type	Latin name	Mode/ feature	Date time
S1	Beijing	ENF	Mixed *Platycladus orientalis* and *Pinus bungeana*	Upward/ Blue	May 26, 2015; Dec. 18, 2015[a]
S2	Beijing	DBF	*Platanus hispanica*	Upward/ Blue	May 29, 2015
S3	Beijing	DBF	*Populus L.*	Upward/ Blue	Jun. 15, 2016[b]
S4	Beijing	Vegetation	*Triticum aestivum*	Upward/ Blue	Jun. 15, 2016
S5	Heilongjiang	Vegetation	*Zea mays L*	Downward/ GI	Jun. 25, 2015
S6	Heilongjiang	Vegetation	*Zea mays L*	Upward/ Blue	Jul. 30, 2015
S7	Hunan	ENF	*Cunninghamia lanceolata*	Upward/ Blue	Jun. 5, 2016
S8	Inner Mongolia	DNF	*Larix gmelinii*	Upward/ Blue	May 25, 2016

ENV evergreen needleleaf forest, *DBF* deciduous broadleaf forest, *DNF* deciduous needleleaf forest
[a]Date on which the images at a shooting angle of 57.5° were also captured
[b]Date on which the images were also captured in multiple directions

Fig. 5 Operation of LAISmart for different vegetation types. (**a**) Upward looking mode for tall forest (left) and short but continuous (right) vegetation. (**b**) Downward looking mode for short and discrete vegetation

looking mode is used in which the greenness index is employed to segment the soil and vegetation.

To validate the performance of LAISmart, the LAI values in every plot were also collected by using the LAI-2000 instrument after measurement was conducted by using the proposed method. When conducting measurement by using LAI-2000, twilight time was selected to reduce the influences of direct sunlight. After one measurement was conducted in totally diffused light outside the forest or above the vegetation canopy, seven or nine radiance values below the canopy of the LAI-2000 instrument were acquired at the approximate location of LAISmart sampling point. The averaged LAI values of LAISmart and the output of the LAI-2000 instrument in the same plot were taken as a pairwise measurement, which were used to validate the performance of LAISmart.

In addition to the aforementioned datasets, which were used for point-to-point comparison of different instruments, we conducted other experiments to evaluate the rationality of using a lower zenith gap rather than a higher zenith to retrieve LAI from images captured by the smartphone. In these experiments, we acquired images at the zenith angle of 57.5° on the point at which LAI-2000 was operated at the ENF site on December 18, 2015 (part of S1 data in Table 1). We also conducted a multidirectional experiment at the DBF site on June 15, 2016 (S3 in Table 1). The calculated 57.5° and multidirectional gaps were used to compare with those of the LAI-2000 instrument acquired at the corresponding direction.

Results

In summary, 114 pairwise data of LAISmart and LAI-2000 were acquired in the eight datasets. For brevity, we arranged the datasets into three groups, i.e., DNF and DBF as Group 1 of deciduous leaf forest, because they share the common feature of leaf growing and defoliating process with time; ENF as Group 2 of evergreen leaf forests; and maize and wheat crops as Group 3. The statistical characteristics of the LAI values in these groups are shown in Table 2.

In general, mean and median were highly consistent between the two different instruments. It was deduced that compared with the commercial LAI-2000 instrument, the smartphone-based LAI measurement, LAISmart, can produce acceptable results with high accuracy by averaging multi-point values. Moreover, a larger variance was observed for Group 1 and Group 3 datasets, which was derived mainly by the larger maximum value of LAI-2000. This indicates that LAISmart might underestimate the LAI value in situations of high-density vegetation cover.

For the ENF and crops datasets, we measured the forest LAI at two different times. It was expected that the LAI values for the ENF dataset would remain at a relatively constant level even though there was a time lag between the two measurements, whereas the LAI values of crops in the process of the growth stage were expected to differ significantly from those of the ENF type. The histogram of

Table 2 Statistics of LAI values of LAISmart and LAI-2000

	Instrument	Samples	Mean	Median	Maximum	Minimum	Variance
Group 1	LAI-2000	19	2.84	2.34	5.11	1.35	1.62
	LAISmart		2.61	2.10	3.92	1.53	0.72
Group 2	LAI-2000	25	2.50	2.38	4.40	1.28	0.62
	LAISmart		2.38	2.20	4.63	1.55	0.59
Group 3	LAI-2000	70	2.09	1.07	4.77	0.19	2.28
	LAISmart		2.07	1.43	4.02	0.23	1.62

the two types of datasets confirmed these expectations, as illustrated in Fig. 6. Although a slight increase was noted for ENF, which was caused mainly by the new needle leaves emerging in 1 year, most of the observation values were near 2.4 m^2/m^2 (Fig. 6a and b). For the crops dataset, as expected, the experiment results revealed that the maize was in the rapid growth stage, which is reflected in the mean value from 0.99 m^2/m^2 (Fig. 6c) to 3.47 m^2/m^2 (Fig. 6d) in about 1 month. It is also indicated in Fig. 6 that the histogram was wider in the early growth stage than in the middle stage, which can be explained by the decrease in landscape heterogeneity during vegetation growth from June to July.

In summary, the LAISmart instrument revealed the ground LAI variation manifested from the different vegetation types and different vegetation growth stages.

We compared the LAISmart values with the measured LAI value of LAI-2000 in the form of a scatter plot (Fig. 7a) and divided the dataset into three groups (Table 2 and Fig. 7b–d). After performing point-by-point comparisons and fitting the results by using an intercept-zero linear regression model, the regression statistical results of

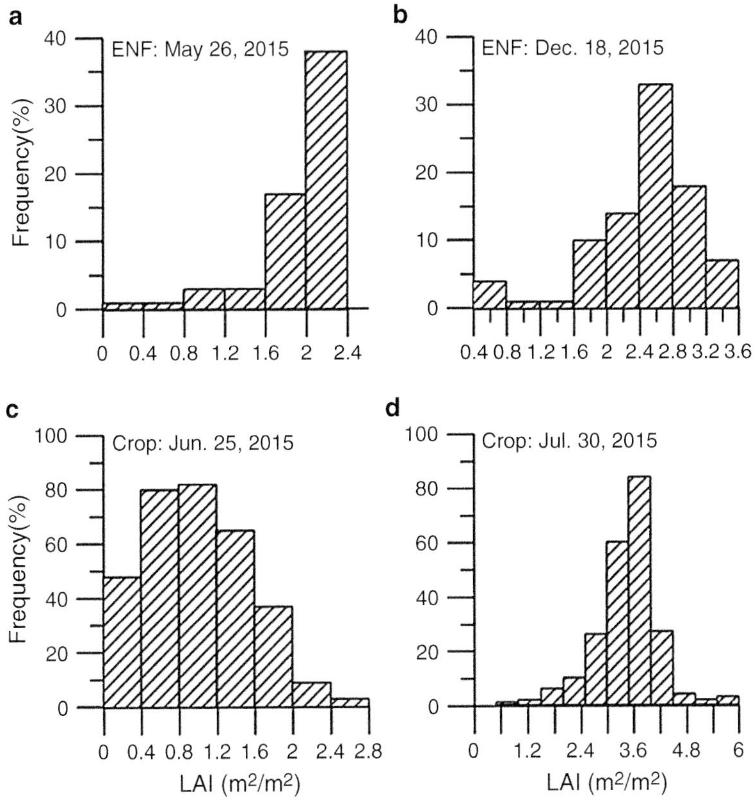

Fig. 6 Frequency distribution histogram of ENF and crops LAI values at different times

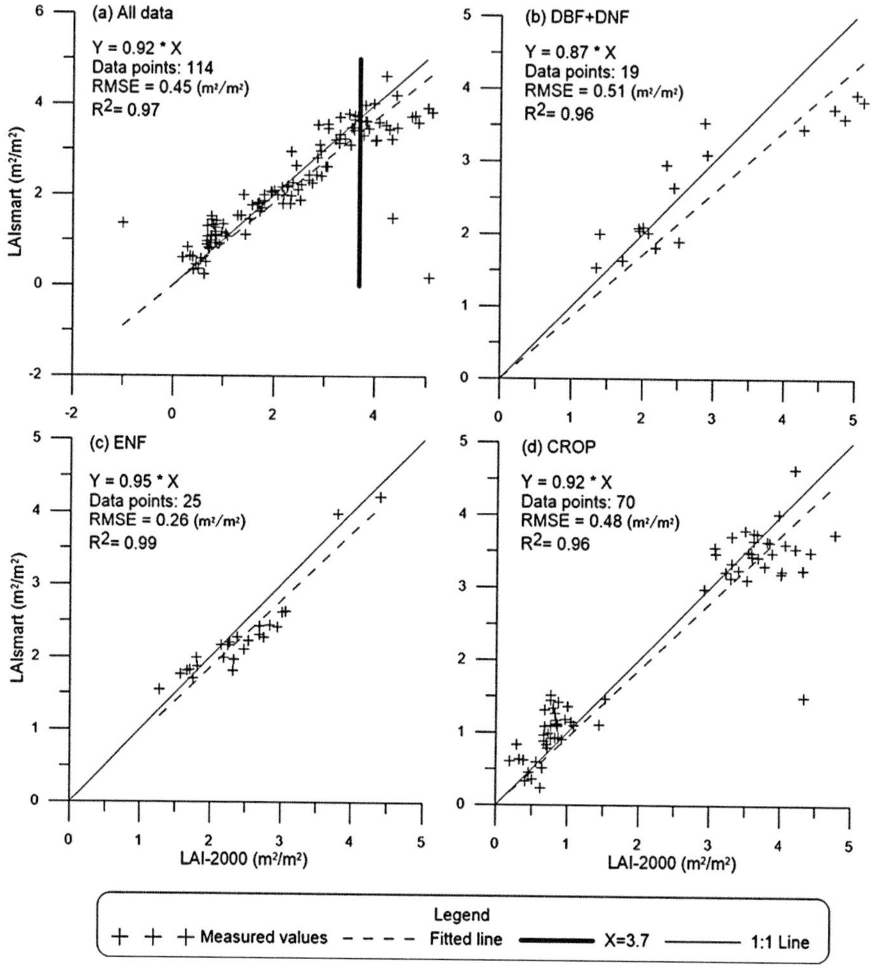

Fig. 7 Scatter diagram of LAISmart and LAI-2000 measurements, where the separation point of 3.7 (m^2/m^2) is also plotted as a black line in subplot (**a**)

the datasets showed a common absolute high determination coefficient (R^2) of 0.97 in total with a range of 0.96 to 0.99 for the three groups. Their root–mean–square error (RMSE) value was 0.45 m^2/m^2 in total and exhibited a relatively higher range of 0.26–0.51 m^2/m^2 for the three groups. The largest biases between LAISmart and LAI-2000 were observed in the datasets of DBF and DNF, although high R^2 values were noted for the DNF dataset. Because there was high correlation between the two instruments, such biases may not be the result of the camera saturation on high-density vegetation leaves.

In general, the measurements of the two instruments were strongly correlated. The high R^2 values suggest that they can reveal the magnitude range of LAI from a minimum 0.19 m^2/m^2 to a maximum 5.11 m^2/m^2. The slopes of the datasets were

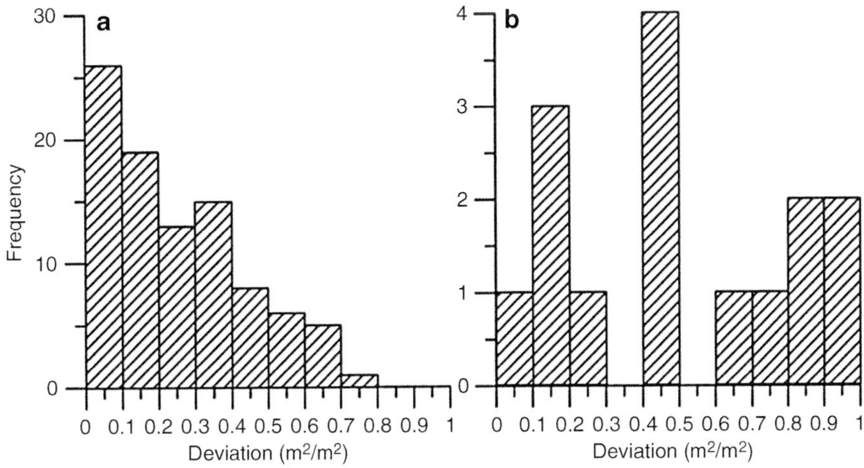

Fig. 8 Frequency of the deviation between the LAI values of LAI-2000 and LAISmart when LAI is divided into two subgroups by the value of 3.7 m^2/m^2. When LAI is lower than the separation point (**a**), the deviation is lower than that in the higher value range (**b**). The RMSEs of the measured values between LAI-2000 and LAISmart in the two groups are (**a**) 0.30 and (**b**) 1.03 m^2/m^2, respectively

close to 1.0 except the case of DNF and DBF vegetation, which had a lower slope of 0.87. This indicates that in most circumstances, the measurements of LAISmart are nearly identical to those of LAI-2000.

However, further analysis showed that the measurements of the two instruments have very different statistical characteristics when the LAI values in all datasets are cataloged into two subgroups by the point of 3.7 m^2/m^2. When the LAI was lower than 3.7 m^2/m^2, the scatter points of the two instruments were distributed around the 1:1 line. This suggests that the measurements of LAISmart are close to those of LAI-2000 in the lower LAI, which can be observed on the left side of the black line in Fig. 7a, and that the frequency of deviation is closest to the left side of the x-axis (Fig. 8a). In such a case, a small RMSE of 0.30 m^2/m^2 was noted. For a different case, however, the deviation was relatively greater, and the RMSE reached 1.03 m^2/m^2 (Fig. 8b), as observed on the right side of the black line in Fig. 7a. We believe that much of the deviation of the two instruments is attributed to the misclassification of background and vegetation when the LAI is in a higher value range.

Prospect on Leaf Area Index Measurement

Ground-based measurement on leaf area index is a hot topic in plant science, ecology, and remote sensing research area. Here we present two new technologies which use modern network communication (LAINet) and mobile computing platform (LAISmart). Though there is a forward step on improving the efficiency of leaf area index data collecting, it should be aware that the current method still facing two types of challenge: the first one is that all the current instruments, either imaging or non-imaging method, are subjected to the weather condition. For example, the LiCor LAI-2000 or LAI-2200C and the DHP (digital hemispherical photography) method are recommended to be operated under diffuse light condition, and the LAISmart instrument is also in this case, while the DEMON (CSIRO, Canberra, Australia) is required to be operated under direct sunlight condition, and the LAINet instrument is also in this case. The limitation on the weather condition reduces the available work time in the field experiment. The second challenge is that in most cases, the performance of the current methods will be decreased in the closed-canopy forests, e.g., LAI mapping from optical data is hindered by the saturation (i.e., LAI does not affect the reflectance or transmittance signal after certain canopy cover). The reasons behind the two above challenges are that the current methods use the passive signal source, i.e., they use the reflected or transmitted solar light to infer vegetation leaf area index.

As an active sensor, light detection and ranging (LIDAR) can provide three-dimensional information from pulses penetrating through canopies and reveal the vertical stratification of a forest (Sasaki et al. 2008). The laser pulses emitted by LIDAR sensors penetrate forest canopies and offer such information on canopy structure and subcanopy topography that is not possible to acquire using passive optical techniques (Heiskanen et al. 2015). The penetration ratio of the airborne LiDAR system (ALS) pulses through the canopy layer represents gap fraction, and this may be converted into LAI based on the Beer–Lambert law, so the LIDAR systems provide alternative data for LAI measurement. There were many successful cases from Terrestrial to airborne LiDAR systems in measuring forest structure parameters (Renslow et al. 2000; Tang et al. 2012; Guang et al. 2017).

However, it should be noted that though the LIDAR sensors can provide complementary advantage compared to traditional passive optical method, the barriers preventing LIDAR to be as popular as passive optical method are the scarce of the portable and the low cost LIDAR instruments. But, we noted that there were some researchers began to use low cost and light weight LIDAR instrument in monitoring vegetation dynamics, e.g., Culvenor et al. (2014) proposed an automated in-situ laser scanner for monitoring forest leaf area index, and then the low cost LIDAR instrument was used on the daily monitoring of forest canopy dynamics (Griebel et al. 2015).

Another method using active signal to measure vegetation leaf area index or other structural parameters is the trial of sensors (Escolà et al. 2011; Llorens et al. 2011). From the principle of the active methods, they all measure the distance between sensors and the target, so there is no much difference between the LIDAR and ultrasonic methods. But, the LIDAR may produce higher accuracy distance measurement, thus the LIDAR many be more suitable on the circumstance where the higher performance is required. But, the most attractive advantage of ultrasonic device is its lower cost compared with LIDAR sensors.

On the independent development of passive and active method, there also emerges the integration of the active and passive types of methods, e.g., Mazzetto et al. (2010) presented a prototype for the monitoring of canopy health and vigor in precision viticulture. Though there is no leaf area index produced in their system, the output of the structural parameter, canopy thickness, may be used as the proxy variables to calculate the leaf area index. On the basis of the preliminary result, it is promising to measure vegetation leaf index area using the integration of passive and active method, which may overcome the disadvantage of individual method.

References

D. Arias, J. Calvo-Alvarado, A. Dohrenbusch, Calibration of LAI-2000 to estimate leaf area index (LAI) and assessment of its relationship with stand productivity in six native and introduced tree species in Costa Rica. For. Ecol. Manag. **247**(1–3), 185–193 (2007)

D.T. Booth, S.E. Cox, C. Fifield, M. Phillips, N. Willlamson, Image analysis compared with other methods for measuring ground cover. Arid Land Res. Manag. **19**(2), 91–100 (2005)

N.J.J. Bréda, Ground-based measurements of leaf area index: a review of methods, instruments and current controversies. J. Exp. Bot. **54**(392), 2403–2417 (2003)

J.M. Chen, Optically-based methods for measuring seasonal variation of leaf area index in boreal conifer stands. Agric. For. Meteorol. **80**(2–4), 135–163 (1996)

J.M. Chen, T.A. Black, Measuring leaf area index of plant canopies with branch architecture. Agric. For. Meteorol. **57**(1–3), 1–12 (1991)

J.M. Chen, T.A. Black, Defining leaf area index for non-flat leaves. Plant Cell Environ. **15**(4), 421–429 (1992)

S.D. Culvenor, J.G. Newnham, A. Mellor, C.N. Sims, A. Haywood, Automated in-situ laser scanner for monitoring forest leaf area index. Sensors **14**(8), 14994–15008 (2014)

L. Di, K. Moe, T.L. van Zyl, Earth observation sensor web: an overview. IEEE J. Sel. Top Appl. Earth Obs. Remote Sens. **3**(4), 415–417 (2010)

A. Escolà, S. Planas, J.R. Rosell, J. Pomar, F. Camp, F. Solanelles, F. Gracia, J. Llorens, E. Gil, Performance of an ultrasonic ranging sensor in apple tree canopies. Sensors **11**(3), 2459–2477 (2011)

S. Fuentes, R.D. Bei, C. Pozo, S. Tyerman, Development of a smartphone application to characterise temporal and spatial canopy architecture and leaf area index for grapevines. Wine Viticult. J. **27**(6), 56–60 (2012)

S. Garrigues, R. Lacaze, F. Baret, J.T. Morisette, M. Weiss, J.E. Nickeson, R. Fernandes, S. Plummer, N.V. Shabanov, R.B. Myneni, Y. Knyazikhin, W. Yang, Validation and intercomparison of global Leaf Area Index products derived from remote sensing data. J. Geophys. Res. Biogeosci. **113**(G2), G02028 (2008)

N.S. Goel, Models of vegetation canopy reflectance and their use in estimation of biophysical parameters from reflectance data. Remote Sens. Rev. **4**(1), 1–212 (1988)

J. Goudriaan, The bare bones of leaf-angle distribution in radiation models for canopy photosynthesis and energy exchange. Agric. For. Meteorol. **43**(2), 155–169 (1988)

A. Griebel, L.T. Bennett, D.S. Culvenor, G.J. Newnham, S.K. Arndt, Reliability and limitations of a novel terrestrial laser scanner for daily monitoring of forest canopy dynamics. Remote Sens. Environ. **166**, 205–213 (2015)

Z. Guang, M. Lixia, J.U.H. Eitel, H. Wei, T.S. Magney, L.M. Moskal, L. Mingshi, Retrieving directional gap fraction, extinction coefficient, and effective leaf area index by incorporating scan angle information from discrete aerial lidar data. IEEE Trans. Geosci. Remote Sens. **55**(1), 577–590 (2017)

V. Haverd, J.L. Lovell, M. Cuntz, D.L.B. Jupp, G.J. Newnham, W. Sea, The canopy semi-analytic Pgap and radiative transfer (CanSPART) model: formulation and application. Agric. For. Meteorol. **160**, 14–35 (2012)

J. Heiskanen, L. Korhonen, J. Hietanen, P.K.E. Pellikka, Use of airborne lidar for estimating canopy gap fraction and leaf area index of tropical montane forests. Int. J. Remote Sens. **36**(10), 2569–2583 (2015)

A.R.G. Lang, Simplified estimate of leaf area index from transmittance of the sun's beam. Agric. For. Meteorol. **41**(3–4), 179–186 (1987)
X. Li, G. Cheng, S. Liu, Q. Xiao, M. Ma, R. Jin, T. Che, Q. Liu, W. Wang, Y. Qi, J. Wen, H. Li, G. Zhu, J. Guo, Y. Ran, S. Wang, Z. Zhu, J. Zhou, X. Hu, Z. Xu, Heihe watershed allied telemetry experimental research (HiWATER): scientific objectives and experimental design. Bull. Am. Meteorol. Soc. **94**(8), 1145–1160 (2013)
J. Llorens, E. Gil, J. Llop, A. Escolà, Ultrasonic and LIDAR sensors for electronic canopy characterization in Vineyards: advances to improve pesticide application methods. Sensors **11**(2), 2177–2194 (2011)
S.N. Martens, S.L. Ustin, R.A. Rousseau, Estimation of tree canopy leaf area index by gap fraction analysis. For. Ecol. Manag. **61**(1–2), 91–108 (1993)
E.G. Mason, M. Diepstraten, G.L. Pinjuv, J.-P. Lasserre, Comparison of direct and indirect leaf area index measurements of *Pinus radiata* D. Don. Agric. For. Meteorol. **166–167**, 113–119 (2012)
F. Mazzetto, A. Calcante, A. Mena, A. Vercesi, Integration of optical and analogue sensors for monitoring canopy health and vigour in precision viticulture. Precis. Agric. **11**(6), 636–649 (2010)
R. Morais, M.A. Fernandes, S.G. Matos, C. Serôdio, P.J.S.G. Ferreira, M.J.C.S. Reis, A ZigBee multi-powered wireless acquisition device for remote sensing applications in precision viticulture. Comput. Electron. Agric. **62**(2), 94–106 (2008)
T. Nilson, A theoretical analysis of the frequency of gaps in plant stands. Agric. Meteorol. **8**, 25–38 (1971)
N. Otsu, A threshold selection method from gray-level histograms. IEEE Trans. Syst. Man Cybern. **9**(1), 62–66 (1979)
J. Pisek, Y. Ryu, M. Sprintsin, L. He, A.J. Oliphant, L. Korhonen, J. Kuusk, A. Kuusk, R. Bergstrom, J. Verrelst, K. Alikas, Retrieving vegetation clumping index from Multi-angle Imaging SpectroRadiometer (MISR) data at 275(m) resolution. Remote Sens. Environ. **138**, 126–133 (2013)
Y. Qu, W. Han, L. Fu, C. Li, J. Song, H. Zhou, Y. Bo, J. Wang, LAINet – a wireless sensor network for coniferous forest leaf area index measurement: design, algorithm and validation. Comput. Electron. Agric. **108**, 200–208 (2014)
Y. Qu, J. Meng, H. Wan, Y. Li, Preliminary study on integrated wireless smart terminals for leaf area index measurement. Comput. Electron. Agric. **129**, 56–65 (2016)
Y. Qu, J. Wang, J. Song, J. Wang, Potential and limits of retrieving conifer leaf area index using smartphone-based method. Forests **8**(6), 217 (2017)
M. Renslow, P. Greenfield, T. Guay, Evaluation of multi-return LIDAR for forestry applications. Report prepared for The Inventory and Monitoring Steering Committee of the USDA Forest Service. US Department of Agriculture Forest Service – Engineering. Remote Sensing Applications Center, November 2000. RSAC-2060/4810-LSP-0001-RPT1. (2000)
P.W. Rundel, E.A. Graham, M.F. Allen, J.C. Fisher, T.C. Harmon, Environmental sensor networks in ecological research. New Phytol. **182**(3), 589–607 (2009)
T. Sasaki, J. Imanishi, K. Ioki, Y. Morimoto, K. Kitada, Estimation of leaf area index and canopy openness in broad-leaved forest using an airborne laser scanner in comparison with high-resolution near-infrared digital photography. Landsc. Ecol. Eng. **4**(1), 47–55 (2008)
M. Schirrmann, A. Hamdorf, A. Giebel, K.H. Dammer, A. Garz, A mobile sensor for leaf area index estimation from canopy light transmittance in wheat crops. Biosyst. Eng. **140**, 23–33 (2015)
H. Tang, R. Dubayah, A. Swatantran, M. Hofton, S. Sheldon, D.B. Clark, B. Blair, Retrieval of vertical LAI profiles over tropical rain forests using waveform lidar at La Selva, Costa Rica. Remote Sens. Environ. **124**, 242–250 (2012)
J.M. Welles, Some indirect methods of estimating canopy structure. Remote Sens. Rev. **5**(1), 31–43 (1990)
K.R. Whitford, I.J. Colquhoun, A.R.G. Lang, B.M. Harper, Measuring leaf area index in a sparse eucalypt forest: a comparison of estimates from direct measurement, hemispherical photography, sunlight transmittance and allometric regression. Agric. For. Meteorol. **74**(3–4), 237–249 (1995)

Radar Remote Sensing of Land Surface Parameters

Alberto Alonso-González and Irena Hajnsek

Contents

Abstract

This chapter describes some of the advantages and limitations of radar remote sensing in the context of land cover and agriculture. In the Introduction section, a short overview is given about the interest and techniques for remote sensing in this context. The motivation to focus on radar and SAR and a brief review of the main techniques are described in order to give a synopsis of the type of

A. Alonso-González
Microwave and Radar Institute, German Aerospace Center (DLR), Wessling, Germany
e-mail: Alberto.Alonso-Gonzalez@dlr.de

I. Hajnsek (✉)
Microwave and Radar Institute, German Aerospace Center (DLR), Wessling, Germany

Institute of Environmental Engineering, Swiss Federal Institute of Technology (ETH), Zurich, Switzerland
e-mail: irena.hajnsek@dlr.de

X. Li, H. Vereecken (eds.), *Observation and Measurement of Ecohydrological Processes*, Ecohydrology, https://doi.org/10.1007/978-3-662-48297-1_12

information that may be extracted from these sensors. Sections "SAR Remote Sensing for Land Cover" and "SAR Remote Sensing of Vegetation" depict some of the main or more relevant applications of SAR within the context of land cover and vegetation. Due to the broad scope of this topic, the focus has been put on four big topics: soil moisture retrieval and land cover classification as examples of SAR remote sensing applications over land and forest and agriculture monitoring as examples over vegetation. Finally, an outlook is given with a summary of the current state of the art, the limitations of remote sensing sensors, and the desired requirements for the different described applications.

Keywords
Remote sensing · Radar · SAR · Polarimetry · Interferometry · Land cover · Agriculture

Introduction

Remote sensing is a technique to achieve information of a distant object or phenomena without having a direct contact with it. This is usually achieved by analyzing the interaction of some propagated signal, as electromagnetic waves, with matter.

The visibility and importance of remote sensing techniques has increased notably during the last decades, in particular due to the ability to gather information about the Earth from space, employing spaceborne sensors. These techniques, then, allow gathering information at global scales with a regular revisit time which makes them exceptionally useful and cost-effective.

Different criteria can be employed for the classification of the different remote sensing techniques. On the one hand, they may be categorized depending on the electromagnetic source employed for illumination. Accordingly, those systems having their own illumination source are also called active systems, while other systems that just measure the radiation emitted or reflected by the target from an external source are called passive systems, as shown in Fig. 1. On the other hand, another classification may be performed according to the electromagnetic spectrum range employed as, for instance, microwave, infrared, or optical systems.

In the following subsections, two of the most commonly used remote sensing techniques, based on optical and microwave sensors, are briefly described.

Optical and Microwave Remote Sensing

Two of the most widely used remote sensing techniques may be considered optical and microwave sensors, depending on the electromagnetic wavelength employed. Optical sensors are passive sensors that measure the light of the Sun reflected on the Earth surface. They have been widely used for extracting information about our environment. Optical images are easily interpreted as they are more comparable to our visual sense, in terms of information and acquisition geometry. Moreover, the

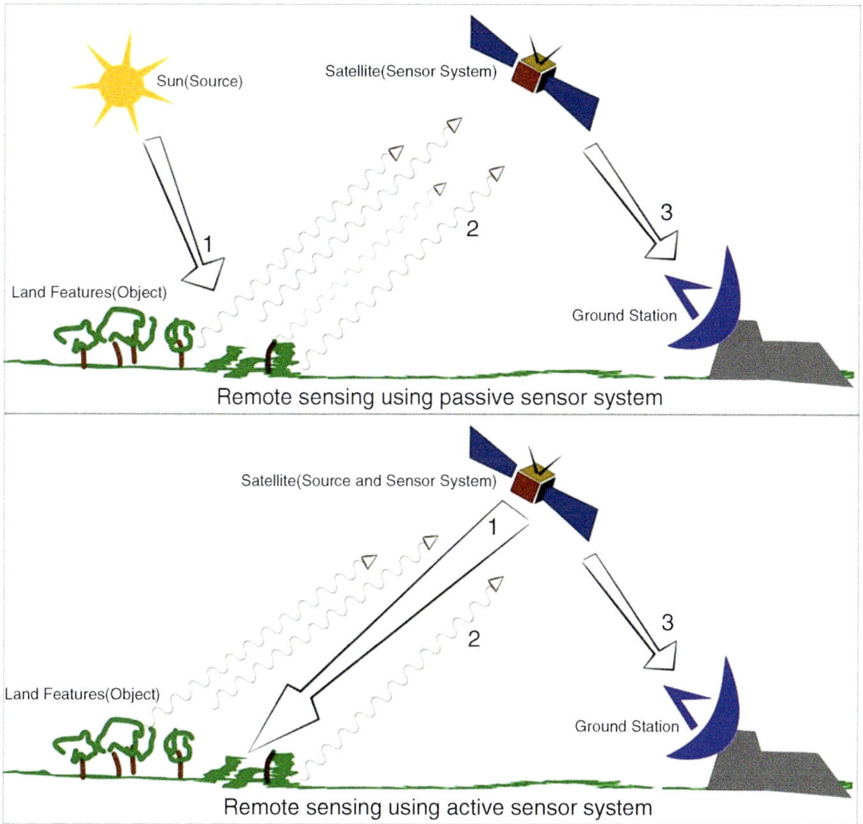

Fig. 1 Illustration of active and passive remote sensing satellites (Illustration of Remote Sensing by Arkarjun, licensed under CC BY-SA 3.0)

wavelengths corresponding to the optical spectrum are located around the maximum energy emission of the spectrum of the Sun. All these properties make optical remote sensing a very attractive technique for Earth monitoring.

However, all these advantages may be outweighed by the strong interaction of the electromagnetic waves at visible frequencies with the atmosphere and especially the troposphere. This is especially important for spaceborne sensors, where the clouds and other atmospheric effects may prevent the acquisition of data from the surface of the Earth. This is particularly critical over some areas that are completely cloud covered most of the time. Additionally, since optical sensors are passive sensors, they can only extract images of the Earth during day conditions, since they need the illumination of the Sun to acquire the data. To partially overcome these limitations, some optical systems acquire information not only on the optical bands, but on several narrow bands (from a few tens to several hundreds) ranging from infrared to ultraviolet. These sensors are called hyperspectral sensors, and, although they cannot overcome the previously mentioned atmospheric limitations, they can acquire valuable information related to some chemical properties of water and vegetation that are related to specific spectral bands.

To elude the previously mentioned atmospheric interaction problems, microwave frequencies might be used. The atmosphere is completely transparent to electromagnetic waves with wavelengths between a few centimeters up to 10 m, as shown in Fig. 2. Moreover, these frequencies are able to penetrate and see through the clouds and other atmospheric effects. These properties make microwave remote sensing a very valuable tool for all-weather Earth monitoring.

However, at these frequencies, the power of the electromagnetic radiation emitted by the Sun is much weaker, making almost impossible to measure the surface-reflected waves from the Sun. Instead, the microwave emission of the surface of the Earth can be measured by radiometers, conforming passive microwave sensors. Nonetheless, since this emission is a very weak signal, the resolution achieved by these sensors is usually in the order of some tens of kilometers, being its main limitation. Another possibility is to use an active system that, having its own source of illumination, transmits an electromagnetic pulse and measures the reflected waves coming from the Earth surface. Different transmitted pulses may be combined coherently to form a synthetic aperture and get a higher-resolution image, in the order of meters. This technique is known as synthetic aperture radar (SAR). SAR sensors are described in more detail in the next section.

Compared to optical and hyperspectral sensors, SAR systems are more sensitive to geometric and dielectric properties (related to the propagation of electromagnetic waves through the medium) of the surface, instead of its chemical properties. They have the advantage of being all-weather and day-night sensors, enabling the acquisition of reliable time series of the Earth surface evolution. However, they are usually more difficult to interpret than optical images, since they are obtained at completely different frequencies and geometry.

Synthetic Aperture Radar (SAR)

In a typical radar system, an electromagnetic pulse is transmitted through an antenna. When this pulse hits a target, located at a certain distance, a part of the energy of this pulse is transmitted toward the receiver. This signal is detected by the receiver and processed. Therefore, radar measures the radar reflectivity of the scene, that is, the fraction of the transmitted power that is reflected back toward the receiver, and the distance between the antenna and the illuminated object.

On synthetic aperture radar (SAR) systems, a higher-resolution image is obtained by coherently processing the radar echoes received by the sensor at different positions (Brown 1967). Therefore, SAR systems are usually mounted on a moving platform, as an airplane or satellite. The flight direction is referred to as the *azimuth* direction in this scenario. The radar beam is slanted from the nadir direction perpendicularly to the flight direction, known as the *range* direction. The length of the strip generated by the movement of the illuminated area of the antenna is known as *swath*. These elements are depicted in Fig. 3.

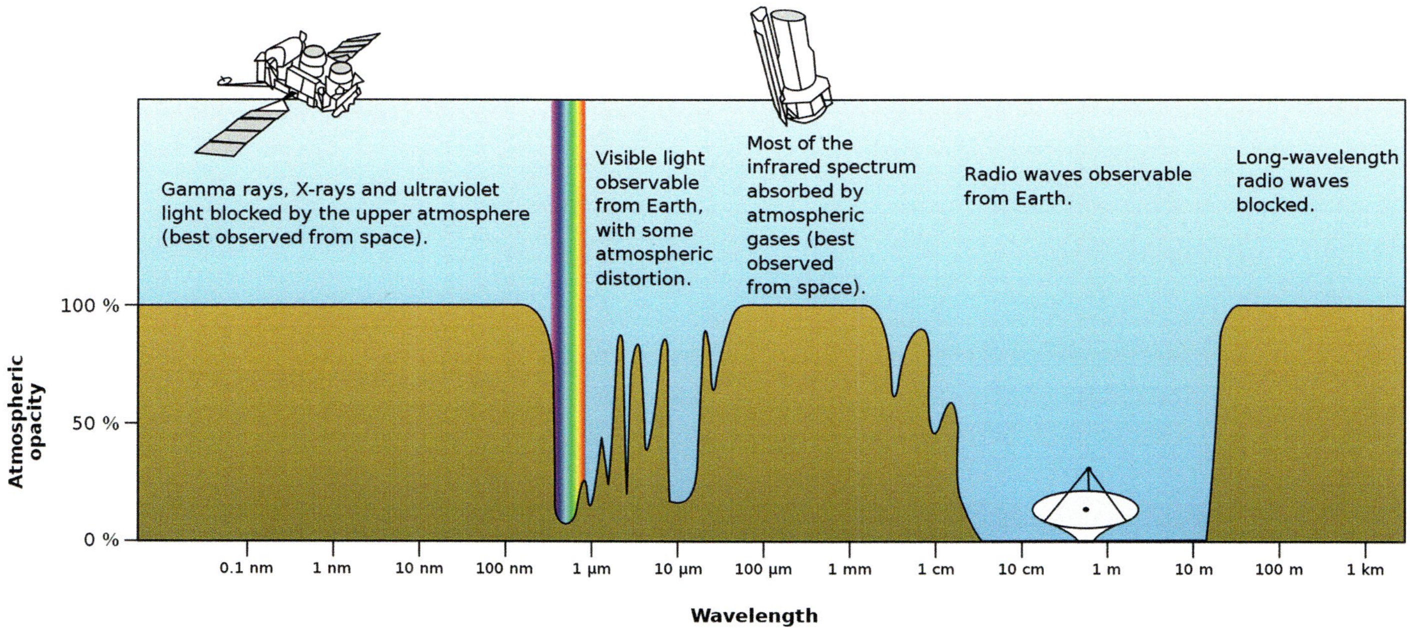

Fig. 2 Atmospheric opacity to different electromagnetic spectrum bands (Electromagnetic transmittance, or opacity, of the Earth's atmosphere. NASA (original); SVG by Mysid)

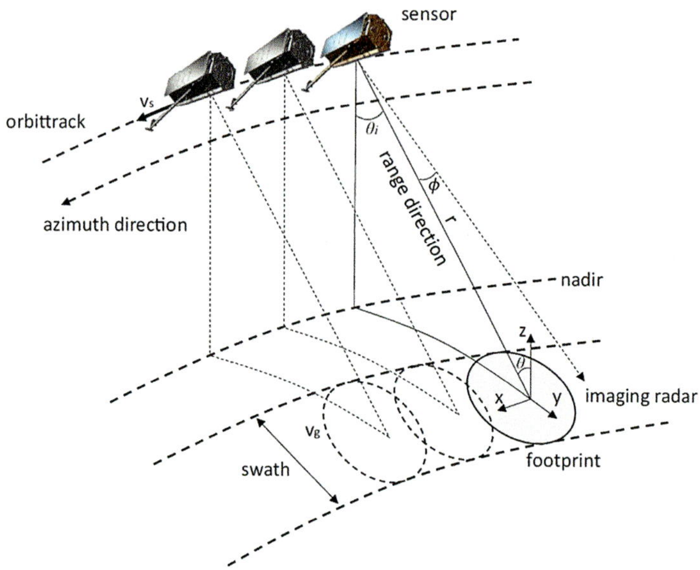

Fig. 3 Synthetic aperture radar acquisition geometry (Alonso-Gonzalez 2014)

It is worth mentioning that, since SAR is an active system, producing its own source of illumination allows having full control of the transmitted pulse, allowing exploiting other techniques as polarimetry or interferometry, which will be described in the following sections. Note that the exploitation of these techniques is much more difficult or impossible for passive sensors, since the light produced by the Sun or the emitted radiation is not a coherent or polarized illumination.

As a counterpart, SAR images are affected by a noisy pattern known as *speckle* (Ulaby et al. 1982). Each resolution cell is the coherent combination of all the individual scatterers within. This combination may be constructive and destructive and depends on their intensity and distribution in the resolution cell. Since this distribution is not known, the speckle is considered as a noise, and the description of the scattering process has to be performed statistically (Lee 1981).

SAR Polarimetry

Polarimetric SAR (PolSAR) systems achieve additional information of the scene by taking advantage of the vector nature of the electromagnetic waves (Balanis 1989). In this case, different polarization states are employed for the transmitted and received radar echoes that interact differently with the scene. PolSAR sensors acquire several SAR images or channels on a single pass, according to the distinct wave polarization states employed for transmission and reception. Therefore, polarimetry allows separating different scattering mechanisms occurring within a single resolution cell.

Another important concept in SAR polarimetry is *polarization synthesis* (Boerner 1981). Accordingly, if two orthogonal polarization states are employed for transmission and reception, the full information of the target response is obtained in the polarimetric space, allowing the exploration of its response at any other polarization state. Therefore, two polarization states, usually vertical and horizontal polarizations, have to be employed in transmission and reception in order to fully characterize the scattered wave E_h^s, E_v^s with respect to the incident wave E_h^i, E_v^i

$$\begin{bmatrix} E_h^s \\ E_v^s \end{bmatrix} = \frac{e^{-jkr}}{r} \begin{bmatrix} S_{hh} & S_{hv} \\ S_{vh} & S_{vv} \end{bmatrix} \begin{bmatrix} E_h^i \\ E_v^i \end{bmatrix} \tag{1}$$

where E_x^s and E_x^i represent the scattered and incident wave at polarization x, respectively, and S_{xy} represents the scattering parameter at transmitted polarization x and received y.

Fully polarimetric SAR systems, also known as quad-polarimetric, measure the full scattering matrix $\boldsymbol{S}$, containing the four S_{xy} parameters. Dual PolSAR sensors, for instance, only measure two of these parameters.

In the case of distributed scattering, as mentioned before, the scattering process has to be defined statistically. Therefore, the second-order statistics are employed by vectoring the scattering matrix $\boldsymbol{S}$ into the scattering vector $\boldsymbol{w}$ and estimating the covariance matrix $\boldsymbol{C}$ as

$$\boldsymbol{C} = E\left\{\boldsymbol{w}_L \boldsymbol{w}_L^H\right\} \tag{2}$$

where $\boldsymbol{w}_L = [S_{hh}, S_{hv}, S_{vh}, S_{vv}]^T$ is the lexicographic scattering vector (Cloude and Pottier 2005) and $\boldsymbol{w}^H$ denotes the complex transpose conjugate operator. Alternatively, instead of the lexicographic basis, the Pauli basis can be employed $\boldsymbol{w}_P = 1/\sqrt{2}[S_{hh} + S_{vv}, S_{hh} - S_{vv}, S_{hv} + S_{vh}, j(S_{hv} - S_{vh})]^T$, resulting in the so-called coherence matrix $\boldsymbol{T}$ (Cloude and Pottier 2005). Additionally, when the receiver is located at the same position than the transmitter, the two cross-polarization channels are statistically equivalent ($S_{hv} = S_{vh}$), and the previously defined scattering vectors are reduced to three dimensions.

Fully polarimetric images are usually represented in the Pauli basis since this representation may be easily related with the elementary or canonical scattering mechanisms. The first component can be related to surface scattering and the second to double-bounce scattering, whereas the third one corresponds to volume scattering. In the typical Pauli RGB representation, the intensity of these channels are assigned to blue, red, and green channels, respectively, in order to generate a color composition of the scattering intensities of these canonical mechanisms. Figure 4 shows the Pauli RGB representation of a fully polarimetric image of the DLR ESAR sensor at L-band over Oberpfaffenhofen, in Germany. As it may be seen, forest areas appear in greenish colors, due to the predominance of volume scattering. Over urban and industrial areas, strong red spots are observed due to double-bounce scattering over human-made buildings. The agricultural fields appear in different combination of colors, especially blue and red, depending on their crop type and phenological stage.

Fig. 4 Pauli RGB composition of an ESAR image over Oberpfaffenhofen, Germany (Alonso-Gonzalez 2014). Surface, double-bounce, and volume scattering components are assigned to blue, red, and green channels, respectively

SAR Interferometry

Since SAR sensors are coherent sensors, they can measure the complex amplitude and phase of the received echo. In two SAR acquisitions with the same acquisition geometry but acquired from slightly different positions, the distance traveled by both of the pulses may differ, generating a relative phase difference between them. The exploitation of this differential phase between two or more SAR acquisitions is called SAR interferometry, or InSAR.

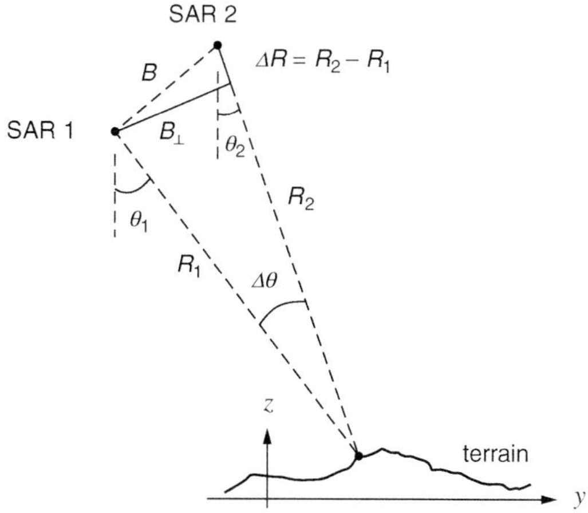

Fig. 5 Across-track interferometric acquisition geometry (Bamler and Hartl 1998)

Figure 5 shows a scheme of the acquisition geometry for across-track InSAR (Bamler and Hartl 1998). In this case, the azimuth direction is perpendicular to the drawing. It may be seen that for two images with a perpendicular baseline B, a difference ΔR is observed in the distance transmitted by the two pulses. This difference is sensitive to different components as geometric parameters (flat earth), height of the terrain, atmospheric propagation effects, deformation, and noise. Therefore, InSAR has been traditionally used for performing digital elevation models (DEMs) (Farr et al. 2007; Krieger et al. 2007) and for deformation monitoring (Bamler and Hartl 1998; Ferreti et al. 2001).

DEMs are very important for the monitoring and modeling of the Earth processes. Several applications rely on precise DEMs of the surface, like geomorphology, hydrology, physical modeling, flight planning, engineering and infrastructure design, etc. When these maps are required on a global scale, SAR interferometry is one of the best and most effective methods to acquire this type of information, as it may be seen in Fig. 6.

For this reason, several spaceborne campaigns have been devoted to DEM generation with InSAR. A good example is the Shuttle Radar Topography Mission (SRTM) (Farr et al. 2007). It was an international effort to generate nearly global DEMs of the Earth (from latitudes 58S to 60 N) with a bistatic radar onboard the Space Shuttle Endeavour in the year 2000. The final resolution over the coverage area is 3 arcseconds (90 m). The SRTM had two receiving antennas separated by 60 m mast in order to generate the across-track baseline required for interferometry in a single pass. Therefore, it was the first single-pass interferometer in space.

It is worth mentioning that DEM generation with InSAR also presents some difficulties. Due to the radar acquisition geometry, some parts of the scene may be highly distorted or not imaged at all, as an effect of the foreshortening and shadowing effects. This is particularly pronounced over high topography areas or areas with low coherence, as deserts. Although this voids only amount for around 0.2% of the total area, this might be a problem in some areas. In the final SRTM product, these voids were filled by interpolation, as it may be seen in Fig. 7.

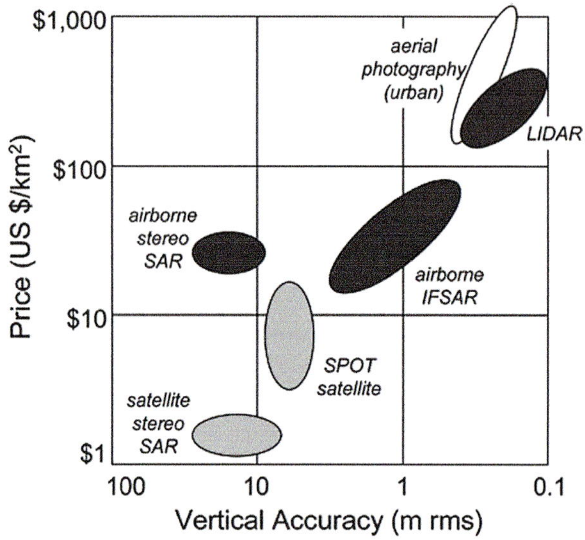

Fig. 6 Relative cost and accuracy of different techniques for DEM generation (Richards 2007)

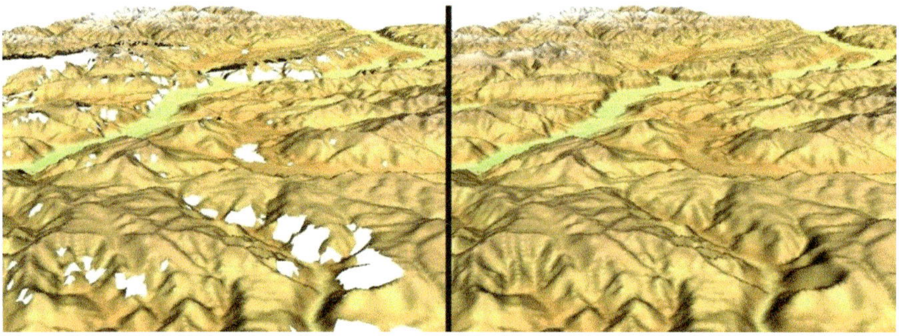

Fig. 7 SRTM generated DEM and processing voids filled by interpolation (Void filling of SRTM DEM data with splines in GRASS GIS by Markus Neteler, licensed under CC BY-SA 2.5)

In order to circumvent the limitations of the SRTM and improve its coverage and resolution, the TanDEM-X mission was launched in 2010 (Krieger et al. 2007). It consists of two SAR satellites orbiting in a close formation. The primary objective of the mission is to generate a detailed DEM covering the whole Earth surface with unprecedented resolution (12 by 12 m) and accuracy (2 m relative accuracy). Figure 8 shows a comparison between the products obtained with these missions.

SAR Tomography

The transmitted electromagnetic wave can penetrate through semitransparent media, as in the vegetation canopy, resulting into volume scattering. As explained in the

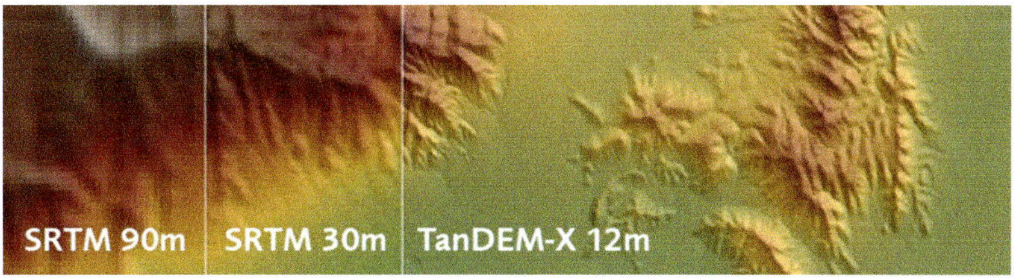

Fig. 8 Comparison between SRTM and TanDEM-X DEMs (From http://www.terraspatium.gr/worlddemtrade.html)

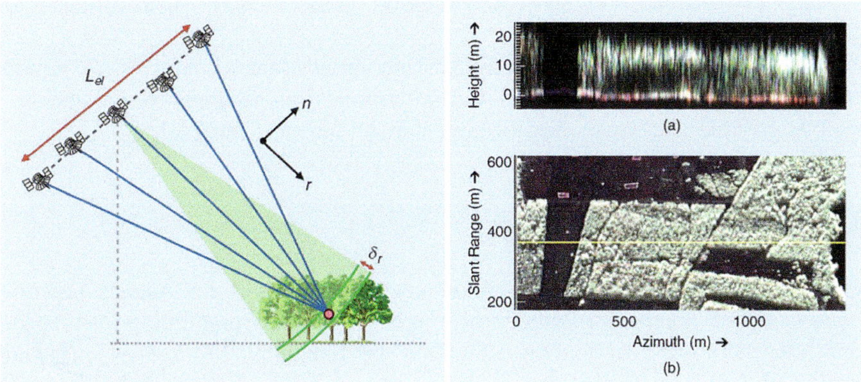

Fig. 9 SAR tomography acquisition geometry representing the synthetic aperture in elevation L_{el} (left) and polarimetric tomographic profile of a forested area using the DLR ESAR system at L-band (right) (Moreira et al. 2013)

previous section, two acquisitions can be combined in order to get height sensitivity. However, when the scatterers are distributed in height, as in the forest scenario for longer wavelengths, this is not enough to image the vertical structure. Several acquisitions with different baselines might be employed to get information about the vertical structure of the scatterers in the scene. This technique, called SAR tomography or TomoSAR, is equivalent to form an additional synthetic aperture in the direction perpendicular to range and azimuth in order to extract a full three-dimensional reflectivity profile (Moreira et al. 2013). This concept is illustrated in the left image of Fig. 9, showing the synthetic aperture in elevation denoted as L_{el}.

One of the applications of SAR tomography is the imaging of volumes in vegetated areas. The right image of Fig. 9 shows one example of the polarimetric vertical profile obtained for a section of a forest area in Dornstetten, Germany. The yellow line in the lower image represents the transect corresponding to the upper image. In this case, a total of 21 tracks were employed with an average separation of 20 m. Different contributions might be clearly identified in the tomogram, especially the ground and canopy layers, appearing in different colors in the Pauli representation. These tomograms have been exploited for different applications as, for instance, estimating the ground topography behind the canopy, for the analysis of structural parameters of the canopy or the detection of objects hidden below the foliage. Some applications of this technique will be briefly described in the following sections.

SAR Remote Sensing for Land Cover

Remote sensing is an important technique for land cover monitoring and classification. The ability to provide information over a large area or even globally makes this technique very valuable for extracting several biophysical or geophysical parameters (Lee and Pottier 2009; Elachi and Van Zyl 2006). Additionally, the excellent repeatability attained with spaceborne sensors is essential for input estimates of land process models.

There are a large number of different land applications for which SAR remote sensing might be especially useful, but in the following sections, special attention will be given to land cover classification and soil moisture retrieval.

Land Cover Classification

Land cover maps might be very useful for several domains as, for instance, land-use classification (Cloude and Pottier 1997), urban settlement monitoring (Henderson and Xia 1997), or forest biomass estimation (Le Toan et al. 2011). These maps are very useful for resource management and planning but also for monitoring and modeling the land use and the environmental changes induced by human beings.

A good example of the importance and interest for a regional land cover map is the CORINE Land Cover (CLC), produced by the European Environment Agency (EEA) (CLC2006 technical guidelines 2007). It is a map covering the whole European Union, and it has been already released three times. The CLC maps are usually performed by a combination of semiautomatic methods and visual inspection employing optical sensors (mainly Landsat and SPOT). Due to the operational limitations of these techniques, the minimum mapping resolution is 25 hectares (roughly around 500 by 500 m), and a release takes around 5 years.

SAR systems can improve significantly these maps since, as explained previously, they provide a high-resolution and all-weather reliable source of information (Balzter et al. 2015). The following sections are devoted to give more details about two of the most interesting properties of SAR in the context of land cover classification and land-use mapping: SAR polarimetry and multi-temporal acquisitions.

Polarimetric Classification

As mentioned in section "SAR Polarimetry," the electromagnetic waves interact differently with the targets according to their polarization state. This property makes SAR polarimetry a very valuable tool for land cover classification. More specifically, the obtained SAR polarimetric response depends on the geometric and dielectric properties at the scale of the wavelength.

Several techniques can be found in the literature in order to exploit SAR polarimetry for land-use classification. To make easier the interpretation of the radar response, usually polarimetric decompositions are employed as, for instance, the entropy (H), anisotropy (A), and alpha angle decomposition (Cloude and Pottier 1997). This technique is based on the eigen-decomposition of the coherence matrix. From this decomposition, it extracts three parameters which may be easily interpreted physically:

1. The entropy (H) parameter gives an idea of the randomness of the scattering process. It is equal to 1 when it is totally random and to 0 when one specific type of scattering is dominant.
2. The anisotropy (A), also in the range from 0 to 1, indicates the dominance of the second most dominant scattering mechanism with respect to the third one.
3. The mean alpha angle reflects the type of scattering. It ranges from 0 to 90 degrees, where 0 indicates surface scattering and 45 corresponds to volume and 90 to multiple scattering.

From the entropy and alpha parameters, a classification can be extracted into eight classes, corresponding to the different zones in the H/alpha plane shown in Fig. 10 (Cloude and Pottier 1997). The anisotropy (A) parameter may also be employed in order to refine the classification into 16 classes.

Figure 11 shows, in the left, the result of applying the H/alpha classification over the image presented in Fig. 4. The color representation and the scatterplot of the H/alpha plane are shown on the right.

The previously described polarimetric classification scheme has the inconvenience that it is completely predefined and lacks some degree of adaptability to the classes present in the scene. To solve these limitations, the Wishart classifier is an iterative classification algorithm based on the initialization provided by the previously described H/alpha classification (Lee et al. 1999).

Fig. 10 Classification based on the entropy (H) – alpha plane

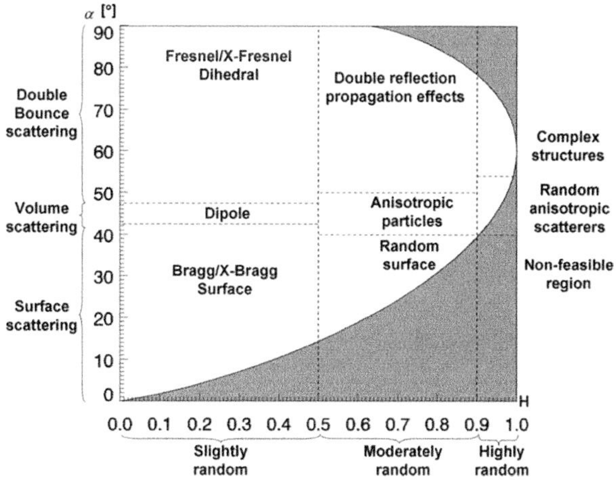

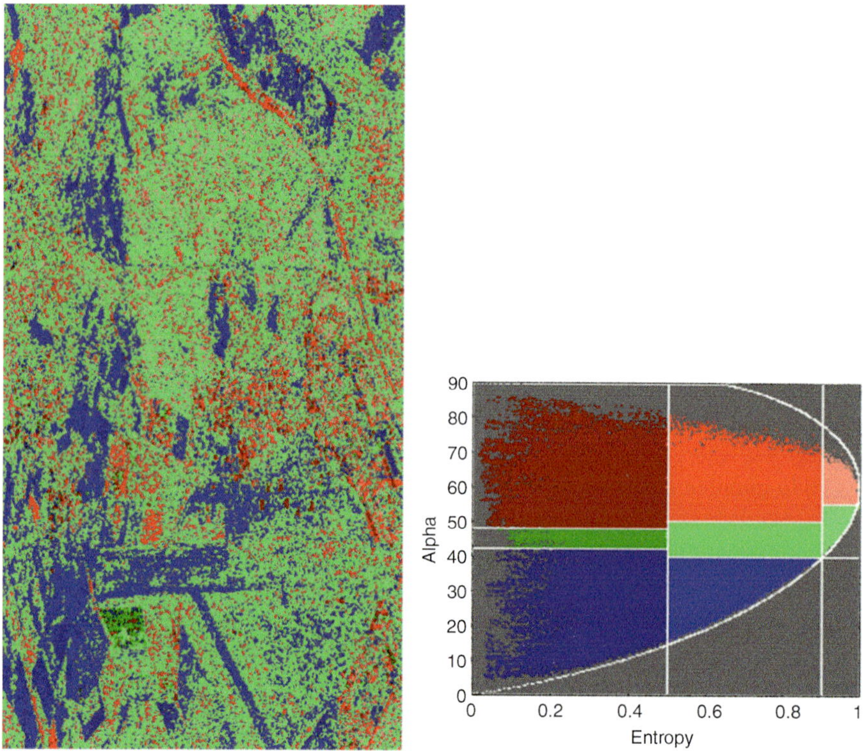

Fig. 11 H/alpha classification of the ESAR image (left) and color code and scatterplot of the H/alpha plane (Alonso-Gonzalez 2014)

The Wishart classifier is based on the statistical distribution of the estimated coherence matrices, the Wishart distribution, in order to assign each sample of the image to closest distribution. Once all the pixels have been assigned to one of the possible classes, the coherence matrix of each class is recomputed. Then, each sample is reassigned again and the process is repeated until convergence:

1. Perform the H/alpha classification to initialize data into eight classes, as shown in Fig. 10.
2. Compute the coherence matrices $\boldsymbol{T}_i$ for each of the classes.
3. Assign each pixel of the image to the most probable coherence matrix $\boldsymbol{T}_i$ according to the Wishart distance (Lee et al. 1999).
4. Recompute the coherence matrices $\boldsymbol{T}_i$ for each of the new classes.
5. Repeat the steps 3 to 5 until convergence.

In the first step, the H/A/alpha classification may be employed to obtain 16 classes instead of the H/alpha. Figure 12 shows the Wishart classification technique for 8 and 16 classes applied over the same ESAR image presented in Fig. 4. As it

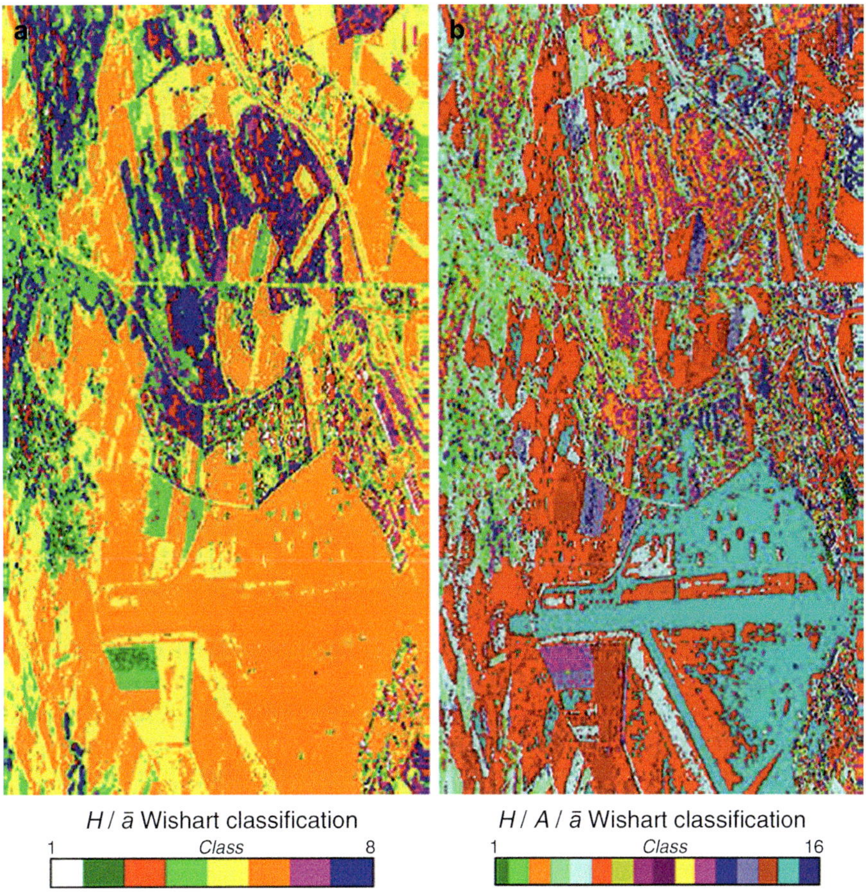

Fig. 12 Wishart classification of the image shown in Fig. 4 into 8 (left) and 16 (right) classes (Alonso-Gonzalez 2014)

may be seen, the Wishart classification can achieve a better adaptation to the image content than the H/alpha.

The previously mentioned approaches correspond to *unsupervised classification* techniques. In this case, no a priori information is given about the classes, and the method tries to find automatically the most differentiated classes according to the image content. On the other hand, in *supervised classification* techniques, some information is provided about the classes to find, usually as a set of training samples for each class. Then, all the image pixels are assigned to the most similar class. However, the same principles explained here are usually used, as the Wishart classifier for measuring the similarity to some predefined classes (Lee et al. 1994). Another possibility is to use polarimetric decomposition, as described previously, in combination with other polarimetric features in order to apply some of the

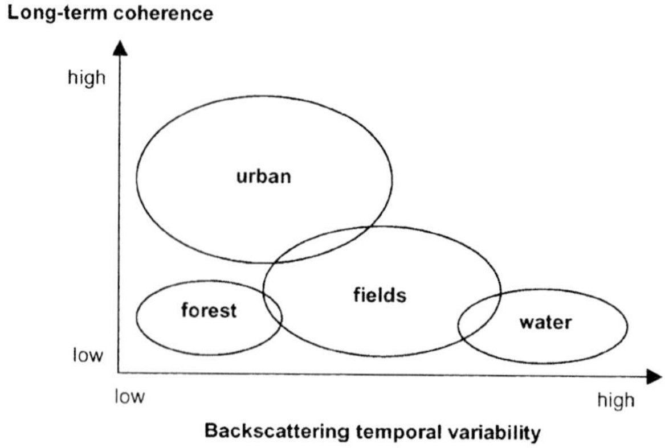

Fig. 13 Distribution of some classes in the backscattering temporal variability and long-term coherence space (Bruzzone et al. 2004)

established data classification and clustering schemes like support vector machines (SVM) (Lardeux et al. 2009) or decision trees (Qi et al. 2012).

Multi-temporal Classification

One of the most important properties of SAR is its ability to acquire images at day-night and all-weather conditions. Since these imaging capabilities may not be achieved by classical optical sensors, SAR sensors are an extremely useful and effective data source for building multi-temporal datasets. This property, combined with the fact that SAR is able to perform coherent observations, is a unique property for performing multi-temporal classification (Bruzzone et al. 2004).

The exploitation of the temporal evolution opens a completely new aspect to explore and significantly improve the target separation and discrimination. The different temporal evolution of the targets is an important property to distinguish them. As an example, Fig. 13 shows that some types of land cover can be separated by the temporal components of the backscattering temporal variability and long-term coherence (Bruzzone et al. 2004). It is worth noticing that some of these classes may be especially difficult to classify properly only with polarimetry, as the urban areas, but the temporal component makes them easily separable. Note also that all this information is additional to the polarimetric information mentioned in the previous section.

One of the areas on which multi-temporal SAR data is especially valuable is agricultural land-use classification. SAR data are particularly sensitive to geometric and dielectric properties of the object structure as, for instance, the size, shape, and distribution of the scatterers, which makes it very useful for crop discrimination (Ulaby et al. 1998). However, since agricultural crops are changing significantly during the different phenological stages, agricultural land-use classification can improve considerably by adding the temporal signatures of the different crops (Skriver 2012). Figure 14 shows the error obtained in agricultural land-use Wishart classification employing only one acquisition (red line) and employing all the previous acquisitions (green line) (Skriver et al. 2011). It may be seen that employing only one acquisition my lead to large classification errors, up to 50%, when the different

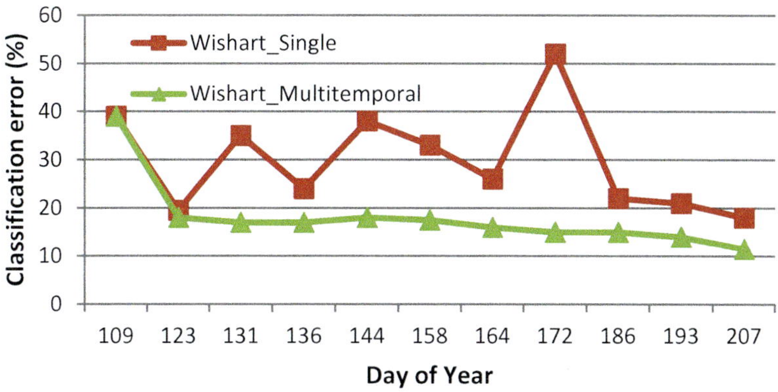

Fig. 14 Average classification error for the AgriSAR06 campaign with the Wishart classifier. In red, only one image is used. In green, all the images previous to the given day of year are used (Skriver et al. 2011)

fields have a similar backscattering response. In the multi-temporal approach, a few images may be combined in order to reduce the error down to 10–15%.

Soil Moisture Retrieval

Soil moisture retrieval with microwave remote sensing has been an active topic in research since it first appeared. It is an important topic from a hydrological perspective, since soil moisture is an important variable in the separation of the rainfall into runoff and infiltration. Therefore, it has an important role in ecological applications as flood forecasting, environmental modeling, or agricultural monitoring (Wagner et al. 2007).

From a physical point of view, soil moisture changes the effective dielectric constant of the soil, which affects its electromagnetic properties (Ulaby et al. 1982). Compared to passive radiometers, SAR sensors can provide a significant improvement by reducing the spatial scale from several kilometers down to meters. This jump in spatial resolution may become especially valuable as it allows a precise separation and identification of soil moisture trends with land cover, allowing its exploitation for hydrological models at field scale, potentially opening new applications.

Polarimetric Decomposition Techniques

Several models can be found in the literature to describe the radar backscattering properties of bare soil, ranging from statistical (Fung et al. 1992; Hajnsek et al. 2003) to empirical approaches (Dubois et al. 1995). These models depend on the dielectric constant of the soil, carrying the information related to soil moisture, but also on other soil parameters like its surface roughness or correlation length. As a consequence, polarimetric data are the first choice for the inversion of the signal into soil moisture (Hajnsek et al. 2003).

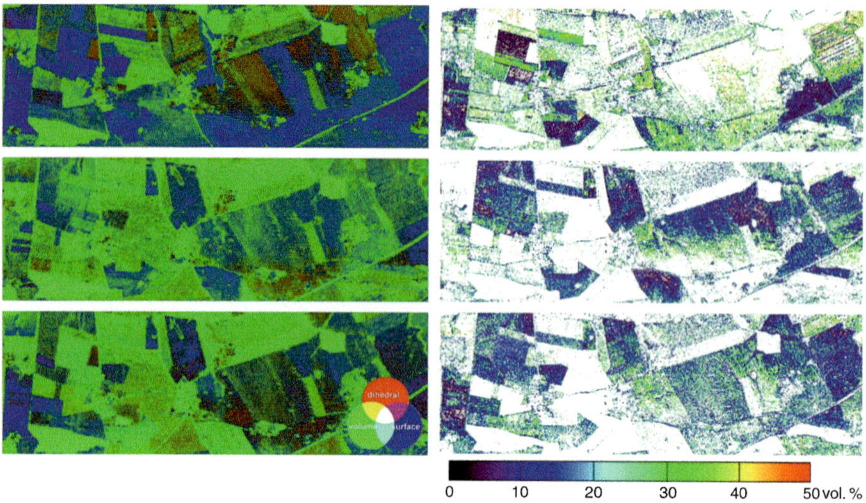

Fig. 15 Left, AgriSAR 2006 RGB normalized power representation of the polarimetric decomposition into volume (green), surface (blue), and double bounce (red). Right, volumetric soil moisture inversion obtained from the surface component (Hajnsek et al. 2009a). From top to bottom, the L-band images were acquired in April 19, June 14, and July 12, 2006, respectively

However, these models only describe the bare soil radar response, and, in a real scenario, usually some kind of vegetation may be found on top of it. In this case, polarimetric decompositions have been proposed in the literature in order to separate the ground and vegetation contributions from the total SAR signal (Hajnsek et al. 2009a). A physical model is employed in order to separate the different components, usually surface, volume, and double-bounce components. This separation normally requires fully polarimetric SAR data due to dimensionality constrains in the decomposition problem.

Figure 15 shows the relative powers of the three components for the polarimetric composition and the volumetric soil moisture obtained by the inversion of the surface component. Note that the soil and vegetation interaction into the scattering process is a very complex problem that has not been completely solved yet. For this reason, these models might fail to describe the obtained response at some pixels, and only a certain degree of inversion rate is obtained with these methods.

However, it is worth mentioning that, in order to be able to perform this polarimetric separation properly, a significant degree of penetration into the vegetation layer is required. For that reason, SAR soil moisture inversion is usually performed at long wavelengths, as L-band SAR. To illustrate this, Fig. 16 shows the simulated PolSAR backscattering for a maize field at L-, C- and X-band with respect to crop height for two different soil moisture values, 5% and 40% volumetric SM. The Tor Vergata vegetation model (Bracaglia et al. 1995) has been employed for the simulation. In these plots it may be seen that the sensitivity to soil moisture depends on the vegetation height but also on the frequency. When the crop height is above 1 m or 60 cm at C- and X-band, the

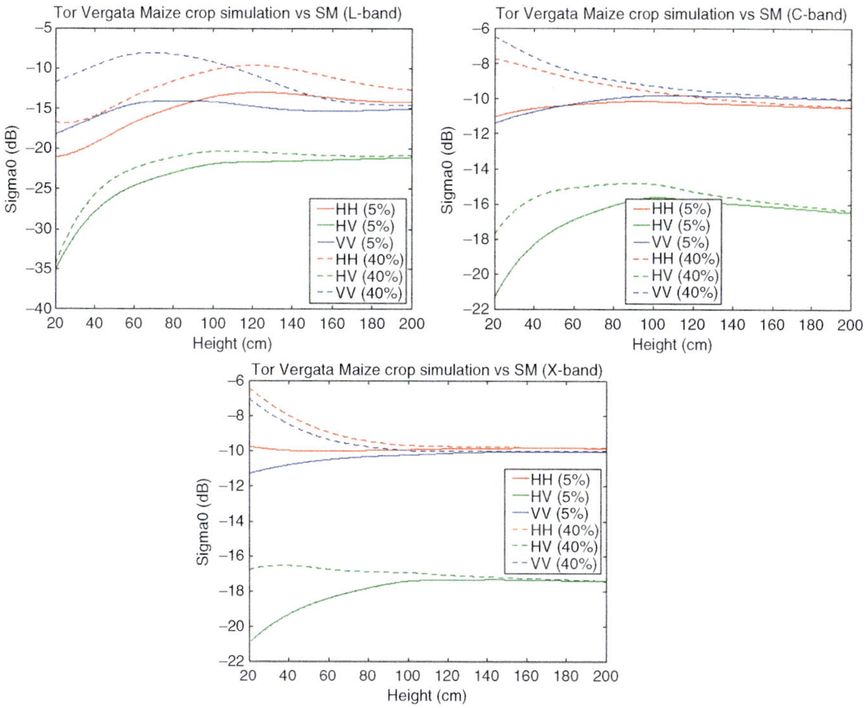

Fig. 16 Tor Vergata model (Bracaglia et al. 1995) PolSAR response at 40° incidence angle at L-, C- and X-band for a maize crop versus height for a very dry soil (5% vol. SM, in solid lines) and a very wet soil (40% vol. SM, dashed lines)

sensitivity to soil moisture of the SAR response is almost completely lost, as the solid and dashed lines come close to each other. On the other hand, at L-band there is still some soil moisture sensitivity even for tall vegetation up to 2 m.

Differential Interferometric Techniques

As mentioned previously, the modeling of the soil and vegetation interaction is a very complex task. Therefore, some recent studies have explored also additional approaches to analyze soil moisture effects over SAR data, especially by exploiting time series datasets, containing several acquisitions of the same scene at different time instants. Some approaches have tried to exploit the backscattering change between acquisitions in order to infer soil moisture changes (Moran et al. 2000) instead of its absolute value. These approaches tend to be more independent of some soil parameters as surface roughness or correlation length, but they might be disturbed by other sources of change, especially due to the vegetation layer.

Another approach to exploit time series is to apply differential interferometry. The effect of soil moisture change into the interferometric coherence and phase had already been observed before. Especially at longer wavelengths, the scattering of bare soil cannot be considered as a pure surface since some degree of penetration is

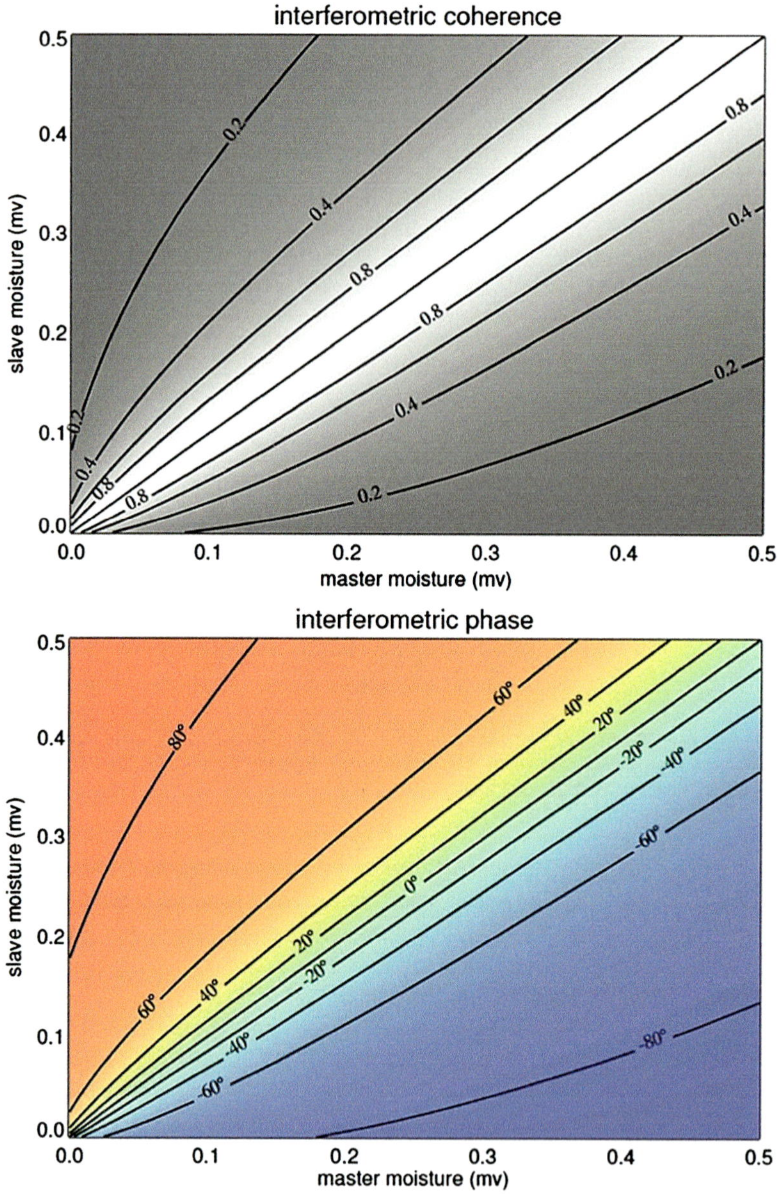

Fig. 17 Interferometric SAR soil moisture model predictions for the interferometric coherence and phase with respect to the master and slave volumetric soil moisture (De Zan et al. 2014)

observed (a few centimeters). As a consequence, when an interferogram is performed between two acquisitions with different soil moisture, an interferometric phase and coherence disturbance is observed. The differential vertical propagation of the electromagnetic waves was employed in (De Zan et al. 2014) to define an interferometric

SAR model for soil moisture. The predicted interferometric coherence and phase for two acquisitions according to this model is shown in Fig. 17. As it may be seen, the interferometric phase difference is correlated with the soil moisture difference between acquisitions. This idea might also be employed to invert the soil moisture changes of the time series, as described in (De Zan et al. 2014). However, this technique has also some limitations, since the changes in the vegetation on top of the soil might also affect the interferometric observables. As a result, the model predictions are not valid when there is tall vegetation on top of the soil or when the surface is very rough, limiting the differential propagation effect (De Zan et al. 2014).

SAR Remote Sensing of Vegetation

In forest and agriculture scenarios, some parameters like height, geometric structure, or dielectric parameters are difficult to measure on the ground extensively in space and time, with accuracies that also depend on vegetation conditions. Therefore, among the different remote sensing techniques, SAR systems are especially valuable in this scenario. They have the ability to image large areas independently of the climate conditions at high spatial and temporal resolutions. Moreover, depending on the frequency band and on the vegetation type, they have some degree of penetration into the vegetation volume, providing unique information about the vertical structure of such media until the ground.

PolInSAR and TomoSAR techniques were introduced at the end of the 1990s, and they were an important step toward developing remote sensing applications relevant to vegetation structure. The sensitivity of the interferometric coherence to the vertical structure of volume scatterers may be exploited together with the potential of SAR polarimetry to interpret and characterize the individual scattering processes at different structural components for the determination of relevant structural parameters of vegetation from a set of SAR measurements. Nowadays, PolInSAR and TomoSAR techniques allow the investigation of the 3D structure of natural volume scatterers, as it will be described in the following sections.

Forest Applications

Since its conception, SAR has been a traditional tool in order to detect the presence of forest and to identify changes or disturbances, like deforestation (Saatchi et al. 1997). For these applications, usually polarimetric acquisitions are needed, since the volume scattering that occurs in the forest canopy is better distinguished in the cross-polarized channels. Therefore, strong disturbances may be detected by the comparison of the backscattered power at these polarimetric channels between acquisitions at different dates. The same concept might be exploited even further, in order to estimate the forest biomass by analyzing the radar backscattering at the different polarimetric channels in longer wavelengths (Beaudoin et al. 1994). These approaches, however, are strongly affected by other environmental or acquisition parameters that might also affect the radar reflectivity, as the incidence angle or local slope and the moisture of the forest and the soil.

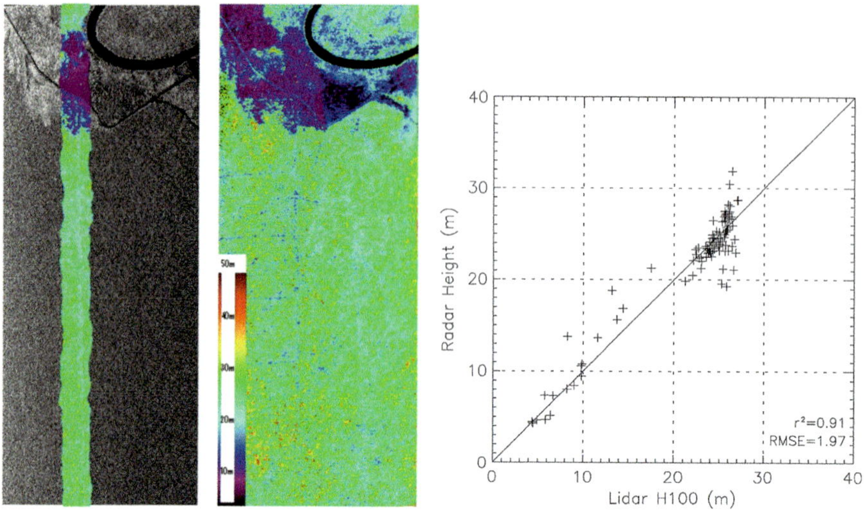

Fig. 18 SAR P-band image of the INDREX-II campaign with LIDAR H100 measurements (left), PolInSAR-derived forest height (center), and height estimates vs LIDAR H100 validation plot (right) (Hajnsek et al. 2009b)

In order to analyze the forest biomass and structure, forest height is one of the most important parameters, along with basal area, stock volume, and species composition. It is also important for the characterization of forest dynamics. Therefore, polarimetric SAR interferometry (PolInSAR) techniques have also been exploited for a more direct and reliable estimation of these parameters (Cloude and Papathanassiou 2003). In this case, a two-layer model is widely used in the literature for modeling the vertical structure of the scatterers. Usually a cloud of small particles is used to model the vegetation layer on top of a narrow impenetrable ground contribution. For this reason, this model is commonly called in the literature the random volume over ground (RVOG). Forest height inversion by exploiting this two-layer model is, probably, the most developed and mature PolInSAR application. When no significant temporal decorrelation exists, the forest height might be estimated with high accuracy (Cloude and Papathanassiou 2003). This technique has been validated over a variety of forest types (temperate, boreal, and tropical) and at different frequencies, ranging from P- up to X-band (Hajnsek et al. 2009b). Figure 18 shows one example of the results and validation activities performed with this technique. In this case the INDREX-II campaign was employed, in order to support and validate the bio-/geophysical parameter inversion techniques from SAR data. It was conducted over tropical forest in Indonesia. The images on the left and center of Fig. 18 show the radar reflectivity image and the extracted forest height with PolInSAR inversion. The plot on the right represents the obtained forest height with respect to the LIDAR-measured forest height. As it may be seen, a good correspondence is obtained between both measurements, indicating the potential of the technique for forest monitoring.

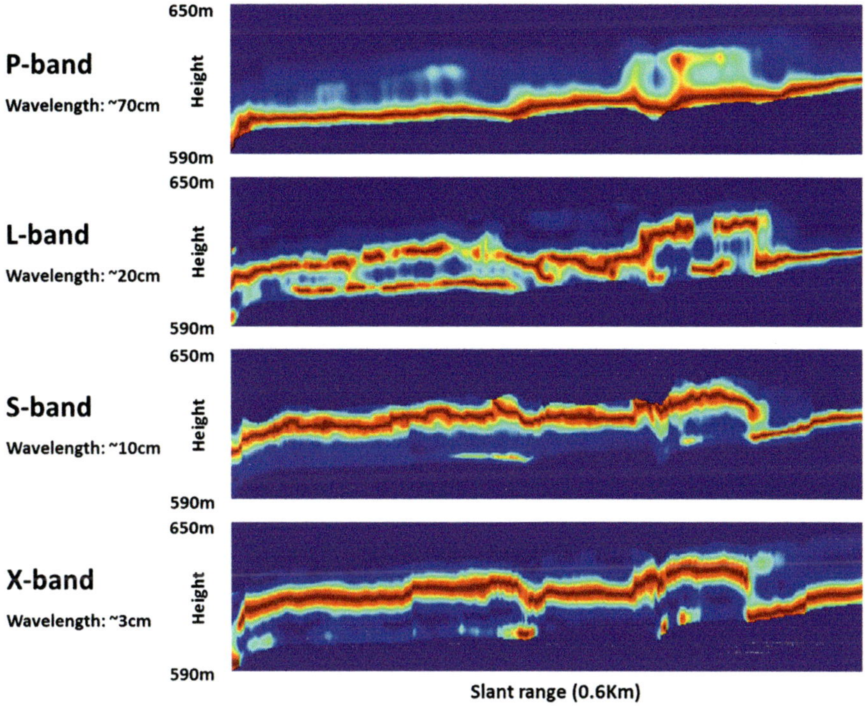

Fig. 19 Tomographic estimation of the backscattered power profile in the slant range vertical profile on the Traunstein forest size, in southern Germany. The profiles are shown for different frequencies from P- to X-band (Pardini et al. 2015)

More information about the forest may be extracted by analyzing its vertical structure, which might be extracted by means of multiple baselines by SAR tomography, as explained before. With this technique, the 3D internal structure of the forest volume might be analyzed with longer wavelengths, in order to have a good penetration into the canopy. Several studies and experiments have analyzed the capabilities of this technique, especially at lower frequencies (P- and L-band) (Tebaldini and Rocca 2012; Neumann et al. 2010). Note, however, that the vertical reflectivity profile depends strongly on the wavelength of observation and, therefore, is not easy to translate this information to physical parameters of the tree structure. Figure 19 illustrates this effect showing the vertical profile of the backscattered power for the same transect of the Traunstein forest at different frequencies from P- to X-band (Pardini et al. 2015). It may be seen that at lower frequencies (P- and L-band) a larger amount of penetration into the forest canopy is observed and, as a consequence, a stronger contribution from the ground is observed in the vertical profile. At higher frequencies (S- and X-band, in this case), most of the observed power in the vertical profile comes from the canopy layer, due to the limited penetration.

Some recent studies have tried to extract more information from this data by defining a more complex coherence model that might fit to the multiple baseline

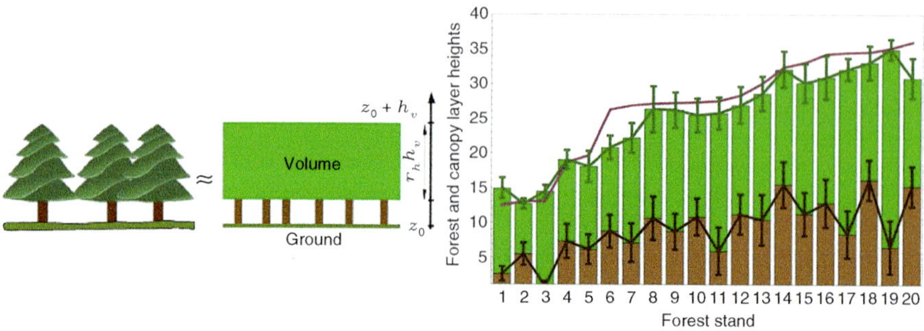

Fig. 20 Physical model employed for the complex coherence (left) and inversion results for different stands of the Traunstein forest (right) (Neumann et al. 2010). The red line corresponds to the measured height

acquisitions (Neumann et al. 2010). In this study, the volume of the canopy is defined as a homogeneous layer between two heights, instead of defining it from ground, as in the RVOG case. Moreover, due to the availability of several baselines, more parameters might be extracted like differential extinction, particle shape and orientation, etc. The model and some inversion results for the Traunstein forest are shown in Fig. 20.

These inversion results, however, are very difficult to validate due to the lack of ground measurements information, despite forest height estimations obtained from LIDAR. Moreover, some parameters like volume particle shape and orientation are very difficult to measure since they depend on the observing frequency, water content, and seasonality of the forest. Therefore, some other studies tend to use different approaches in order to measure some relative metrics that could be more physically grounded as, for instance, vertical and horizontal structure of the forest (Pardini et al. 2015). In any case, the extraction of information from SAR tomography profiles and translation into physical parameters of the forest is still not fully understood, and active research is performed in the field. Probably when future SAR missions more focused on tomography are deployed (Moreira et al. 2015) and a higher amount of data and validation measurements are available, a deeper knowledge and comprehension of this problem will be achieved, allowing further exploitation of SAR tomography for this application.

Agricultural Monitoring

The ability of SAR systems to map large areas of the Earth independently of the weather conditions is a valuable property for agricultural monitoring. Note that the phenological processes of agricultural crops induce changes significantly faster than in the forest scenario, sometimes requiring a weekly revisiting time to properly follow the plant evolution. These revisiting time constraints might be very difficult to follow with optical sensors, for instance, which are always subject to atmospheric state in order to achieve valuable acquisitions. Additionally, since SAR is working at completely different frequencies of the electromagnetic spectrum, it is sensitive to different properties compared to optical systems. In general, optical and hyper-spectral sensors are more sensitive to chemical properties of the vegetation, while SAR is more susceptible to its geometric and dielectric properties. Moreover, the high resolution acquired by SAR allows performing measurements at field or within field scale, making it an important tool in agricultural monitoring.

Polarimetric Change Analysis

Due to the sensitivity of SAR and especially polarimetric SAR to the structure of the vegetation, it is a relevant tool for detecting and monitoring the temporal changes of the agricultural crops when evolving through the different phenological stages. Several approaches may be found in the literature for the analysis and interpretation of the changes in PolSAR time series for agricultural monitoring (Alonso-Gonzalez et al. 2016). In this case, although some of the changes in the evolution of agricultural crops might be detected even with single polarization data, it is much more convenient to employ fully polarimetric data in order to improve change sensitivity and especially to allow the interpretation of the changes in terms of simple scattering mechanisms, as in the Pauli basis.

The change analysis technique proposed in (Alonso-Gonzalez et al. 2016) employs the polarimetric contrast (Boerner 1981), that is, the ratio between back-scattered powers, at the different polarization states. By exploring this ratio for all the polarimetric space, a set of polarization states that maximize and minimize this contrast might be obtained that conform a basis of the fully polarimetric state. Moreover, the contrast allows separating the changes into increase and decrease, and the polarization states that optimize this contrast might be represented into the Pauli basis for an easier interpretation of the observed changes. Figure 21 shows a crop of the Pauli RGB images of the CROPEX 2014 campaign for different dates and the corresponding land-use map (Alonso-Gonzalez et al. 2016). This campaign was conducted over the Wallerfing agricultural test site, in southern Germany with the DLR's F-SAR airborne sensor. It consists of several multi-baseline acquisitions

Fig. 21 C-band Pauli RGB images of the CROPEX 2014 campaign and land-use map. From left to right, Pauli RGB image on May 22, June 12, July 3, and July 24, 2014, and land-use map representing maize, wheat, and barley in yellow, green, and red colors, respectively (Alonso-Gonzalez et al. 2016)

at different dates and frequencies during the growing and senescence stages of several crops, including maize, wheat, and barley. In parallel with the acquisitions, ground measurements were collected over some fields.

As it may be seen in Fig. 21, some changes might be observed over the fields, but, in order to analyze them systematically, the polarimetric change analysis described in (Alonso-Gonzalez et al. 2016) might be applied to obtain the representation of the polarization states increasing and decreasing, as shown in Fig. 22. From these representations, it is clearer that a strong increase might be observed between May 22 and June 12, appearing in different colors in maize and barley fields. These changes correspond to the growing of the maize plants and the fruit maturation of the barley crop. On the other hand, a decrease might be observed in wheat and barley crops from July 3 to July 24, produced by the harvesting and plant senescence, respectively.

This change analysis technique might be applied to every pair of images of the time series, and, in order to represent all this information, the *change matrix* might be employed (Alonso-Gonzalez et al. 2016). In the change matrix, both axes represent the acquisition times, and the values in the upper diagonal represent the increase, whereas values in the lower diagonal represent the decrease. The changes obtained for the three crop types described before are represented in the change matrices in Fig. 23 with different colors. As it may be seen, this technique might be useful for monitoring the agricultural crops temporal evolution with polarimetric SAR time series.

Inversion of Crop Height

The same concept of the two component models employed in the RVOG model, as explained before, might also be applied over vegetation. However, in contrast to forest applications, in agriculture vegetation, the scattering particles are typically assumed to have a nonrandom distribution. Therefore, the propagation through the volume is not isotropic, which complicates significantly the model inversion, and

Fig. 22 Pauli RGB polarimetric change representation of the change analysis. From left to right, Pauli RGB representation of the decrease and increase for the acquisitions from May 22 to June 12 and decrease and increase from July 3 to July 24, respectively (Alonso-Gonzalez et al. 2016)

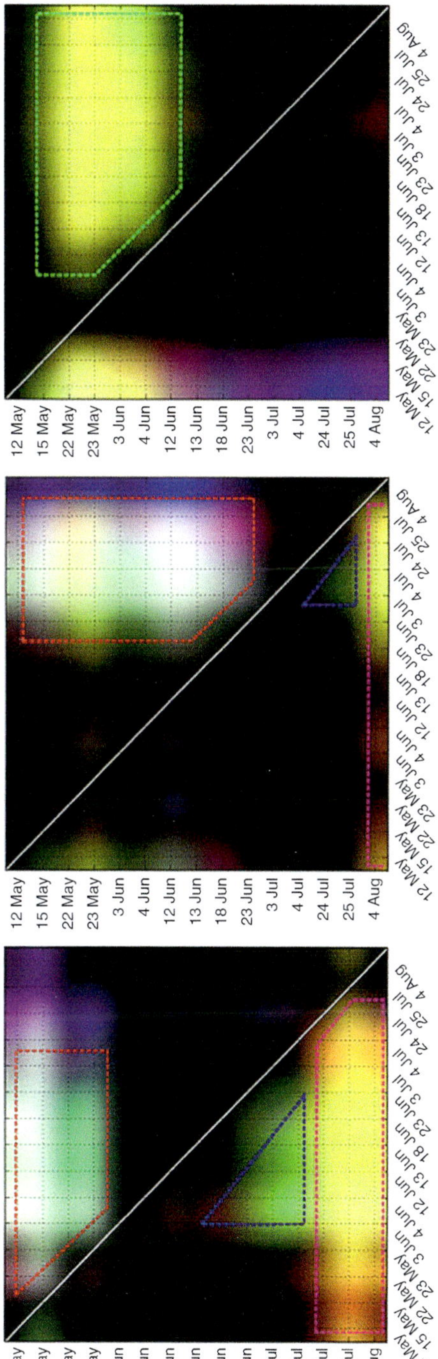

Fig. 23 Change matrices for barley (left), wheat (center), and maize (right) crops. The areas marked with dashed lines correspond to plant growth (yellow), fruit maturation (red), plant drying and senescence (blue), and harvest (magenta) phenological changes (Alonso-Gonzalez et al. 2016)

some assumptions need to be considered that might not be true in a real scenario (Joerg et al. 2016). Additionally, the height values and the accuracy needed for monitoring changes in agricultural vegetation require much longer baselines than in the forest scenario. This involves a more complex interferometric processing and needs for a precise digital elevation model, since the height of ambiguity at these long baselines is too small to extract this information directly from the data (Alonso-González et al. 2016). Moreover, the vegetation in agriculture changes much faster than the forest, resulting in a higher temporal decorrelation. These two reasons make the practical adoption of the technique from spaceborne sensors more complicated, since specific spatial baselines and very short temporally spaced acquisitions are needed for this application.

For these reasons, the development of PolInSAR applications for monitoring the structural and dielectric changes in agricultural vegetation is at an early stage. The absence of suitable multi-baseline data acquired in the same day, as explained before, and the corresponding ground measurements during all the crop phenological stages is one of the major reasons. This gap has been partially filled only in the last years by means of purposely designed airborne campaigns (Lopez-Sanchez et al. 2012a) that enabled also SAR tomography analyses (Joerg et al. 2016). The crop height inversion results obtained with the AgriSAR 2006 campaign are shown in Fig. 24 (Lopez-Sanchez et al. 2012a). Four acquisitions with three baselines at L-band were

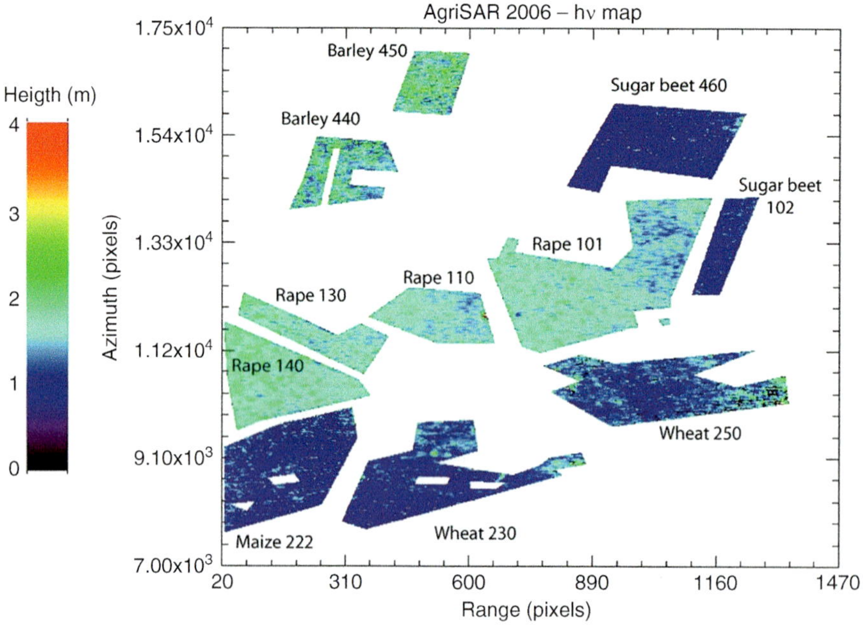

Fig. 24 Crop height inversion obtained over some fields of the AgriSAR 2006 campaign (Lopez-Sanchez et al. 2012a)

employed in order to generate these height inversion maps. Despite the fact that the frequency is not the optimum for agricultural monitoring, the comparison with ground measurements shows a relative error around 10% in the height estimates for maize and rape fields (Lopez-Sanchez et al. 2012a). In some other fields, the errors obtained were higher.

During the Science Phase of the TanDEM-X mission, longer baselines were maintained during the second and third trimesters of 2015 for experimental purposes. The acquisitions collected during this period might be employed to analyze the temporal evolution of agricultural fields with interferometry. Since this data is collected at X-band, a small degree of penetration into the crop canopy is observed, and the full inversion of the two component models might not be generally performed, but an analysis of the interferometric phase center evolution was performed in (Alonso-González et al. 2016), which shows the potential of this technique from a spaceborne sensor. Figure 25 shows the retrieved phase center differences between some acquisitions. In the left image, the phase center height difference between May 25 and June 5 may be seen. Since these two acquisitions are only 11 days apart, no significant change in height is observed. However, when the two acquisitions in May 25 and July 30 are compared (in the middle image), a significant change might be observed. Comparing these changes with the land-use map shown in the right image, it may be seen that an increase is observed over the maize fields and a decrease on the rape fields. These changes correspond to the growing of the maize and the harvest of the rape crops, respectively, between these dates. Even a smaller decrease might also be seen over the wheat fields, corresponding to its harvest, which corresponds to a phase center height change smaller than 1 meter, showing the sensitivity of these interferometric techniques even from a spaceborne sensor like TanDEM-X. Figure 26 illustrates the complete TanDEM-X image geocoded over an optical image, while Fig. 27 shows the temporal evolution of the relative phase center heights compared to the first acquisition, showing the potential of the technique for monitoring large agricultural areas from space (Alonso-González et al. 2016).

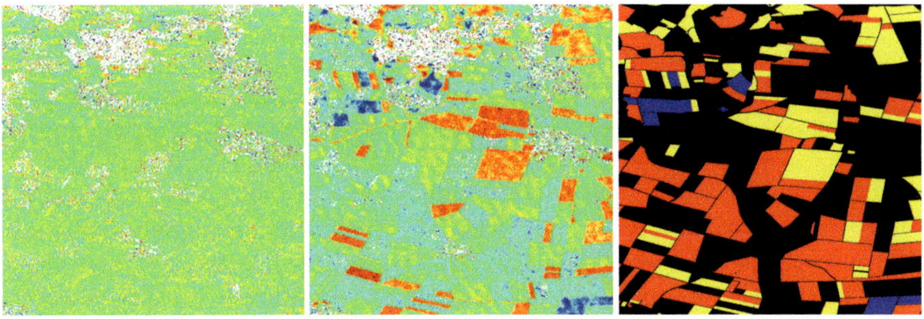

Fig. 25 Retrieved phase center height difference between acquisitions in May 25–June 5 (left) and May 25–July 30, 2015 (center), and land use of the area, representing maize, wheat, and rape in yellow, red, and blue, respectively (Alonso-González et al. 2016)

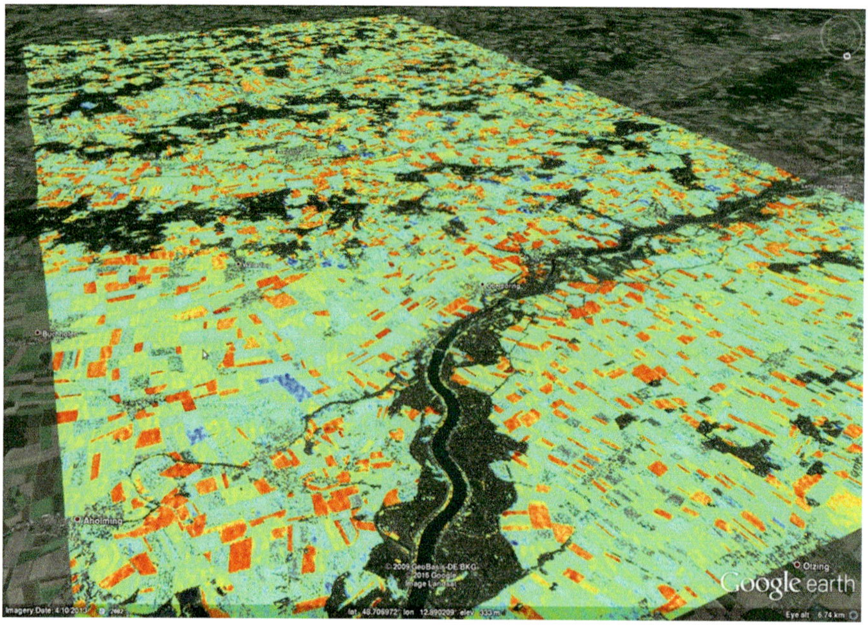

Fig. 26 Geocoded phase center height difference obtained between the acquisitions in May 25–June 5, 2015, for the complete TanDEM-X acquisition over the Wallerfing test site

Phenology Retrieval

The next step in order to deliver higher-level products that might be directly useful for agricultural management and planning would be to derive the crop phenological information from these acquisitions or even estimations of the crop yield. Some studies in the literature have tried to address this issue employing fully or dual polarimetric SAR data at different frequencies (Lopez-Sanchez et al. 2012b; Yang et al. 2014).

However, this is a difficult task since the observed radar response temporal evolution for the different crops depends strongly on the crop type but also on other parameters like the local incidence angle, the row orientation and plant density, etc. For this reason, the studies have focused in the extraction of the phenological information for a few particular species and test sites. One of the crop types more studied in the SAR literature is rice, due to its importance as a basic food resource for many countries, especially in Asia, and also due to its global climate importance for the methane emissions associated to its cultivation (Lopez-Sanchez et al. 2012b). Moreover, it has some important properties from a radiometric perspective, since it is cultivated on flat areas with the soil covered by water, which makes the processing easier due to the elimination of local incidence angle variations induced by topography and produces a strong backscattering response at the ground level produced by the double bounce, an interaction between the plant stocks and the water surface.

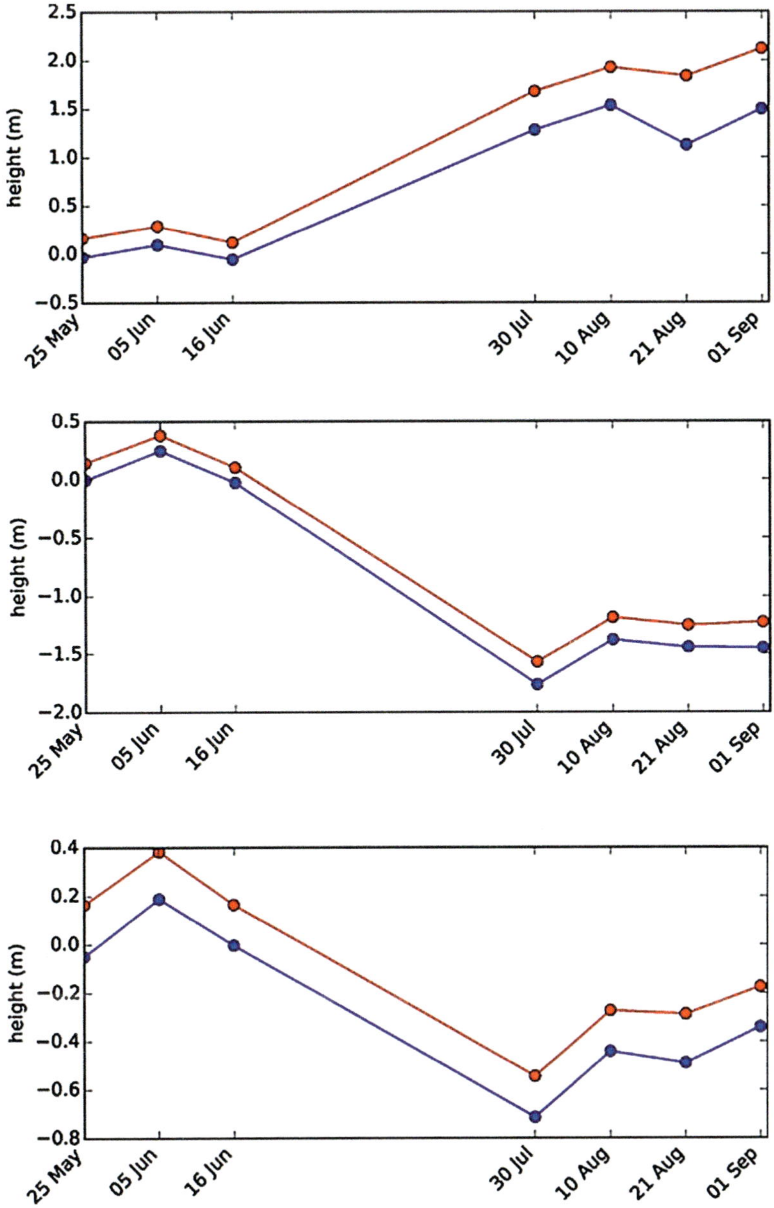

Fig. 27 Plots of the interferometric phase center height difference evolution with respect to the first acquisition for maize (left), rape (center), and wheat (right) (Alonso-González et al. 2016)

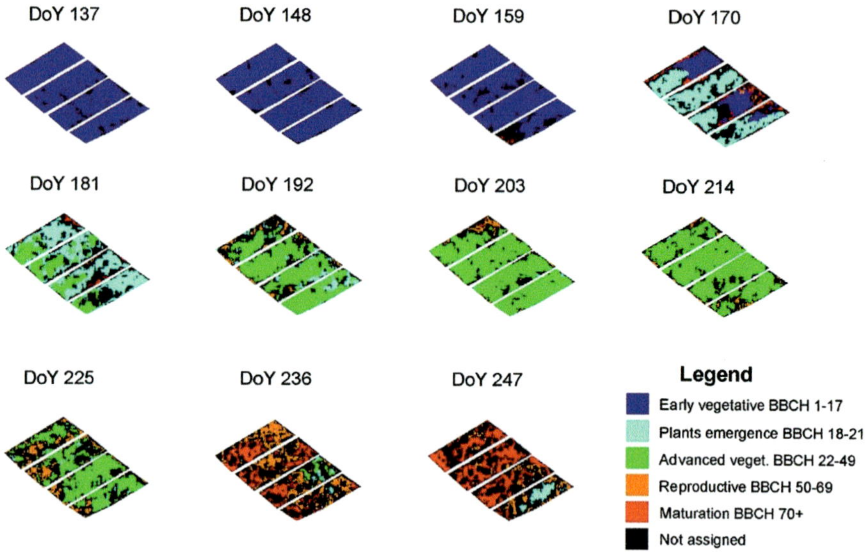

Fig. 28 Rice phenology estimated for different fields at different days of the year (DoY) employing dual polarimetric TerraSAR-X data (Lopez-Sanchez et al. 2012b)

In (Lopez-Sanchez et al. 2012b) some polarimetric decomposition parameters were employed to define a phenology extraction algorithm for rice employing dual polarimetric TerraSAR-X data. The results are shown in Fig. 28.

Some other studies have employed fully polarimetric C-band SAR data from RADARSAT-2 data to perform a more detailed rice phenology extraction and validation (Yang et al. 2014) employing decision trees. An accuracy of around 85% or better might be achieved according to the authors. Note, however, that in order to achieve this accuracy, two different models have been trained for two different rice subspecies and local adjustments have to be performed for every test site. This fact reveals the difficulties in the generalization of these phenological extraction models.

Vertical Structure Monitoring

SAR tomography might also be applied to monitor the vertical distribution of backscattered power over agricultural crops. However, as in the crop height inversion case, the technique is still at an early stage of development due to the lack of suitable campaigns with several multi-baseline acquisitions within the same day. Nevertheless, some specific airborne campaigns have been recently devoted to perform these types of analysis.

One example of such specific datasets is the CROPEX 2014 dataset, as explained before. The vertical reflectivity profiles of different crops were analyzed in (Joerg et al. 2016) with polarimetric SAR tomography at different frequencies, and some results are shown in Fig. 29.

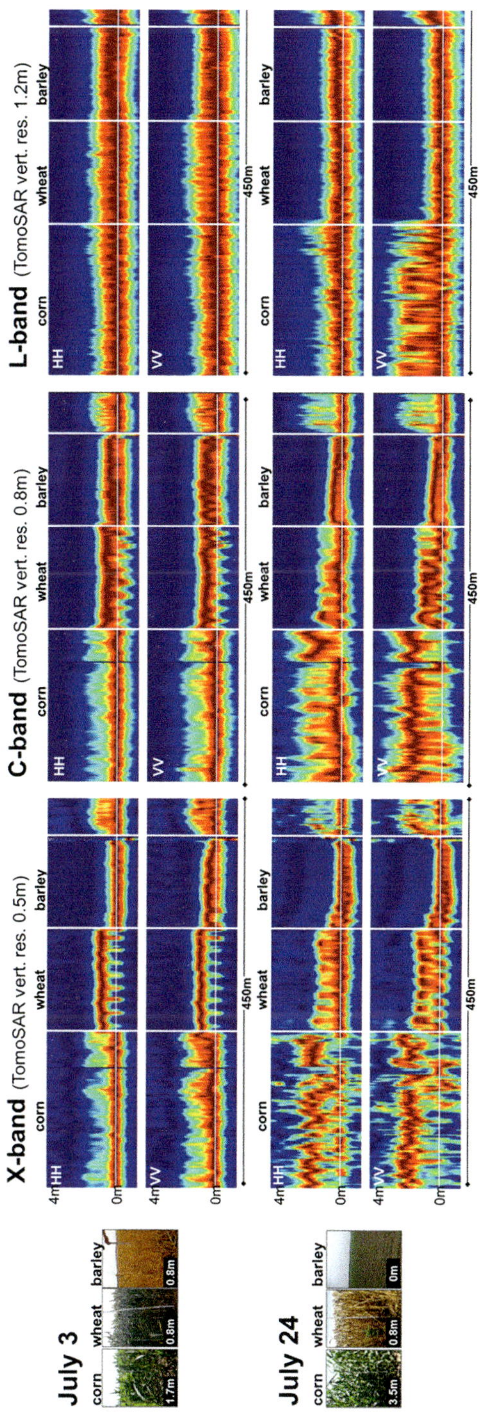

Fig. 29 SAR tomography normalized vertical reflectivity profiles for two different dates at different frequencies (Joerg et al. 2016)

At L-band the smaller vertical resolution limits the sensitivity to smaller plants. No significant difference is observed at this band for the different crop types on July 3, and only the corn can be differentiated at VV polarization on July 24. Comparing X- and C-band results, similar scattering profiles might be observed, although a higher volume scattering is present at X-band. Moreover, the vertical profile of corn is noisier at X-band on July 24. As discussed in (Joerg et al. 2016), this may be caused by additional temporal decorrelation between the different passes. The difference between HH and VV polarization on some vertical profiles at all the frequencies is also clear, which suggests the presence of significant orientation effects in the vegetation volume.

Between these dates, the changes due to the plant growing in the corn plants may be clearly seen in the vertical profile. Additionally, the plant drying and senescence of the wheat field appears as a movement of scattering phase center toward the soil, even if the plant height does not change between these dates. This shows the sensitivity of the radar response to the geometric and dielectric properties of the plant (Joerg et al. 2016).

Furthermore, from these results, it may be concluded that the choice of frequency depends on the particular application. L-band, for instance, is more sensitive to the ground and to the bigger plant components. On the other hand, at X-band the smaller plant components contribute to a higher volume component than at C-band. For this reason the scattering phase center is mainly on the top of the vegetation volume. Therefore, in some cases, a detailed analysis of the whole vegetation volume might be hindered by the limited penetration capabilities at X-band. Conversely, the use of C-band enables a better analysis of the whole vegetation volume. Even in the fully growth corn canopy, the ground component can still be identified.

Outlook

The previous sections have given a brief introduction of some of the concepts, techniques, and applications for vegetation and land cover monitoring with SAR. This section gives a short summary of the relevant techniques and requirements for the different applications described.

In the context of land cover classification, as described in section "Land Cover Classification," SAR images are a valuable tool due to its ability to monitor large areas with high resolution independently of the weather conditions. However, single reflectivity images generally do not provide enough information for a reliable differentiation between different land cover types. Moreover, the effect of speckle strongly affects the backscattering information. Therefore, in order to achieve better separation between the scene classes, SAR polarimetry might be employed. In this way, different elemental scattering processes like surface, double-bounce, or volume scattering might be distinguished in order to improve classification. Additionally temporal series are valuable sources of information for boosting classification accuracy. Interferometric coherence might be employed to distinguish reliably human-made targets from natural scenarios, and the temporal evolution dynamics is an important source to separate the different agricultural crops and forested areas.

SAR data can also be employed for soil moisture retrieval at much higher resolution than passive sensors like radiometers. However, SAR data is more affected by other soil parameters like surface roughness or the presence of vegetation. In order to invert for all these parameters normally, fully polarimetric data is needed. Moreover, in order to mitigate the effect of these parameters and being able to achieve significant response from the ground, even below the vegetation layer, longer wavelengths are required, usually L- or P-band. Additionally, some recent studies have demonstrated the effect of soil moisture in the interferometric phase and coherence at these frequencies. This demonstrates that some degree of soil penetration might be observed, and, in a complementary way, this information might be exploited to detect soil moisture variations.

Polarimetric SAR interferometry (PolInSAR) has been widely used in the last decades for the analysis of forested areas. It exploits the coherent nature of SAR in order to extract information in the vertical dimension. Two-layer models might be used to extract information related to the forest height and the ground topography underneath. Note, however, that some degree of penetration is required in order to penetrate through the forest canopy to the ground, requiring longer wavelengths as L-band or P-band. Additionally, the interferometric baselines have to be appropriate for having a good sensitivity to forest height while preventing ambiguities. When several baselines properly distributed are available, SAR tomography can be exploited to image the 3D reflectivity profile of the forest. For this technique special attention needs to be put to temporal decorrelation, which degrades the obtained results due to small changes in the vertical distribution of scatterers during the acquisitions. It is worth mentioning that some future SAR missions like ESA Biomass or Tandem-L (Moreira et al. 2015) will put strong focus on PolInSAR and SAR tomography over forest.

As opposed to optical systems, SAR sensors are more sensitive to geometric and dielectric properties of the plant, allowing them to provide complementary information for agricultural monitoring. The analysis of the polarimetric temporal evolution might be employed in order to detect phenological changes of different crop types. Additionally, PolInSAR techniques might also be employed to extract plant height and structural parameters. However, for agriculture the constraints in terms of baseline distribution and temporal decorrelation are much stronger than in forest scenario, since the crops change much faster than forest. Some studies have also demonstrated that TomoSAR might also be applied when appropriate resolution and baselines are available. For this application, due to the smaller size of the plants and its canopy volume components with respect to forest, medium wavelengths are usually preferred, as C- or X-band.

References

A. Alonso-Gonzalez, Multidimensional and temporal SAR data representation and processing based on binary partition trees, PhD Thesis, Universitat Politecnica de Catalunya (UPC), Signal Theory and Communications Department, 2014

A. Alonso-Gonzalez, H. Joerg, K. Papathanassiou, I. Hajnsek, Change analysis and interpretation in polarimetric time series over agricultural fields at C-band, in *Proceedings of EUSAR 2016: 11th European Conference on Synthetic Aperture Radar*, Hamburg, 2016

A. Alonso-González, H. Joerg, K. Papathanassiou, I. Hajnsek, Dual-polarimetric agricultural change analysis of long baseline TanDEM-X time series data, in *Geoscience and Remote Sensing Symposium (IGARSS), 2016 I.E. International*, Beijing, 2016

C. Balanis, *Advanced Engineering Electromagnetics* (Wiley, New York, 1989)

H. Balzter, B. Cole, C. Thiel, C. Schmullius, Mapping CORINE land cover from sentinel-1A SAR and SRTM digital elevation model data using random forests. Remote Sens. **7**(11), 14876–14898 (2015)

R. Bamler, P. Hartl, Synthetic aperture radar interferometry. Inverse Prob. **14**, R1–R54 (1998)

A. Beaudoin, T. Le Toan, S. Goze, E. Nezry, A. Lopes, E. Mougin, C.C. Hsu, H.C. Han, J.A. Kong, R.T. Shin, Retrieval of forest biomass from SAR data. Int. J. Remote Sens. **15**(14), 2777–2796 (1994)

W. Boerner, Optimal polarization concept in radar imaging, *Technical report,* 1981

M. Bracaglia, P. Ferrazzoli, L. Guerriero, A fully polarimetric multiple scattering model for crops. Remote Sens. Environ. **54**(3), 170–179 (1995)

W. Brown, Synthetic aperture radar. IEEE Trans. Aerosp. Electron. Syst. **3**(2), 217–229 (1967)

L. Bruzzone, M. Marconcini, U. Wegmueller, A. Wiesmann, An advanced system for the automatic classification of multitemporal SAR images. IEEE Trans. Geosci. Remote Sens. **42**(6), 1321–1334 (2004)

CLC2006 technical guidelines, vol. 17, EEA Technical Report, 2007

S.R. Cloude, K.P. Papathanassiou, Three-stage inversion process for polarimetric SAR interferometry. IEE ProcRadar. Sonar Navig. **150**(3), 125–134 (2003)

S. Cloude, E. Pottier, An entropy based classification scheme for land applications of polarimetric SAR. IEEE Trans. Geosci. Remote Sens. **35**(1), 68–78 (1997)

S. Cloude, E. Pottier, Concept of polarization entropy in optical scattering. Opt. Eng. **34**(6), 1599–1610 (2005)

F. De Zan, A. Parizzi, P. Prats-Iraola, P. Lopez-Dekker, A SAR interferometric model for soil moisture. IEEE Trans. Geosci. Remote Sens. **52**(1), 418–425 (2014)

P.C. Dubois, J. Van Zyl, T. Engman, Measuring soil moisture with imaging radars. IEEE Trans. Geosci. Remote Sens. **33**(4), 915–926 (1995)

C. Elachi, J. Van Zyl, *Introduction to the Physics and Techniques of Remote Sensing. Wiley Series in Remote Sensing and Image Processing* (Wiley, Hoboken, 2006)

T.G. Farr, P.A. Rosen, E. Caro, R. Crippen, R. Duren, S. Hensley, D. Seal, et al., The shuttle radar topography mission. Rev. Geophys. **45**(2), RG2004+ (2007)

A. Ferreti, C. Prati, F. Rocca, Permanent scatterers in SAR interferometry. IEEE Trans. Geosci. Remote Sens. **39**(1), 8–20 (2001)

A.K. Fung, Z. Li, K.S. Chen, Backscattering from a randomly rough dielectric surface. IEEE Trans. Geosci. Remote Sens. **30**(2), 356–369 (1992)

I. Hajnsek, E. Pottier, S.R. Cloude, Inversion of surface parameters from polarimetric SAR. IEEE Trans. Geosci. Remote Sens. **41**(4), 727–744 (2003)

I. Hajnsek, T. Jagdhuber, H. Schon, K.P. Papathanassiou, Potential of estimating soil moisture under vegetation cover by means of PolSAR. IEEE Trans. Geosci. Remote Sens. **47**(2), 442–454 (2009a)

I. Hajnsek, F. Kugler, S.K. Lee, K.P. Papathanassiou, Tropical-forest-parameter estimation by means of pol-InSAR: The INDREX-II campaign. IEEE Trans. Geosci. Remote Sens. **47**(2), 481–493 (2009b)

F. Henderson, Z. Xia, SAR applications in human settlement detection, population estimation and urban land use pattern analysis: A status report. IEEE Trans. Geosci. Remote Sens. **35**(1), 79–85 (1997)

H. Joerg, M. Pardini, I. Hajnsek, K. P. Papathanassiou, Analysis of orientation effects of crop vegetation volumes by means of SAR tomography at different frequencies, in *Proceedings of 11th European Conference on Synthetic Aperture Radar (EUSAR)*, Hamburg, 2016

G. Krieger, A. Moreira, H. Fiedler, I. Hajnsek, M. Werner, M. Younis, M. Zink, TanDEM-X: A satellite formation for high-resolution SAR interferometry. IEEE Trans. Geosci. Remote Sens. **45**(11), 3317–3341 (2007)

C. Lardeux, P. Frison, C. Tison, J. Souyris, B. Stoll, B. Fruneau, J. Rudant, Support vector machine for multifrequency SAR polarimetric data classification. IEEE Trans. Geosci. Remote Sens. **47**(12), 4143–4152 (2009)

T. Le Toan, S. Quegan, M. Davidson, H. Balzter, P. Paillou, K. Papathanassiou, S. Plummer, F. Rocca, S. Saatchi, H. Shugart, L. Ulander, The BIOMASS mission: Mapping global forest biomass to better understand the terrestrial carbon cycle. Remote Sens. Environ. **15**(11), 2850–2860 (2011)

J. Lee, Speckle analysis and smoothing of synthetic aperture radar images. Comput. Graphics Image Process. **1**(17), 24–32 (1981)

J. Lee, E. Pottier, *Polarimetric Radar Imaging: From Basics to Applications* (CRC Press, Boca Raton, 2009)

J. Lee, M. Grunes, R. Kwok, Classification of multi-look polarimetric SAR imagery based on complex Wishart distribution. Int. J. Remote Sens. **15**(11), 2299–2311 (1994)

J. Lee, M. Grunes, T. Ainsworth, L. Du, D. Schuler, S. Cloude, Unsupervised classification using polarimetric decomposition and the complex Wishart classifier. IEEE Trans. Geosci. Remote Sens. **37**(5), 2249–2258 (1999)

J.M. Lopez-Sanchez, I. Hajnsek, J.D. Ballester-Berman, First demonstration of agriculture height retrieval with PolInSAR airborne data. IEEE Geosci. Remote Sens. Lett. **9**(2), 242–246 (2012a)

J.M. Lopez-Sanchez, S.R. Cloude, J.D. Ballester-Berman, Rice phenology monitoring by means of SAR polarimetry at X-band. IEEE Trans. Geosci. Remote Sens. **50**(7), 2695–2709 (2012b)

M.S. Moran, D.C. Hymer, J. Qi, E.E. Sano, Soil moisture evaluation using multi-temporal synthetic aperture radar (SAR) in semiarid rangeland. Agric. For. Meteorol. **105**(1), 69–80 (2000)

A. Moreira, P. Prats-Iraola, M. Younis, G. Krieger, I. Hajnsek, K.P. Papathanassiou, A tutorial on synthetic aperture radar. IEEE Geosci. Remote Sens. Mag. **1**(1), 6–43 (2013)

A. Moreira, G. Krieger, I. Hajnsek, K. Papathanassiou, M. Younis, P. Lopez-Dekker, S. Huber, M. Villano, M. Pardini, M. Eineder, F. De Zan, A. Parizzi, Tandem-L: a highly innovative bistatic SAR mission for global observation of dynamic processes on the Earth's surface. IEEE Geosci. Remote Sens. Mag. **3**(2), 8–23 (2015)

M. Neumann, L. Ferro-Famil, A. Reigber, Estimation of forest structure, ground, and canopy layer characteristics from multibaseline polarimetric interferometric SAR data. IEEE Trans. Geosci. Remote Sens. **48**(3), 1086–1104 (2010)

M. Pardini, M. Tello Alonso, A. Torano Caicoya, M. Heym, K. Papathanassiou, A comparison of P- and L-Band PolInSAR 3-D forest structure estimates: a study case in the Traunstein Forest, in *Proceedings of the ESA PolInSAR Workshop*, Frascati (Rome), 2015

Z. Qi, A. Yeh, X. Li, Z. Lin, A novel algorithm for land use and land cover classification using RADARSAT-2 polarimetric SAR data. Remote Sens. Environ. **118**, 21–39 (2012)

M. Richards, A beginner's guide to interferometric SAR concepts and signal processing. IEEE Aerosp. Electron. Syst. Mag. **22**(9), 5–29 (2007)

S.S. Saatchi, J.V. Soares, D.S. Alves, Mapping deforestation and land use in Amazon rainforest by using SIR-C imagery. Remote Sens. Environ. **59**(2), 191–202 (1997)

H. Skriver, Crop classification by multitemporal C- and L-band single- and dual-polarization and fully polarimetric SAR. IEEE Trans. Geosci. Remote Sens. **50**(6), 2138–2149 (2012)

H. Skriver, F. Mattia, G. Stalino, A. Balenzano, V. Pauwels, N. Verhoest, M. Davidson, Crop classification using short-revisit multitemporal SAR data. IEEE J. Sel. Top. Appl. Earth Obs. Remote Sens. **4**(2), 423–431 (2011)

S. Tebaldini, F. Rocca, Multibaseline polarimetric SAR tomography of a boreal forest at P-and L-bands. IEEE Trans. Geosci. Remote Sens. **50**(1), 232–246 (2012)

F. Ulaby, R. Moore, A. Fung, *Microwave Remote Sensing: Active and Passive. Volume 2 – Radar Remote Sensing and Surface Scattering and Emission Theory* (Artech House Publishers, 1982)

F. Ulaby, R. Moore and A.K. Fung, *Microwave Remote Sensing, Active and Passive. Vol. III: From Theory to Applications* (Artech House Publishers, 1986)

W. Wagner, G. Blöschl, P. Pampaloni, J.C. Calvet, B. Bizzarri, J.P. Wigneron, Y. Kerr, Operational readiness of microwave remote sensing of soil moisture for hydrologic applications. Hydrol. Res. **38**(1), 1–20 (2007)

Z. Yang, K. Li, L. Liu, Y. Shao, B. Brisco, W. Li, Rice growth monitoring using simulated compact polarimetric C band SAR. Radio Sci. **49**(12), 1300–1315 (2014)

Root Processes Affecting Soil Moisture Patterns in Ecohydrology

Axelle Koch, Félicien Meunier, Harry Vereecken, and Mathieu Javaux

Contents

Abstract

Soil moisture patterns arise from the combined processes induced by vegetation, soil properties, climate, topography, parent material, and time. In this chapter, we focus on how vegetation induces soil moisture patterns, particularly how plant

A. Koch · F. Meunier
Earth and Life Institute/Environmental Sciences, Université catholique de Louvain, Louvain-la-Neuve, Belgium
e-mail: Axelle.koch@uclouvain.be; felicien.meunier@uclouvain.be

H. Vereecken
Agrosphere Institute (IBG-3), Forschungszentrum Jülich GmbH, Jülich, Germany
e-mail: h.vereecken@fz-juelich.de

M. Javaux (✉)
Earth and Life Institute/Environmental Sciences, Université catholique de Louvain, Louvain-la-Neuve, Belgium

Agrosphere Institute (IBG-3), Forschungszentrum Jülich GmbH, Jülich, Germany
e-mail: Mathieu.javaux@uclouvain.be

X. Li, H. Vereecken (eds.), *Observation and Measurement of Ecohydrological Processes*, Ecohydrology, https://doi.org/10.1007/978-3-662-48297-1_13

root processes affect the soil moisture distribution. Four different mechanisms were identified as potential drivers of soil moisture variability: root growth, root water uptake and transpiration, plant competitions, and rhizosphere properties. High transpiration, root growth, and root water uptake generally increase the soil moisture variability for drying conditions. On the other hand, other mechanisms reduce the soil moisture variation under drying condition including (1) compensation, which plants extract water in the wettest part of the soil; (2) hydrotropism, which roots tend to grow toward wetter zone of the soil; and (3) plant competition, which different plants try to segregate the depths at which they take up water. In addition, rhizosphere-specific properties tend to increase the variability when the soil is wetted from dry condition or to decrease it under wet conditions. We used a plant architecture model to illustrate how soil and root properties combine to generate or destroy soil moisture relations.

Keywords

Root · Soil · Soil-root interactions · Functional-structural plant model · Soil moisture · Rhizosphere · R-SWMS

Introduction

Ecohydrological patterns are being found in desert and arid regions, wetlands, riverine landscapes, land use systems, and soil-plant-atmosphere systems, among others. The presence of patterns in soil-plant-atmosphere systems may be the result of system's properties that cause self-organization leading to typical ecohydrological patterns, or it is the result of emergent properties that appear due to changes in scale and the presence of patterns in controls acting on the system (Vereecken et al. 2016). Self-organization of a system implies that it reaches in an autonomous manner a state of higher organization or order over time characterized by initiation and formation of patterns. As stated in Vereecken et al. (2016), "The formation of patterns via emergence can [then] be defined as the result of dynamic interactions between the system components on the micro-level, which create a pattern (or an emergent property) on the macro-level that provides novel information on the system as a whole." In general, patterns in any system can be identified and characterized using (1) images and maps that show the spatial distribution of the process, (2) statistical measures that capture structure in temporal and spatial data, and (3) models that capture the dynamics of states in a system. A review on methods and applications of pattern characterization in soil-vegetation-atmosphere systems was recently published by Vereecken et al. (2016).

Ecological and hydrological patterns are closely connected. For example, processes such as infiltration, overland flow, and evapotranspiration may lead to the formation of vegetation patterns in dry regions (Schröder 2006). Meinders and Breemen (2005) showed that soil-vegetation feedback processes might introduce self-organized formation of patterns in semiarid systems through nonlinear interactions between water and vegetation.

Numerous authors have worked on the drivers of soil water heterogeneity. McBratney et al. (2003) extended Jenny's equation (Jenny 1941) and identified six factors that impact soil moisture distribution (s, c, o, r, p, a): (s) soil properties (Jawson and Niemann 2007), (c) climate (Cho and Choi 2014), (o) vegetation (Hupet and Vanclooster 2002), (r) topography (Zhu and Lin 2011; Sur et al. 2013), (p) parent material (Markoski et al. 2016), and (a) time (Gerke et al. 2016). These factors can either increase soil moisture variability and accentuate soil moisture patterns or destroy them (Teuling and Troch 2005). It has been established that soil hydraulic property variability is a driver of soil moisture variability (Yoo et al. 1998; Jawson and Niemann 2007; Martinez Garcia et al. 2014). It is also known that a general link exists between soil moisture spatial characteristics and its state (i.e., whether the soil is dry or wet). In general, precipitation acts as a pattern moisture destroyer (Yoo et al. 1998). Western et al. (2003) showed that soil moisture spatial variability is the highest under intermediate conditions and diminishes under extreme conditions. Yet when vegetation comes into play, different variability patterns might arise due to additional dynamic and spatially variable processes.

The objective of this chapter is to emphasize the role of plants and in particular of root-related processes in creating or destroying spatial and temporal soil moisture patterns in contrasted pedo-climatic conditions. We will first review literature case studies to identify a list of root-related processes and discuss if they promote or destroy soil moisture patterns (respectively, positive/negative feedbacks). Then, we will make use of a plant-soil model to quantify how these processes combine with soil and climate to create/destroy spatial and temporal patterns of soil moisture. These will be investigated using the relationship between the coefficient of variation (CV) of the soil moisture and its mean value.

Plants as Drivers for Soil Moisture Pattern

It was long thought that soil properties varied more intensively than root water uptake (RWU) over the field and therefore were the key factors controlling soil moisture pattern (Clothier 1988; White 1988). However, some studies emphasized the role of RWU on soil moisture spatial and temporal variability. For example, Hupet and Vanclooster (2002) highlighted that soil moisture spatiotemporal variability was depth dependent in a maize field during summer. Indeed, soil moisture variations at deeper layers (>75 cm) were less affected by RWU and precipitation dynamics. The RWU is a dominant driver of soil moisture distribution under dry conditions, whereas soil properties play a more important role under wet conditions (Katul et al. 1997; Hupet and Vanclooster 2002). Moreover, Baroni et al. (2013) revealed that soil-vegetation interaction also affected the soil moisture pattern through a principal component analysis (PCA). However it was not clear how RWU influences soil moisture variance and hence the soil water content CV. Indeed, Katul et al. (1997) showed that RWU processes tended to amplify soil moisture variance under wet conditions and to reduce it under dry conditions, while Hupet and Vanclooster (2002) observed an increase in soil moisture variability under dry conditions. However, this apparent controversy could be explained by the fact that soil types were different and so was soil water redistribution. For understanding the impact of vegetation on the relationship between coefficient of variation of soil moisture and its mean, Martinez Garcia et al. (2014) conducted an experiment with different land uses: bare soil and vegetation covered. They found that moisture spatial variability under vegetation was higher than for bare soil. Moreover, they pointed out that the slope of that relationship was half as much than for bare soil conditions. This implies that soil moisture variability under vegetation is always high, even under extreme moisture conditions. Korres et al. (2013) underlined that plant type and the development stage are other important parameters to take into account and also lead to temporal variability over the growing season.

Hupet and Vanclooster (2005) studied the impact of spatial variability of hydrological processes (rainfall, RWU, and evapotranspiration) on soil moisture patterns under maize crops in Louvain-la-Neuve (well-drained silty-loam) during summer 2003 (unusually dry). Different mechanisms were shown to control soil moisture patterns. First, they observed that the distribution of plants together with their aboveground architecture affected the distribution of rainfall at the soil surface through throughfall and stemflow. These phenomena induced patterns in infiltration and hence soil moisture. The spatial distribution of soil moisture depletion evolved during the season. At the beginning, depletion occurred mainly in shallow layers near the maize feet, thereby increasing soil moisture variability between row and inter-row. Later, deeper soil layers (45–65 cm depth) were depleted due to root growth, and soil moisture contrast between row/inter-row were even more pronounced. Finally, at the end of the dry season, lateral redistribution of water occurred from soil below inter-row (wetter) to the previous locations of RWU (drier), thereby providing more water to plants. This redistribution phenomenon decreased soil moisture spatial variability. Thus, the variability of infiltration (due to vegetation cover) together with the variability of RWU generated moisture patterns (start and middle of the season), while lateral soil water redistribution (due to relatively high soil hydraulic conductivity of silt loam) decreased the moisture variability.

Zhang et al. (2016) investigated the dynamic of soil properties and soil moisture over successive vegetation restoration states on the Loess Plateau in China. This restoration followed five stages, namely, two grass restoration stages, one shrub restoration stage, and two forest restoration stages. From grassland to forest climax, soil bulk density decreased, total porosity increased, and clay and silt contents increased. All these variations were depth dependent and occurred firstly in the topsoil. A diminution in vertical soil moisture variation was also noted. This water redistribution could be explained by an increased soil hydraulic conductivity or by extended root systems that allowed RWU all over the soil profile. Both processes tended to homogenize soil water potential. Recent studies seem to confirm the impact of vegetation cover on soil moisture dynamic (Niether et al. 2017; Özkan and Gökbulak 2017). Niether et al. (2017) studied soil moisture dynamics under different cocoa production, namely, monoculture and agroforestry. Their research confirmed that under agroforestry system (where roots explore a wider depth than in monoculture), the variability of soil moisture was reduced with soil depth. Özkan and

Gökbulak (2017) compared soil moisture dynamics under herbaceous vegetation and forest cover. Firstly, they observed that mean soil moisture was higher under herbaceous vegetation than under forest cover. This is probably due to the absence of canopy interception (i.e., higher net rainfall) and to smaller transpiration losses. Secondly, they noted that soil moisture increased with depth in winter. During vegetation season, they noticed that soil moisture depleted in the subsoil under forest cover, while it depleted in the topsoil under herbaceous vegetation. These soil moisture changes are driven by RWU and so linked to the rooting depths. Tree roots, which are able to extract water from the whole depth profile, decrease the vertical moisture variability by taking water from where the potential is the highest.

Under semiarid and arid conditions, several studies demonstrated that vegetation increased infiltration thereby promoting the development of ecohydrological patterns. Thompson et al. (2010) investigated the vegetation-infiltration relationship under different ecosystems and confirmed a strong correlation between biomass and infiltration under arid conditions. They hypothesize that not only the passive impact of the vegetation on the prevention of surface crust or the creation of habitat for microorganisms and microfauna but also active mechanisms related to root growth were the drivers of this relationship.

Root Processes Affecting the Soil Moisture Pattern

Through the previous examples, numerous plant-related processes were pointed out, which affect soil water heterogeneity and thus destroy or generate soil moisture patterns. In particular, belowground plant traits might control patterns (Guswa 2012) and ecosystem processes (Bardgett et al. 2014). Recent advances in the understanding of root physiology might help in understanding the occurrence and relevance of these processes.

Root Growth

By driving the development of the root system architecture with time and thus the potential distribution of root water uptake, root growth is the principal mechanism, which shapes the soil moisture patterns. Gravitropism, i.e., the tendency of roots to grow downward, will generally induce a vertical soil moisture distribution. However, root growth is also impacted by soil water distribution through direct and indirect mechanisms. Hydrotropism (Moriwaki et al. 2013) is the process by which root tips grow toward wetter soil zones. Hydro-patterning (Bao et al. 2014) is another mechanism, which has been discovered recently, by which primary roots locate their lateral roots toward available water. Root growth is also indirectly affected by soil water content via its impact on soil mechanical resistance (Whalley and Bengough 2013; Ruiz et al. 2015). Indeed soil strength is inversely proportional to soil moisture, and higher soil resistance impedes elongation and root invasion (Yapa et al. 1988). These processes will lead to a negative feedback between growth and soil moisture patterns. By growing toward wetter zones, roots will tend to homogenize soil moisture distribution under medium soil moisture conditions. Under dry conditions, in addition, low available soil water will induce plant water stress and then close stomata and decrease carbon acquisition and limit growth.

Root Water Uptake and Transpiration

The uptake of soil water by roots is a passive mechanism, which depends on the gradient of water potential between bulk soil and root xylem and the hydraulic resistances in between. Root water uptake fluxes therefore vary in both space and time as a function of root system development (de Dorlodot et al. 2007), root hydraulic properties (Zarebanadkouki et al. 2014) that depend in turn on root type and age (Vetterlein and Doussan 2016), soil water availability (Couvreur et al. 2014b), and root water potential distributions (Lobet et al. 2014). In addition, soil water heterogeneity also impacts hormonal production (Puértolas et al. 2015), which partially controls the distribution of root lateral conductivity, the root conductance and leaf growth (Caldeira et al. 2014), and eventually the root water uptake distribution.

In general, high transpiration rate will tend to generate drying patterns proportional to root distribution (Singh et al. 1998). However, compensation process, i.e., the mechanism by which root extracts water from the wetter zones of the soil (Lobet et al. 2014), will on the other hand tend to homogenize soil moisture distribution. Hydraulic lift is an extreme case of compensation process where water is moved upward through the root system (Prieto et al. 2012).

It has been demonstrated that the impact of compensation increases with soil water potential heterogeneity and with lower transpiration (Javaux et al. 2013). This typically occurs when the soil is already relatively dry, when soil hydraulic conductivity is low, and when plant is under water stress. Important soil water potential gradients might then appear, and root compensation could then play an important role. It has been shown that compensation was promoted for root systems with large radial conductivity in particular in wet zones (Meunier et al. 2017). Yet the magnitude and relevance of compensation is still a matter of debate (Burgess 2011).

Rhizosphere and Root Zone Properties

Plants are known to affect infiltration and soil water retention directly or indirectly. The vegetation aboveground organs affect the amount of precipitation that infiltrates the soil through interception and crust prevention, and they diminish the evaporation losses. In addition, during plant growth, roots physically and chemically alter their surrounding environment, i.e., the rhizosphere (Bengough 2012). It has been demonstrated that root zones under arid conditions represent resource islands for microorganism activity, which affect soil structure, pore connectivity, and infiltration capacity (Bhark and Small 2003). For instance, Dunkerley (2002) showed that soil infiltration was gradually increasing toward trunks of Australian mulga, independently of canopy coverage, demonstrating a change in soil hydraulic properties induced by roots.

Root elongation and development may compact the soil around it (Aravena et al. 2011) but also may generate air gaps between root and soil in particular under drying conditions (Carminati et al. 2009), which might promote preferential flow and therefore increase infiltration heterogeneity (Johnson and Lehmann 2006; Bargués Tobella et al. 2014).

Mucilage exudation has been shown to make rhizosphere more hydrophobic under drying conditions (Ahmed et al. 2016), which might stabilize the patterns. It has been hypothesized that this hydrophobicity would be a positive trait of deep-rooted plants to avoid competition with shallow-rooted plants (Zeppenfeld et al. 2017), which would also stabilize soil moisture patterns. On the other hand, some authors claim that changes in root hydraulic properties will lead to a more homogeneous distribution of the uptake (Schwartz et al. 2016). In general, rhizosphere properties will induce heterogeneous infiltration patterns and will tend to increase soil moisture variability under dry conditions and decrease it when the soil is under medium conditions, by preferential infiltration in drier zone.

Multiple Plants/Competition

The location of plants impacts soil water variability, an extreme example being observed in agricultural fields, where row-inter-row patterns determine soil moisture patterns (Hupet and Vanclooster 2002; Beff et al. 2012; Couvreur et al. 2014a). Plant competition is affected by soil water distribution (Gordon et al. 1989), but the consequences of competition for water for the evolution of plants and the functioning of ecosystems are poorly explored (Craine and Dybzinski 2013). Weigelt et al. (2000) showed that the belowground competition was affected by soil water status and that structure of plant communities depends on the seasonal change of soil water. In principle competition between plants should lead ultimately to a more homogeneous distribution of RWU and thus acts as a destroyer of soil moisture variability. Yet this effect should depend on plant-specific traits like stomatal, regulation, rooting depths, and their spatial distribution.

Combination of Processes

Root water uptake in wet soils generates patterns under high transpiration conditions and heterogeneous root distribution. The magnitude and temporal stability of these patterns will depend on the soil water redistribution, controlled by soil hydraulic conductivity. For low transpiration situations, dry conditions, and more heterogeneous water distribution, mechanisms like hydrotropism, competition between roots and plants, or compensated root water uptake will come into play and tend to homogenize soil water content distributions. These plant-related homogenizing mechanisms particularly take place when soil moisture redistribution is impeded due to low soil hydraulic conductivity. On the opposite, when rainfall occurs under dry soil conditions, rooted soils will most probably generate very heterogeneous infiltration patterns due to rhizosphere-related processes coupled with canopy interception and stemflow. Figure 1 summarizes the different processes in a conceptual plot between soil moisture variability and mean soil moisture.

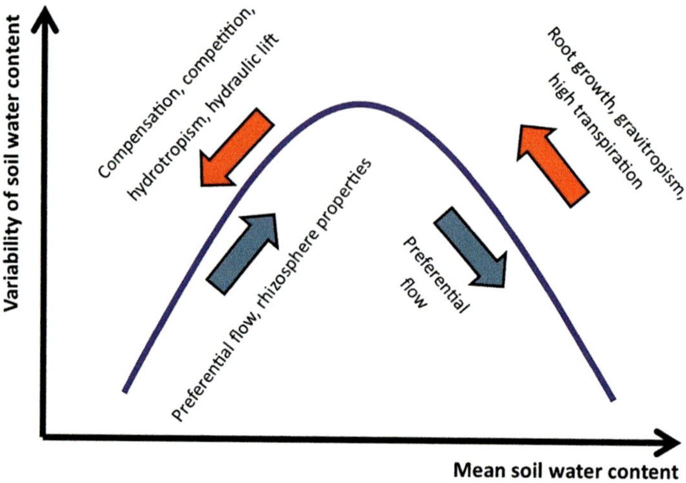

Fig. 1 Soil moisture variability vs averaged soil moisture and associated root processes, which control variability

Using Quantitative Models to Investigate Feedback Between Root Properties and Soil Moisture

How the root-related processes combine with heterogeneous soil properties to maintain or destroy patterns has hardly been quantified. Indeed while, under wet conditions, plants will create patterns potentially counterbalanced by soil water redistribution, the opposite takes place under dry conditions, when soil water redistribution decreases but root compensation increases. Soil hydraulic conductivity, heterogeneous soil properties, root distribution, and hydraulic properties but also transpiration are some of the factors that will influence the appearance of soil moisture patterns. Kuhlmann et al. (2012) were probably the first ones to investigate root water uptake compensation in heterogeneous soils and showed that it tended to increase transpiration. However, no studies used a mechanistic model to investigate these interactions. Hereafter we use a process-based root-soil model to illustrate how such tools might help understand and quantify root impacts on soil moisture patterns.

Root-Soil Functional Structural Model

We used R-SWMS (see Javaux et al. 2008 for an exhaustive presentation of the model), a functional-structural root-soil model to simulate root water uptake and soil water flow. In this model, the distribution of root water uptake is not imposed a priori, but rather the result of the continuous energy minimization required to fulfill the evaporative demand. The root axial and radial water flow equations through roots are solved iteratively with the Richards equation in soil to estimate the distribution of the RWU based on the water potential distribution in soil and roots. Stress appears when the evaporative demand cannot be fulfilled by the plant without dropping its leaf water potential below a limiting value fixed as threshold.

We simulated changes of soil moisture distribution by running different scenarios with one root system located in a wet soil profile of 1.25 m depth and subject to different levels of transpiration (Fig. 2). No flux top and bottom soil boundary conditions were imposed (i.e., no rain nor drainage). Periodic boundary conditions were used for lateral fluxes of water in roots and soil (i.e., root and water fluxes leaving from one border come in at the opposite side to represent the presence of identical neighbors), mimicking a homogeneous distribution of plants. Among the different factors affecting soil moisture distribution, six were investigated (see Table 1): soil hydraulic properties (sandy loam vs silty clay), soil property distribution (homogeneous vs random), root system conductance (low vs normal), evaporative demand (low vs high), plant density (low vs high; see Fig. 7 in "Appendix B"), and root hydraulic property distribution (homogeneous vs heterogeneous). In total, combinations of these 6 factors lead to 64 (2^6) scenarios.

Figure 2 shows the soil properties used in the study. Heterogeneity in soil properties was generated using randomly distributed scaling factors (mean = 1, and s.d. = 0.1). All parameters for reproducing this study are given in appendix A. In Fig. 2, homogeneous soil hydraulic properties are represented by the dark dashed and solid lines for the silty clay and the sandy loam soil types, respectively. The

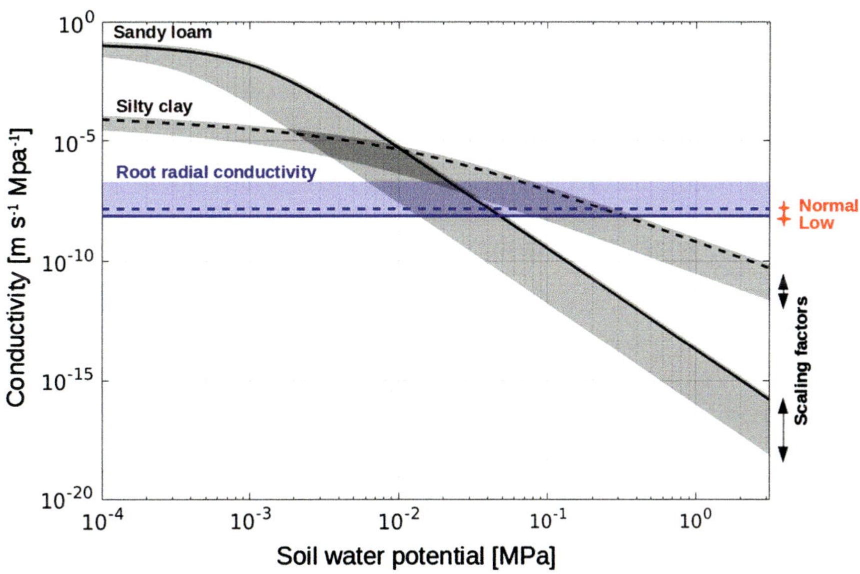

Fig. 2 Soil and root conductivity used in the scenario analysis. In red the crosses indicate the total root system conductivities (total root system conductance divided by the total root surface area). Shaded zones represent variability due to scaling factors for soils and due to heterogeneous distribution of root properties for root systems

Table 1 Summary of the parameters varied in the scenario analysis

Parameters	Scenario 1	Scenario 2	Illustration
Soil type	Sandy loam	Silty clay	Fig. 2
Scaling factors	None	Random	Fig. 2
Root system conductance	Low	High	Fig. 2
Evaporative demand	Low	High	Fig. 3
Plant density	Low	High	Fig. 3
Root conductivity	Heterogeneous	Homogeneous	Fig. 3

variability in soil properties induced by scaling factors is represented by the gray envelopes around the mean values. As the y-axis is in log scale, the symmetry of the soil hydraulic conductivities results in asymmetric gray envelopes. Heterogeneity in root properties was imposed by making them function of root age just as in Doussan et al. (2006). The homogeneous root conductivity distributions are represented in Fig. 2 as well by the blue dashed and solid lines for the normal and low root system conductances, respectively. The blue envelope encompasses the range of hydraulic conductivities in both low and normal conductance scenarios in case of heterogeneous root hydraulic distribution. The red crosses indicate the total root system conductivities (total root system conductance divided by the total root surface area) for the two root system conductance scenarios.

Figure 3 represents the two levels of plant density in the two respective subplots: high (a) and low (b). The two top-boundary conditions are indicated as well at the soil surface. Finally, the homogeneous and heterogeneous root hydraulic properties are mapped in subplots a and b, respectively, for the normal root system conductance.

Note that no feedback between soil water content and root growth was modeled in these simulations, i.e., we did not consider root growth at all. The results were analyzed by plotting the coefficient of variation (CV) of the soil water content vs the average value for all depths by depth increment $\Delta z = 3$ cm. The absolute value of CV is generally low as compared to field measurements (Vereecken et al. 2014) but should only be seen as an illustration of the potential of such models to investigate the shape of such relations, rather than an exact representative value.

Model Application

Figure 4 illustrates the CV and mean water content obtained for scenarios with high evaporative demand, low plant conductance (K_{rs}), homogeneous distribution of plant root properties, and heterogeneous soil property distribution (scaling factors) for two soil types: sandy loam (top) and silty clay (down). A typical bell curve appears for the silty clay soil, while only the increasing trend is observed for sandy loam. This is because sandy loam has a much higher hydraulic conductivity than silty clay in the explored range of soil moisture for these scenarios (Fig. 2). This generates much higher soil hydraulic redistribution for sandy loam,

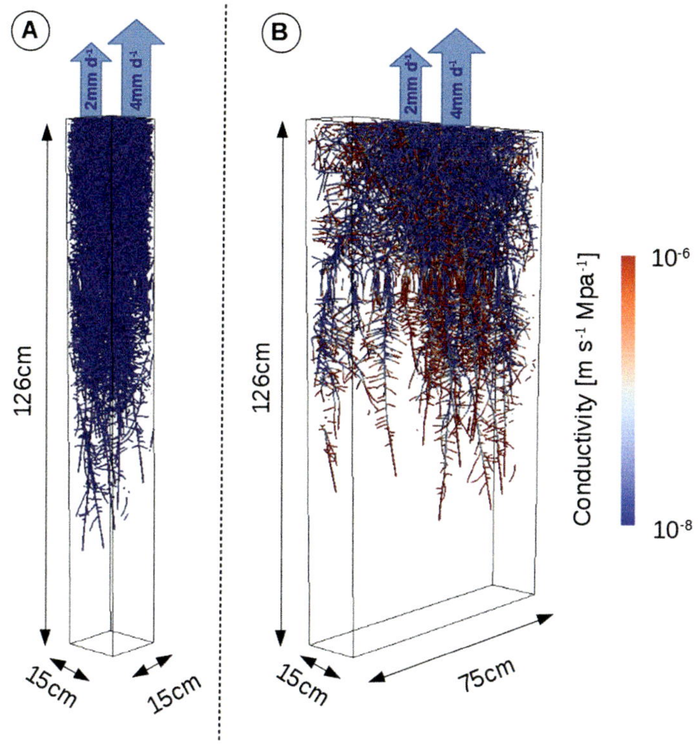

Fig. 3 Plant root system hydraulic architectures in the high (A) and low (B) plant density scenarios. The root hydraulic conductivity distribution is either homogeneous (A) or heterogeneous (B) for the normal root system conductance. The two levels of evaporative demand are indicated by the blue arrows

which decreases soil moisture variability. The compensation process, which is expected to decrease the variability at medium soil moisture, seems to affect the relation only for silty clay, as hydraulic gradients in sandy loam are lower. Probably that for sandy loam, compensation would occur at lower soil moisture values, but this was not explored with these simulations. An interesting feature that appears is that the bell shape of the CV-mean relationship is independent of the depth, and a unique curve seems to arise from the interactions between soil and roots.

Figure 5 gives an overview of soil moisture relations in the silty clay soil type for the two root conductances and the two soil-type heterogeneities. An interesting feature is that the same bell shape is observed for all cases, even with homogeneous soil properties, indicating the key role of soil-plant interactions. Obviously, soil heterogeneity properties increase the CV of soil moisture but to a lower extent than to the total variability induced by the root water uptake. In addition, root and soil heterogeneity do not affect the shape of the relations. A similar observation was made for sandy loam (not shown). When root system conductance is higher (high K_{rs}

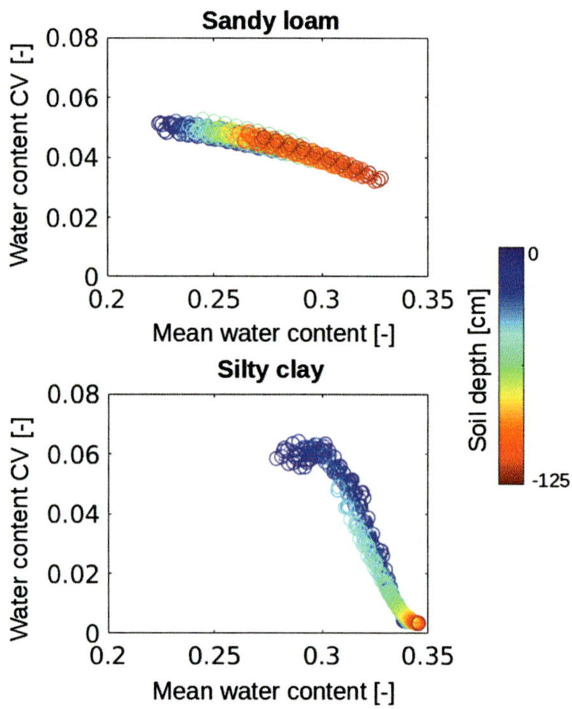

Fig. 4 Coefficient of variation (CV) vs average water content for the two considered soils. Simulations with random scaling factors, high evaporative demand, low root conductance, homogeneous uptake, and low plant density. Soil depth is represented with a color scale (3-cm depth increments, from blue to red, from top to bottom)

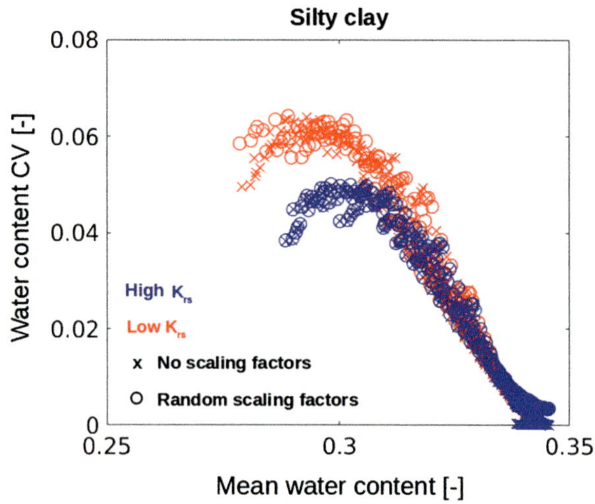

Fig. 5 Coefficient of variation (CV) vs average water content for the silty clay soil under high transpiration demand (4 mmd^{-1}). Red is for low root system conductance, blue for high root system conductance, the open circle for random scaling factors, and the crosses for the absence of scaling factors

scenario), compensation process is enhanced and tends to decrease the soil moisture variability at higher soil moisture level.

In Fig. 6, two evaporative demands were compared. It is observed that a higher transpiration rate generates higher soil water content variability, which can only be

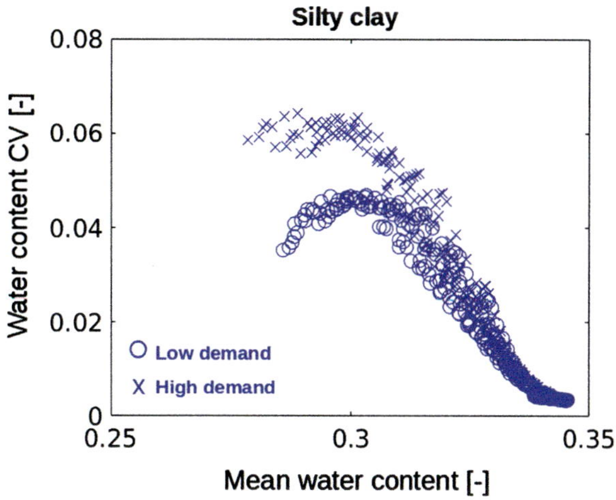

Fig. 6 Impact of evaporative demand on soil variability. Scenario with random scaling factors, high evaporative demand, low K_{rs}, the silty clay soil, homogeneous uptake, low plant density, and depth by depth (3 cm step, no color)

balanced by root compensation mechanism at lower soil moisture. Indeed, higher transpiration rates tend to follow the root distribution profile, while lower transpiration rates increase the impact of compensation and smooth out soil moisture variability.

Conclusions and Outlook

Vegetation is one of key driving factors for generating soil moisture patterns. In general, transpiration and root water uptake tend to increase soil moisture variability. Patterns are generated in soils with low hydraulic conductivity, where soil hydraulic redistribution cannot take place and cannot reduce vegetation-induced soil moisture heterogeneity. However, several plant root mechanisms exist, which may decrease soil moisture variability under drying conditions: growth, compensation, and competition. On the other hand, the change in rhizosphere properties by roots induces more heterogeneous infiltration patterns and preferential flow. Despite the fact that the relation between vegetation and infiltration has been demonstrated in literature, the specific impact of roots on preferential infiltration still needs to be proven. Identifying and quantifying how roots alter directly or indirectly rhizosphere hydraulic conductivity remain challenging open questions in plant science.

A process-based model for root water uptake was used to investigate and quantify the impact of soil and root properties on the soil moisture dynamics. The shape of the relationship between the coefficient of variation of the water content and its mean was affected by root and soil mean hydraulic properties and their variability. It was shown that a unique relation between averaged and standard error of soil moisture could be identified for a given set of root and soil properties. However, numerous additional processes like canopy interception, root growth, or change of rhizosphere were not accounted for in this example. The quantification of the impact of these intricate plant-related processes on soil moisture patterns could still be improved.

Future research should focus on developing process-based plant-soil models able to quantify water fluxes within and between soil/plant systems. How stable is the shape of soil moisture relation with additional processes and boundary conditions should be investigated.

In general, the functional understanding and modeling of soil-vegetation feedbacks within the water and carbon cycles remains a challenge. Combinations of tracers and noninvasive and modeling techniques should allow us to better quantify water and carbon fluxes in the soil pant system and to identify processes generating or destroying soil moisture patterns.

Appendix A

Table of model parameters used for the simulations

Soil type	θ_r	θ_s	α (cm^{-1})	n	K_s (cmd^{-1})	λ
Sandy loam	0.065	0.41	0.075	1.89	106	0.5
Silty clay	0.07	0.36	0.005	1.09	0.48	0.5

Scaling factor	Range of a	Range of b
None	1	1
Random	0.5–1.5	0.5–1.5

Root system conductance	Krs [cm^3.hPa^{-1}.d^{-1}]
Low	0.08
High	0.16

Evaporative demand	Tact [mm.d^{-1}]
Low	2
High	4

Plant density	Nplants [m^{-2}]
Low	9
High	45

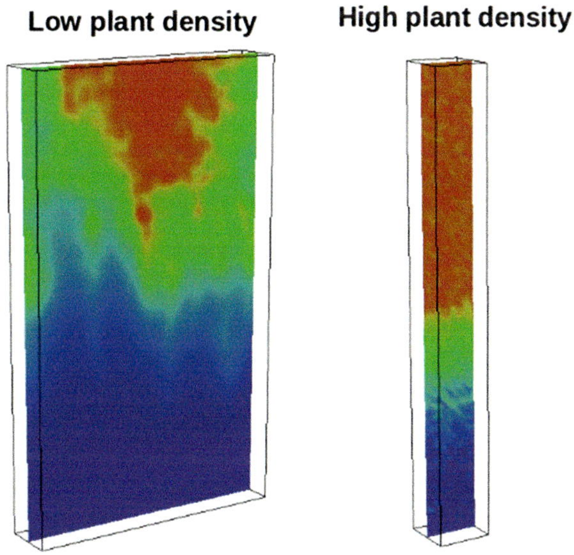

Fig. 7 Impact of sowing density on the water content distribution at the very end of the simulation. The represented scenario is for the high evaporative demand, the silty clay soil, the heterogeneous root hydraulic properties, and the random scaling factors

Root hydraulic conductivity	Low conductance	High conductance
	Range of k_r (10^{-8} m.s^{-1}. MPa^{-1})	Range of k_r (10^{-8} m.s^{-1}. MPa^{-1})
Heterogeneous	1–10	2–20
Homogeneous	0.75	1.5

Appendix B

See Fig. 7.

References

M.A. Ahmed, E. Kroener, P. Benard, M. Zarebanadkouki, A. Kaestner, A. Carminati, Plant Soil **407**, 161 (2016)

J.E. Aravena, M. Berli, T.A. Ghezzehei, S.W. Tyler, Environ. Sci. Technol. **45**, 425 (2011)

Y. Bao, P. Aggarwal, N.E. Robbins, C.J. Sturrock, M.C. Thompson, H.Q. Tan, C. Tham, L. Duan, P.L. Rodriguez, T. Vernoux, S.J. Mooney, M.J. Bennett, J.R. Dinneny, Proc. Natl. Acad. Sci. **111**, 9319 (2014)

R.D. Bardgett, L. Mommer, F.T.D. Vries, Trends Ecol. Evol. **29**, 692 (2014)
A. Bargués Tobella, H. Reese, A. Almaw, J. Bayala, A. Malmer, H. Laudon, U. Ilstedt, Water Resour. Res. **50**, 3342 (2014)
G. Baroni, B. Ortuani, A. Facchi, C. Gandolfi, J. Hydrol. **489**, 148 (2013)
L. Beff, T. Günther, B. Vandoorne, V. Couvreur, M. Javaux, Hydrol. Earth Syst. Sci. Discuss. **9**, 8535 (2012)
A. G. Bengough, Vadose Zone J. **11**(2) (2012). https://doi.org/10.2136/vzj2011.0111
E.W. Bhark, E.E. Small, Ecosystems **6**, 185 (2003)
S.S.O. Burgess, Plant Soil **341**, 25 (2011)
C.F. Caldeira, L. Jeanguenin, F. Chaumont, F. Tardieu, Nat. Commun. **5**, 5365 (2014)
A. Carminati, D. Vetterlein, U. Weller, H.-J. Vogel, S.E. Oswald, Vadose Zone J. **8**, 805 (2009)
E. Cho, M. Choi, J. Hydrol. **516**, 317 (2014)
B. E. Clothier, in *Flow and Transport in the Natural Environment: Advances and Applications*, ed. by W.L. Steffen, O.T. Denmead (Springer, New York, 1988)
V. Couvreur, J. Vanderborght, X. Draye, M. Javaux, Water Resour. Res. (2014a). https://doi.org/10.1002/2014WR015608
V. Couvreur, J. Vanderborght, L. Beff, M. Javaux, Hydrol. Earth Syst. Sci. **18**, 1723 (2014b)
J.M. Craine, R. Dybzinski, Funct. Ecol. **27**, 833 (2013)
S. de Dorlodot, B. Forster, L. Pagès, A. Price, R. Tuberosa, X. Draye, Trends Plant Sci. **12**, 474 (2007)
C. Doussan, A. Pierret, E. Garrigues, L. Pages, Plant Soil **283**, 99 (2006)
D. Dunkerley, Hydrol. Process. **16**, 119 (2002)
H.H. Gerke, H. Rieckh, M. Sommer, Soil Tillage Res. **156**, 230 (2016)
D.R. Gordon, J.M. Menke, K.J. Rice, Oecologia **79**, 533 (1989)
A.J. Guswa, Vadose Zone J. **11** (2012). https://doi.org/10.2136/vzj2011.0159
F. Hupet, M. Vanclooster, J. Hydrol. **261**, 86 (2002)
F. Hupet, M. Vanclooster, J. Hydrol. **303**, 247 (2005)
M. Javaux, T. Schroeder, J. Vanderborght, H. Vereecken, Vadose Zone J. **7**, 1079 (2008)
M. Javaux, V. Couvreur, J. Vanderborght, H. Vereecken, Vadose Zone J. **12** (2013). https://doi.org/10.2136/vzj2013.02.0042
S.D. Jawson, J.D. Niemann, Adv. Water Resour. **30**, 366 (2007)
H. Jenny, *Factors of Soil Formation. A System of Quantitative Pedology* (McGraw-Hill, New York, 1941)
M.S. Johnson, J. Lehmann, Ecoscience **13**, 324 (2006)
G. Katul, P. Todd, D. Pataki, R. Oren, Water Resour. **33**, 611 (1997)
W. Korres, T.G. Reichenau, K. Schneider, J. Hydrol. **498**, 89 (2013)
A. Kuhlmann, I. Neuweiler, S.E.A.T.M. van der Zee, R. Helmig, Water Resour. Res. **48**, W02534 (2012)
G. Lobet, V. Couvreur, F. Meunier, M. Javaux, X. Draye, Plant Physiol. **164**, 1619 (2014)
M. Markoski, T. Mitkova, V. Tanaskovik, V. Spalevic, Z. Zgorelec, Agric. For. **62**, 175 (2016)
G. Martinez Garcia, Y.A. Pachepsky, H. Vereecken, J. Hydrol. **516**, 154 (2014)
A.B. McBratney, M.L. Mendonça Santos, B. Minasny, On digital soil mapping. Geoderma **117**, 3 (2003)
M. Meinders, N.V. Breemen, *Ecosystem Function in Heterogeneous Landscapes* (Springer, New York, 2005), pp. 207–227
F. Meunier, Y. Rothfuss, T. Bariac, P. Biron, P. Richard, J.-L. Durand, V. Couvreur, J. Vanderborght, M. Javaux, Vadose Zone J. (2017). https://doi.org/10.2136/vzj2016.12.0134
T. Moriwaki, Y. Miyazawa, A. Kobayashi, H. Takahashi, Am. J. Bot. **100**, 25 (2013)
W. Niether, U. Schneidewind, L. Armengot, N. Adamtey, M. Schneider, G. Gerold, Catena **158**, 340 (2017)
U. Özkan, F. Gökbulak, Catena **149**, 158 (2017)
I. Prieto, C. Armas, F.I. Pugnaire, New Phytol. **193**, 830 (2012)
J. Puértolas, M.R. Conesa, C. Ballester, I.C. Dodd, J. Exp. Bot. **66**, 2325 (2015)

S. Ruiz, D. Or, S.J. Schymanski, PLoS One **10**, e0128914 (2015)
B. Schröder, Hydrol. Earth Syst. Sci. **10**, 967 (2006)
N. Schwartz, A. Carminati, M. Javaux, Water Resour. Res. **52**, 264 (2016)
J.S. Singh, D.G. Milchunas, W.K. Lauenroth, Plant Ecol. **134**, 77 (1998)
C. Sur, Y. Jung, M. Choi, Geoderma **207–208**, 234 (2013)
A.J. Teuling, P.A. Troch, Geophys. Res. Lett. **32**, 1 (2005)
S.E. Thompson, C.J. Harman, P. Heine, G.G. Katul, J. Geophys. Res. Biogeosci. **115**, n/a (2010)
H. Vereecken, J.A. Huisman, Y. Pachepsky, C. Montzka, J. van der Kruk, H. Bogena, L. Weihermüller, M. Herbst, G. Martinez, J. Vanderborght, J. Hydrol. **516**, 76 (2014)
H. Vereecken, Y. Pachepsky, C. Simmer, J. Rihani, A. Kunoth, W. Korres, A. Graf, H.J.-H. Franssen, I. Thiele-Eich, Y. Shao, J. Hydrol. **542**, 63 (2016)
D. Vetterlein, C. Doussan, Plant Soil **407**, 145 (2016)
A. Weigelt, M. Röttgermann, T. Steinlein, W. Beyschlag, Folia Geobot. **35**, 169 (2000)
A. W. Western, R. B. Grayson, G. Blöschl, D. J. Wilson, in *Scaling Methods in Soil Physics*, ed. by Y. Pachepsky, D.E. Radcliffe, H.M. Selim (CRC Press, Boca Raton, 2003), pp. 119–140
W.R. Whalley, A.G. Bengough, *Plant Roots*: the hidden half, ed. by A. Eshel, T. Beeckman (CRC Press, New York, 2013), pp. 37-1–37-15
I. White, in *Flow and Transport in the Natural Environment: Advances and Applications*, ed. by W.L. Steffen, O.T. Denmead (Springer, New York, 1988), pp. 59–85
L.G.G. Yapa, D.D. Fritton, S.T. Willatt, Plant Soil **109**, 9 (1988)
C. Yoo, J.B. Valdés, G.R. North, Adv. Water Resour. **21**, 375 (1998)
M. Zarebanadkouki, E. Kroener, A. Kaestner, A. Carminati, Plant Physiol. **166**, 487 (2014)
T. Zeppenfeld, N. Balkenhol, K. Kóvacs, A. Carminati, PLoS One **12**, e0182188 (2017)
Y. Zhang, L. Deng, W. Yan, Z. Shangguan, Catena **137**, 52 (2016)
Q. Zhu, H. Lin, Geoderma **163**, 45 (2011)

Upscaling Issues in Ecohydrological Observations

Harry Vereecken, Yakov Pachepsky, Heye Bogena, and Carsten Montzka

Contents

Abstract

Ecohydrological processes are strongly controlled by complex interactions between the subsurface or vadose zone, the vegetation, and the atmosphere. Upscaling of ecohydrological processes requires an understanding of the fundamental processes and states controlling water-related fluxes in vegetation and soils as well as the characterization of the inherent spatial variability occurring in these systems from the local to the catchment scale and beyond. In this chapter we address upscaling of soil water processes and hydraulic properties in the vadose

H. Vereecken (✉) · H. Bogena · C. Montzka
Agrosphere Institute (IBG-3), Forschungszentrum Jülich GmbH, Jülich, Germany
e-mail: h.vereecken@fz-juelich.de; h.bogena@fz-juelich.de; c.montzka@fz-juelich.de

Y. Pachepsky
Environmental Microbial and Food Safety Laboratory, USDA ARS Beltsville Agricultural Research Center, Beltsville, MD, USA
e-mail: Yakov.Pachepsky@ARS.USDA.GOV

X. Li, H. Vereecken (eds.), *Observation and Measurement of Ecohydrological Processes*, Ecohydrology, https://doi.org/10.1007/978-3-662-48297-1_14

zone, the upscaling of soil water-plant processes, the use of data assimilation techniques to estimate ecohydrologically relevant parameters, and the use of novel sensing techniques and observational platforms. The integration of novel upscaling approaches and novel sensing techniques will provide a unique opportunity to improve our understanding of ecohydrological processes.

Keywords
Evapotranspiration · Ecohydrology · Scaling approaches · Pedotransfer functions

Introduction

Recently, several papers have been published that review the state of the art in characterizing and quantifying ecohydrological processes. Asbjornsen et al. (2011) identified several key challenges in studying plant-water relationships in the context of ecohydrological processes including the need to better account for the spatial heterogeneity and variability in the controls on water fluxes across sales, a better understanding of the interaction and feedbacks between soil moisture dynamics and vegetation, and the complex interaction between vegetation dynamics and soil hydraulic properties, hydrological fluxes, and catchment properties. Porporato et al. (2002) identified the importance of quantifying soil moisture dynamics to understand a set of ecological processes such as occurrence, frequency, and duration of water stress in water-limited ecosystems and its impact of vegetation dynamics. Guswa et al. (2002) analyzed the role of soil water dynamics in relation to evapotranspiration (ET) of an African savanna. (Eamus et al. 2015) analyzed recent insights obtained from satellite- and field-based studies with respect to groundwater-dependent ecosystems. They identified three groups of methods that can be used to identify these ecosystems such as remote sensing techniques, groundwater level measurements, and stable isotope techniques. Svoray et al. (2015) pointed out that there is a lack of limited laboratory and field experimental data about ecohydrological processes and their impact on biota in drylands.

Upscaling Ecohydrological Processes

Parameters are typically defined in the framework of a continuum description of physical, chemical, and biological processes and are valid for a representative elementary volume. Application of these parameters at scales beyond the local or elementary volume is pruned with errors as the larger-scale systems such as fields, catchments, and river basins are heterogeneous by nature, and this heterogeneity is not captured in these parameters. In this section we describe the various approaches that can be used to parametrize water flow in the vadose zone. A discussion about approaches to upscale water-plant processes from leaf to single canopy, canopy to field or stand, and to catchment scale and beyond has been provided in chapter ▶ "Upscaling Issues in Ecohydrological Observations" (Asbjornsen et al. 2011).

There is an abundant literature available on how to upscale processes from the local to the field scale. In the field of vadose zone hydrology, upscaling concepts based on first principles have mainly focused on bare field systems (Vereecken et al. 2008). Vereecken et al. (2007) predicted the spatial variability of soil moisture variability versus mean soil water content for different soils using a closed-form equation derived from a stochastic analysis of Richards equation using first-order perturbation theory derived by Zhang et al. (1998). They showed that knowledge of the statistical properties of soil hydraulic parameters describing the Brooks-Corey model provided a good estimation of the $\sigma(<\theta>)$ function where σ is the standard deviation of the mean soil moisture content and $<\theta>$ the mean soil moisture content (-). Qu et al. (2015a) derived the closed-form equation for $\sigma(<\theta>)$ using first-order perturbation theory of water flow in a one-dimensional heterogeneous flow domain and the van Genuchten parameter. They used pedotransfer functions to estimate the $\sigma(<\theta>)$ for different field sites located in Germany, China, and Australia. This indicates that for field to small catchment scales, the spatial variability of soil hydraulic properties is key in determining the spatial variability of soil moisture.

Various methods have been used to derive effective flow equations and parameters describing water flow in the vadose zone which were reviewed by Vereecken et al. (2007). These methods include first-order perturbation theory, homogenization, scaling theory based on Miller-Miller scaling, the aggregation approach, inverse estimation, and data fusion or data assimilation in porous media among others. Here we will give a short description of the most frequently used methods. The data assimilation or data fusion approach will be presented in section "Estimating Key Parameters in Describing Ecohydrological Processes." For a comprehensive review, we refer the reader to Vereecken et al. (2007). In the first-order perturbation theory, the soil hydraulic properties are described by a mean value and a fluctuation around the mean value. In order to characterize the spatial variability of the fluctuations, parameters are assumed to be realization of a second-order stationary process allowing for a simpler mathematical treatment. Effective parameters of soil hydraulic parameters can then be formulated as closed-form expressions depending on the small-scale variability of these parameters.

The Miller-Miller scaling (Miller and Miller 1956) is based on the similarity assumption of the pore space geometry. Based on physical analysis of capillary flow phenomena, they derived scaling relationships for matrix potential or pressure head and hydraulic conductivity, K, among others, using the microscopic laws for capillary pressure forces and viscous flow in porous media. The scale factors in general are defined as conversion factors that relate the characteristics of one system to the corresponding characteristics of another system (Tillotson and Nielsen 1984). In terms of water flow at the field scale, Miller-Miller scaling can be used to derive a mean moisture retention, hydraulic conductivity function, and a set of scaling factors that allows to retrieve the local-scale properties. A global data set of upscaled soil hydraulic parameters for global simulations based on Miller-Miller scaling has recently been published by (Montzka et al. 2017). Two approaches in deriving scaling factors of soil water processes are typically distinguished (Vereecken et al. 2007): (i) the dimensional analysis, called similitude analysis, which is based on the physical characteristics of a soil, and (ii) the functional normalization approach, which is based on regression analysis.

The concept of aggregation refers to estimating aggregated soil properties such as hydraulic parameters that describe relevant soil water processes and soil moisture states of soil map units or numerical grids in land surface atmosphere or climate models that typically consist of a mixture of soil types (Vereecken et al. 2007). The aggregation approaches require the definition of the fine-scale heterogeneity elements and finding weights to create a weighted sum of their functioning that will represent the next level of hierarchy.

Contrary to the above methods, inverse modeling uses information contained in state parameters to derive effective parameter values. Vereecken et al. (2007) distinguished three different types of inverse modeling approaches to estimate effective parameters: (i) using methods that observe the system at a larger scale (e.g., remote sensing from satellite platforms), (ii) using smaller-scale data as calibration targets for a system described (and defined) at a larger scale (e.g., water level measurements, average soil water tension, or soil water contents), or (iii) using spatial statistics and structures of local-scale state variables to infer the spatial structure of smaller-scale process parameters, which can subsequently be used in forward upscaling.

The main advantage of this approach is that effective parameters can be estimated for complex natural systems such as soil-plant-atmosphere systems which cannot be treated mathematically to derive closed-form expressions for these effective parameters. The main disadvantage is that these parameters are not universally valid as they strongly depend on the system properties and the formulation of the boundary conditions. Inverse parameter estimation has been used, e.g., by Vrugt et al. (2001) to estimate root water uptake parameters.

Estimating Key Parameters in Describing Ecohydrological Processes

In this section we will address methods that allow estimating key parameters that control ecohydrological processes. The first part deals with the estimation of soil hydraulic properties using pedotransfer functions. The second part addresses data assimilation methods that can be used to estimate vegetation parameters that are time dependent.

Pedotransfer Functions

Ecohydrology, as any environmental science, encounters the need of estimating parameters at coarse scales where direct measurements are not possible. This appears to be true for all compartments of ecohydrological systems. Soil is one of such compartments. Pedotransfer functions were defined as relationships to derive soil properties that we need from properties that we have (Wosten et al. 2001). Inputs for pedotransfer functions are in most cases data available from soil survey such as contents of soil textural fractions and organic matter content. The majority of pedotransfer functions provide such conversion at the soil horizon level of soil organization. Soil horizon is the smallest soil organization level. Pedotransfer functions are needed for coarser scales corresponding to field or stand, catena, watershed, or region. Development of such functions is the challenging field of current active research.

The following problems are currently being addressed:

1. How to develop a PTF if soil cover within the spatial computational cell is composite, i.e., encompasses different soil textures and/or consists of soils from different taxonomic units?
2. How to account for spatial variation of environmental covariates, such as land use or topography, that are used as pedotransfer inputs at coarse scales?
3. How to make sure that the PTF output is the correct parameter for a given scale and given model?
4. How to evaluate the performance of pedotransfer functions?

Each of these problems is model- and scale-specific. Different ecohydrological models may require different approaches and lead to different solutions.

Two methods of the PTF application for composite soil covers have been proposed to date. One assumes that the PTF input soil properties have to be aggregated first over all components within the composite cover and PTF output soil parameters have to be computed with these aggregated inputs. Another approach consists in estimating PTF output soil parameters for each of soil cover components and then aggregating the PTF application results. Several aggregation procedures were considered. Miller and White (1998) developed a conterminous US multilayer soil characteristics data set for regional climate and hydrology modeling. The continuous variables were aggregated over all components of a given map unit of the composite using the percentages of the component area as weights. For categorical variables, the relative amounts of each category in each component were multiplied by percentage of the component and summed; the category having the largest total was chosen as the category for the map unit, even if it represented less than half the area of the map unit. For parameters such as available water content that were represented by maximum and minimum values, average values were computed and then aggregated by their areal weight. Romanowicz et al. (2005) worked on hydrological modeling for a catchment in Belgium and demonstrated that the aggregation results depend on the threshold value set for the smallest spatial unit (subbasin) used in the model as a single entity. Shangguan et al. (2014) hypothesized that, because the relationship between the soil basic properties and the derived soil hydraulic parameters is nonlinear, the better way is to first calculate the soil hydraulic parameters for each component and then aggregate them.

Other authors proposed to take advantage of the availability of ranges of basic soil properties for individual soil cover components. Haskett et al. (1995) considered crop simulations at the county level and proposed creating frequency distributions based on the recorded properties of the individual soils within soil association and then approximating this distribution by the use of a beta function.

It needs to be noted that aggregation will filter out the differences of soil properties between soil components within a given mapping unit no matter what is done first – aggregation of basic properties or parameter estimation for each component of the composed soil cover. It was suggested that a palliative solution can be doing simulations for each soil cover component separately and then aggregating simulation results to the composite soil cover spatial unit, i.e., simulation cell of remote sensing footprint (Dai et al. 2003). This approach is realistic only if predominantly vertical flow and transport are of interest. Another important feature of aggregation is that it is applicable only to mass-conserving soil parameters, i.e., parameters that are not dependent on scale. Parameters that control flow and transport, such as hydraulic conductivity or dispersivity, show scale dependence in a range of scales at the horizon scale and therefore cannot be aggregated simultaneous. These parameters seem to become scale-independent after the scale crosses a threshold value, but it is currently not known whether there is another range of scale – dependence for even larger scales.

The effect of the aggregation methods can be observed best when soil maps of different scales are used. Livneh et al. (2015) compared results obtained with soil textural information obtained from the digital general soil map of the US (STATSGO2) soil database (1: 250,000) and from FAO-based harmonized world soil database (1: 5,000,000) over the Mississippi river basin. Their grid-based distributed hydrological model simulated canopy interception, snow accumulation and melt, root zone soil moisture and ET, infiltration, surface and subsurface runoff, percolation, and baseflow and flood routing. Model simulations using STATSGO2 consistently led to greater separation between fast and slow runoff responses (by approximately 24–39%) and, generally, more extreme responses to historic flood and drought events. Not unexpectedly, the choice of soil database altered the partitioning of precipitation between ET and runoff and affected the correlation structure between forced and modeled fluxes by up to 0.2 points.

Using environmental covariates can be advantageous in estimating soil parameters on coarse scales in at least one of two ways. First, the applicability of pedotransfer functions for soil horizon scale may be extended if attributes of land use and/or topography are included in the pedotransfer relationships. Pachepsky and Rawls (2004) proposed the topotextural grouping that provided reasonable accurate estimates of water content at 33 kPa (Figs. 1 and 2). Complementing field-determined textural classes with field-recorded topographic descriptors brought the accuracy of water retention estimates to the level characteristically used for determining textural class from laboratory textural analysis. Fuka et al. (2016) demonstrated that using topographic attributes can improve soil information at the watershed scale. They incorporated the topographically adjusted soils into an initialization of the Soil and Water Assessment Tool model for 15 research watersheds to investigate how

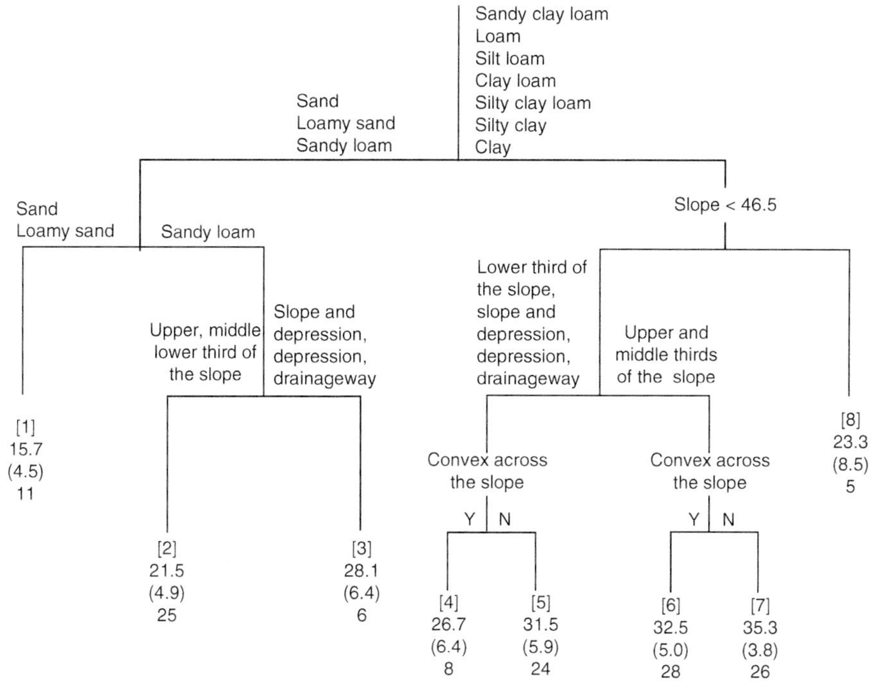

Fig. 1 Regression tree to estimate water retention at −33 kPa from field textural class, genetic horizon, and topographic descriptors; "y" and "n" mean "yes" and "no" answers to the parameter definition in the box above; terminal nodes contain the node number in brackets, the average water retention at −33 kPa for the group, the standard deviation of water retention within groups in parentheses, and the number of samples in the group

changes in soil properties influence modeled hydrological responses at the field scale. The results showed that the topographically adjusted soils produced better runoff predictions in 50% of the fields.

The important aspect of using environmental covariates in pedotransfer consists in their map resolution dependence. The scale dependence exists also when the environmental covariates are used to interpolate results of PTF application. Poggio et al. (2010) showed that the spatial interpolation of the PTF-derived available soil water capacity appears to be scale dependent in that models fitted have different significant covariates and different estimated values according to the region considered. Their results suggested that the estimates from the model fitted at the appropriate extent are the most accurate.

Meaning of the terms used for soil hydraulic properties appears to be model-specific. The coarse scale modeling employs various modes of soil flow in soils. Examples are Richards equation as used, e.g., in JULES (Best et al. 2011), and soil moisture accounting models such as FAO56 (Allen et al. 1998) and MWBUS (Pachepsky et al. 2006). The parameter of field capacity appears to be present in different soil models, and its meaning actually depends on the way how a specific

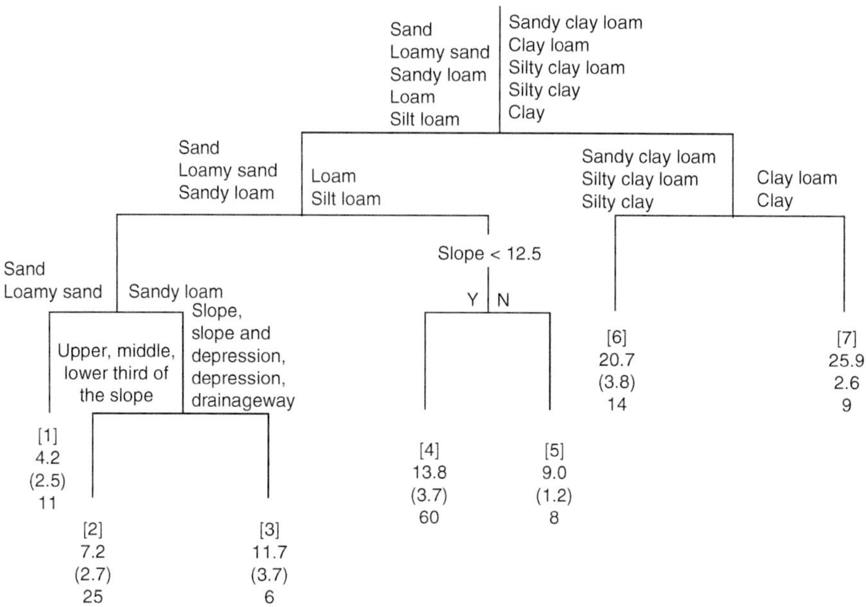

Fig. 2 Regression tree to estimate water retention at −1500 kPa from field textural class, genetic horizon, and topographic descriptors; "Y" and "N" mean "yes" and "no" answers to the parameter definition in the box above; terminal nodes contain the node number in brackets, the average water retention at −1500 kPa for the group, the standard deviation of water retention within groups in parentheses, and the number of samples in the group

model simulates infiltration. For example, a pure bucket model will have the value of the "field capacity" different from the model that has a kinetic equation for the infiltration losses.

Using pedotransfer to estimate parameters in these soil models is far from trivial. For example, the FAO56 method involves assuming that storage capacity available for plant uptake can be found from the equation $(\theta_{FC} - \theta_{WP})L_r$, where L_r is the rooting depth and θ_{FC} and θ_{WP} are the water contents of the soil at field capacity and wilting point, respectively (Allen et al. 1998). It is often assumed that θ_{FC} can be found from water retention curves obtained at a particular soil water potential at the horizon scale. Under such assumption, PTF developed at the soil horizon scale can be used to find the water contents at specific soil water potential and therefore can be used at coarser scales. However, the assumption about the existence of the specific soil water potential corresponding to the field capacity at the pedon and coarser scales appears to be inaccurate. Nemes et al. (2011) demonstrated that an additional pedotransfer correction is needed to convert water contents at specific potentials at the horizon scale to field capacity at the pedon scale. Recently it was proposed to determine parameters of the SMAP from solutions of the Richards equation found for various textures and then to relate the parameters of the simplified model to soil texture (Mathias et al. 2015). The Richards equation included the ET term. Caution needs to be exercised in applying this approach, though, since under this approach Richards equation gets its parameters from the horizon-scale pedotransfer functions that can be inaccurate outside of the area of their development.

Evaluation of pedotransfer functions on coarse scales appears to be exclusively functional, i.e., dependent of the performance of a model in which soil parameters are found from PTFs. For example, Gutmann and Small (2007) compared the performance of PTF and inverse modeling in Noah land surface model used to estimate the latent heat flux. They showed that inverse modeling reduces errors between measured and modeled latent heat flux from 88 to 28 W/m^2; with increasing vegetation cover, the sensitivity to and requirements of accuracy of importance of soil hydraulic properties decrease. Fuka et al. (2016) reported improvements in PTFs based on the comparison of computed and measured runoff. Model-independent PTF evaluation at coarse scales remains elusive.

Overall, scale effect on the soil model parameterization with PTFs came to the research forefront fairly recently. The value of PTFs in ecohydrological research, assessment, and predictions cannot be overestimated. Whenever PTFs are applied, their use should include analysis of the effects of scale on the PTF inputs, on model in which PTFs are used, and on spatiotemporal dynamics of environmental variables used for calibration and validation of those models.

The Remote Sensing Contribution to Ecohydrological Scaling

Remote sensing typically records reflectances or emissions from Earth's surface, which are often not directly the specific key parameter of interest. However, remote sensing measurements show direct relationships to the occurrence of spatial ecohydrological patterns. Empirical regression analyses as well as radiative transfer models are available to retrieve a large number of parameters from raw sensor data. Spaceborne sensors are able to extend point-scale measurements beyond the capabilities of proximal sensing methods with the drawback of investigation of the top soil layer only. In situ measurements are often necessary to estimate the regression statistics or to calibrate and validate retrieval models. Up- and downscaling have also been performed by utilizing different remote sensors at different spatiotemporal resolutions. Two steps to gain information about ecohydrological processes exist: to use remote sensing data as complementary source of information to derive spatial patterns from point-scale measurements and to use this area-wide data product to inform models. Key parameters for ecohydrological processes can be extended from point-scale in situ measurements to larger scales by remote sensing. This is important to identify spatial patterns. Statistical regression methods such as ordinary least square regression, regression with correlated errors (maximum likelihood), and ordinary kriging describe the required relationships (Castaldi et al. 2014).

One important input for PTFs is organic carbon content of the soil. Remote sensing is able to identify the organic carbon content of the top soil during unvegetated periods by hyperspectral measurements (Vaudour et al. 2016). Similarly, clay content has been successfully estimated by hyperspectral data (Middleton et al. 2013). Recently, Ouerghemmi et al. (2016) developed a semi-blind source separation method to separate vegetation spectra from soil spectra in order to retrieve the clay content over semi-vegetated surfaces. Wu et al. (2015) classified degraded land based on the silt content with the multispectral BJ-1 satellite. Similarly, Zhai et al. (2006) classified soil textural differences by soil spectra. A remaining problem of identifying PTF input data by hyperspectral remote sensing is the variable soil moisture content, because it reduces the strength of the relevant absorption features in the soil spectra. Castaldi et al. (2015) analyzed two methods to consider a priori known soil moisture during clay content estimation. Jana and Mohanty (2011) used a Bayesian neural network applied across spatial scales to approximate fine-scale soil hydraulic properties. In addition to the input of PTFs, hydraulic properties can also be directly estimated by remote sensing. For example, Babaeian et al. (2015) analyzed the ability of current satellite systems to retrieve soil hydraulic properties by a stepwise multiple linear regression method directly from spectral information. Levi et al. (2015) directly link remotely sensed soil properties as PTF inputs with the pre- and post-monsoon vegetation response. They identified that the antecedent precipitation was more important for explaining the relationships between modeled soil properties and vegetation growth than the amount of monsoon precipitation. With the current spaceborne hyperspectral sensor – Hyperion on the EO-1 satellite – retrievals often fail due to the poor signal-to-noise ratio (Castaldi et al. 2014). Future spaceborne sensors such as the Environmental Mapping and Analysis Program (EnMap), the Hyperspectral Imager Suite (HISUI), the PRecursore IperSpettrale della Missione Applicativa (PRISMA), the China Commercial Remote-sensing Satellite System (CCRSS), and the NASA Hyperspectral Infrared Imager (HyspIRI) may overcome this problem and improve the spatial characterization of point-scale measurements.

Applications of remotely sensed PTF inputs or PTF results are rare and the full potential not exploited so far.

For example, Jana et al. (2012) scaled in situ measurements of soil hydraulic properties to maps of 30 m and 1 km resolution by a neural net and implemented this in the HYDRUS-3D model. The results perform reasonably well for soil moisture as compared to standard high-resolution soil maps. Babaeian et al. (2016) used diffuse soil spectral information in the visible, near-infrared, and shortwave-infrared range from soil samples to estimate area-wide soil hydraulic parameters and predict soil moisture profiles using a 1D mechanistic water flow model. Three different types of parametric transfer functions (spectrotransfer functions, pedotransfer functions, and spectral pedotransfer functions) were aggregated from the point scale to 1 km^2 pixel size to provide coarse scale estimates of van Genuchten-Mualem (VGM) hydraulic parameters. Simulated subsurface soil water dynamics matched well with the ground-truth measurements. These findings indicate the feasibility of using spectral data to predict VGM hydraulic parameters and, ultimately, to predict soil water dynamics at the larger scales. Future hyperspectral satellite systems such as EnMap are able to deliver the required observation accuracies.

The most significant feedbacks between ecological processes and the hydrological cycle are featured by transpiration and plant water use. These are also related to canopy and atmosphere exchange of carbon dioxide (CO_2) through vegetation photosynthesis and autotrophic respiration (Zhang et al. 2016). Currently, in situ measurements of ET are difficult to scale up for regional applications, so remote sensing methods have been increasingly used to estimate ET (Marshall et al. 2016). In general, three groups of methods exist: (1) vegetation-based methods, (2) radiometric land surface temperature-based methods, and (3) triangle or scatterplot inversion methods. Vegetation-based methods typically need an index of vegetation biomass or leaf area index (LAI) as input to calculate ET. Choudhury et al. (1994) already identified the relationship between vegetation indices and ET. Nagler et al. (2013) used water balance or flux tower measurements to calibrate this relationship for a season-long estimation of ET from MODIS. Guerschman et al. (2009) additionally incorporated a vegetation moisture index to improve the results. Radiometric land surface temperature-based (LST) methods utilize remote thermal measurements of the land surface within a surface energy balance (SEB) expression. Land surface flux equations are combined with remotely sensed temperatures, e.g., within the one-source models SEBAL and METRIC or within the two-source models TSMN and ALEXI. The difference is that two-source models do not treat the Earth's surface as a whole, but allow for energy and vapor fluxes between the soil surface and the canopy. Song et al. (2016) improved the two-source model's performance over densely vegetated and advection conditions by advancing the surface temperature partitioning method in TSEB. ALEXI and its disaggregated version DisALEXI additionally couple the LST method with an atmospheric boundary layer model to improve the already encouraging results. Triangle methods that utilize the relationship between multispectral vegetation indices (VI) and LST differences have been reviewed by Petropoulos et al. (2009). The analysis of the scatterplot generated in the triangular (or trapezoidal) LST-VI feature space of a heterogeneous region identifies the limits of ET for different vegetation conditions. With this information gradual values of ET can be estimated including the separation into transpiration and evaporation (Zhang et al. 2016). A combination of LST-VI methods with land surface models provides further increase in performance, because instead of the linear, a nonlinear interpretation of the LST-VI space is possible (Biggs et al. 2013). Additionally, also water balance models have been used to estimate larger-scale ET (Wan et al. 2015). Where polar orbiting satellites are able to provide ET estimates for snapshots only, geostationary satellites provide diurnal cycles of ET. For example, Sun et al. (2012) calculated ET from Meteosat Second Generation SEVIRI data for Africa. The remaining problem is the lack of information during cloudy conditions. Solutions have been developed by using the sine function, the constant evaporative fraction, constant reference evaporative fraction method, or a combination of polar orbiting and geostationary satellites to estimate daily ET (Sun et al. 2012; Xu et al. 2015a).

In addition to the parameter introduction into ecohydrological models, a quantification of ecohydrological processes at different scales can also be done by state (and parameter) update in data assimilation systems; see the following section.

Data Assimilation Techniques

Ecohydrological models differ from one scale to another, and multiple models are used at the same scales. Parameters of those models are essential input data in ecohydrological projects. The common source of the model parameters is the model calibration that consists in fitting the models to monitoring data collected over long periods of time. Being the well-developed field in hydrological modeling, calibration methods tend to have inherent weaknesses in their treatment of uncertainty as being primarily attributable to the model parameters without explicit treatment of input, output, and model uncertainties (Vrugt et al. 2008).

Data assimilation (DA) consists of using observations to periodically correct both modeling results available at the observation times and future modeling results. This is achieved with data assimilation that strives for improvements in both model parameters and state variables of the hydrological model (Evensen 2009). The general idea of data assimilation is to estimate uncertainties in model parameters, state variables, and data and to use observations to update parameters and state variables based on how their uncertainties compare to data uncertainties (Reichle 2008). The less certain data are, the less will be changed in parameters and state variables and vice versa. Originally data assimilation presented hardly tractable problems since it required the model to estimate changes of uncertainties in parameters and state variables in time. The applicability of DA was radically expanded with the development of the ensemble Kalman filter (EnKF) method (Evensen 1994) that allowed deriving the time series of uncertainties from running an ensemble of models with equiprobable random inputs.

Comparisons of parameter estimation with data assimilation and traditional calibration favored the data assimilation for the Richards equation as a model of soil water dynamics. Li and Ren (Li and Ren 2011) estimated parameters of soil hydraulic functions by assimilating observations of soil water pressure dynamics using the EnKF in infiltration and drainage experiments. The results of the synthetic experiments on 12 soils with different textures indicated that EnKF estimates can quickly approach stable estimates. In contrast to the batch calibration process that used the least squares objective function, the EnKF reduced the risk of obtaining suboptimal estimates. In this work, the EnKF also performed well in the multiparameter estimation scenarios in a heterogeneous soil profile with in situ field observation.

DA appeared to be a good choice in parameterization of models containing lumped representations of both processes and spatial domain heterogeneities that are referred as unresolved structure or structural error in models. Erdal et al. (2014) applied data assimilation to the data of soil water dynamics on synthetic and real lysimeter with layered soil under a crop. They showed that DA can provide reasonable estimates of parameters of a homogeneous soil profile model with bias corrections in the case of its application to the layered soil profile.

The EnKF assumes the normal distributions of model and observation errors. That may introduce errors in data assimilation for highly nonlinear systems. This is especially true for soil water content modeling with Richards equation. Using the empirical cumulative error distribution functions rather than presuming normal distribution of errors appears to be more appropriate. This has been implemented using particle filtering method (Gordon et al. 1993). Another possibility of dealing with nonlinearities consists in applying EnKF to transformed state and parameter variables when the transformation reduces the nonlinearity.

At coarser scales, remote sensing data serve as an essential input for both mechanistic Richards equation-based models and simple semiempirical water flow models. Alvarez-Garreton et al. (2014) applied DA to parameterization of rainfall runoff model and showed that this application may in turn help to improve the satellite-based input data. Xu et al. (2015b) improved the turbulent heat flux estimates via a dual EnKF DA scheme by optimizing model parameters and state variables. This can be an additional advantage of using DA, given the uncertainty in accuracy of remote sensing moisture retrieval algorithms. Another advantage may stem from parameterization of the spatial structure present in EnKF spatial patterns. For more complex mechanistic models, application of DA allows one to quantify scale effects on properties of ecohydrological systems (Pachepsky and Hill 2016). Further potential for multi-scale DA in terrestrial systems has been reviewed by Montzka et al. (2012).

Hierarchically Scaled Observation Systems and Platforms

In terrestrial ecosystems, one faces the challenge of large spatial and temporal variability of soil moisture from the soil aggregate-to-plot-to-field and stand-to-landscape scales (Vereecken et al. 2014). For instance, factors controlling soil moisture include atmospheric forcing, topography, soil properties, and vegetation and human impacts (e.g., irrigation, groundwater withdrawal), which all interact in complex, nonlinear ways. For better addressing the key research questions in ecohydrology, e.g., understanding threshold behaviors, positive feedback mechanisms, and complex interactions between the different environmental compartments, long-term environmental observation at multiple scales is needed (Asbjornsen et al. 2011). Typically experimental designs tend to focus on a specific scale. In order to address the scaling issue properly, we need a hierarchical design to study ecohydrological processes. New developments of hierarchically scaled observation systems are now underway to be used in long-term monitoring networks such as CZO, LTER, TERENO, and HiWATER that enable the analysis of scale-dependent processes. In the following we present key developments in multi-scale ecohydrological measurement techniques.

Multivariate and Multi-scale Remote Sensing Techniques

Remote sensing allows for an area-wide analysis of ecohydrological processes, e.g., the functional connection between soil properties, soil moisture, and vegetation patterns. Multivariate remote sensing refers to the simultaneous observation of multiple state variables. In these studies, the measurement data are either of the same scale, or, more commonly, scale disparities are neglected. Multi-scale remote sensing operates at significantly different resolutions and the application of a scaling technique. A combination of multivariate and/or multi-scale remote observations within a data assimilation framework typically improves the knowledge of further ecohydrological states and fluxes (Montzka et al. 2012).

Lausch et al. (2013) used highly resolved spatial and spectral resolution data from airborne imaging spectrometer (AISA) in combination with ground-based geophysical measurements (electromagnetic induction, EMI, and gamma ray) to study the spectral response of vegetation as a function of soil properties and soil moisture patterns. Similarly, Rudolph et al. (2015) demonstrated the vegetation patterns (e.g., leaf area index, LAI) derived from satellite remote sensing which can be combined with EMI measurements to detect subsurface structures of soil textures at larger scales. Vegetation monitoring is typically performed in the VIS and NIR bands. Here, multispectral optical instruments, such as Landsat, RapidEye, and Sentinel-2, have been used extensively to characterize vegetation status. This includes not only the provision of spectral indices but also the derivation of higher-level data such as leaf area index (Qu et al. 2015b) and physiognomic land cover maps (Homer et al. 2015). Canopy water content and leaf water potential are highly dynamic quantities and are important variables for determining water deficit and drought effects. Recently, remote sensing methods for determining leaf equivalent water and canopy water content from different combinations of spectral bands have been introduced (Ustin et al. 2012).

Finer-scale environmental characteristics compared to spaceborne systems can be delivered by unmanned aerial vehicles (UAVs). In addition, UAVs are very attractive for campaign studies due to their high flexibility and availability at reduced costs. The large variety of different remote sensing platforms enables the provision of ecohydrological information from local to global scale and the development of scaling strategies.

Multi-scale Isotope Techniques to Unravel Evapotranspiration Processes

Stable isotope analysis (δ18O and δ2H) of water in soils and plants as well as of water fluxes (e.g., precipitation, ET, runoff) can be used to for the partitioning of hydrological water fluxes (Stockinger et al. 2015; Wen et al. 2016; Vereecken et al. 2016). The mass differences of the major and minor water isotopologues lead to thermodynamic and kinetic isotopic fractionation, causing measurable differences in the isotopic composition of the water pools of different terrestrial ecosystem compartments (e.g., soil, groundwater, surface water, plant, and atmosphere). These differences can be used for water balance closure and flux partitioning in the soil-plant-atmosphere continuum from the plot (Rothfuss et al. 2010) to the field and regional scales (Moreira et al. 1997). For instance, very recently Jasechko et al. (2016) analyzed seasonal cycles of δ18O in rain, snow, and streamflow from 254 catchments around the world and found that a substantial proportion of global streamflow less than 3 months old indicates fast water transport. In another study, Evaristo et al. (2015) used δ2H and δ18O data from 47 globally distributed sites to show that in most cases water supplying groundwater recharge and streamflow is different from the water that supplies plant transpiration, suggesting evidence of ecohydrological separation across different ecosystems. Yet, there is still an ongoing debate on how global terrestrial water fluxes are partitioned (Coenders-Gerrits et al. 2014). Coenders-Gerrits et al. (2014) concluded that although isotopes are useful for partitioning evaporation at smaller scales, results at the larger scales can be highly sensitive to data uncertainties and that improvement in observation techniques is needed to enhance the application of isotopes for partitioning of hydrological fluxes.

Wireless Sensor Networks

Wireless sensor networks consist of a multitude of small sensor nodes embedded in the environment that can quantify spatially heterogeneous phenomena, e.g., temperature or soil moisture fields and vegetation conditions, with high temporal and spatial resolution (Bogena et al. 2010; Jin et al. 2014; Qu et al. 2014). Such distributed data enables the determination of covariance structures of auto-correlated drivers to assist in upscaling and downscaling of state variables in time and space (Loescher et al. 2007). Slower changes in system states might best be observed at broader scales than the local, i.e., field, stand, and ecosystem scales. Such would be the example of measuring the below-canopy microclimate of many ecosystems within the landscape. The physical structure of these systems does not change rapidly and thus can be assessed on a campaign basis through the use of wireless ad hoc sensor networks (Inagaki et al. 2011). Ad hoc sensor networks are extremely flexible and can be used temporarily to assess spatial heterogeneity of ecosystem states in areas that are not part of the "fixed" sensor network. In addition, these ad hoc networks can be deployed in a chronosequence design to substitute time for space. Moreover, these ad hoc sensor networks can operate periodically on a regular or event-driven basis thus allowing to capture other nonlinear behaviors in the landscape that otherwise could not be measured, i.e., heat islands, shifting agriculture, flooding, and wildfires.

Another promising application for wireless sensor networks is collaborative catchment-scale management. To this end, Zia et al. (2013) propose a multi-scale wireless monitoring approach, which should consider three scale levels. At the field scale, monitoring is aimed at improved understanding of small-scale processes such as leaching of nutrients in soil and groundwater and retention of fertilizers by soil. At the medium scale, monitoring should be focused on assessing agricultural practices such as improved irrigation strategies and use of alternative fertilizers or reduced application rates while maintaining same crop yields. Finally, the basin-scale monitoring also covers groundwater and stream water quality and regional land use assessments for whole catchments.

Outlook

In this chapter we analyzed different approaches and methods to upscale ecohydrological processes from the local to the catchment scale and beyond. Many new techniques have emerged in the last year that show a huge potential for upscaling as they either allow a high spatial and/or temporal measurement of key state variables of the ecohydrological system. Recent progress has shown that, e.g., spatiotemporal patterns of soil moisture can be used to derive soil hydraulic properties at field and catchment scale in combination with pedotransfer functions and stochastic theories. In addition these patterns provide essential information that enables deriving the role of ecohydrological processes such as root water uptake and rainfall partitioning. Patterns derived from ground-based measurements can also be combined with remotely sensed patterns to improve our understanding of underlying ecohydrological processes, or they may be used for calibration and validation of remotely sensed data. With respect to the estimation of soil hydraulic properties, several pedotransfer research directions have to be pursued to address the issues of scale. Soil data aggregation methods need to be better documented and compared. Digital soil mapping research presents additional prospects for that. The role of environmental covariates has to be understood better, given their importance can be different in different geographic settings. Parent material has to be actively used as the environmental covariate. Class PTFs as well as categorical attributes of other soil-vegetation-atmosphere compartments have a substantial role to play at coarser scales. Better correspondence has to be established between specific PTFs and specific complex and simpler semiempirical models. It remains to be determined how regionally important soil processes, such as salinization, tropical soil mineralogy, etc., have to be accounted at coarse-scale PTFs. It may be important to find out if time-(seasonally) dependent PTFs can be beneficial at coarser scales. The concurrent use and comparison of several PTFs will be beneficial in most projects.

Remote sensing can help upscaling ecohydrological processes, variables, and parameters. Often regression methods are used to advance point-scale measurements to larger regions. Here, more physical approaches need to be developed that describe the reflectance characteristics toward a physical parameter retrieval.

Data assimilation is a promising methodology of ecohydrological model parameterization. It is relatively young approach in this field, and more needs to be learned about its efficiency as affected by the ensemble size, the observation and model errors, the assimilation interval, the types and ranges of water dynamics, the variability of the estimated parameters, the spatial patterns of ecohydrological parameters, and the structural errors in models. Data assimilation for models of different complexity needs to be evaluated at coarser scales. In summary, due to the rapid developments in technological and modeling capabilities, novel understanding of ecohydrological processes will continue to emerge in the next years to come.

References

R.G. Allen, L.S. Pereira, D. Raes, M. Smith, Crop evapotranspiration-guidelines for computing crop water requirements-FAO irrigation and drainage paper 56, FAO, Rome 300, D05109 (1998)

C. Alvarez-Garreton, D. Ryu, A.W. Western, W.T. Crow, D.E. Robertson, The impacts of assimilating satellite soil moisture into a rainfall-runoff model in a semi-arid catchment. J. Hydrol. **519**, 2763 (2014)

H. Asbjornsen et al., Ecohydrological advances and applications in plant-water relations research: a review. J. Plant. Ecol. **4**, 3 (2011)
E. Babaeian, M. Homaee, C. Montzka, H. Vereecken, A.A. Norouzi, Towards retrieving soil hydraulic properties by hyperspectral remote sensing. Vadose Zone J. **14**(3), (2015). https://doi.org/10.2136/vzj2014.07.0080
E. Babaeian, M. Homaee, C. Montzka, H. Vereecken, A.A. Norouzi, M.T. van Genuchten, Soil moisture prediction of bare soil profiles using diffuse spectral reflectance information and vadose zone flow modeling. Remote Sens. Environ. **187**, 218 (2016)
M.J. Best et al., The joint UK land environment simulator (JULES), model description – part 1: energy and water fluxes. Geosci. Model Dev. **4**, 677 (2011)
T.W. Biggs, G.P. Petropoulos, N.M. Velpuri, M. Marshall, E.P. Glenn, P. Nagler, A. Messina, Remote sensing of energy fluxes and soil moisture content, G.P. Petropoulos (Taylor and Francis, New York, 2013), p. 59
H.R. Bogena, M. Herbst, J.A. Huisman, U. Rosenbaum, A. Weuthen, H. Vereecken, Potential of wireless sensor networks for measuring soil water content variability. Vadose Zone J. **9**, 1 (2010)
F. Castaldi, R. Casa, A. Castrignano, S. Pascucci, A. Palombo, S. Pignatti, Estimation of soil properties at the field scale from satellite data: a comparison between spatial and non-spatial techniques. Eur. J. Soil Sci. **65**, 842 (2014)
F. Castaldi, A. Palombo, S. Pascucci, S. Pignatti, F. Santini, R. Casa, Reducing the influence of soil moisture on the estimation of clay from hyperspectral data: a case study using simulated PRISMA data. Remote Sens-Basel **7**, 15561 (2015)
B.J. Choudhury, N.U. Ahmed, S.B. Idso, R.J. Reginato, C.S.T. Daughtry, Relations between evaporation coefficients and vegetation indexes studied by model simulations. Remote Sens. Environ. **50**(1) (1994)
A.M.J. Coenders-Gerrits, R.J. van der Ent, T.A. Bogaard, L. Wang-Erlandsson, M. Hrachowitz, H. H.G. Savenije, Uncertainties in transpiration estimates. Nature **506**, E1 (2014)
Y.J. Dai et al., The common land model. B. Am. Meteorol. Soc. **84**, 1013 (2003)
D. Eamus, S. Zolfaghar, R. Villalobos-Vega, J. Cleverly, A. Huete, Groundwater-dependent ecosystems: recent insights from satellite and field-based studies. Hydrol. Earth. Syst. Sci. **19**, 4229 (2015)
D. Erdal, I. Neuweiler, U. Wollschlager, Using a bias aware EnKF to account for unresolved structure in an unsaturated zone model. Water Resour. Res. **50**, 132 (2014)
J. Evaristo, S. Jasechko, J.J. McDonnell, Global separation of plant transpiration from groundwater and streamflow. Nature **525**, 91 (2015)
G. Evensen, Sequential data assimilation with a nonlinear quasi-geostrophic model using Monte-Carlo methods to forecast error statistics. J. Geophys. Res. Oceans **99**, 10143 (1994)
G. Evensen, The ensemble Kalman filter for combined state and parameter estimation Monte Carlo techniques for data assimilation in large systems. IEEE Control. Syst. Mag. **29**, 83 (2009)
D.R. Fuka, A.S. Collick, P.J.A. Kleinman, D.A. Auerbach, R D. Harmel, Z M. Easton, Improving the spatial representation of soil properties and hydrology using topographically derived initialization processes in the SWAT model. Hydrol. Process., n/a (2016)
N.J. Gordon, D.J. Salmond, A.F.M. Smith, Novel-approach to nonlinear non-Gaussian Bayesian state estimation. IEEE Proc. F Radar Signal Process. **140**, 107 (1993)
J.P. Guerschman, A.I.J.M. Van Dijk, G. Mattersdorf, J. Beringer, L.B. Hutley, R. Leuning, R.C. Pipunic, B.S. Sherman, Scaling of potential evapotranspiration with MODIS data reproduces flux observations and catchment water balance observations across Australia. J. Hydrol. **369**, 107 (2009)
A.J. Guswa, M.A. Celia, I. Rodriguez-Iturbe, Models of soil moisture dynamics in ecohydrology: a comparative study. Water Resour. Res. **38**, 1166 (2002)
E.D. Gutmann, E.E. Small, A comparison of land surface model soil hydraulic properties estimated by inverse modeling and pedotransfer functions. Water Resour. Res. **43** (2007.) W05418
J.D. Haskett, Y.A. Pachepsky, B. Acock, Use of the beta-distribution for parameterizing variability of soil properties at the regional level for crop yield estimation. Agric. Syst. **48**, 73 (1995)

C. Homer et al., Completion of the 2011 national land cover database for the conterminous United States – representing a decade of land cover change information. Photogramm. Eng. Remote. Sens. **81**, 345 (2015)

M.N. Inagaki, T. Fukatsu, M. Hirafuji, M.M. Nachit, Monitoring soil moisture under wheat growth through a wireless sensor network in dry conditions. J. Environ Sci. Eng. **5**, 428 (2011)

R.B. Jana, B.P. Mohanty, Enhancing PTFs with remotely sensed data for multi-scale soil water retention estimation. J. Hydrol. **399**, 201 (2011)

R.B. Jana, B.P. Mohanty, Z. Sheng, Upscaling soil hydraulic parameters in the picacho mountain region using bayesian neural networks. T Asabe **55**, 463 (2012)

S. Jasechko, J. Kirchner, J. Welker and J.J. McDonnell Substantial Portion of global river discharge less than 3 months old, Nat. Geosci. **9**, 9, 126 (2016)

R. Jin et al., A nested ecohydrological wireless sensor network for capturing the surface heterogeneity in the midstream areas of the Heihe River basin, China. Ieee Geosci Remote S **11**, 2015 (2014)

A. Lausch, S. Zacharias, C. Dierke, M. Pause, I. Kuhn, D. Doktor, P. Dietrich, U. Werban, Analysis of vegetation and soil patterns using hyperspectral remote sensing, EMI, and gamma-ray measurements. Vadose Zone J. **12(4)**, (2013). https://doi.org/10.2136/vzj2012.0217

M.R. Levi, M.G. Schaap, C. Rasmussen, Application of spatial pedotransfer functions to understand soil modulation of vegetation response to climate. Vadose Zone J. **14**(9), (2015). https://doi.org/10.2136/vzj2014.09.0126

C. Li, L. Ren, Estimation of unsaturated soil hydraulic parameters using the ensemble Kalman filter. Vadose Zone J. **10**, 1205 (2011)

B. Livneh, R. Kumar, L. Samaniego, Influence of soil textural properties on hydrologic fluxes in the Mississippi river basin. Hydrol. Process. **29**, 4638 (2015)

H.W. Loescher et al., Enhancing water cycle measurements for future hydrologic research. B. Am. Meteorol. Soc. **88**, 669 (2007)

M. Marshall, P. Thenkabail, T. Biggs, K. Post, Hyperspectral narrowband and multispectral broadband indices for remote sensing of crop evapotranspiration and its components (transpiration and soil evaporation). Agric. For. Meteorol. **218**, 122 (2016)

S.A. Mathias, T.H. Skaggs, S.A. Quinn, S.N.C. Egan, L.E. Finch, C.D. Oldham, A soil moisture accounting-procedure with a Richards' equation-based soil texture-dependent parameterization. Water Resour. Res. **51**, 506 (2015)

E.M. Middleton, S.G. Ungar, D.J. Mandl, L. Ong, S.W. Frye, P.E. Campbell, D.R. Landis, J.P. Young, N.H. Pollack, The earth observing one (EO-1) satellite mission: over a decade in space. IEEE J-Stars **6**, 243 (2013)

E.E. Miller, R.D. Miller, Physical theory for capillary flow phenomena. J. Appl. Phys. **27**, 324 (1956)

D.A. Miller, R.A. White, A conterminous United States multilayer soil characteristics dataset for regional climate and hydrology modeling. Earth Interact. **2**, 1 (1998)

C. Montzka, V.R.N. Pauwels, H.J.H. Franssen, X.J. Han, H. Vereecken, Multivariate and multiscale data assimilation in terrestrial systems: a review. Sensors-Basel **12**, 16291 (2012)

C. Montzka, M. Herbst, L. Weihermüller, A. Verhoef, H. Vereecken, A global data set of soil hydraulic properties and sub-grid variability of soil water retention and hydraulic conductivity curves. Earth Syst. Sci. Data Discuss. **2017**, 1 (2017)

M.Z. Moreira, L.D.L. Sternberg, L.A. Martinelli, R.L. Victoria, E.M. Barbosa, L.C.M. Bonates, D. C. Nepstad, Contribution of transpiration to forest ambient vapour based on isotopic measurements. Glob. Chang. Biol. **3**, 439 (1997)

P.L. Nagler, E.P. Glenn, U. Nguyen, R.L. Scott, T. Doody, Estimating riparian and agricultural actual evapotranspiration by reference evapotranspiration and MODIS enhanced vegetation index. Remote Sens-Basel **5**, 3849 (2013)

A. Nemes, Y.A. Pachepsky, D.J. Timlin, Toward improving global estimates of field soil water capacity. Soil Sci. Soc. Am. J. **75**, 807 (2011)

W. Ouerghemmi, C. Gomez, S. Naceur, P. Lagacherie, Semi-blind source separation for the estimation of the clay content over semi-vegetated areas using VNIR/SWIR hyperspectral airborne data. Remote Sens. Environ. **181**, 251 (2016)
Y. Pachepsky, R.L. Hill, Scale and scaling in soils. Geoderma 287, 4-30 (2017)
Y. Pachepsky, W.J. Rawls, *Development of Pedotransfer Functions in Soil Hydrology*, vol 30 (Elsevier, Amsterdam, 2004)
Y. Pachepsky, A. Guber, D. Jacques, J. Simunek, M.T. Van Genuchten, T. Nicholson, R. Cady, Information content and complexity of simulated soil water fluxes. Geoderma **134**, 253 (2006)
G. Petropoulos, T.N. Carlson, M.J. Wooster, S. Islam, A review of T-s/VI remote sensing based methods for the retrieval of land surface energy fluxes and soil surface moisture. Prog. Phys. Geogr. **33**, 224 (2009)
L. Poggio, A. Gimona, I. Brown, M. Castellazzi, Soil available water capacity interpolation and spatial uncertainty modelling at multiple geographical extents. Geoderma **160**, 175 (2010)
A. Porporato, P. D'Odorico, F. Laio, L. Ridolfi, I. Rodriguez-Iturbe, Ecohydrology of water-controlled ecosystems. Adv. Water Resour. **25**, 1335 (2002.) Pii s0309-1708(02)00058-1
Y.H. Qu, Y.Q. Zhu, W.C. Han, J.D. Wang, M.G. Ma, Crop leaf area index observations with a wireless sensor network and its potential for validating remote sensing products. IEEE J-Stars **7**, 431 (2014)
W. Qu, H.R. Bogena, J.A. Huisman, J. Vanderborght, M. Schuh, E. Priesack, H. Vereecken, Predicting subgrid variability of soil water content from basic soil information. Geophys. Res. Lett. **42**, 789 (2015a)
Y.H. Qu, W.C. Han, M.G. Ma, Retrieval of a temporal high-resolution leaf area index (LAI) by combining MODIS LAI and ASTER reflectance data. Remote Sens-Basel **7**, 195 (2015b)
R.H. Reichle, Data assimilation methods in the earth sciences. Adv. Water Resour. **31**, 1411 (2008)
A.A. Romanowicz, M. Vanclooster, M. Rounsevell, I. La Junesse, Sensitivity of the SWAT model to the soil and land use data parametrisation: a case study in the Thyle catchment, Belgium. Ecol. Model. **187**, 27 (2005)
Y. Rothfuss, P. Biron, I. Braud, L. Canale, J.L. Durand, J.P. Gaudet, P. Richard, M. Vauclin, T. Bariac, Partitioning evapotranspiration fluxes into soil evaporation and plant transpiration using water stable isotopes under controlled conditions. Hydrol. Process. **24**, 3177 (2010)
S. Rudolph et al., Linking satellite derived LAI patterns with subsoil heterogeneity using large-scale ground-based electromagnetic induction measurements. Geoderma **241**, 262 (2015)
W. Shangguan, Y.J. Dai, Q.Y. Duan, B.Y. Liu, H. Yuan, A global soil data set for earth system modeling. J. Adv. Model. Earth Syst **6**, 249 (2014)
L.S. Song et al., Applications of a thermal-based two-source energy balance model using Priestley-Taylor approach for surface temperature partitioning under advective conditions. J. Hydrol. **540**, 574 (2016)
M.P. Stockinger, A. Lucke, J.J. McDonnell, B. Diekkruger, H. Vereecken, H.R. Bogena, Interception effects on stable isotope driven streamwater transit time estimates. Geophys. Res. Lett. **42**, 5299 (2015)
Z. Sun, M. Gebremichael, J. Ardo, A. Nickless, B. Caquet, L. Merboldh, W. Kutschi, Estimation of daily evapotranspiration over Africa using MODIS/Terra and SEVIRI/MSG data. Atmos. Res. **112**, 35 (2012)
T. Svoray, S. Assouline, G. Katul, Introduction to a special section on ecohydrology of semiarid environments: confronting mathematical models with ecosystem complexity. Water Resour. Res. **51**, 8677 (2015)
P.M. Tillotson, D.R. Nielsen, Scale factors in soil science. Soil Sci. Soc. Am. J. **48**, 953 (1984)
S.L. Ustin, D. Riano, E.R. Hunt, Estimating canopy water content from spectroscopy. Israel J. Plant Sci. **60**, 9 (2012)
E. Vaudour, J.M. Gilliot, L. Bel, J. Lefevre, K. Chehdi, Regional prediction of soil organic carbon content over temperate croplands using visible near-infrared airborne hyperspectral imagery and synchronous field spectra. Int. J. Appl. Earth Obs. **49**, 24 (2016)

H. Vereecken, T. Kamai, T. Harter, R. Kasteel, J. Hopmans, J. Vanderborght, Explaining soil moisture variability as a function of mean soil moisture: a stochastic unsaturated flow perspective. Geophys. Res. Lett. **34 (22)** L22402 (2007)

H. Vereecken, J.A. Huisman, H. Bogena, J. Vanderborght, J.A. Vrugt, J.W. Hopmans, On the value of soil moisture measurements in vadose zone hydrology: a review. Water Resour. Res. **44** W00D06 (2008)

H. Vereecken et al., On the spatio-temporal dynamics of soil moisture at the field scale. J. Hydrol. **516**, 76 (2014)

H. Vereecken, A. Schnepf, J.W. Hopmans, M. Javaux, D. Or, T. Roose, J. Vanderborght, M.H. Young, W. Amelung, M. Aitkenhead, S.D. Allison, S. Assouline, P. Baveye, M. Berli, N. Brüggemann, P. Finke, M. Flury, T. Gaiser, G. Govers, T. Ghezzehei, P. Hallett, H.J. Hendricks Franssen, J. Heppell, R. Horn, J.A. Huisman, D. Jacques, F. Jonard, S. Kollet, F. Lafolie, K. Lamorski, D. Leitner, A. McBratney, B. Minasny, C. Montzka, W. Nowak, Y. Pachepsky, J. Padarian, N. Romano, K. Roth, Y. Rothfuss, E.C. Rowe, A. Schwen, J. Šimůnek, A. Tiktak, J. Van Dam, S.E.A.T.M. van der Zee, H.J. Vogel, J.A. Vrugt, T. Wöhling, I.M. Young, Modeling Soil Processes: Review, Key Challenges, and New Perspectives. Vadose Zone J. **15**(5), (2016). https://doi.org/10.2136/vzj2015.09.01310

J.A. Vrugt, M.T. van Wijk, J.W. Hopmans, J. Simunek, One-, two-, and three-dimensional root water uptake functions for transient modeling. Water Resour. Res. **37**, 2457 (2001)

J.A. Vrugt, P.H. Stauffer, T. Wohling, B.A. Robinson, V.V. Vesselinov, Inverse modeling of subsurface flow and transport properties: a review with new developments. Vadose Zone J. **7**, 843 (2008)

Z.M. Wan, K. Zhang, X.W. Xue, Z. Hong, Y. Hong, J.J. Gourley, Water balance-based actual evapotranspiration reconstruction from ground and satellite observations over the conterminous United States. Water Resour. Res. **51**, 6485 (2015)

X. Wen, B. Yang, X. Sun, X. Lee, Evapotranspiration partitioning through in-situ oxygen isotope measurements in an oasis cropland. Agric. For. Meteorol. **230–231**, 89 (2016)

J.H.M. Wosten, Y.A. Pachepsky, W.J. Rawls, Pedotransfer functions: bridging the gap between available basic soil data and missing soil hydraulic characteristics. J. Hydrol. **251**, 123 (2001)

J.J. Wu, Z.Y. Li, Z.H. Gao, B.Y. Wang, L.N. Bai, B. Sun, C.L. Li, X.Y. Ding, Degraded land detection by soil particle composition derived from multispectral remote sensing data in the Otindag Sandy Lands of China. Geoderma **241**, 97 (2015)

T.R. Xu, S.M. Liu, L. Xu, Y.J. Chen, Z.Z. Jia, Z.W. Xu, J. Nielson, Temporal upscaling and reconstruction of thermal remotely sensed instantaneous evapotranspiration. Remote Sens-Basel **7**, 3400 (2015a)

T.R. Xu, S.M. Liu, Z.W. Xu, S.L. Liang, L. Xu, A dual-pass data assimilation scheme for estimating surface fluxes with FY3A-VIRR land surface temperature. Sci. China Earth Sci. **58**, 211 (2015b)

Y.S. Zhai, J.A. Thomasson, J.E. Boggess, R.X. Sui, Soil texture classification with artificial neural networks operating on remote sensing data. Comput. Electron. Agr. **54**, 53 (2006)

D.X. Zhang, T.C. Wallstrom, C.L. Winter, Stochastic analysis of steady-state unsaturated flow in heterogeneous media: comparison of the Brooks-Corey and Gardner-Russo models. Water Resour. Res. **34**, 1437 (1998)

K. Zhang, J.S. Kimball, S.W. Running, A review of remote sensing based actual evapotranspiration estimation. Wiley Interdiscip. Rev. Water **3**, 834 (2016)

H. Zia, N.R. Harris, G.V. Merrett, M. Rivers, N. Coles, The impact of agricultural activities on water quality: a case for collaborative catchment-scale management using integrated wireless sensor networks. Comput. Electron Agric. **96**, 126 (2013)

Field Experimental Design for a Watershed Observing System

Xin Li, Guodong Cheng, Chunfeng Ma, Yingchun Ge, and Zhonghao Zhang

Contents

X. Li (✉)
Institute of Tibetan Plateau Research, Chinese Academy of Sciences, Beijing, China

CAS Center for Excellence in Tibetan Plateau Earth Sciences, Chinese Academy of Sciences, Beijing, China
e-mail: xinli@itpcas.ac.cn

G. Cheng
Cold and Arid Regions Environmental and Engineering Research Institute, Chinese Academy of Sciences, Lanzhou, China

Institute of Urban Studies, Shanghai Normal University, Shanghai, China

C. Ma
Key Laboratory of Remote Sensing of Gansu Province, Cold and Arid Regions Environmental and Engineering Research Institute, Chinese Academy of Sciences, Lanzhou, China

Y. Ge
Cold and Arid Regions Environmental and Engineering Research Institute, Chinese Academy of Sciences, Lanzhou, China

Z. Zhang
Institute of Urban Studies, Shanghai Normal University, Shanghai, China

X. Li, H. Vereecken (eds.), *Observation and Measurement of Ecohydrological Processes*, Ecohydrology, https://doi.org/10.1007/978-3-662-48297-1_15

Abstract

A watershed, regarded as the ideal unit for practicing earth system science, possesses all of the complexities of the land surface system. Thus, building a watershed observing system and conducting a field experiment for the observation of the ecohydrological processes at watershed scale is the best way to understand the complexities of the land surface system. This chapter presents an overview on the designing and conducting of a field experiment and establishing a watershed observing system. Several key scientific problems are addressed: (1) why we need a watershed observing system, (2) what are the characteristics of a watershed observing system, and (3) how can we design and establish a watershed observing system. We believe that a watershed observing system is the prerequisite of the watershed science development and helping to improve the understanding of the ecohydrological processes. The watershed observing system possesses the multidiscipline and multi-scale characteristics. It must be able to capture the spatiotemporal heterogeneity and to quantify the uncertainty of the key ecohydrological variables. Finally, by taking two sequential watershed-scale observation experiments as example, we furthermore illustrate how to design and establish a watershed observing system.

Keywords

Watershed observing system · Field experiment · Ecohydrology · Heterogeneity · Uncertainty · Scaling · HiWATER

Why a Watershed Observing System

Traditionally, watershed science refers to the subjects of hydrology and water resource management; however, in recent years, watershed science has experienced a revolutionary change that can be attributed to the forceful interactions among the multiple disciplines of earth science and the rigorous advancements in earth observing and information gathering (Cheng et al. 2014; Cheng and Li 2015). Water and its interactions with other systems in a watershed are increasingly becoming the focus of scientific communities. New subjects such as ecohydrology, hydroecology, hydroeconomics, environmental hydrogeology, and hydroinformatics are emerging. The integrated perspective of an "atmosphere-water-ecosystem-soil-human" continuum, which regards a river basin as a whole, has rapidly evolved and is premature – this new science is called watershed science (Cheng 2009; Cheng and Li 2015). A river basin or watershed in its totality contains most of the dynamics of the earth system, including the atmosphere, mountain cryosphere, rivers, lakes, groundwater, ecosystems, land use, and human activities, all of which provide a vivid portrayal of the earth system. The water cycle, which is one the most important processes on earth, can generally be closed within a river basin or watershed.

Therefore, a river basin can be considered as a miniature model of ecohydrology or as an atomic unit of nature. From the viewpoint of fundamental research, watershed science is the study of ecohydrological processes at the watershed scale. Considering the integrated approach of watershed management, watershed science is an applied science that investigates sustainable development through the optimal allocation of natural resources and human activities in a river basin (NRC 1997; Li and Cheng 2008). In recent years, discussion of watershed science has been very active and has stressed the interdisciplinary nature of watershed science (Aylward 2005; Uhlenbrook 2006). The National Research Council (NRC)'s report titled "Opportunities in the Hydrologic Sciences" (NRC 1991) played a critical role in the gradual formation of watershed science and cultivated a climax in the comprehensive study of hydrology. Thereafter, the NRC successively published "Watershed Research in the U.S. Geological Survey," "New Strategies for America's Watersheds," and "River Science at the U. S. Geological Survey" (NRC 1997, 1999, 2007). These three advisory reports on watershed and river research systematically addressed and thoroughly discussed watershed science, launching the new field into its premature stage.

Without an earth observing system (EOS), there would be no earth system science. Similarly, the development of watershed science also requires the development of a watershed observing system. Earth observation has helped with the advancement of watershed science. Remote sensing, wireless sensor networks (WSNs), GPS hydrology, and other modern earth observing techniques are revolutionizing the hydrology, ecology, and other disciplines of earth science. For example, it was stated that "Through some remarkable technological accomplishments, some important variables and processes associated with the water cycle and ecological process can be retrieved now from satellite observations" (NRC 2008). Major components of the water cycle and ecohydrological processes including precipitation, evapotranspiration, runoff, soil moisture, snow, glaciers and ice, soil freeze/thaw, gross primary productivity (GPP), net primary productivity (NPP), net ecosystem exchange (NEE), biomass, carbon flux, and even groundwater dynamics can now be measured from space at various spatial and temporal scales. Earth observing techniques can also be applied in river basins; water resource management significantly benefits from improved measurement of watershed topography, land use, ecosystem processes, and geophysical and geochemical processes (GEO 2005).

This variety of techniques greatly improves our understanding of natural processes by improving our ability to observe ecohydrology and to more clearly and accurately quantify water cycle elements and/or ecological processes. However, some questions remain unanswered regarding building an effective watershed observing system; a few are listed below.

1. How can we establish a comprehensive observing platform that can more effectively support research on ecohydrological processes in a river basin and implement integrated watershed sciences?

2. How well can remote sensing data be used to improve our understanding of the ecohydrological processes in a river basin?
3. What types of ecohydrological processes can be monitored through remote sensing? How accurate are these measurements? How can we develop more reliable and robust ecohydrological remote sensing methods through field experiments?
4. Can the ecohydrological processes observed (obtained) at a point scale be upscaled to larger spatial scales (e.g., a catchment scale) through remote sensing? If this is possible, how can we achieve this transformation?
5. How can we design an in situ observation network? What are the optimum density and scale of individual in situ observations? How can we capture the spatiotemporal variations of each observation item at the catchment scale?
6. How can we design in situ sampling strategies aimed at remote sensing validation?
7. How can we apply scale transformation to effectively use multisource, multiscale remote sensing data and ground-based measurements in hydrological, ecological, and integrated watershed models?
8. How can we integrate remote sensing observations, in situ measurements, and model simulations to accurately estimate the state variables and fluxes in the water cycle and ecological processes?
9. How can we merge multisource, multiscale observations into a watershed-scale ecohydrological assimilation system that improves the accuracy and predictability of hydrological and ecological models?
10. How can we effectively use multisource data in integrated river basin management?
11. How might observing and modeling mutually interact?
12. How can the heterogeneity be captured?

This chapter presents the authors' thoughts on building a watershed observing system for ecohydrological processes.

Characteristics of a Watershed Observing System

Watershed-scale observing and modeling systems for ecohydrological processes are prerequisites for watershed science. A watershed observing system should be capable of observing water, energy, ecology, and geochemical cycles as well as socioeconomic activities using state-of-the-art, high-resolution, high-accuracy, high-efficacy, multiscale integrated techniques. The system should be developed to capture the spatial and temporal variabilities, heterogeneity, and connectivity of the ecohydrological processes at the watershed scale and test the strength of modern observing techniques, and it should be easy to incorporate into integrated watershed models.

Multidisciplinary Characteristics

As watershed science has distinct multidisciplinary, interdisciplinary, and even transdisciplinary characteristics, it requires the participation and collaboration of all disciplines related to the earth surface system. However, the quite different paradigms and traditions of different subject areas may confer different perspectives and encourage different ideas. For example, meteorologists, hydrologists, or ecologists performing an observation experiment within a watershed may consider different measurement methods, instrumentation, observation items, spatial scales, temporal scales, and experiment controllability even though atmospheric science, hydrology, and ecology are all closely related to watershed science. Based on our observations, we provide the following analysis to illustrate why the integration of different paradigms is important.

First: the land surface experiment. Traditionally, land surface science has been considered as a subdiscipline of atmospheric science, and its goal is to understand land-atmosphere interactions. In many land surface experiments, the main observation instruments are weather stations and flux towers. The observed variables include the meteorological states and fluxes of the near-surface and boundary layers. In terms of spatial scale, since land surface parameterization schemes are designed for general circulation models, a grid scale ranging from tens to hundreds of kilometers has received the most focus. Land surface experiments therefore assume that the observed surface is homogeneous (or only has macroscale heterogeneity) in order to use a general circulation model (GCM) grid. In terms of temporal scale, land surface process experiments are performed on a very short time scale and are characterized by an intensive observation period (IOP). Land surface experiments are rarely controllable because the nature of the boundary layer is uncontrollable. Therefore, monitoring experiments contain only a series of observations that focus on the different natural characteristics of climate and landscapes. Generally, these experiments depend on micrometeorological and flux measurements and can monitor only the expected, natural formation of different climate and landscape characteristics.

Second: the hydrological experiment. Different from land surface experiments, a hydrological experiment stresses not only the measurements of the water budget in the vertical dimension but also the horizontal interactions among the observed objects, e.g., different hydrological response units. All components of the water cycle, such as rainfall, evaporation, transpiration, runoff, infiltration, snow cover, soil moisture, freeze/thaw states, and groundwater table, should be measured. In terms of spatial scale, hydrological observation experiments hierarchically use watershed, catchment, or hill slope (10^3 to 1 m) as the basic observing unit. Therefore, hydrological observations require much denser spatial instrumentation than do land surface experiments. In terms of temporal scale, hydrological experiments also try to capture fast processes: 1 h or 1 day is defined as the basic time step, and the observation period often continues for at least 1 hydrological year, especially in cold regions, where soil freezing and thawing as well as snow melting and other hydrological processes occur. Hydrological experiments are also mainly based on monitoring, but some experiments are controllable or semi-controllable, such as measurements of runoff at a hill slope.

Third: the ecological experiment. Ecological studies always focus on the processes and inner relationships among the components of an ecosystem. Therefore, the spatial scale in ecological experiments is even smaller, and the basic observation unit is the organ, individual, or sampling plot. With the exception of some short-term process observations, long-term monitoring over many years or even centuries is highly valued. Both field campaigns and laboratory analyses are involved in observation approaches. An advantage of ecological experiments is that the variables (e.g., light, heat, water, gas, and fertilizer) may be highly controlled in order to observe a corresponding ecological response.

Understanding ecohydrological processes requires an interdisciplinary approach that utilizes the three different spatial and temporal scales described above. In addition, watershed sciences should use the river basin or watershed as the basic unit of study, giving priority to water; however, equal attention should be given to approaches that utilize monitoring or a controllable experiment. Figure 1 depicts the multidisciplinary nature of a watershed observing system.

Multiscale Perspectives

When determining the spatial scale to be observed, a homogeneous land surface is traditionally selected as the target of observation because it is relatively easy and cheap to carry out measurements, analyses, and modeling on a homogeneous surface. However, heterogeneity is one of the essential characteristics of natural processes. Strictly speaking, there is no completely homogeneous surface on earth. Therefore, a homogeneous land surface can no longer be viewed as the primary observing target, and the heterogeneities of the measured object, both in time and

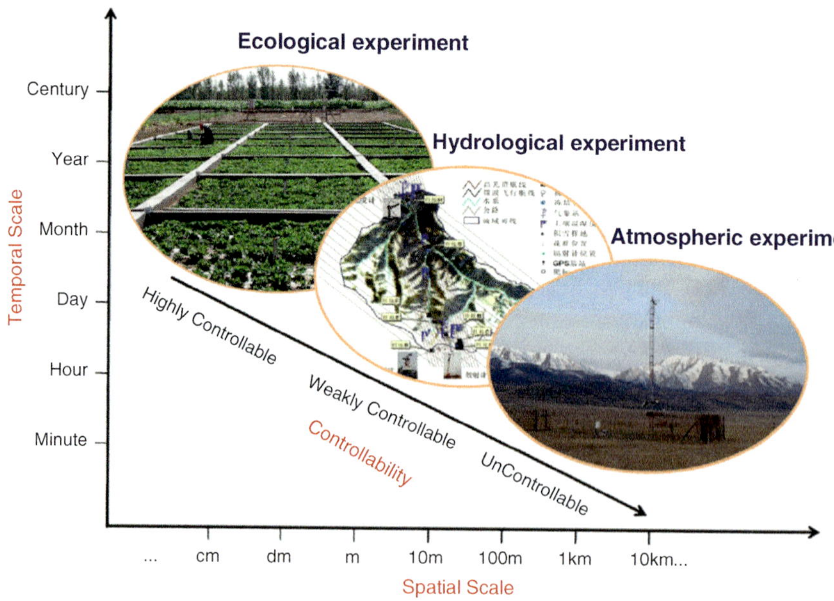

Fig. 1 Multidisciplinary nature of a watershed observing system

space, should be captured (McDonnell et al. 2007). Since observing heterogeneity is very complicated and deeply connected to scale problems, the question arises of how to capture the spatiotemporal heterogeneity through multiscale observations.

Figure 2 demonstrates our view of the multiscale observations used to describe land surface heterogeneities across several moderate-resolution pixels (i.e., 50–200 m). Traditionally, the remote sensing observations that are generated from sources ranging from satellite to airborne to ground-based are nested in their resolution, with resolutions ranging from coarse to fine. However, when relying on the greatly enhanced resolution of satellite remote sensing data, the development of large-scale ground observation methods and the emergence of sensor networks, i.e., satellite, airborne, and ground hierarchy, is no longer a nesting observation structure; rather, it is integrated to comprise a multiscale observation system. In this system, heterogeneities can be captured by diverse observing methods, e.g., sensor networks, large-scale observations, and remote sensing. In this context, the scale issues become dramatically more complex because the effective sensing ranges are inconsistent among different observations. Therefore, to quantitatively measure heterogeneity

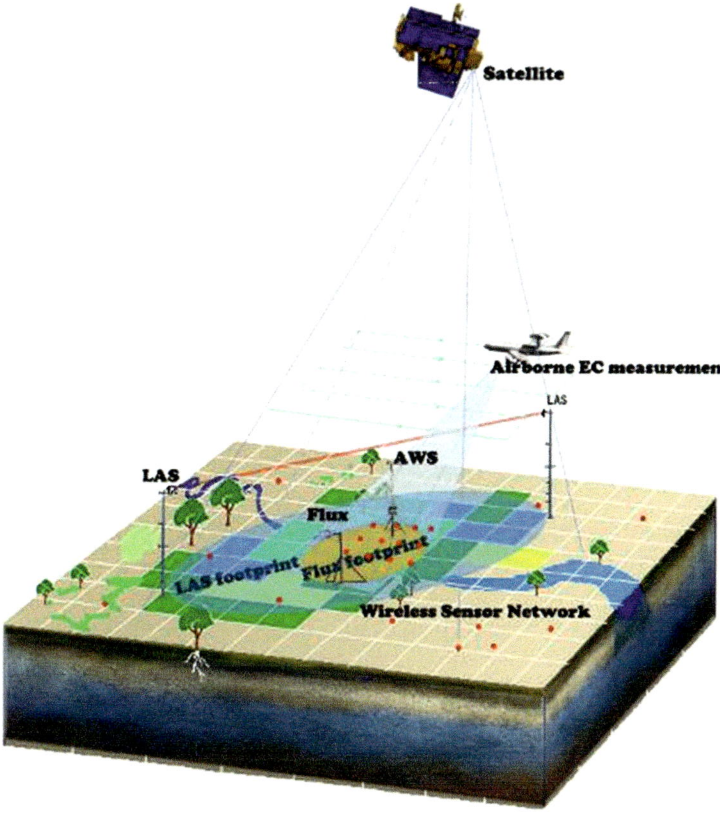

Fig. 2 Multiscale characteristics of a watershed observing system

and error and to compare various observations, a multiscale observation system should consider the following:

- The representative areas or footprints from different observation methods should be overlapped, e.g., remote sensing, eddy covariance (EC), or large-aperture scintillometer (LAS).
- The spatial representative errors of the observations should be statistically unbiased, and the variance should be as small as possible. In addition, the statistical spatial distribution of the observational errors (i.e., the probability distribution function) should be known and should be quantitatively estimated according to the source weight function or the spatial arrangement of the sampling points.
- Based on the characteristics of the heterogeneity in the study area, the sampling scheme can be designed in advance, but it must also consider the transport accessibility, security, and other practical factors.

Monitoring Experiment Versus Controllable Experiment

Traditionally, earth science is believed to initiate "from observation, rather than controllable experiment." However, as earth science has become an experimental science, controllable experiments have been increasingly proposed to validate scientific hypotheses (McDonnell et al. 2007). Thus, by the dual actuations of scientific questioning and technological advancement, a synthesized observation experiment should become another key characteristic of a watershed observing system used to observe ecohydrological processes; additionally, equal attention should be given to monitoring and controllable experiments. Therefore, the following ideas must be considered.

(1) The goal of a monitoring experiment is to observe a natural process or to acquire forcing data, parameters, and validation data sets that can be used to run and validate an existing model. Meanwhile, the state variables should be collected with high resolution so that these observations can be assimilated into hydrological and ecological models. Because the scales of heterogeneity are inconsistent among the different variables/parameters, specific observing and sampling schemes should be designed for each observation item. The monitoring experiment should capture the heterogeneities if they can be observed and measured within the simulation and observation scales of the equipment.
(2) The controllable experiment intends to observe natural processes and validate scientific hypotheses, which are often manifested by models under development or by new theories being formulated. Traditionally, the execution of a controllable experiment is difficult because the subject of geoscience is nature. Once the observation capability is enhanced, a controllable or semi-controllable experiment becomes feasible. The most typical example of a controllable experiment is the Biosphere 2 experiment. In addition, according to the gradient characteristics of the climate and landscape, comparative observation experiments can also be viewed as a type of controlled experiment.

Therefore, a key step in the development of a mature watershed observing system is to conduct a monitoring experiment that is supported by a controllable experiment. As mentioned in the NRC's "Opportunities in the Hydrologic Sciences" (NRC 1991), "large-scale experiments can provide a basic platform for various researchers to carry out their research work." If no variable large-scale observations and experiments were completed, earth surface system sciences would not continue to mature.

Uncertainties Associated with Observations

Evaluation of the validity and accuracy of remotely sensed and ground-based observations is integral to watershed measurement; therefore, the importance of understanding the uncertainties associated with observations should be emphasized.

The boundary conditions and parameters that affect the hydrological and ecological processes are highly heterogeneous both in time and space. Therefore, considerable uncertainties can appear during the upscaling procedure, which involves a transition of known hydrological and ecological processes from the microscale to the macroscale (i. e., from a grid scale to a watershed scale). This uncertainty is usually caused not by an unknown process but rather by spatial and temporal heterogeneities.

Ground measuring can only discern the "true value" of the observed object at the exact moment and scale of measurement (at least when the instrument error is very small). Much uncertainty surrounds spatial representativeness. If the watershed-scale spatiotemporal variations in the main state variables, parameters, and forcing data that impact the hydrological and ecological processes can be captured through well-designed intensive observations, used in multidisciplinary models as an input to perform an ensemble forecast, or assimilated into hydrological and ecological models, it is possible to significantly improve the accuracy of hydrological and ecological modeling and forecasting (especially for short-term prediction), and the errors of the model can be measured simultaneously.

Remote sensing is a macro and quantitative observational approach for measuring hydrological and ecological processes in terrestrial conditions, but it is also the tool that bridges observations and model simulations. However, because remote sensing can only provide indirect observations, greater uncertainty exists in both the inversion (e.g., soil moisture) and indirect estimation of ecohydrological variables that rely on these models (e.g., evapotranspiration (ET) and ecosystem productivity). Therefore, the remotely sensed observing capability of the hydrological and ecological processes cannot be significantly improved without the existence of well-designed validation experiments.

Building a Watershed Observing System

In watershed science, the advancement of water cycle and ecological process modeling and the reinforcement of integrated watershed management rely on modern observation techniques and a comprehensive watershed observing system.

Further, a watershed observing system is established by the dual actuations of specific watershed science questions and practical watershed management problems. The aim of observation may be (1) to test a hypothesis in the form of hydrological (Savenije 2009) and ecological models or (2) to rectify the parameters and variables that are used in the hydrological and ecological models in order to improve watershed management in terms of sustainable development. The former can be considered a type of controlled experiment, and the latter is type of monitoring experiment.

Supporting Hypothesis

Modeling must be based on scientific hypotheses. These hypotheses, which are manifested during the development of models and new theories, can be validated through controlled experiments that occur in a watershed observing system. However, since heterogeneity always exists in hydrology and ecology, both in situ and remote sensing observations must try to capture and quantify the spatial and temporal heterogeneity. At least three basic hypotheses should be considered in a watershed observing system.

1. The first hypothesis. The effects of surface heterogeneity are adequately considered during the selection of an observing position. Though the microscale water cycle and ecological process have been thoroughly studied, there are great uncertainties in how this information can be applied to the macroscale, such as from a grid to a watershed. However, high-spatiotemporal heterogeneity of boundary conditions, parameters, and variables in the water cycle and ecological processes themselves, rather than these unknown factors, tend to play a vital role in bringing about those uncertainties.
2. The second hypothesis. Ground-based observations obtain true values that can only represent the current conditions, even when using instruments with high accuracies. In other words, the amount of uncertainty increases with the spatial scale. However, if the spatiotemporal characteristics of primary variables, parameters, and forcing data in the water cycle and ecological processes are acquired through intensive observations and are used as inputs or assimilated into the hydrological and ecological models, the predictability of hydrological and ecological simulation and the ecohydrological processes are more likely to be improved, especially in the short term.
3. The third hypothesis. As a macroscale and quantitative observation approach of the terrestrial water cycle and ecological processes, remote sensing is a bridge between point observations and model simulations. However, as an indirect variable, hydrological and ecological modeling with remote sensing generates great uncertainties, both in inversion (e.g., soil moisture) and in the indirect estimation of the water cycle and ecological variables (e.g., evapotranspiration). Therefore, water cycle and ecological process observations through remote sensing approaches cannot be improved without well-designed and validated experiments.

Components of a Watershed Observing System

The watershed observing system should competently measure all aspects of energy, water, and biogeochemical cycles and ecological processes, as well as social and economic activities. Our view of a watershed observing system can be depicted as shown in Fig. 3, which refers to the experience of the constitution of EOS. The system consists of the following sections:

1. **Satellite remote sensing** incorporates traditional visible and near-infrared (VNIR), thermal infrared (TIR), and passive/active microwave methods; in addition, new methods, including LiDAR and the gravity technique, are being developed.
2. **Airborne remote sensing and telemetry** is an essential experimental approach and an important component of this observing system. Airborne remote sensing is usually superior to satellite remote sensing in terms of spatial resolution and can accomplish some of the observational tasks that satellite remote sensing cannot, e. g., mapping of high-resolution DEM and land cover/vegetation functional type. In addition, other common telemetry observations for watershed hydrology and ecology can be investigated, e.g., airborne eddy covariance and ground penetrating radar (GPR) measurements.
3. **Ground-based measurements** consist of the following components:
   - An automatic and advanced observing system that usually comprises an automatic meteorological station (AMS), a monitoring network and other automatic instruments used to measure fluxes (i.e., energy, water, gas and biochemistry), solar radiation (including the effective energy of photosynthesis) and longwave radiation, soil moisture and temperature, ecology, surface runoff and groundwater, GPP, NPP, NEE, biomass, etc.
   - A WSN is a modern information technique that integrates sensor technology, automatic control, and data transmission/storage/processing/analysis. Based on the acquisition of high-density observations, the WSN provides superior, integrated, standardized, and automated observation capabilities for hydrological and ecological processes researched at the watershed scale. Concentrated observations can be made of some specific land surface processes, e.g., snow hydrology, frozen soil hydrology, forest hydrology, and canopy hydrological processes.
   - An operational observing network built by operating departments from various fields, including meteorology, hydrology, ecology, hydrogeology, and water resources, is also a major component in developing a watershed observing system. The system consists of ordinary and automatic meteorological stations, hydrological stations, leaf area indices (LAIs), groundwater measurements, and records on irrigation and water allocation in canal systems.
   - Ground-based remote sensing involves specific instruments, e.g., weather radar, microwave radiometer, scatterometer, and spectrometer.
   - Geochemistry observations should include carbon cycle, isotope, and surface/underground water and air quality measurements.
   - Atmospheric aerosol and temperature and humidity profiles should be measured.

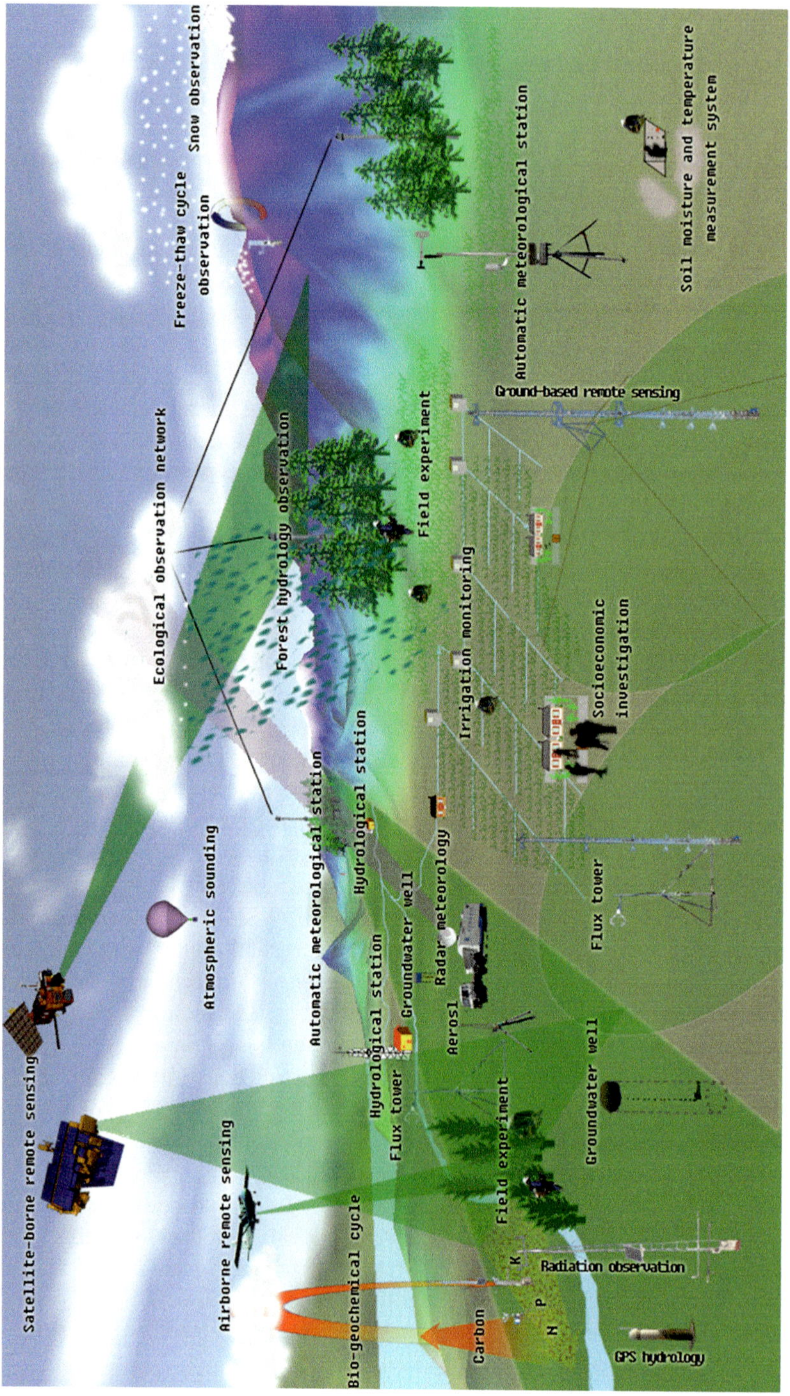

Fig. 3 Composition of a watershed observing system

- Social-economic surveys, e.g., using economic methods to evaluate the spatiotemporal patterns of ecosystem service value for cropland and forest in the process of urbanization.

Strategies for Building a Watershed Observing System

The following strategies are important for building a watershed observing system:

(1) Proposing watershed-specific scientific questions, identifying watershed-specific practical problems, and selecting appropriate physically based hydrological, ecological, and other environmental models. Observation experiments should be conducted based on (a) the state variables and parameters that are needed in models, e.g., forcing data, (b) the understanding of ecohydrological processes, and (c) the development of novel methods. Meanwhile, it is also important to validate new models and theories through observing systems.
(2) Capturing the heterogeneity in space and time. For parameters, even if we cannot determine the spatial distribution function, observations can provide information on the constraints used in the hydrological model, e.g., the upper and lower limits of the parameters. Homogeneous land surface should no longer be the sole target in observations; rather, heterogeneous surfaces should be treated. Moreover, the uncertainties associated with these observations should be considered during the entire process of devising multiscale observing schemes to capture the heterogeneities of various variables and parameters at different scales. Thus, a scale transformation is expected to be achieved, and a relevant method should be developed.
(3) Observe at different (stratified) scales. Various spatiotemporal scales in hydrology, ecology, and land surface observations should be considered, and a multiscale nested observing scheme should be implemented through synergistic satellite, airborne, and ground-based observations.
(4) Observations should be able to characterize the heterogeneity. We can utilize the concept of representativeness error to unify the errors caused by scale representativeness during observations. Based on the integration of field experiments and observations, we can develop a data assimilation method to evaluate complex land surface systems from a stochastic perspective.
"Current physically based models generally fail to explicitly incorporate these aspects of spatial variability" (Seyfried and Wilcox 1995). This is partially because of the driving mechanisms and partially because we do not have enough available spatially distributed data, with the exceptions of land cover maps, soil maps, and DEMs. In the process from transforming point data to large-scale data, common methods ignore the spatial correlation among sampling points. The statistics of the spatiotemporal characteristics of ecohydrological observations and their related environmental variables should be used to improve the accuracy of modeling. It should be noted that spatial variability is not going to be characterized only by mean values or variance, which have been traditionally used in statistics; in contrast, spatial variability should be characterized by spatial correlation functions,

correlation range, and spatial distribution functions. Fundamentally, capturing the probability density function (pdf) of variables, parameters, and forcing data is important. Meteor-hydrological, hydrogeological, and ecohydrological processes should be observed.

(5) A capable and affordable observing system is crucial to our observation of ecohydrological processes. In detail, the optimal selection of observation density and scale is an essential task in designing an in situ observation network that provides information on ecohydrological variables. A scheme that integrated kriging theory, spatial stratification, and temporal stability analysis was determined to be effective in scale selection. Based on the relationships among ecohydrological variables and human activities; meteorological, environmental, and hydrogeological conditions; and other factors, the optimized patterns of observation can be determined by expert knowledge, ecohydrological modeling, and historical observations. To estimate the means of certain properties and their estimated variances for complicated surfaces, the method for surfaces with nonhomogeneity (MSN) theory (Wang et al. 2009) combined spatial stratification and a kriging method based on the assumption that a nonhomogeneous surface can be conveniently and rapidly applied to optimize sampling sites in watershed regions.

(6) Observation of human activities is challenging. Accompanied by a complementary system of observation of human activities (NRC 2007), validation is essential and important for producing remote sensing products. Pixel-scale validation sites with nested multiscale observation systems can be used to solve some key scientific problems associated with remote sensing product validation (e.g., the scale effect and scale transformation) (Ma et al. 2015). Multi-year in situ measurements will ensure the high-accuracy and interannual validity of the remote sensing products, which will provide dynamic regional monitoring and simulation capabilities. Potential linkages with information system, modeling and data assimilation should be considered. It has been suggested that the ecohydrological observations, models, and data assimilations need to produce more "usable" information with enhanced reliability to ensure the adoption of research results; however, "usable" information should still be utilized by the user or the decision-maker. The information system plays a key role in linking information producers and users. The integration of useful and relevant scientific information is necessary and critical to enable informed decision-making.

New, cutting-edge technology should be encouraged because it can provide the current and fully developed technological features required for ecohydrological observation, modeling, and data assimilation.

Case Studies of Building a Watershed Observing System

A watershed observing system can be demonstrated using two comprehensive watershed-scale observation experiments, i.e., the Watershed Allied Telemetry Experimental Research (WATER) (Li et al. 2009) and the Heihe Watershed Allied Telemetry Experimental Research (HiWATER) (Li et al. 2013).

Overview of WATER

The WATER experiment, which was a simultaneous airborne, satellite-borne, and ground-based remote sensing observation approach used to improve the observability, understanding, and predictability of hydrological and related ecological processes at the catchment scale, took place from 2007–2009 in the Heihe River Basin, which is China's second largest inland river basin (Li et al. 2009, 2012a). Observation items were separated into five categories, including hydrological and ecological variables, atmospheric driving data, vegetation parameters, soil parameters, and aerodynamic parameters, which corresponded to observations from four scales that contributed to the river basin, focus test area, dense observation area, and observation plot (Li et al. 2009); these four scales included the river basin, the key experimental area, the foci experimental area, and the experiment site. First, as a multiscale land surface/hydrological experiment in a cold and arid region, the experiment provided dynamic monitoring methods and cases at the watershed scale, and it developed a land surface data assimilation system based on remote sensing observation; additionally, it established scaling transformation approaches to inverse land surface hydroecological variables by using active and positive microwave remote sensing production. Finally, this approach generated a basic data set for watershed science by carrying out synchronous experiments comprising satellite remote sensing and in situ measurements.

Airborne sensors, including microwave radiometers at L, K, and Ka bands, imaging spectrometers, thermal imagers, CCD, and LiDAR, were used. Multiscale and multisource satellite data were collected. Ground measurements were carried out at various scales. Several remote sensing methods for the estimation of key ecohydrological variables were used (Li et al. 2012b). (1) The high-accuracy observations of watershed-scale precipitation were carried out using a truck-mounted dual-polarized Doppler radar. (2) The evapotranspiration estimation models were developed and improved. (3) A snow depth retrieval algorithm was developed in the mountainous area. (4) The influence of roughness on soil moisture estimation was identified and eliminated by using multiangle SAR data. Generally, the WATER experiment applied various airborne remote sensors to develop and improve the models and algorithms for key ecohydrological variables/parameters, illustrating the irreplaceable role of airborne remote sensing in obtaining high-quality and high-resolution data. However, the scaling approach and methodology referring to the application of remote sensing in ecohydrology were not emphasized.

The HiWATER

Based on the existing observational infrastructure in the Heihe River Basin and the outcomes of the WATER experiment, a new program called the Heihe Watershed Allied Telemetry Experimental Research (HiWATER) (Li et al. 2013) was proposed and successfully carried out to establish a comprehensive research platform that integrated observations, data management, and model simulations to foster twenty-first-century watershed science in China. The program performed airborne, satellite-borne, and ground-based remote sensing experiments at various scales.

The HiWATER Watershed Observing System.
The HiWATER established several multiscale, nested hydrometeorological observing networks (Fig. 4), flux observing matrices (Xu et al. 2013), and ecohydrological WSNs (Jin et al. 2014; Qu et al. 2014; Ma et al. 2015; Wang et al. 2016) throughout the upstream, midstream, and downstream portions of the Heihe River Basin. Additionally, a large number of airborne- and ground-based remote sensing experiments were simultaneously conducted.

In 2012, multiscale observation experiments using surface evapotranspiration measurements were conducted in a heterogeneous oasis and in an oasis-desert transition zone in the midstream area of the basin. The experimental setup dozens of automatic weather stations (AWSs), eddy covariance systems, and large-aperture scintillations within and outside the oasis. A large matrix consisting of six ECs and AWSs was set up outside the oasis, and a small matrix consisting of 18 ECs, 17 AWSs, 4 pairs of LASs, a trace isotope system, and more than 250 WSN nodes (200 soil moisture and soil temperature nodes and 50 LAI nodes) were set up within the oasis.

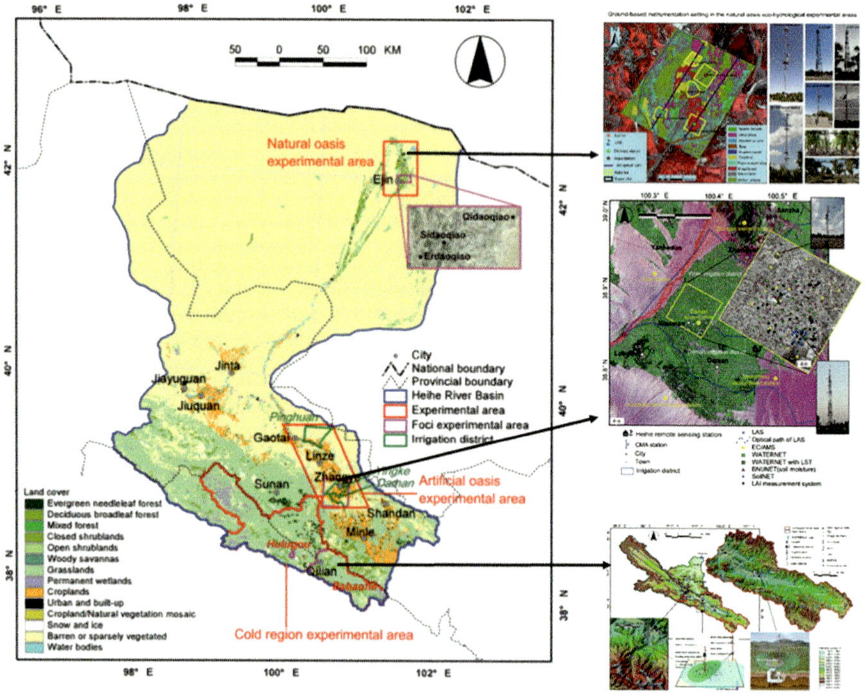

Fig. 4 Multiscale observing system in Heihe River Basin

In the upstream portion of the basin, a multiscale nested observing network was established in 2013. The network consisted of 8 AWSs and more than 40 WSN nodes, as well as a superstation for observing the freezing and thawing cycles. Several key ecohydrological variables and fluxes, including precipitation, ET, soil moisture, soil temperature, snow, freezing and thawing cycle, and photosynthesis, were obtained from the network.

In the downstream portion of the basin, five ECs and AWS towers, two pairs of LASs, and a small WSN for soil moisture measurements were established in different ecosystems, including bare soil, cropland, and wood land.

The Airborne Remote Sensing Experiment

The objective of the airborne remote sensing experiment was to observe the key ecohydrological variables and parameters, namely, snow melt (SM; including freeze/thaw status), snow cover, amount of irrigation water, land surface temperature (LST), emissivity, LAI, roughness, albedo, vegetation canopy structure, and high-resolution DEM. All of these variables/parameters were applied to facilitate the understanding of the inland river basin ecohydrological processes; the development, improvement, and validation of these ecohydrological models; the transformation of the scales between the ground measurements and the satellite remote sensing observations; and the support of the integration of multiscale models and their corresponding applications.

Aircraft was equipped with various sensors, including microwave radiometers, an imaging spectrometer, a thermal imaging camera, LiDAR, and other measuring systems that targeted key ecohydrological processes (Table 1). The goal of these missions was to improve the remote sensing methods for observing key ecohydrological variables/parameters, to develop scale methods by acquiring multi-spatial resolution remote sensing data sets, and to reserve some flights to satisfy the needs of relevant projects.

The airborne remote sensing experiment was performed over several key areas and transects. Based on the core scientific questions addressed in each experimental area and the corresponding specific targets, several relevant items were recorded, including flight path, endurance time, airborne remote sensor types, platforms, flight frequency, and attitude, and the experiment was designed to be spatiotemporally consistent with the simultaneous surface sampling strategy, the station, and WSN observations and the satellite data.

To ensure the credibility of airborne data, the flight attitude, airborne sensor calibration, and data product quality control were carefully maintained. In addition, the rationally designed flight lines and the flying time were closely matched with the ground-based and satellite observations. A multi-sensor configuration with a single flight strategy was complied with to enhance the efficiency of the aerial remote sensing data acquisition and to increase the temporal comparability of various data sets.

Table 1 Airborne sensors used in HiWATER (Li et al. 2013)

	Instrument	Observation items
	LiDAR+CCD	Canopy structure, NPP, DEM, forest height, land cover type, tree crown size, LUCC, LST, reflectance, albedo
	CASI-1500 VNIR and SASI-600 SWIR imaging spectrometer	LUCC, LST, reflectance, albedo, vegetation canopy temperature
	TASI-600 push broom hyperspectral thermal sensor system	Soil moisture, LST, vegetation canopy temperature
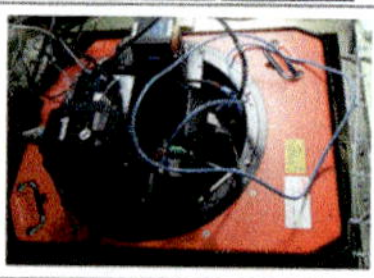	Wide-angle infrared dual-mode line/area array scanner	Soil moisture, LST, canopy temperature, snow surface temperature, emissivity, biogeophysical parameters, biogeochemical parameters, reflectance, and albedo
	PLMR	Soil moisture, soil freeze/thaw status, and snow moisture, VWC, LST, vegetation canopy temperature
	L band passive microwave radiometer	Soil moisture, soil freeze/thaw status, and snow moisture
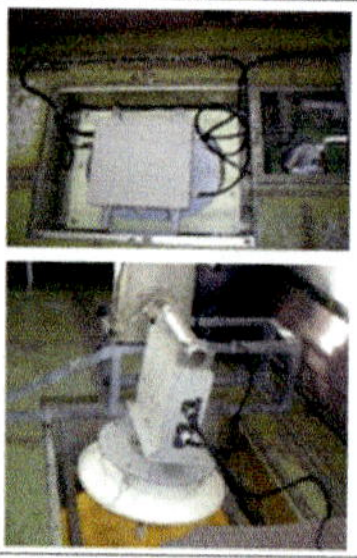	K and Ka band passive microwave radiometers	Soil freezing /thawing states, snow depth, and snow moisture

Thematic Experiments

The thematic experiment was concentrated on specific ecohydrological processes. The thematic experiment of HiWATER was initiated to conduct multiscale evapotranspiration observations over nonhomogeneous land surfaces (Liu et al. 2016; Xu et al. 2013).

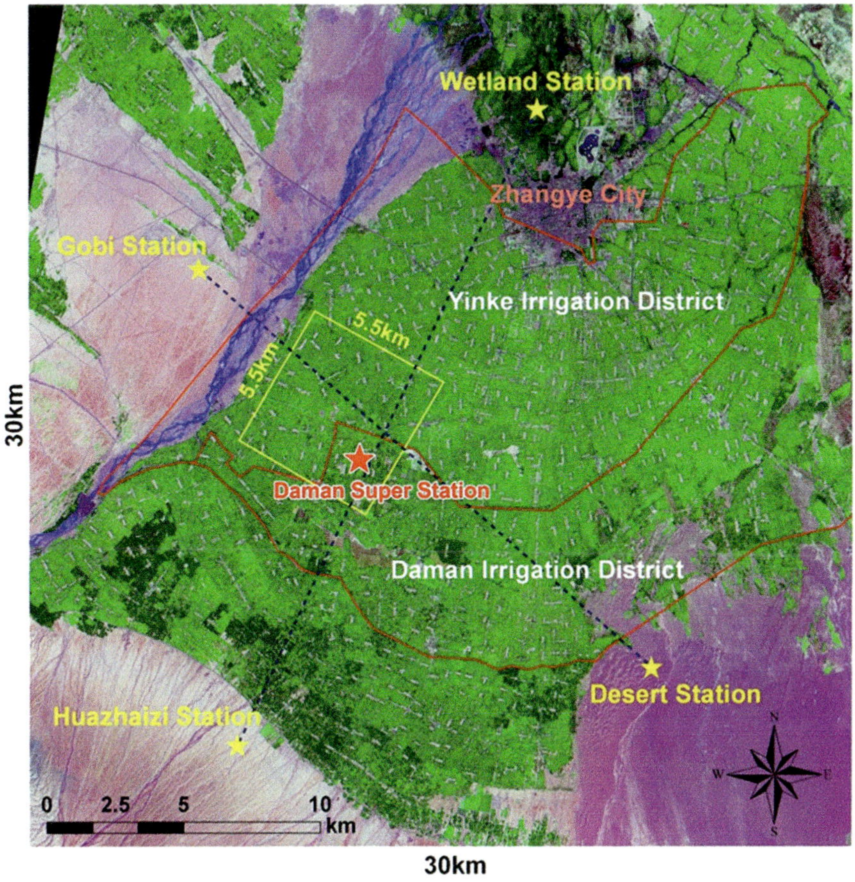

Fig. 5 The schematic map of the 30 km × 30 km experimental area

Two nested experimental areas were set up near Zhangye City. The larger site covered an area of 30 km × 30 km and contained a cross-shaped observation arrangement including one superstation (i.e., the Daman Superstation) and four ordinary stations. The smaller (kernel) experimental area located inside the aforementioned experimental area had a size of 5.5 × 5.5 km (Fig. 5), and ground-based measurements were performed by the EC and LAS compositional observation matrix, AWSs, and WSN. Combined with spaceborne and airborne remote sensing (VNIR, TIR, microwave and LiDAR, etc.), and the airborne EC system, a three-dimensional observation network was formulated to determine areal surface ET, key ecohydrological variables, and hydrometeorological elements at the pixel scale over typical landscapes in the middle reaches of the Heihe River Basin. Simultaneously, stable isotope measurements were performed for the partitioning of soil evaporation and plant transpiration.

Remote Sensing Products of Ecohydrological Variables

The HiWATER has developed algorithms for producing remote sensing products for seven key ecohydrological variables, and it has generated high-spatiotemporal-resolution remote sensing products for ecohydrological variables of the Heihe River Basin (Li et al. 2017), including 14 years (2000–2013) of high-spatiotemporal precipitation products for the HRB with 0.05°spatial and hourly resolutions (Pan et al. 2015, 2017), 16 years (2000–2015) of daily snow cover products with 500-m spatial resolution, 8 years (2008–2015) of daily soil moisture products with 1-km spatial resolution (Ma et al. 2013; Kang et al. 2017), and 5 years (2011–2015) of monthly precision land cover products with 30-m spatial resolution (Zhong et al. 2015) and multi-spatiotemporal-resolution vegetation fraction products (2011–2015) (Mu et al. 2015) phenological products (Xia et al. 2012) with 1-km resolution and daily NPP products with 1-km resolution from 2012 to 2015. The precisions of all products were better than those of the products available at that time.

Scaling Issues

The scaling issue has always been a key problem in remote sensing, but the development of theory and methodology is still seriously hampered by the scarcity of multiscale observations. One of the core objectives of the HiWATER was to carry out multiscale observations to support scaling transformation studies (Jin et al. 2017; Li et al. 2016; Liu et al. 2016).

- The scaling-related concepts, including spatial averaging, spatial upscaling, observing footprint, representativeness error, and observation truth, have been mathematically defined.
- The multiscale and multivariable sampling methods were proposed based on the theory of geo-statistics, which could not only obtain globally unbiased optimal estimates but also capture the spatial heterogeneity of the variable.
- The representativeness errors of radiation, carbon flux, and soil moisture were investigated based on the multipoint footprint observations from HiWATER. The results showed that the representativeness error of a single-point observation over a heterogeneous surface could not be ignored. Thus, the representativeness error should be carefully considered in remote sensing validation activities.
- An upscaling method for single-point and footprint-scale observations was proposed, and the kriging approach was extended to the regression kriging, area-area kriging, and spatiotemporal regression kriging. In addition, a nonlinear upscaling approach under the Bayesian theorem framework was proposed.
- A general research framework for upscaling, which included aspects of sampling design, multiscale observations, metrics of representativeness error, and validation activities, was preliminarily formed.

Data Quality Control and Sharing

The sharing of scientific experimental data is not only an important part of the experiment but also an important obligation to the scientific community. It is an

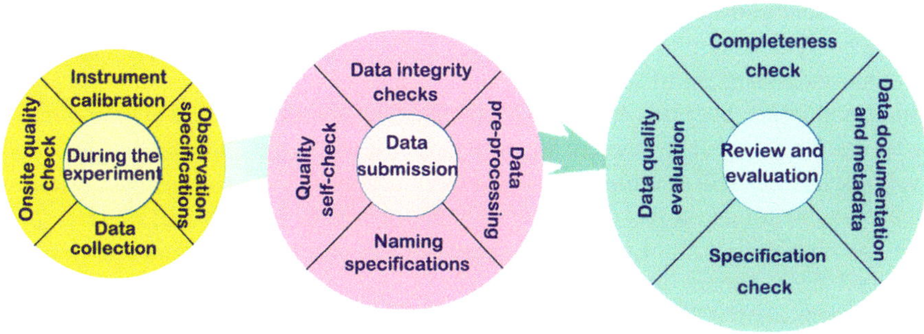

Fig. 6 HiWATER data collection, processing, archiving, and sharing process (Li et al. 2017)

important mean by which to test the experimental results and improve the quality of the data. During and after the HiWATER, almost all experimental data were released to the public after systematic processing, metadata and document writing, development of data information systems, and data quality control were performed (Guo and Liu 2015; Li et al. 2010, 2017).

An automatic data assembling information system was developed, and the data quality control rules were determined during HiWATER. All the data observed by the observation system were assembled into the data center via the wireless transformation. The main point of the data quality control was to establish a working mechanism that included a quality control algorithm that was compatible and coordinated with a variety of quality control algorithms to complete the automatic quality control of observation data in certain processes. After this reprocessing and data quality control, the data were then evaluated by holding related evaluation conferences or by having the data producers and experts review and discuss the results. Finally, the data were released, and the evaluation continued. Furthermore, improvements were made based on the feedback from the users. The whole data quality control process can be described in Fig. 6.

Both the HiWATER and WATER experiments embraced policies of "full and open" sharing. All the data that were not confidential were made fully available to the public. The user could freely download the data after being granted permission by the data management sector.

Summary

Watershed science has been identified as an emerging new science. Watershed observing systems have the capacity to promote the development of ecohydrological processes at the watershed scale by means of accurately measuring a series of state variables and fluxes from the natural and human processes at multiple spatiotemporal scales. In addition, it is very important that new hypotheses can be validated, and new models and approaches can be potentially proposed by making use of the abundant data resources made available by the observing system.

Our understanding of watershed science will depend on our observation abilities and capabilities, both of which have significantly improved and are evolving into a new era. Revolutionary changes are occurring in hydrological and ecological observations, and this will also bring a revolutionary change in watershed science. Most of the changes are driven by the powerful tools of remote sensing and other revolutionary observing technologies. Many previously unobservable variables can now be derived from remote sensing techniques. Though many of them still cannot be readily observed, a few new techniques, such as GPM, SMOS, SMAP, new-generation gravity, and LiDAR, enable the improved observation ability of ecohydrological processes. New techniques such as WSNs will play a critical role because they provide promising methods to fill the gap between point measurements and remote sensing. Challenges in the next decade come from the expectation of more reliable and higher-resolution observations, particularly for those ecohydrological variables that have not been captured, e.g., soil moisture, snow water equivalent in mountainous areas (NRC 2007, 2008).

In this chapter, we envisioned a future watershed observing system for assessing ecohydrological processes, and it strongly relies on multiscale observations. The advancement of techniques is providing us good opportunities; however, we should prioritize what we are going to observe, what the strategy of measurement should be, and what we should do first. This priority setting should come from different disciplines – hydrology, ecology, remote sensing, and information science. Experimental watersheds should act as innovation platforms to test new ideas, instruments, models, and integrated approaches. Comprehensive experiments (platforms) are required to test both the observation techniques and their applicability in hydrology and ecology.

However, it is still challenging to quantify the key water cycle and ecological process variables and fluxes and to improve their accuracy and precision. Uncertainties are closely related to heterogeneity and therefore are an inherent part of measurements in hydrology, ecology, and other land surface sciences. To serve the next generation of integrated watershed models, watershed observing systems should have the ability to express the heterogeneous behavior of ecohydrological processes.

Most importantly, raw observations must be translated into useful information and knowledge for managing watersheds. This goal further depends on integrating multiscale observations, whether they are in situ, remotely sensed, or from modeling. It is difficult to use observation by itself to construct the spatiotemporal evolution of the water cycle and related ecological processes in a river basin. Models should be used to extend this purpose so that assimilating observations in watershed models is of particular importance because a better estimation can be achieved from model-data fusion.

The development of a watershed observing system for ecohydrological processes relies on the mutual interaction of modeling and measurement. In summary, watershed observing systems will act as the most important infrastructure in the future decades of earth surface system science. Modelers and experimentalists should cooperate to achieve this goal.

Acknowledgments This work is supported by the National Natural Science Foundation of China (grant No. 91425303), and the Strategic Priority Research Program of the Chinese Academy of Sciences, Grant No. XDA20100104.

References

B. Aylward, Towards watershed science that matters. Hydrol. Process. **19**(13), 2643–2647 (2005)

G.D. Cheng, *Integrated management of the water-ecology-economy system in the Heihe River basin* (Science Press, Beijing, 2009), p. 581

G. Cheng, X. Li, Integrated research methods in watershed science. Sci. China-Earth Sci. **58**, 1159–1168 (2015)

G.D. Cheng, X. Li, W.Z. Zhao, Z.M. Xu, Q. Feng, S.C. Xiao, H.L. Xiao, Integrated study of the water-ecosystem-economy in the Heihe River basin. Natl. Sci. Rev. **1**, 413–428 (2014)

GEO (Group on Earth Observations), Global Earth Observation System of Systems GEOSS 10-Year Implementation Plan Reference Document (2005), pp. 209

J.W. Guo, F. Liu, Automatic data quality control of observations in wireless sensor network. IEEE Geosci. Remote Sens. Lett. **12**(4), 716–720 (2015)

R. Jin, X. Li, B.P. Yan, X.H. Li, W.M. Luo, M.G. Ma, J.W. Guo, J. Kang, Z.L. Zhu, a.S.J. Zhao, A nested ecohydrological wireless sensor network for capturing the surface heterogeneity in the midstream area of the Heihe River basin, China. IEEE Geosci. Remote Sens. Lett. **11**(11), 2015–2019 (2014)

R. Jin, X. Li, S.M. Liu, Understanding the heterogeneity of soil moisture and evapotranspiration using multiscale observations from satellites, airborne sensors, and a ground-based observation matrix. IEEE Geosci. Remote Sens. Lett. **14**(11), 2132–2136 (2017)

J. Kang, R. Jin, X. Li, C.F. Ma, J. Qin, Y. Zhang, High spatio-temporal resolution mapping of soil moisture by integrating wireless sensor network observations and MODIS apparent thermal inertia in the Babao River basin, China. Remote Sens. Environ. **191**, 232–245 (2017)

X. Li, G.D. Cheng, On the watershed observing and modeling systems. Adv. Earth Sci. **23**(7), 756–764 (2008)

X. Li, X.W. Li, Z.Y. Li, M.G. Ma, J. Wang, Q. Xiao, Q. Liu, T. Che, E.X. Chen, G.J. Yan, Z.Y. Hu, L.X. Zhang, R.Z. Chu, P.X. Su, Q.H. Liu, S.M. Liu, J.D. Wang, Z. Niu, Y. Chen, R. Jin, W.Z. Wang, Y.H. Ran, X.Z. Xin, H.Z. Ren, Watershed allied telemetry experimental research. J. Geophys. Res. Atmos. **114** (2009). https://doi.org/10.1029/2008JD011590

X. Li, X.W. Li, Z.Y. Li, Watershed allied Telemetry Experimental research (WATER) datasets are available for open access (Chinese with English abstract). Remote Sens. Technol. Appl. **25**(6), 761–764 (2010)

X. Li, X.W. Li, Z.Y. Li, J. Wang, M.G. Ma, Q. Liu, Q. Xiao, Z.Y. Hu, T. Che, J.M. Wang, Q.H. Liu, E.X. Che, G.J. Yan, S.M. Liu, W.Z. Wang, L.X. Zhang, J.D. Wang, Z. Niu, R. Jin, Y.H. Ran, L. X. Wang, Progress on the watershed allied telemetry experimental research (WATER). Remote Sens. Technol. Appl. **27**(5), 637–649 (2012a.) (in Chinese with English Abstract)

X. Li, Q. Liu, Q.H. Liu, J. Wang, M.G. Ma, Q. Xiao, T. Che, R. Jin, Y.H. Ran, Progress on the watershed allied telemetry experimental research (WATER): Remote sensing of key hydrological and ecological parameters. Remote Sens. Technol. Appl. **27**(5), 650–662 (2012b.) in Chinese with English Abstract

X. Li, G.D. Cheng, S.M. Liu, Q. Xiao, M.G. Ma, R. Jin, T. Che, Q.H. Liu, W.Z. Wang, Y. Qi, J.G. Wen, H.Y. Li, G.F. Zhu, J.W. Guo, Y.H. Ran, S.G. Wang, Z.L. Zhu, J. Zhou, X.L. Hu, Z.W. Xu, Heihe watershed allied telemetry experimental research (HiWATER): Scientific objectives and experimental design. Bull. Am. Meteorol. Soc. **94**, 1145–1160 (2013)

X. Li, R. Jin, S.M. Liu, Y. Ge, Q. Xiao, Q.H. Liu, M.G. Ma, Y.H. Ran, Upscaling research in HiWATER: Progress and prospects. J. Remote Sens. **20**(5), 1993–2002 (2016)

X. Li, S.M. Liu, Q. Xiao, M.G. Ma, R. Jin, T. Che, W.Z. Wang, X.L. Hu, Z.W. Xu, J.G. Wen, L.X. Wang, A multiscale dataset for understanding complex ecohydrological processes in a heterogeneous oasis system. Sci. Data **4**, 170083 (2017). https://doi.org/10.1038/sdata.2017.83

S.M. Liu, Z.W. Xu, L.S. Song, Q.Y. Zhao, Y. Ge, T.R. Xu, Y.F. Ma, Z.L. Zhu, Z.Z. Jia, F. Zhang, Upscaling evapotranspiration measurements from multi-site to the satellite pixel scale over heterogeneous land surfaces. Agric. For. Meteorol. **230–231**, 97–113 (2016)

C.F. Ma, W.Z. Wang, X.J. Han, X. Li, Soil moisture retrieval in the Heihe River basin based on the real thermal inertia method. IEEE J. Sel. Top. Appl. Earth Obs. Remote Sens. **6**(3), 1460–1467 (2013)

M.G. Ma, T. Che, X. Li, Q. Xiao, K. Zhao, X.P. Xin, A prototype network for remote sensing validation in China. Remote Sens. **7**(5), 5187–5202 (2015). https://doi.org/10.3390/rs70505187

J.J. McDonnell, M. Sivapalan, K. Vache, S. Dunn, G. Grant, R. Haggerty, C. Hinz, R. Hooper, J. Kerchner, M.L. Roderick, J. Selker, M. Weiler, Moving beyond heterogeneity and process complexity: a new vision for watershed hydrology. Water Resour. Res. **43**, W07301 (2007). https://doi.org/10.1029/2006WR005467

X.H. Mu, S. Huang, H.Z. Ren, G.J. Yan, W. Song, G. Ruan, Validating GEOV1 fractional vegetation cover derived from coarse-resolution remote sensing images over croplands. IEEE J. Sel. Top. Appl. Earth Obs. Remote Sens. **8**(2), 439–446 (2015)

NRC (Committee on Opportunities in the Hydrologic Sciences Water Science and Technology Board Commission on Geosciences, Environment, and Resources), *Opportunities in the Hydrologic Sciences* (National Academy Press, Washington, DC, 1991), p. 368

NRC (Committee on River Science at the U.S. Geological Survey), *River Science at the U.S. Geological Survey* (National Academies Press, Washington, DC, 2007), p. 206

NRC (Committee on Scientific Accomplishments of Earth Observations from Space), *Earth Observations from Space: The First 50 Years of Scientific Achievements* (National Academies Press, Washington, DC, 2008), p. 142

NRC (Committee on U.S. Geological Survey), *Watershed Research in the U.S. Geological Survey* (National Academies Press, Washington, DC, 1997), p. 96

NRC (Committee on Watershed Management), *New Strategies for America's Watersheds* (National Academies Press, Washington, DC, 1999), p. 328

X.D. Pan, X. Li, G.D. Cheng, H.Y. Li, X.B. He, Development and evaluation of a river-basin-scale high spatio-temporal precipitation data set using the WRF model: a case study of the Heihe River basin. Remote Sens. **7**(7), 9230–9252 (2015)

X.D. Pan, X. Li, G.D. Cheng, Y. Hong, Effects of 4D-Var data assimilation using remote sensing precipitation products in a WRF model over the complex terrain of an arid region river basin. Remote Sens. **9**(9), 963 (2017). https://doi.org/10.3390/rs9090963

Y.H. Qu, Y.Q. Zhu, W.C. Han, J.D. Wang, M.G. Ma, Crop leaf area index observations with a wireless sensor network and its potential for validating remote sensing products. IEEE J. Sel. Top. Appl. Earth Obs. Remote Sens. **7**(2), 431–444 (2014)

H. Savenije, HESS opinions "the art of hydrology". Hydrol. Earth Syst. Sci. **13**(2), 157–161 (2009)

M.S. Seyfried, B.P. Wilcox, Scale and the nature of spatial variability: field examples having implications for hydrological modeling. Water Resour. Res. **31**, 173–184 (1995)

S. Uhlenbrook, Catchment hydrology – a science in which all processes are preferential. Hydrol. Process. **20**(16), 3581–3585 (2006)

J. Wang, G. Christakos, M. Hu, Modeling spatial means of surfaces with stratified non-homogeneity. IEEE Trans. Geosci. Remote Sens. **47**(12), 4167–4174 (2009)

S.G. Wang, X. Li, Y. Ge, R. Jin, M.G. Ma, Q.H. Liu, J.G. Wen, S.M. Liu, Validation of regional-scale remote sensing products in China: from site to network. Remote Sens. **8**(12), 980 (2016). https://doi.org/10.3390/rs8120980

C.F. Xia, J. Li, Q.H. Liu, Monitoring vegetation phenology in China using time-series MODIS LAI data. IEEE Geosci. Remote Sens. Symp., 48–51 (2012). https://doi.org/10.1109/IGARSS.2012.6350912. Munich, Germany

Z.W. Xu, S.M. Liu, X. Li, S.J. Shi, J.M. Wang, Z.L. Zhu, T.R. Xu, W.Z. Wang, M.G. Ma, Intercomparison of surface energy flux measurement systems used during the HiWATER-MUSOEXE. J. Geophys. Res. Atmos. **118**, 13140–13157 (2013)

B. Zhong, A. Yang, A. Nie, Y. Yao, H. Zhang, S. Wu, Q. Liu, Finer resolution land-cover mapping using multiple classifiers and multisource remotely sensed data in the Heihe River basin. IEEE J. Sel. Top. Appl. Earth Obs. Remote Sens. **8**(10), 4973–4992 (2015)

Index

X. Li, H. Vereecken (eds.), *Observation and Measurement of Ecohydrological Processes*, Ecohydrology, https://doi.org/10.1007/978-3-662-48297-1

T

Printed by Printforce, the Netherlands